Computational methods in structural dynamics

Monographs and textbooks on mechanics of solids and fluids

editor-in-chief: G. Æ. Oravas

Mechanics: Dynamical systems

editor: L. Meirovitch

Computational methods in structural dynamics

Leonard Meirovitch

Reynolds Metals Professor
Department of Engineering Science and Mechanics
Virginia Polytechnic Institute and State University

SIJTHOFF & NOORDHOFF   1980
Alphen aan den Rijn, The Netherlands
Rockville, Maryland, U.S.A.

ISBN 90 286 0580 0

Printed in The Netherlands

To Jo Anne

To Jo Anne

Preface

The last several decades have witnessed impressive progress in structural dynamics. This progress can be attributed to a happy confluence of new methods of analysis, such as the finite element method and substructure synthesis, and the digital computer. The newly developed methods of analysis permit the structural analyst to formulate increasingly complex problems and the computer enables him to obtain numerical results to these problems. As a result, the emphasis in the field of structural dynamics has been shifting steadily toward computational methods. But, a good understanding of the techniques for deriving the equations of motion for complex structures and of the computational algorithms for solving these equations demands an increasing mathematical sophistication on the part of the structural dynamicist. This book presents the latest developments in structural dynamics and it also provides the mathematical background necessary for the understanding of the various modelling techniques and computational algorithms. The book is intended as a text for a one-year graduate course in structural dynamics as well as a reference book for structural dynamicists. It assumes some prior knowledge of the field of vibrations, such as that normally acquired in an elementary course in mechanical and structural vibrations.

An important aspect of structural dynamics is mathematical modelling. Mathematical models can be classified according to the spatial distribution of the parameters describing the system properties. There are two major classes, namely, discrete and distributed-parameter models. Less often one encounters hybrid systems, i.e., systems that are part discrete and part distributed. Discrete systems are commonly described by ordinary differential equations and distributed systems by partial differential equations. Most dynamical systems are actually distributed and, except for some simple classical examples, exact solutions for the response are difficult, if not impossible to produce. This points to approximate solutions as an alternative, which almost invariably amounts to the representation of the distributed model by a discrete one in a process known as spatial discretization. In approximating a distributed system by a discrete one, the question of truncation of the order of the system plays an important part. Of course, the subject is to produce a discrete model of relatively low order that is capable of simulating the behavior of the distributed system with the desired degree of accuracy.

The book can be regarded as consisting of three parts: discrete systems, distributed systems and discretized systems. The contents of the book can be best reviewed through a brief discussion of the individual chapters. Any modern treatment of the vibration of discrete systems must rely heavily on linear algebra. In Ch. 1, concepts from linear algebra of particular interest in vibrations, such as vector spaces and matrices, are introduced. Chapter 2 is devoted to the free vibration of discrete systems. In this chapter, the equations of motion for a large variety of dynamical systems are derived. The free vibration problem leads naturally to the characteristic-value problem, or the eigenvalue problem, one of the most basic problems in linear algebra. A widely used method for deriving the response, known as modal analysis, is based on the solution of the eigenvalue problem. Chapter 2 sets the stage for the following three chapters. Indeed, Ch. 3 presents a general discussion of the algebraic eigenvalue problem and Ch. 4 examines the behavior of the eigensolution in a qualitative manner. In Ch. 4, some very powerful and far-reaching principles, such as Rayleigh's principle, the maximum-minimum principle and the inclusion principle, are presented. In Ch. 5, the emphasis changes from qualitative considerations to quantitative methods. The chapter is devoted almost exclusively to computational algorithms for the solution of the algebraic eigenvalue problem including the most modern ones. The response of discrete systems to arbitrary excitations is discussed in Ch. 6. In this chapter, various techniques from linear system theory for solving sets of simultaneous ordinary differential equations are presented. Consistent with the expectation that the response to arbitrary excitations must be ultimately simulated on a digital computer, the idea of discretization in time of the equations of motion is introduced. In Ch. 7, the attention is shifted to the vibration of distributed-parameter systems. The chapter is devoted to exact solutions to vibration problems and has a classical appearance. One of its objectives is to highlight properties shared by all solutions to vibration problems for a large class of distributed-parameter systems, namely, self-adjoint systems. These properties can be established regardless of whether solutions in terms of known functions can be derived or not. Knowledge of the solution properties plays an important role in producing approximate solutions, as demonstrated in Ch. 8. Among the approximate methods discussed in Ch. 8, the Rayleigh–Ritz method occupies a special place in structural dynamics. In fact, as pointed out in Ch. 9, the finite element method itself can be regarded as a Rayleigh–Ritz method. Note that the methods discussed in Chs. 8 and 9 are essentially discretization procedures. Complex structures require complex mathematical models, in the sense that the mathematical simulation is in terms of a large number of differential equations. Chapter 10 is devoted to techniques for reducing the order of the system. A method that can be regarded as both a modelling and a reduction method is referred to as substructure synthesis. A number of variants of the method are presented in Ch. 11.

The author wishes to acknowledge the help received from many colleagues and students during various stages of the manuscript. In particular, he wishes to express his appreciation to Professor Robert J. Melosh for his many thoughtful and valuable comments. Many thanks are also due to Professor Earl H. Dowell, Mr. Arthur L. Hale, Dr. Dewey H. Hodges and Professor Harold D. Nelson, as well as to Mr. Haim Baruh, Dr. Hayrani Öz, Mr. Garnett Ryland and Mr. Lawrence M. Silverberg. Finally, the author would like to thank Mrs. Peggy Epperly, Miss Carol Ann Sowder and Mrs. Marlene A. Taylor for their patience in typing various portions of the manuscript.

Contents

Contents

Contents

Concepts from linear algebra

1.1 Introduction

As shown in subsequent chapters, the vibration of linear discrete systems is governed by a set of simultaneous linear ordinary differential equations. The solution of such sets of equations can be obtained most conveniently by means of a linear transformation rendering the set of equations independent. In seeking this linear transformation, the problem is converted from that of a set of simultaneous differential equations to that of a set of simultaneous algebraic equations. The latter problem is known as the eigenvalue problem and it represents an important problem in linear algebra. Hence, a brief discussion of certain pertinent concepts of linear algebra is in order. In particular, a discussion of vector spaces, as well as of closely related topics such as determinants and matrices, should prove most rewarding.

1.2 Linear vector spaces

In discussing vector spaces, it proves convenient to introduce the concept of a field. A *field* is defined as a set of scalars possessing certain algebraic properties. The real numbers constitute a field, and so do the complex numbers.

Let us consider a set of elements F such that for any two elements α and β in F it is possible to define another two unique elements belonging to F, the first denoted by $\alpha + \beta$ and called the *sum* of α and β, and the second denoted by $\alpha\beta$ and called the *product* of α and β. The set F is called a field if these two operations satisfy the five field postulates:

1. *Commutative laws.* For all α and β in F,

   (i) $\alpha + \beta = \beta + \alpha$, (ii) $\alpha\beta = \beta\alpha$

2. *Associative laws.* For all α, β and γ in F.

   (i) $(\alpha + \beta) + \gamma = \alpha + (\beta + \gamma)$, (ii) $(\alpha\beta)\gamma = \alpha(\beta\gamma)$

3. *Distributive laws.* For all α, β and γ in F,

$$\alpha(\beta + \gamma) = \alpha\beta + \alpha\gamma$$

4. *Identity elements.* There exist in F elements 0 and 1 called the zero and the unity elements, respectively, such that $0 \neq 1$, and for all α in F,

   (i) $\alpha + 0 = \alpha$,   (ii) $1\alpha = \alpha$

5. *Inverse elements.*

   (i) For every element α in F there exists a unique element $-\alpha$, called the additive inverse of α, such that $\alpha + (-\alpha) = 0$.
   (ii) For element $\alpha \neq 0$ in F there exists a unique element α^{-1}, called the multiplicative inverse of α, such that $\alpha\alpha^{-1} = 1$.

It is clear that the set of all real numbers satisfy all five postulates.

Next, we wish to define the concept of *linear vector space*, also referred to as *linear space* and *vector space*. Let L be a set of elements called *vectors* and F a field of *scalars*. Then, if L and F are such that two operations, namely *vector addition* and *scalar multiplication*, are defined for L and F, the set of vectors together with the two operations are called a *linear vector space L over a field F.* For every two elements x and y in L, it satisfies the postulates:

1. *Commutativity.* $\mathbf{x} + \mathbf{y} = \mathbf{y} + \mathbf{x}$
2. *Associativity.* $(\mathbf{x} + \mathbf{y}) + \mathbf{z} = \mathbf{x} + (\mathbf{y} + \mathbf{z})$
3. There exists a unique vector $\mathbf{0}$ in L such that $\mathbf{x} + \mathbf{0} = \mathbf{0} + \mathbf{x} = \mathbf{x}$
4. For every vector $\mathbf{x}$ in L there exists a unique vector $-\mathbf{x}$ such that $\mathbf{x} + (-\mathbf{x}) = (-\mathbf{x}) + \mathbf{x} = \mathbf{0}$.

Hence, the rules of vector addition are similar to those of ordinary algebra. Moreover, for any vector $\mathbf{x}$ in L and any scalar α in F, there is defined a unique *scalar product* $\alpha\mathbf{x}$ which is also an element of L. The scalar multiplication must be such that, for all α and β in F and all $\mathbf{x}$ and $\mathbf{y}$ in L, it satisfies the postulates:

5. *Associativity.* $\alpha(\beta\mathbf{x}) = (\alpha\beta)\mathbf{x}$
6. *Distributivity.* (i) $\alpha(\mathbf{x} + \mathbf{y}) = \alpha\mathbf{x} + \alpha\mathbf{y}$,   (ii) $(\alpha + \beta)\mathbf{x} = \alpha\mathbf{x} + \beta\mathbf{x}$
7. $1\mathbf{x} = \mathbf{x}$, where 1 is the unit scalar, and $0\mathbf{x} = \mathbf{0}$.

We have considerable interest in a vector space L possessing n elements of the field F, i.e., in a vector space of n-tuples. We shall write any two such vectors in L as

$$
\mathbf{x} = \begin{bmatrix} x_1 \\ x_2 \\ \cdot \\ \cdot \\ \cdot \\ x_n \end{bmatrix} \qquad \mathbf{y} = \begin{bmatrix} y_1 \\ y_2 \\ \cdot \\ \cdot \\ \cdot \\ y_n \end{bmatrix} \tag{1.1}
$$

and refer to them as *n-vectors*. The set of all *n*-vectors is called the *vector space L^n*. Then, the addition of these two vectors is defined as

$$
\mathbf{x} + \mathbf{y} = \begin{bmatrix} x_1 + y_1 \\ x_2 + y_2 \\ \cdots \cdots \\ x_n + y_n \end{bmatrix} \tag{1.2}
$$

Moreover, if α is a scalar in F, then the product of a scalar and a vector is defined as

$$
\alpha \mathbf{x} = \begin{bmatrix} \alpha x_1 \\ \alpha x_2 \\ \cdots \\ \alpha x_n \end{bmatrix} \tag{1.3}
$$

Let S be a subset of the vector space L. Then, S is a *subspace* of L if the following statements are true:

1. If $\mathbf{x}$ and $\mathbf{y}$ are in S, then $\mathbf{x} + \mathbf{y}$ is in S.
2. If $\mathbf{x}$ is in S and α is in F, then $\alpha \mathbf{x}$ is in S.

1.3 Linear dependence

Let us consider a set of vectors $\mathbf{x}_1, \mathbf{x}_2, \ldots, \mathbf{x}_n$ in a linear space L and a set of scalars $\alpha_1, \alpha_2, \ldots, \alpha_n$ in F. Then the vector $\mathbf{x}$ given by

$$
\mathbf{x} = \alpha_1 \mathbf{x}_1 + \alpha_2 \mathbf{x}_2 + \cdots + \alpha_n \mathbf{x}_n \tag{1.4}
$$

is said to be a *linear combination* of $\mathbf{x}_1, \mathbf{x}_2, \ldots, \mathbf{x}_n$ with coefficients $\alpha_1, \alpha_2, \ldots, \alpha_n$. The vectors $\mathbf{x}_1, \mathbf{x}_2, \ldots, \mathbf{x}_n$ are said to be *linearly independent* if the relation

$$
\alpha_1 \mathbf{x}_1 + \alpha_2 \mathbf{x}_2 + \cdots + \alpha_n \mathbf{x}_n = \mathbf{0} \tag{1.5}
$$

can be satisfied only for the trivial case, i.e., only when all the coefficients $\alpha_1, \alpha_2, \ldots, \alpha_n$ are identically zero. If the relation (1.5) is satisfied and at least one of the coefficients $\alpha_1, \alpha_2, \ldots, \alpha_n$ is different from zero, then the vectors $\mathbf{x}_1, \mathbf{x}_2, \ldots, \mathbf{x}_n$ are said to be *linearly dependent*, with the implication that one vector is a linear combination of the remaining $n-1$ vectors.

The subspace S of L consisting of all the linear combinations of the vectors $\mathbf{x}_1, \mathbf{x}_2, \ldots, \mathbf{x}_n$ is called a subspace *spanned* by the vectors $\mathbf{x}_1, \mathbf{x}_2, \ldots, \mathbf{x}_n$. If $S = L$, then $\mathbf{x}_1, \mathbf{x}_2, \ldots, \mathbf{x}_n$ are said to *span L*.

Example 1.1

Consider the two independent vectors

$$\mathbf{x}_1 = \begin{bmatrix} 1 \\ 2 \\ 3 \end{bmatrix}, \qquad \mathbf{x}_2 = \begin{bmatrix} 2 \\ -1 \\ 1 \end{bmatrix} \tag{a}$$

in a three-dimensional space. The set of all linear combinations of $\mathbf{x}_1$ and $\mathbf{x}_2$ span a plane passing through the origin and the tips of $\mathbf{x}_1$ and $\mathbf{x}_2$. The three vectors

$$\mathbf{x}_1 = \begin{bmatrix} 1 \\ 2 \\ 3 \end{bmatrix}, \qquad \mathbf{x}_2 = \begin{bmatrix} 2 \\ -1 \\ 1 \end{bmatrix}, \qquad \mathbf{x}_3 = \begin{bmatrix} 5 \\ 0 \\ 5 \end{bmatrix} \tag{b}$$

span the same plane, because $\mathbf{x}_3$ lies in the plane spanned by $\mathbf{x}_1$ and $\mathbf{x}_2$. Hence, the three vectors are linearly dependent. Indeed, it can be easily verified that

$$\mathbf{x}_1 + 2\mathbf{x}_2 - \mathbf{x}_3 = \mathbf{0} \tag{c}$$

so that $\mathbf{x}_3$ is really a linear combination of $\mathbf{x}_1$ and $\mathbf{x}_2$ (see Fig. 1.1).

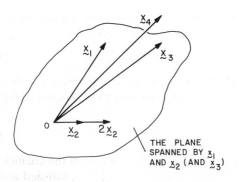

Figure 1.1

4

On the other hand, the three vectors

$$\mathbf{x}_1 = \begin{bmatrix} 1 \\ 2 \\ 3 \end{bmatrix}, \qquad \mathbf{x}_2 = \begin{bmatrix} 2 \\ -1 \\ 1 \end{bmatrix}, \qquad \mathbf{x}_4 = \begin{bmatrix} 5 \\ 1 \\ 5 \end{bmatrix} \tag{d}$$

are linearly independent because

$$\alpha_1 \mathbf{x}_1 + \alpha_2 \mathbf{x}_2 + \alpha_4 \mathbf{x}_4 \neq \mathbf{0} \tag{e}$$

for all cases other than the trivial one. The three vectors $\mathbf{x}_1$, $\mathbf{x}_2$, and $\mathbf{x}_4$ span a three-dimensional space.

1.4 Bases and dimension of a vector space

A vector space L over F is said to be finite dimensional if there exists a finite set of vectors $\mathbf{x}_1, \mathbf{x}_2, \ldots, \mathbf{x}_n$ which span L, i.e., such that every vector in L is a linear combination of $\mathbf{x}_1, \mathbf{x}_2, \ldots, \mathbf{x}_n$.

Let L be a vector space over F. A set of vectors $\mathbf{x}_1, \mathbf{x}_2, \ldots, \mathbf{x}_n$ which span L is called a *generating system* of L. If $\mathbf{x}_1, \mathbf{x}_2, \ldots, \mathbf{x}_n$ are linearly independent and span L, then the generating system is called a *basis* for L. If L is a finite-dimensional vector space, any two bases for L contain the same number of vectors. The basis can be regarded as the generalization of the concept of a coordinate system.

Let L be a finite dimensional vector space over F. The *dimension* of L is defined as the number of vectors in any basis for L. This integer is denoted by dim L. The vector space L^n is spanned by n linearly independent vectors, so that dim $L^n = n$.

Consider an arbitrary n-vector $\mathbf{x}$ in L^n with components $x_1, x_2, \ldots, x_n$ and introduce a set of n-vectors given by

$$\mathbf{e}_1 = \begin{bmatrix} 1 \\ 0 \\ \cdot \\ \cdot \\ \cdot \\ 0 \end{bmatrix}, \qquad \mathbf{e}_2 = \begin{bmatrix} 0 \\ 1 \\ \cdot \\ \cdot \\ \cdot \\ 0 \end{bmatrix}, \ldots, \qquad \mathbf{e}_n = \begin{bmatrix} 0 \\ 0 \\ \cdot \\ \cdot \\ \cdot \\ 1 \end{bmatrix} \tag{1.6}$$

The vector $\mathbf{x}$ can be written in terms of the vectors $\mathbf{e}_i$ $(i = 1, 2, \ldots, n)$ as follows:

$$\mathbf{x} = x_1 \mathbf{e}_1 + x_2 \mathbf{e}_2 + \cdots + x_n \mathbf{e}_n = \sum_{i=1}^{n} x_i \mathbf{e}_i \tag{1.7}$$

It follows that L^n is spanned by the set of vectors $\mathbf{e}_i$ $(i = 1, 2, \ldots, n)$, so

5

that the vectors $\mathbf{e}_i$ constitute a generating system of L^n. The set of vectors $\mathbf{e}_i$ can be verified as being linearly independent and they are generally referred to as the *standard basis for* L^n.

Example 1.2

The vectors $\mathbf{x}_1$, $\mathbf{x}_2$, and $\mathbf{x}_4$ of Example 1.1 form a basis for a three-dimensional vector space. Any vector $\mathbf{x}$ in L^3 can be written as a unique linear combination of $\mathbf{x}_1$, $\mathbf{x}_2$, and $\mathbf{x}_4$. For example, it can be verified that the vector

$$\mathbf{x} = \begin{bmatrix} 3 \\ 0 \\ 4 \end{bmatrix} \tag{a}$$

can be represented in the form

$$\mathbf{x} = 2\mathbf{x}_1 + 3\mathbf{x}_2 - \mathbf{x}_4 \tag{b}$$

The same vector $\mathbf{x}$ can be also represented in the terms of the standard basis $\mathbf{e}_1, \mathbf{e}_2, \mathbf{e}_3$ for L^3. Indeed, it is easy to see that

$$\mathbf{x} = 3\mathbf{e}_1 + 0\mathbf{e}_2 + 4\mathbf{e}_3 \tag{c}$$

1.5   Inner products and orthogonal vectors

Various concepts encountered in two- and three-dimensional spaces, such as the length of a vector and orthogonality, can be generalized to n-dimensional spaces. This requires the introduction of additional definitions.

Let L^n be an n-dimensional vector space defined over the field F of scalars. If to each pair of vectors $\mathbf{x}$ and $\mathbf{y}$ in L^n is assigned a unique scalar in F, called the *inner product* of $\mathbf{x}$ and $\mathbf{y}$, then L^n is said to be an *inner product space*. The vectors $\mathbf{x}$ and $\mathbf{y}$ can be complex, in which case $\bar{\mathbf{x}}$ and $\bar{\mathbf{y}}$ denote their complex conjugates. The inner product is denoted by $(\mathbf{x}, \mathbf{y})$ and must satisfy the following postulates:

1. $(\mathbf{x}, \mathbf{x}) \geq 0$ for all $\mathbf{x}$ in L^n and $(\mathbf{x}, \mathbf{x}) = 0$ if and only if $\mathbf{x} = \mathbf{0}$.
2. $(\mathbf{x}, \mathbf{y}) = \overline{(\mathbf{y}, \mathbf{x})}$
3. $(\lambda\mathbf{x}, \mathbf{y}) = \lambda(\mathbf{x}, \mathbf{y})$ and $(\mathbf{x}, \lambda\mathbf{y}) = \bar{\lambda}(\mathbf{x}, \mathbf{y})$ for all λ in F
4. $(\mathbf{x}, \mathbf{y} + \mathbf{z}) = (\mathbf{x}, \mathbf{y}) + (\mathbf{x}, \mathbf{z})$ for all $\mathbf{x}$, $\mathbf{y}$, and $\mathbf{z}$ in L^n.

The most common definition of the *complex inner product* is

$$(\mathbf{x}, \mathbf{y}) = x_1\bar{y}_1 + x_2\bar{y}_2 + \cdots + x_n\bar{y}_n \tag{1.8}$$

which represents a complex number. An inner product space defined over the field of complex numbers is called a *unitary space*.

When **x** and **y** are real vectors, Eq. (1.8) reduces to

$$(\mathbf{x}, \mathbf{y}) = x_1 y_1 + x_2 y_2 + \cdots + x_n y_n \tag{1.9}$$

which defines the *real inner product*, a real number. A finite-dimensional inner product space defined over the real scalar field is called a *Euclidean space*.

It is often desirable to have a measure of the size of a vector. Such a measure is called the *norm*. It is designated by the symbol $\|\mathbf{x}\|$ and is required to possess the following properties:

1. $\|\mathbf{x}\| \geq 0$ and $\|\mathbf{x}\| = 0$ if and only if $\mathbf{x} = \mathbf{0}$
2. $\|\lambda \mathbf{x}\| = |\lambda|\, \|\mathbf{x}\|$ for any scalar λ
3. $\|\mathbf{x} + \mathbf{y}\| \leq \|\mathbf{x}\| + \|\mathbf{y}\|$

where property 3 is known as the triangle inequality. Note that $|\lambda|$ denotes the absolute value, or modulus of λ.

A commonly used norm is the *quadratic norm*

$$\|\mathbf{x}\| = (\mathbf{x}, \mathbf{x})^{1/2} \tag{1.10}$$

which defines the *length* of the vector **x**. In the case of real vector spaces, Eq. (1.10) reduces to

$$\|\mathbf{x}\| = \left(\sum_{i=1}^{n} x_i^2 \right)^{1/2} \tag{1.11}$$

which defines the *Euclidean norm*. Equation (1.11) can be recognized as the extension to n dimensions of the ordinary concept of length of a vector in two and three dimensions.

A vector whose norm is equal to unity, $\|\mathbf{x}\| = (\mathbf{x}, \mathbf{x})^{1/2} = 1$, is called a *unit vector*. Any nonzero vector can be *normalized* so as to form a unit vector by simply dividing the vector by its norm

$$\hat{\mathbf{x}} = \frac{\mathbf{x}}{\|\mathbf{x}\|} \tag{1.12}$$

It is easy to verify that the vectors $\mathbf{e}_i$ defined by Eqs. (1.6) are unit vectors.

When the vectors **x** and **y** are real, the inner product is sometimes referred to as the *dot product*. We recall from ordinary vector analysis that the dot product of two vectors in the two- and three-dimensional space can be used to define the cosine of the angle between the two vectors.

This concept can be generalized to the n-dimensional space by writing

$$\cos \theta = \frac{(\mathbf{x}, \mathbf{y})}{\|\mathbf{x}\| \, \|\mathbf{y}\|} = (\hat{\mathbf{x}}, \hat{\mathbf{y}}) \tag{1.13}$$

Any two vectors $\mathbf{x}$ and $\mathbf{y}$ in L^n are said to be *orthogonal* if and only if

$$(\mathbf{x}, \mathbf{y}) = 0 \tag{1.14}$$

which represents a generalization of the ordinary concept of perpendicularity. If each pair of vectors in a given set are mutually orthogonal, then the set is said to be an *orthogonal set*. If, in addition, the vectors have unit norms, the vectors are said to be *orthonormal*. *Any set of mutually orthogonal nonzero vectors in L^n is linearly independent.* To show this, let us assume that the orthogonal set of vectors $\mathbf{x}_1, \mathbf{x}_2, \ldots, \mathbf{x}_n$ satisfies a relation of the type (1.5) and form the inner products

$$\begin{aligned}
0 = (\mathbf{x}_i, \mathbf{0}) &= (\mathbf{x}_i, \alpha_1 \mathbf{x}_1 + \alpha_2 \mathbf{x}_2 + \cdots + \alpha_n \mathbf{x}_n) \\
&= \alpha_1 (\mathbf{x}_i, \mathbf{x}_1) + \alpha_2 (\mathbf{x}_i, \mathbf{x}_2) + \cdots + \alpha_n (\mathbf{x}_i, \mathbf{x}_n) \\
&= \alpha_i (\mathbf{x}_i, \mathbf{x}_i), \quad i = 1, 2, \ldots, n
\end{aligned} \tag{1.15}$$

Because $(\mathbf{x}_i, \mathbf{x}_i) \neq 0$, it follows that Eqs. (1.15) can be satisfied if and only if all the coefficients α_i are identically zero, so that the set of vectors $\mathbf{x}_1, \mathbf{x}_2, \ldots, \mathbf{x}_n$ must be linearly independent. Owing to the independence property, orthogonal vectors, and in particular orthonormal vectors, are convenient choices for basis vectors. A classical example of an orthonormal set of vectors used as a basis are the unit vectors $\mathbf{e}_i$, which explains why these vectors are referred to as a standard basis for L^n.

1.6   The Gram-Schmidt orthogonalization process

Orthogonal vectors are by definition independent, but independent vectors are not necessarily orthogonal. A set of independent vectors, however, can be rendered orthogonal. In computational work, it is often desirable to work with a set of orthogonal vectors, so that the procedure for rendering independent vectors orthogonal is of special interest. The procedure is known as the *Gram-Schmidt orthogonalization process*.

Let us consider the set of independent vectors $\mathbf{x}_1, \mathbf{x}_2, \ldots, \mathbf{x}_n$ and denote the desired orthogonal vectors by $\mathbf{y}_1, \mathbf{y}_2, \ldots, \mathbf{y}_n$. These latter vectors can be normalized by dividing each of the vectors by its norm, so that the orthonormal vectors $\hat{\mathbf{y}}_1, \hat{\mathbf{y}}_2, \ldots, \hat{\mathbf{y}}_n$ are given by

$$\hat{\mathbf{y}}_i = \mathbf{y}_i / \|\mathbf{y}_i\|, \quad i = 1, 2, \ldots, n \tag{1.16}$$

The first vector of the desired orthonormal set is simply

$$\hat{\mathbf{y}}_1 = \hat{\mathbf{x}}_1 = \mathbf{x}_1 / \|\mathbf{x}_1\| \tag{1.17}$$

The second vector, $\mathbf{y}_2$, must be orthogonal to $\hat{\mathbf{y}}_1$. A vector $\mathbf{y}_2$ satisfying this condition can be taken in the form

$$\mathbf{y}_2 = \mathbf{x}_2 - (\mathbf{x}_2, \hat{\mathbf{y}}_1)\hat{\mathbf{y}}_1 \tag{1.18}$$

Indeed, we have

$$(\mathbf{y}_2, \hat{\mathbf{y}}_1) = (\mathbf{x}_2, \hat{\mathbf{y}}_1) - (\mathbf{x}_2, \hat{\mathbf{y}}_1) = 0 \tag{1.19}$$

Of course, the vector $\mathbf{y}_2$ can be normalized by using the second of Eqs. (1.16) to obtain $\hat{\mathbf{y}}_2$. The third vector, $\mathbf{y}_3$, can be written in the form

$$\mathbf{y}_3 = \mathbf{x}_3 - (\mathbf{x}_3, \hat{\mathbf{y}}_1)\hat{\mathbf{y}}_1 - (\mathbf{x}_3, \hat{\mathbf{y}}_2)\hat{\mathbf{y}}_2 \tag{1.20}$$

which is orthonormal to $\hat{\mathbf{y}}_1$ and $\hat{\mathbf{y}}_2$, as

$$\begin{aligned}
(\mathbf{y}_3, \hat{\mathbf{y}}_1) &= (\mathbf{x}_3, \hat{\mathbf{y}}_1) - (\mathbf{x}_3, \hat{\mathbf{y}}_1) = 0 \\
(\mathbf{y}_3, \hat{\mathbf{y}}_2) &= (\mathbf{x}_3, \hat{\mathbf{y}}_2) - (\mathbf{x}_3, \hat{\mathbf{y}}_2) = 0
\end{aligned} \tag{1.21}$$

The vector $\mathbf{y}_3$ can be normalized to obtain $\hat{\mathbf{y}}_3$. Generalizing, we can write

$$\mathbf{y}_i = \mathbf{x}_i - \sum_{j=1}^{i-1} (\mathbf{x}_i, \hat{\mathbf{y}}_j)\hat{\mathbf{y}}_j, \quad i = 1, 2, \ldots, n \tag{1.22}$$

which can be used to compute $\hat{\mathbf{y}}_i$. Clearly, $\hat{\mathbf{y}}_i$ is orthonormal to $\hat{\mathbf{y}}_1, \hat{\mathbf{y}}_2, \ldots, \hat{\mathbf{y}}_{i-1}$ ($i = 2, 3, \ldots, n$). The process is concluded with the computation of $\hat{\mathbf{y}}_n$.

The Gram-Schmidt process described above can often yield computed vectors that are far from being orthogonal (N1, p. 148). An orthogonalization process that is mathematically equivalent but computationally superior to the Gram-Schmidt process is the *modified Gram-Schmidt process*. In the ordinary Gram-Schmidt process, an orthonormal basis $\hat{\mathbf{y}}_i$ ($i = 1, 2, \ldots, n$) is computed in successive steps without altering the original vectors $\mathbf{x}_i$ ($i = 1, 2, \ldots, n$). In the modified Gram-Schmidt process, however, upon computing $\mathbf{y}_i$ the vectors $\mathbf{x}_{i+1}, \mathbf{x}_{i+2}, \ldots, \mathbf{x}_n$ are also changed by insisting that they be orthogonal to $\hat{\mathbf{y}}_i$, as well as to $\hat{\mathbf{y}}_1, \hat{\mathbf{y}}_2, \ldots, \hat{\mathbf{y}}_{i-1}$. The first step is as given by Eq. (1.17), but in addition the vectors $\mathbf{x}_i$ ($i = 2, 3, \ldots, n$) are modified by writing

$$\mathbf{x}_i^{(1)} = \mathbf{x}_i - (\hat{\mathbf{y}}_1, \mathbf{x}_i)\hat{\mathbf{y}}_1, \quad i = 2, 3, \ldots, n \tag{1.23}$$

Forming the inner product $(\hat{\mathbf{y}}_1, \mathbf{x}_i^{(1)})$, we conclude that $\mathbf{x}_i^{(1)}$ ($i = 2, 3, \ldots, n$) are all orthogonal to $\hat{\mathbf{y}}_1$. The next step consists of normalizing $\mathbf{x}_2^{(1)}$ to

1 Concepts from linear algebra

produce $\hat{\mathbf{y}}_2$ as well as of modifying $\mathbf{x}_i^{(1)}$ $(i = 3, 4, \ldots, n)$ by writing

$$\mathbf{x}_i^{(2)} = \mathbf{x}_i^{(1)} - (\hat{\mathbf{y}}_2, \mathbf{x}_i^{(1)})\hat{\mathbf{y}}_2, \quad i = 3, 4, \ldots, n \tag{1.24}$$

It is not difficult to verify that $\mathbf{x}_i^{(2)}$ $(i = 3, 4, \ldots, n)$ are all orthogonal to both $\hat{\mathbf{y}}_1$ and $\hat{\mathbf{y}}_2$. Of course, $\hat{\mathbf{y}}_3$ is obtained by normalizing $\mathbf{x}_3^{(2)}$. Generalizing, the jth step consists of computing $\mathbf{x}_j^{(j-1)}$ and normalizing it to produce $\hat{\mathbf{y}}_j$, or

$$\hat{\mathbf{y}}_j = \mathbf{x}_j^{(j-1)}/\|\mathbf{x}_j^{(j-1)}\| \tag{1.25}$$

and then computing the modified vectors

$$\mathbf{x}_i^{(j)} = \mathbf{x}_i^{(j-1)} - (\hat{\mathbf{y}}_j, \mathbf{x}_i^{(j-1)})\hat{\mathbf{y}}_j, \quad i = j+1, j+2, \ldots, n \tag{1.26}$$

which renders $\mathbf{x}_i^{(j)}$ orthogonal to $\hat{\mathbf{y}}_j$ as well as to $\hat{\mathbf{y}}_1, \hat{\mathbf{y}}_2, \ldots, \hat{\mathbf{y}}_{j-1}$, or

$$(\hat{\mathbf{y}}_k, \mathbf{x}_i^{(j)}) = 0, \quad k = 1, 2, \ldots, j \tag{1.27}$$

The process is completed with the computation and normalization of $\mathbf{x}_n^{(n-1)}$, yielding $\hat{\mathbf{y}}_n$.

When the vectors $\mathbf{x}_i$ $(i = 1, 2, \ldots, n)$ are independent, the ordinary and the modified Gram-Schmidt processes yield the same results. When the vectors $\mathbf{x}_i$ are nearly dependent, however, the ordinary Gram-Schmidt process fails to yield orthonormal vectors, but the modified Gram-Schmidt process does yield vectors that are nearly orthonormal (N1, p. 149).

Example 1.3

Consider the vectors

$$\mathbf{x}_1 = \begin{bmatrix} 1 \\ 1 \\ 1 \end{bmatrix}, \quad \mathbf{x}_2 = \begin{bmatrix} 1 \\ 1 \\ -1 \end{bmatrix}, \quad \mathbf{x}_3 = \begin{bmatrix} 1 \\ -1 \\ 1 \end{bmatrix} \tag{a}$$

and obtain an orthonormal basis in terms of these vectors by the modified Gram-Schmidt process. Use Euclidean norms for the vectors.

From Eq. (1.17), we obtain the first normalized vector

$$\hat{\mathbf{y}}_1 = \hat{\mathbf{x}}_1 = \mathbf{x}_1/\|\mathbf{x}_1\| = \frac{1}{\sqrt{3}}\begin{bmatrix} 1 \\ 1 \\ 1 \end{bmatrix} \tag{b}$$

so that, from Eqs. (1.23), we can write

$$\mathbf{x}_2^{(1)} = \mathbf{x}_2 - (\hat{\mathbf{y}}_1, \mathbf{x}_2)\hat{\mathbf{y}}_1 = \begin{bmatrix} 1 \\ 1 \\ -1 \end{bmatrix} - \left(\frac{1}{\sqrt{3}}\begin{bmatrix} 1 \\ 1 \\ 1 \end{bmatrix}, \begin{bmatrix} 1 \\ 1 \\ -1 \end{bmatrix} \right)\frac{1}{\sqrt{3}}\begin{bmatrix} 1 \\ 1 \\ 1 \end{bmatrix} = \frac{2}{3}\begin{bmatrix} 1 \\ 1 \\ -2 \end{bmatrix}$$

(c)

$$\mathbf{x}_3^{(1)} = \mathbf{x}_3 - (\hat{\mathbf{y}}_1, \mathbf{x}_3)\hat{\mathbf{y}}_1 = \begin{bmatrix} 1 \\ -1 \\ 1 \end{bmatrix} - \left(\frac{1}{\sqrt{3}}\begin{bmatrix} 1 \\ 1 \\ 1 \end{bmatrix}, \begin{bmatrix} 1 \\ -1 \\ 1 \end{bmatrix} \right)\frac{1}{\sqrt{3}}\begin{bmatrix} 1 \\ 1 \\ 1 \end{bmatrix} = \frac{2}{3}\begin{bmatrix} 1 \\ -2 \\ 1 \end{bmatrix}$$

The second vector of the orthonormal set is obtained by simply normalizing $\mathbf{x}_2^{(1)}$, or

$$\hat{\mathbf{y}}_2 = \mathbf{x}_2^{(1)}/\|\mathbf{x}_2^{(1)}\| = \frac{1}{\sqrt{6}}\begin{bmatrix} 1 \\ 1 \\ -2 \end{bmatrix}$$

(d)

Finally, from Eq. (1.24), we can write

$$\mathbf{x}_3^{(2)} = \mathbf{x}_3^{(1)} - (\hat{\mathbf{y}}_2, \mathbf{x}_3^{(1)})\hat{\mathbf{y}}_2 = \frac{2}{3}\begin{bmatrix} 1 \\ -2 \\ 1 \end{bmatrix}$$

$$- \left(\frac{1}{\sqrt{6}}\begin{bmatrix} 1 \\ 1 \\ -2 \end{bmatrix}, \frac{2}{3}\begin{bmatrix} 1 \\ -2 \\ 1 \end{bmatrix} \right)\frac{1}{\sqrt{6}}\begin{bmatrix} 1 \\ 1 \\ -2 \end{bmatrix} = \begin{bmatrix} 1 \\ -1 \\ 0 \end{bmatrix} \quad \text{(e)}$$

so that

$$\hat{\mathbf{y}}_3 = \mathbf{x}_3^{(2)}/\|\mathbf{x}_3^{(2)}\| = \frac{1}{\sqrt{2}}\begin{bmatrix} 1 \\ -1 \\ 0 \end{bmatrix}$$

(f)

It can be verified easily that the vectors $\hat{\mathbf{y}}_1$, $\hat{\mathbf{y}}_2$, and $\hat{\mathbf{y}}_3$ are orthonormal.

1.7   Matrices

A *matrix* is a rectangular array of scalars of the form

$$A = \begin{bmatrix} a_{11} & a_{12} & \cdots & a_{1n} \\ a_{21} & a_{22} & \cdots & a_{2n} \\ \multicolumn{4}{c}{\dotfill} \\ a_{m1} & a_{m2} & \cdots & a_{mn} \end{bmatrix}$$

(1.28)

1 Concepts from linear algebra

The scalars a_{ij} $(i = 1, 2, \ldots, m; j = 1, 2, \ldots, n)$, called the *elements* of A, belong to a given field F. The field is assumed to be either the real field R or the complex field C. Because the matrix A has m rows and n columns it is referred to as an $m \times n$ *matrix*. It is customary to say that the *dimensions* of A are $m \times n$. The position of the element a_{ij} in the matrix A is in the ith row and jth column, so that i is referrred to as the row index and j as the column index.

If $m = n$, the matrix A reduces to a *square matrix of order* n. The elements a_{ii} in the square matrix are called the *main diagonal elements* of A. The remaining elements are referred to as the *off-diagonal elements* of A. In the special case in which all the off-diagonal elements of A are zero, the matrix A is said to be a *diagonal matrix*. If A is diagonal and if all its diagonal elements are unity, $a_{ii} = 1$, then the matrix is called the *unit matrix*, or *identity matrix*, and denoted by I. Introducing the Kronecker delta symbol δ_{ij} defined as

$$
\begin{aligned}
\delta_{ij} &= 1 \quad \text{if} \quad i = j \\
\delta_{ij} &= 0 \quad \text{if} \quad i \neq j
\end{aligned}
\tag{1.29}
$$

the identity matrix can be regarded as a matrix with every element equal to the Kronecker delta and can be written in the form $I = [\delta_{ij}]$. Similarly, a diagonal matrix D can be written in terms of the Kronecker delta in the form $D = [a_{ij}\delta_{ij}]$.

An $m \times n$ matrix A is said to be *upper* (*lower*) *trapezoidal* if $a_{ij} = 0$ for $i > j$ $(i < j)$. A square upper (lower) trapezoidal matrix is said to be *upper* (*lower*) *triangular*. If the diagonal elements of an upper (lower) triangular matrix are unity, then the matrix is referred to as *unit upper* (*lower*) *triangular*.

A square matrix A is said to be *upper* (*lower*) *Hessenberg* if $a_{ij} = 0$ for $i > j + 1$ $(i < j - 1)$. If A is upper and lower Hessenberg simultaneously, then it is said to be *tridiagonal*. Clearly, a tridiagonal matrix has nonzero elements only on the main diagonal and on the diagonals immediately above and below the main diagonal.

A matrix obtained from A by interchanging all its rows and columns is referred to as the *transpose* of A and is denoted by A^T. Hence,

$$
A^T = \begin{bmatrix}
a_{11} & a_{21} & \cdots & a_{m1} \\
a_{12} & a_{22} & \cdots & a_{m2} \\
\multicolumn{4}{c}{\dotfill} \\
a_{1n} & a_{2n} & \cdots & a_{mn}
\end{bmatrix}
\tag{1.30}
$$

It is obvious that if A is an $m \times n$ matrix, then A^T is an $n \times m$ matrix.

Next, let us consider a square matrix A. If the elements of A are such that $a_{ij} = a_{ji}$, then the matrix A is said to be *symmetric*. Hence, a matrix is symmetric if $A = A^T$. On the other hand, if the elements of A are such that $a_{ij} = -a_{ji}$ for $i \neq j$ and $a_{ii} = 0$, then the matrix A is said to be *skew symmetric*. It follows that A is skew symmetric if $A = -A^T$.

Let A be an $n \times n$ matrix with real or complex elements and define the *conjugate* of A as the $n \times n$ matrix $\bar{A}$ whose elements are given by $\bar{a}_{ij}$. Moreover, let A^H be the *conjugate transpose* of A, $A^H = \bar{A}^T$. The matrix A^H is sometimes referred to as the *adjoint matrix*. If the matrix A is such that

$$A = A^H \tag{1.31}$$

then the matrix A is said to be *Hermitian*. Because Eq. (1.31) states that the matrix A is equal to its adjoint, Hermitian matrices are called *self-adjoint*. Note that *a real Hermitian matrix is symmetric* and *a pure imaginary Hermitian matrix is skew symmetric*.

A matrix consisting of one column and n rows is called a *column matrix* and denoted by

$$\mathbf{x} = \begin{bmatrix} x_1 \\ x_2 \\ \cdot \\ \cdot \\ \cdot \\ x_n \end{bmatrix} \tag{1.32}$$

The column matrix $\mathbf{x}$ can clearly be identified with a vector in L^n and is also known as a *column vector*. The transpose of the column matrix $\mathbf{x}$ is the *row matrix*

$$\mathbf{x}^T = [x_1 \ x_2 \cdots x_n] \tag{1.33}$$

and is also called a *row vector*.

A matrix with all its elements equal to zero is called the *zero matrix* or the *null matrix* and is denoted by 0, $\mathbf{0}$, or $\mathbf{0}^T$, depending on whether it is a rectangular, a column, or a row matrix, respectively.

1.8 Basic matrix operations

Two matrices A and B are said to be equal if and only if they have the same number of rows and columns and $a_{ij} = b_{ij}$ for all pairs of subscripts i and j.

If A and B are two $m \times n$ matrices, then the sum of A and B is

1 Concepts from linear algebra

defined as a matrix C whose elements are

$$c_{ij} = a_{ij} + b_{ij}, \quad i = 1, 2, \ldots, m; \quad j = 1, 2, \ldots, n \quad (1.34)$$

Clearly, $C = A + B$ is also an $m \times n$ matrix. Matrix addition is *commutative* and *associative*. Indeed, if A, B and C are arbitrary $m \times n$ matrices, then

$$A + B = B + A, \quad (A + B) + C = A + (B + C) \quad (1.35)$$

The *product of a matrix A and a scalar* α implies that every element of A is multiplied by α. Hence, if A is an $m \times n$ matrix, then the statement $C = \alpha A$ implies

$$c_{ij} = \alpha a_{ij}, \quad i = 1, 2, \ldots, m; \quad j = 1, 2, \ldots, n \quad (1.36)$$

Next, let us define the *product of two matrices*. If A is an $m \times n$ matrix and B is an $n \times p$ matrix, then the product $C = AB$ of the two matrices is a matrix with the elements

$$c_{ij} = a_{i1}b_{1j} + a_{i2}b_{2j} + \cdots + a_{in}b_{nj} = \sum_{k=1}^{n} a_{ik}b_{kj} \quad (1.37)$$

It is clear from the above that the matrix product is defined only if the number of columns of A is equal to the number of rows of B. In this case, the matrices A and B are said to be *conformable* in the order stated. The matrix product AB can be described as B *premultiplied by* A or A *postmultiplied by* B. It can also be described as B *multiplied on the left by* A or A *multiplied on the right by* B. Matrix multiplication is in general *not commutative*

$$AB \neq BA \quad (1.38)$$

In fact, unless $m = p$ the matrix product BA is not even defined. One notable exception is the case in which one of the matrices is the identity matrix because then

$$AI = IA = A \quad (1.39)$$

Clearly, the order of I must be such that the product is defined.

The matrix product

$$AB = 0 \quad (1.40)$$

does not imply that either A or B, or both A and B, are null matrices. Indeed, an illustration of this statement is provided by

$$\begin{bmatrix} 1 & 1 \\ 1 & 1 \end{bmatrix} \begin{bmatrix} 1 & -1 \\ -1 & 1 \end{bmatrix} = \begin{bmatrix} 0 & 0 \\ 0 & 0 \end{bmatrix} \quad (1.41)$$

14

The matrix product satisfies *associative laws*. If A, B, and C are $m \times n$, $n \times p$ and $p \times q$ matrices, respectively, then it can be verified that

$$D = (AB)C = A(BC) \tag{1.42}$$

is an $m \times q$ matrix whose elements are given by

$$d_{ij} = \sum_{l=1}^{p} \sum_{k=1}^{n} a_{ik} b_{kl} c_{lj} = \sum_{k=1}^{n} \sum_{l=1}^{p} a_{ik} b_{kl} c_{lj} \tag{1.43}$$

The matrix product satisfies *distributive laws*. If A and B are $m \times n$ matrices, C is a $p \times m$ matrix, and D is an $n \times q$ matrix, then it can be shown that

$$C(A + B) = CA + CB, \qquad (A + B)D = AD + BD \tag{1.44}$$

If A is an $m \times n$ matrix and B is an $n \times p$ matrix, so that the product $C = AB$ is given by Eq. (1.37), then

$$C^{T} = (AB)^{T} = B^{T} A^{T} \tag{1.45}$$

To show this, we recognize that to any element a_{ik} in A corresponds the element a_{ki} in A^{T}, and to any element b_{kj} in B corresponds the element b_{jk} in B^{T}. Then, the product

$$\sum_{k=1}^{n} b_{jk} a_{ki} = c_{ji} \tag{1.46}$$

establishes the validity of Eq. (1.45). In words, *the transpose of a product of two matrices is equal to the product of the transposed matrices in reversed order.* As a corollary, it can be verified that if

$$C = A_{1} A_{2} \cdots A_{s-1} A_{s} \tag{1.47}$$

then

$$C^{T} = A_{s}^{T} A_{s-1}^{T} \cdots A_{2}^{T} A_{1}^{T} \tag{1.48}$$

Example 1.4

Calculate the matrix product AB, where

$$A = \begin{bmatrix} 2 & -3 \\ 1 & 5 \end{bmatrix}, \qquad B = \begin{bmatrix} 1 & 3 & 7 \\ -1 & 4 & 2 \end{bmatrix} \tag{a}$$

What can be said about the matrix product BA?

The matrix product AB is formed as follows:

$$AB = \begin{bmatrix} 2(1) - 3(-1) & 2(3) - 3(4) & 2(7) - 3(2) \\ 1(1) + 5(-1) & 1(3) + 5(4) & 1(7) + 5(2) \end{bmatrix} = \begin{bmatrix} 5 & -6 & 8 \\ -4 & 23 & 17 \end{bmatrix} \tag{b}$$

15

The matrix product BA is not defined because B is a 2×3 matrix and A a 2×2 matrix and hence the matrices are not conformable in that order.

Example 1.5

Calculate the matrix products AB and CA, where

$$
A = \begin{bmatrix} 0 & 1 \\ 1 & 0 \end{bmatrix}, \quad
B = \begin{bmatrix} 2 & 4 & 1 & -2 \\ -1 & 5 & 7 & 3 \end{bmatrix}, \quad
C = \begin{bmatrix} 1 & 3 \\ 7 & 4 \\ -2 & 2 \end{bmatrix} \tag{a}
$$

The matrix products AB and CA are as follows:

$$
AB = \begin{bmatrix} -1 & 5 & 7 & 3 \\ 2 & 4 & 1 & -2 \end{bmatrix}, \quad
CA = \begin{bmatrix} 3 & 1 \\ 4 & 7 \\ 2 & -2 \end{bmatrix} \tag{b}
$$

We observe that the matrix AB is obtained from B by interchanging the rows. Similarly, CA is obtained from C by interchanging the columns. Hence, the effect of premultiplying a matrix by A is to permute its rows and the effect of postmultiplying a matrix by A is to permute its columns. For this reason, A is called a *permutation matrix*. In general, a permutation matrix is a matrix obtained by interchanging rows or columns of the identity matrix.

1.9   Determinants

If A is any square matrix of order n with elements in the field F, then it is possible to associate with A a number in F called the *determinant* of A, and denoted by $\det A$ or $|A|$. The determinant of A is said to be of *order* n and can be exhibited in the form

$$
\det A = |A| = \begin{vmatrix}
a_{11} & a_{12} & \cdots & a_{1n} \\
a_{21} & a_{22} & \cdots & a_{2n} \\
\multicolumn{4}{c}{\dotfill} \\
a_{n1} & a_{n2} & \cdots & a_{nn}
\end{vmatrix} \tag{1.49}
$$

Determinants have many interesting and useful properties. We shall examine only those properties pertinent to our future study.

Unlike the matrix A, which represents a given array of numbers, the determinant of A represents a single number with a unique value that can be calculated by following rules for the expansion of a determinant. The expansion rules can be most conveniently discussed by introducing the concept of minor determinants. The *minor determinant* $|M_{rs}|$ corresponding to the element a_{rs} is the determinant obtained from $|A|$ by striking out the rth row and sth column. Clearly, the order of $|M_{rs}|$ is $n-1$. The signed minor determinant corresponding to the element a_{rs} is called the *cofactor* of a_{rs} and is given by

$$\det A_{rs} = |A_{rs}| = (-1)^{r+s} |M_{rs}| \tag{1.50}$$

The value of the determinant of A can be obtained by expanding the determinant in terms of cofactors by the rth row as follows:

$$|A| = \sum_{s=1}^{n} a_{rs} |A_{rs}| \tag{1.51}$$

The determinant can also be expanded by the sth column in the form

$$|A| = \sum_{r=1}^{n} a_{rs} |A_{rs}| \tag{1.52}$$

The value of the determinant is unique, regardless of whether it is expanded by a row or by a column, and regardless of by which row or column. The expansion by cofactors is known as a *Laplace expansion*. There are other ways of expanding determinants, but there is not sufficient reason to be concerned with them. The cofactors $|A_{rs}|$ are determinants of order $n-1$. If $n=2$, then these cofactors are simply scalars. If $n>2$, then these cofactors can be expanded in terms of their own cofactors, and the process repeated until the minor determinants are of order 2. As an illustration, let $n=3$ and expand $\det A$ by the first row as follows:

$$\det A = \begin{vmatrix} a_{11} & a_{12} & a_{13} \\ a_{21} & a_{22} & a_{23} \\ a_{31} & a_{32} & a_{33} \end{vmatrix} = a_{11}|A_{11}| + a_{12}|A_{12}| + a_{13}|A_{13}|$$

$$= a_{11} \begin{vmatrix} a_{22} & a_{23} \\ a_{32} & a_{33} \end{vmatrix} - a_{12} \begin{vmatrix} a_{21} & a_{23} \\ a_{31} & a_{33} \end{vmatrix} + a_{13} \begin{vmatrix} a_{21} & a_{22} \\ a_{31} & a_{32} \end{vmatrix}$$

$$= a_{11}(a_{22}a_{33} - a_{23}a_{32}) - a_{12}(a_{21}a_{33} - a_{23}a_{31})$$
$$+ a_{13}(a_{21}a_{32} - a_{22}a_{31}) \tag{1.53}$$

Because the value of $\det A$ is the same, regardless whether the determinant is expanded by a row or by a column, it follows that

$$\det A = \det A^{T} \tag{1.54}$$

or *the determinant of a matrix is equal to the determinant of the transposed*

17

matrix. It is easy to verify that *the determinant of a triangular matrix is equal to the product of the main diagonal elements*. It follows immediately that *the determinant of a diagonal matrix is equal to the product of the diagonal elements*, and *the determinant of the identity matrix is equal to* 1.

If $\det A = 0$, then the matrix A is said to be *singular*, and if $\det A \neq 0$, the matrix is said to be *nonsingular*. Clearly, a matrix with an entire row or an entire column equal to zero is singular. The evaluation of determinants can be greatly simplified by invoking certain properties of determinants. In fact, it is often possible to establish that a determinant is zero without actually expanding it. From Eq. (1.51), one can deduce the following properties:

1. If two rows (or two columns) are interchanged, then the determinant changes sign.
2. If all the elements of one row (or of one column) are multiplied by a scalar α, then the determinant is multiplied by α.
3. The value of a determinant does not change if one row (or one column) multiplied by a scalar α is added or subtracted from another row (or another column).
4. If every element in one row (or one column) is the sum of two terms, then the determinant is equal to the sum of two determinants, each of the two determinants being obtained by splitting every sum so that one term is in one determinant and the remaining term is in the other determinant.

The above properties permit us to make two observations. Property 2 implies that $\det(\alpha A) = \alpha^n \det A$, where α is a scalar and n is the order of the matrix. On the other hand, Property 3 implies that a determinant with two proportional rows, or two proportional columns, is equal to zero. Property 3 can be used in general to simplify the evaluation of a determinant, and in particular to show that its value is zero, if indeed this is the case.

Example 1.6

Calculate the value of the determinant

$$|A| = \begin{vmatrix} 3 & 2 & -1 \\ 1 & 5 & 2 \\ 3 & -1 & 2 \end{vmatrix} \tag{a}$$

Expanding by the first row, we obtain

$$A = 3 \begin{vmatrix} 5 & 2 \\ -1 & 2 \end{vmatrix} - 2 \begin{vmatrix} 1 & 2 \\ 3 & 2 \end{vmatrix} - 1 \begin{vmatrix} 1 & 5 \\ 3 & -1 \end{vmatrix}$$

$$= 3[5(2) - 2(-1)] - 2[1(2) - 2(3)] - [1(-1) - 5(3)] = 60 \qquad \text{(b)}$$

On the other hand, subtracting three times the second row from the first and third rows, we can write

$$A = \begin{vmatrix} 0 & -13 & -7 \\ 1 & 5 & 2 \\ 0 & -16 & -4 \end{vmatrix} \qquad \text{(c)}$$

Next, expanding by the first column, we obtain

$$A = - \begin{vmatrix} -13 & -7 \\ -16 & -4 \end{vmatrix} = -[(-13)(-4) - (-7)(-16)] = 60 \qquad \text{(d)}$$

which is the same value as that given by (b).

1.10   Inverse of a matrix

If A and B are two $n \times n$ matrices such that

$$AB = BA = I \qquad \text{(1.55)}$$

then B is said to be the *inverse* of A and is denoted by

$$B = A^{-1} \qquad \text{(1.56)}$$

Note that at the same time A is the inverse of B, $A = B^{-1}$.

Next, let us derive a formula for the inverse of a matrix. To this end, let us consider Eq. (1.50) and introduce the *cofactor matrix*

$$\text{cof } A = [(-1)^{r+s} |M_{rs}|] \qquad \text{(1.57)}$$

Then, we can write

$$A (\text{cof } A)^T = \left[\sum_{s=1}^{n} (-1)^{r+s} a_{ps} |M_{rs}| \right] \qquad \text{(1.58)}$$

Recalling Eq. (1.51), we conclude that every element of the matrix A $(\text{cof } A)^T$ can be regarded as a determinantal expansion. When $p = r$ the element is simply equal to det A. On the other hand, when $p \neq r$ the result is zero. This can be explained by recognizing that the determinant corresponding to $p \neq r$ is obtained from the matrix A by replacing the rth row by the pth row and keeping the pth row intact. Because the

19

corresponding determinant has two identical rows, its value is zero. In view of this, Eq. (1.58) can be rewritten as

$$A(\text{cof } A)^T = (\det A)I \tag{1.59}$$

where I is the identity matrix of order N. Multiplying Eq. (1.59) on the left by A^{-1} and dividing through by $\det A$, we obtain

$$A^{-1} = \frac{(\text{cof } A)^T}{\det A} \tag{1.60}$$

Note that the transpose of the cofactor matrix is often referred to as the *adjoint* of A, $(\text{cof } A)^T = \text{adj } A$. It should not be confused with the adjoint matrix A^H defined in Sec. 1.7.

If $\det A = 0$, then no matrix B exists such that Eq. (1.55) is satisfied. To show this, we invoke the following theorem (M16, p. 134): *If A and B are $n \times n$ matrices, then*

$$\det AB = \det A \det B \tag{1.61}$$

But, from Eq. (1.55), we conclude that $\det AB = I$ if $B = A^{-1}$ exists, so that $\det A \neq 0$. Hence, if $\det A = 0$, Eq. (1.55) cannot be satisfied, so that $B = A^{-1}$ does not exist. Recalling that when $\det A = 0$ the matrix is singular, it follows that *an $n \times n$ matrix A has an inverse if and only if it is nonsingular.*

To calculate the inverse of a matrix by means of formula (1.60) it is necessary to evaluate a large number of determinants if the order of the matrix is large. For example, if A is of order n, then the calculation of $\det A$ requires the evaluation of $n!/2$ determinants of order 2. Hence, as n increases, this implies an increasingly large number of multiplications coupled with a progressive loss of accuracy, so that the use of the formula is not recommended. Later in this text we shall study more efficient and more accurate methods for the calculation of the inverse of a matrix.

Next, let us consider the product of matrices (1.47). Multiplying both sides of Eq. (1.47) on the right by $A_s^{-1}, A_{s-1}^{-1}, \ldots, A_1^{-1}$, in sequence, and then on the left by C^{-1}, we obtain

$$C^{-1} = A_s^{-1} A_{s-1}^{-1} \cdots A_2^{-1} A_1^{-1} \tag{1.62}$$

or *the inverse of a product of matrices is equal to the product of the inverse matrices in reversed order.* Of course, Eq. (1.62) implies that all the inverse matrices in question exist.

Example 1.7

Calculate the inverse of the matrix

$$A = \begin{bmatrix} 3 & 2 & -1 \\ 1 & 5 & 2 \\ 3 & -1 & 2 \end{bmatrix} \tag{a}$$

First, we evaluate the minor determinants

$$|M_{11}| = \begin{vmatrix} 5 & 2 \\ -1 & 2 \end{vmatrix} = 12, \qquad |M_{12}| = \begin{vmatrix} 1 & 2 \\ 3 & 2 \end{vmatrix} = -4, \qquad |M_{13}| = \begin{vmatrix} 1 & 5 \\ 3 & -1 \end{vmatrix} = -16$$

$$|M_{21}| = \begin{vmatrix} 2 & -1 \\ -1 & 2 \end{vmatrix} = 3, \qquad |M_{22}| = \begin{vmatrix} 3 & -1 \\ 3 & 2 \end{vmatrix} = 9, \qquad |M_{23}| = \begin{vmatrix} 3 & 2 \\ 3 & -1 \end{vmatrix} = -9 \tag{b}$$

$$|M_{31}| = \begin{vmatrix} 2 & -1 \\ 5 & 2 \end{vmatrix} = 9, \qquad |M_{32}| = \begin{vmatrix} 3 & -1 \\ 1 & 2 \end{vmatrix} = 7, \qquad |M_{33}| = \begin{vmatrix} 3 & 2 \\ 1 & 5 \end{vmatrix} = 13$$

Using Eq. (1.45), we obtain the cofactor matrix

$$\text{cof } A = \begin{bmatrix} 12 & -(-4) & (-16) \\ -(3) & 9 & -(-9) \\ 9 & -(7) & 13 \end{bmatrix} = \begin{bmatrix} 12 & 4 & -16 \\ -3 & 9 & 9 \\ 9 & -7 & 13 \end{bmatrix} \tag{c}$$

Recalling from Example 1.5 that det $A = 60$, we can use Eq. (1.60) and write

$$A^{-1} = \frac{1}{60} \begin{bmatrix} 12 & -3 & 9 \\ 4 & 9 & -7 \\ -16 & 9 & 13 \end{bmatrix} \tag{d}$$

1.11 Partitioned matrices

On many occasions it is convenient to partition matrices into submatrices. Then, under proper circumstances, certain matrix operations can be performed by treating the submatrices as if they were single elements. As an example, let us consider a 3×4 matrix A and partition it as follows:

$$A = \begin{bmatrix} a_{11} & a_{12} & a_{13} & a_{14} \\ a_{21} & a_{22} & a_{23} & a_{24} \\ a_{31} & a_{32} & a_{33} & a_{34} \end{bmatrix} = \begin{bmatrix} A_{11} & A_{12} \\ A_{21} & A_{22} \end{bmatrix} \tag{1.63}$$

where

$$A_{11} = \begin{bmatrix} a_{11} & a_{12} \\ a_{21} & a_{22} \end{bmatrix}, \qquad A_{12} = \begin{bmatrix} a_{13} & a_{14} \\ a_{23} & a_{24} \end{bmatrix} \qquad (1.64)$$

$$A_{21} = [a_{31} \quad a_{32}], \qquad A_{22} = [a_{33} \quad a_{34}]$$

are the submatrices of A. Next, let us consider a 4×4 matrix B partitioned in the form

$$B = \begin{bmatrix} b_{11} & b_{12} & \vdots & b_{13} & b_{14} \\ b_{21} & b_{22} & \vdots & b_{23} & b_{24} \\ \cdots & \cdots & \vdots & \cdots & \cdots \\ b_{31} & b_{32} & \vdots & b_{33} & b_{34} \\ b_{41} & b_{42} & \vdots & b_{43} & b_{44} \end{bmatrix} = \begin{bmatrix} B_{11} & \vdots & B_{12} \\ \cdots & \vdots & \cdots \\ B_{21} & \vdots & B_{22} \end{bmatrix} \qquad (1.65)$$

where

$$B_{11} = \begin{bmatrix} b_{11} & b_{12} \\ b_{21} & b_{22} \end{bmatrix}, \qquad B_{12} = \begin{bmatrix} b_{13} & b_{14} \\ b_{23} & b_{24} \end{bmatrix}$$

$$B_{21} = \begin{bmatrix} b_{31} & b_{32} \\ b_{41} & b_{42} \end{bmatrix}, \qquad B_{22} = \begin{bmatrix} b_{33} & b_{34} \\ b_{43} & b_{44} \end{bmatrix} \qquad (1.66)$$

It is not difficult to verify that the matrix product AB can be obtained by treating the submatrices A_{ik} and B_{kj} as if they were ordinary matrices. Indeed, the elements of the product $C = AB$ are

$$C_{ij} = \sum_{k=1}^{2} A_{ik} B_{kj}, \quad i, j = 1, 2 \qquad (1.67)$$

It should be pointed out, however, that products such as (1.67) are possible only if the matrix A_{ik} has as many columns as the matrix B_{kj} has rows, which is clearly true in the particular case at hand.

   If the off-diagonal submatrices of a square matrix are null matrices, then the matrix is said to be *block-diagonal*. For block-diagonal matrices the determinant of the matrix is equal to the product of the determinants of the submatrices on the main diagonal. For example, if B_{12} and B_{21} in Eq. (1.65) are null matrices, then

$$\det B = \det B_{11} \det B_{22} \qquad (1.68)$$

Actually the above statement is true even if the matrix is only *block-triangular*, i.e., if only the submatrices above (or below) the main diagonal are null matrices.

1.12   Systems of linear equations

Let us consider a system of m linear equations in n unknowns $x_1, x_2, \ldots, x_n$ of the form

$$a_{11}x_1 + a_{12}x_2 + \cdots + a_{1n}x_n = c_1$$
$$a_{21}x_1 + a_{22}x_2 + \cdots + a_{2n}x_n = c_2$$
$$\cdots\cdots\cdots\cdots\cdots\cdots\cdots\cdots\cdots\cdots\cdots\cdots\cdots\cdots \qquad (1.69)$$
$$a_{m1}x_1 + a_{m2}x_2 + \cdots + a_{mn}x_n = c_m$$

The system of equations can be written as the compact matrix equation

$$A\mathbf{x} = \mathbf{c} \qquad (1.70)$$

where $A = [a_{ij}]$ is an $m \times n$ matrix known as the matrix of the coefficients, $\mathbf{x} = [x_1 \, x_2 \cdots x_n]^T$ is the n-vector of the unknowns, and $\mathbf{c} = [c_1 \, c_2 \cdots c_m]^T$ is the m-vector of the constants on the right side of Eqs. (1.69). Our interest is in deriving conditions under which Eq. (1.70) has a solution.

The matrix A can be partitioned into n m-dimensional column vectors of the form

$$A = [\mathbf{a}_1 \, \mathbf{a}_2 \cdots \mathbf{a}_n] \qquad (1.71)$$

where $\mathbf{a}_1 = [a_{11} \, a_{21} \cdots a_{m1}]^T$, etc., are the column vectors. In view of this, the matrix product $A\mathbf{x}$ can be looked upon as a linear combination of the columns of A. Hence, Eq. (1.70) can be written as

$$x_1\mathbf{a}_1 + x_2\mathbf{a}_2 + \cdots + x_n\mathbf{a}_n = \mathbf{c} \qquad (1.72)$$

Equation (1.72) implies that the set of all products $A\mathbf{x}$ is the same as the set of linear combinations of the columns of A. The subspace of L^m spanned by the columns of A is called the *column space* of A and denoted by $\mathcal{R}(A)$. If $\mathbf{y}$ is an m-vector, then $\mathbf{y}^T$ is a row vector with m components. Now, if A is partitioned into m row vectors, then the product $\mathbf{y}^T A$ is a linear combination of the rows of A whose coefficients are the components of $\mathbf{y}$. Hence, the *row space* of A, written $\mathcal{R}(A^T)$, is the subspace of L^n spanned by the row vectors of A.

Next, let us define the *rank* of a matrix A, denoted by rank A, as the dimension of the linear space spanned by its columns. Because the latter is simply the dimension of $\mathcal{R}(A)$, we have

$$\text{rank } A = \dim \mathcal{R}(A) \qquad (1.73)$$

It would appear that rank A should have been more properly referred to as the *column rank* of A, which would have naturally called for the introduction of a *row rank* of A as the dimension of $\mathcal{R}(A^T)$. It turns out, however, that the column rank and row rank of any matrix A are equal (M16, p. 110), so that no such distinction is necessary. In view of the definition of the dimension of a linear space, it follows that *the rank of a matrix A is equal to the maximum number of linearly independent columns of A and it is also equal to the maximum number of linearly independent rows of A, where the two numbers must be the same.*

There is one more vector space associated with any $m \times n$ matrix A. This is the *null space* of A, denoted by $\mathcal{N}(A)$ and defined as the space of all the solutions $\mathbf{x} \neq \mathbf{0}$ satisfying the homogeneous equation $A\mathbf{x} = \mathbf{0}$. The dimension of the null space $\mathcal{N}$ is called the *nullity* of A, dim $\mathcal{N}$ = null A.

Let us return now to Eqs. (1.69) and introduce the *augmented matrix* of the system defined by

$$B = [A, \mathbf{c}] = \begin{bmatrix} a_{11} & a_{12} & \cdots & a_{1n} & c_1 \\ a_{21} & a_{22} & \cdots & a_{2n} & c_2 \\ \cdots\cdots\cdots\cdots\cdots\cdots\cdots\cdots\cdots \\ a_{m1} & a_{m2} & \cdots & a_{mn} & c_m \end{bmatrix} \tag{1.74}$$

Then, Eqs. (1.69) *have a solution* $\mathbf{x}$ *if and only if the rank of the augmented matrix B is equal to the rank of A.* If a solution $\mathbf{x}$ exists, then $\mathbf{c}$ is a linear combination of the columns of A and hence lies in $\mathcal{R}(A)$. It follows that $\mathcal{R}(B) = \mathcal{R}(A)$, and rank B = rank A.

The rank of an arbitrary matrix can be connected with the order of its nonsingular square submatrices. Indeed, according to a theorem of linear algebra, *the rank of any matrix A is equal to the order of the square submatrix of A of greatest order whose determinant does not vanish* (M16, p. 140). It follows that:

a. If $m \geq n$, then the largest possible rank of A is n. If rank A = rank B = n, then Eqs. (1.69) have a unique solution.

b. If $m < n$, then the largest possible rank of A is m. If rank A = rank $B = m$, then Eqs. (1.69) have an infinity of solutions. A unique solution can be chosen in the form of the solution with the minimum norm

$$\mathbf{x} = A^H (AA^H)^{-1} \mathbf{c} \tag{1.75}$$

where AA^H is an $m \times m$ matrix of rank m and is therefore non-singular.

The case in which the number of equations is equal to the number of unknowns is of particular interest. If A is a square matrix of order n, then the following statements are equivalent:

1. The rank of A is n, rank $A = n$.

2. The system $A\mathbf{x} = \mathbf{c}$ has a unique solution for arbitrary vectors $\mathbf{c}$.
3. The system $A\mathbf{x} = \mathbf{0}$ has only the trivial solution $\mathbf{x} = \mathbf{0}$, which implies that null $A = 0$.

The implication of statements 1 and 2 is that the matrix A is nonsingular, so that A possesses an inverse. Considering the case in which the matrix A in Eq. (1.70) is square and premultiplying both sides of the equation by A^{-1}, we obtain

$$\mathbf{x} = A^{-1}\mathbf{c} \tag{1.76}$$

Hence, when A is nonsingular the solution of Eq. (1.70) can be produced by simply calculating the inverse of A. We have shown in Sec. 1.10 that A^{-1} can be obtained by dividing the adjoint of A by the determinant of A; this method for solving sets of simultaneous equations is generally known as *Cramer's rule*. This approach is mainly of academic interest and in practice the procedure is seldom used, especially for large order matrices A, because it involves the evaluation of a large number of determinants, which is time consuming and leads to loss of accuracy. In Sec. 5.2, we shall discuss a more efficient method for deriving the solution of Eq. (1.70), namely, the Gaussian elimination.

Next, let us turn our attention to the homogeneous system $A\mathbf{x} = \mathbf{0}$. We have pointed out earlier that the matrix product $A\mathbf{x}$ represents a linear combination of the column vectors of A. Because this linear combination must equal zero, it follows from Sec. 1.3 that the columns of A are not independent. Hence, the rank of A must be less than n, so that $\det A = 0$. This conclusion can be stated in a more formal manner by means of the well-known theorem of linear algebra: *If A is an $n \times n$ matrix, then the equation $A\mathbf{x} = \mathbf{0}$ has a nontrivial solution $\mathbf{x} \neq \mathbf{0}$ if and only if $\det A = 0$.*

As an application of the above theorem, let us devise a test for the dependence of a set of n-vectors $\mathbf{y}_1, \mathbf{y}_2, \ldots, \mathbf{y}_n$. If the vectors are to be linearly dependent, then they must satisfy a relation of the type

$$\alpha_1 \mathbf{y}_1 + \alpha_2 \mathbf{y}_2 + \cdots + \alpha_n \mathbf{y}_n = \mathbf{0} \tag{1.77}$$

where $\alpha_1, \alpha_2, \ldots, \alpha_n$ are constant scalars. Next, let us form the inner products $(\mathbf{y}_i, \mathbf{y}_j)$. We have shown in Sec. 1.7, however, that vectors can be represented by column matrices. In view of this, the inner product can be written in the matrix form

$$(\mathbf{y}_i, \mathbf{y}_j) = \bar{\mathbf{y}}_j^T \mathbf{y}_i \tag{1.78}$$

Hence, premultiplying Eq. (1.77) by $\bar{\mathbf{y}}_1^T, \bar{\mathbf{y}}_2^T, \ldots, \bar{\mathbf{y}}_n^T$, in sequence, we

25

obtain

$$\alpha_1 \bar{\mathbf{y}}_1^T \mathbf{y}_1 + \alpha_2 \bar{\mathbf{y}}_1^T \mathbf{y}_2 + \cdots + \alpha_n \bar{\mathbf{y}}_1^T \mathbf{y}_n = 0$$
$$\alpha_1 \bar{\mathbf{y}}_2^T \mathbf{y}_1 + \alpha_2 \bar{\mathbf{y}}_2^T \mathbf{y}_2 + \cdots + \alpha_n \bar{\mathbf{y}}_2^T \mathbf{y}_n = 0 \qquad (1.79)$$
$$\cdots\cdots\cdots\cdots\cdots\cdots\cdots\cdots\cdots\cdots\cdots\cdots$$
$$\alpha_1 \bar{\mathbf{y}}_n^T \mathbf{y}_1 + \alpha_2 \bar{\mathbf{y}}_n^T \mathbf{y}_2 + \cdots + \alpha_n \bar{\mathbf{y}}_n^T \mathbf{y}_n = 0$$

Equations (1.79) represent a set of n homogeneous simultaneous equations in the unknown $\alpha_1, \alpha_2, \ldots, \alpha_n$. By the theorem just presented, Eqs. (1.79) have a nontrivial solution if and only if the determinant of the coefficients vanishes

$$|G| = \begin{vmatrix} \bar{\mathbf{y}}_1^T \mathbf{y}_1 & \bar{\mathbf{y}}_1^T \mathbf{y}_2 & \cdots & \bar{\mathbf{y}}_1^T \mathbf{y}_n \\ \bar{\mathbf{y}}_2^T \mathbf{y}_1 & \bar{\mathbf{y}}_2^T \mathbf{y}_2 & \cdots & \bar{\mathbf{y}}_2^T \mathbf{y}_n \\ \cdots\cdots\cdots\cdots\cdots\cdots\cdots \\ \bar{\mathbf{y}}_n^T \mathbf{y}_1 & \bar{\mathbf{y}}_n^T \mathbf{y}_2 & \cdots & \bar{\mathbf{y}}_n^T \mathbf{y}_n \end{vmatrix} = 0 \qquad (1.80)$$

where $|G|$ is known as the *Gramian determinant*. Hence, a necessary and sufficient condition for the set of vectors $\mathbf{y}_1, \mathbf{y}_2, \ldots, \mathbf{y}_n$ to be linearly dependent is that the Gramian determinant be zero.

As a simple illustration, let us consider the unit vectors $\mathbf{e}_i$, as given by Eqs. (1.6). In this case, the *Gramian matrix* G is equal to the identity matrix, $G = I$, so that $|G| = 1$. Hence, the unit vectors $\mathbf{e}_i$ are linearly independent.

Example 1.8

Determine the rank and nullity of the matrix

$$A = \begin{bmatrix} 2 & -1 & 4 & 3 \\ 1 & 5 & -2 & 4 \\ 5 & 3 & 6 & 10 \\ -1 & 6 & -6 & 1 \end{bmatrix} \qquad (a)$$

It is not difficult to verify that $\det A = 0$, so that rank $A < 4$. Hence, at least one of the columns (rows) of A is a linear combination of the other. By inspection, we observe that adding twice the first row to the second we obtain the third row. Moreover, subtracting the first row from the second we obtain the fourth row. Further search will reveal no other combinations of rows, so that two rows of A are linearly independent. It follows that rank $A = 2$.

To determine the nullity of A, we wish to determine first the null space of A by solving the equation

$$A\mathbf{x} = \mathbf{0} \tag{b}$$

which can be written in the explicit form

$$2x_1 - x_2 + 4x_3 + 3x_4 = 0$$
$$x_1 + 5x_2 - 2x_3 + 4x_4 = 0$$
$$5x_1 + 3x_2 + 6x_3 + 10x_4 = 0 \tag{c}$$
$$-x_1 + 6x_2 - 6x_3 + x_4 = 0$$

Using the method of elimination from high-school algebra, the above four equations can be reduced to the two equations

$$x_1 + (18/11)x_3 + (19/11)x_4 = 0$$
$$x_2 - (8/11)x_3 + (5/11)x_4 = 0 \tag{d}$$

while the remaining two equations are identically zero. It can be verified that every solution of Eqs. (d) can be written in the form

$$\mathbf{x} = \alpha_1 \mathbf{u}_1 + \alpha_2 \mathbf{u}_2 \tag{e}$$

where

$$\mathbf{u}_1 = [-18 \quad 8 \quad 11 \quad 0]^T, \qquad \mathbf{u}_2 = [-19 \quad -5 \quad 0 \quad 11]^T \tag{f}$$

The vectors $\mathbf{u}_1$ and $\mathbf{u}_2$ are clearly independent and they span the null space. Hence, they form a basis for the space. The dimension of the null space $\mathcal{N}(A)$ is two, dim $\mathcal{N} = 2$, so that the nullity of A is two, null $A = 2$.

1.13   Matrix norms

As with vectors, it is useful to assign a single number to a matrix which in some sense gives a measure of the magnitude of the matrix. Such a measure is provided by the norm. The *norm of a square matrix A* is a nonnegative number $\|A\|$ which satisfies the conditions

1. $\|A\| \geq 0$, $\|A\| = 0$ if and only if $A = 0$.
2. $\|kA\| = |k| \|A\|$ for any complex scalar k.
3. $\|A + B\| \leq \|A\| + \|B\|$.
4. $\|AB\| \leq \|A\| \cdot \|B\|$.

Corresponding to any vector norm, one can associate with any matrix A a nonnegative quantity defined by max $\|A\mathbf{x}\|/\|\mathbf{x}\|$, $\|\mathbf{x}\| \neq 0$. This quantity is a function of the matrix A and it satisfies the conditions of a matrix norm.

27

It is called the matrix norm *subordinate* to the vector norm. Because

$$\|A\| = \max \frac{\|A\mathbf{x}\|}{\|\mathbf{x}\|}, \qquad \|\mathbf{x}\| \neq 0 \tag{1.81}$$

we have

$$\|A\mathbf{x}\| \leq \|A\| \cdot \|\mathbf{x}\| \tag{1.82}$$

where inequality (1.82) is true for $\|\mathbf{x}\| \neq 0$ or for $\|\mathbf{x}\| = 0$. Matrix and vector norms satisfying an inequality of the type (1.82) for all A and $\mathbf{x}$ are said to be *compatible*. Hence, a vector norm and its subordinate matrix norm are always compatible.

Another matrix norm of particular importance is the Euclidean norm, denoted by $\|A\|_E$ and defined as

$$\|A\|_E = \left(\sum_{i=1}^{n} \sum_{j=1}^{n} |a_{ij}|^2 \right)^{1/2} \tag{1.83}$$

The Euclidean norm has the advantage that it is easy to compute. Moreover, it has the important property that its value is invariant under a certain type of transformation known as a unitary transformation (N1, p. 287).

Free vibration of discrete systems

2.1 Introduction

Mathematical models of dynamical systems can be divided into two large classes: discrete and distributed. The reference is to the system parameters, such as mass, damping and stiffness. Discrete systems are described by variables depending on time alone, whereas distributed systems are described by variables depending on time and space. Accordingly, the motion of discrete systems is governed by ordinary differential equations and the motion of distributed systems by partial differential equations.

The minimum number of dependent variables required to fully describe the motion of a discrete system is known as the *number of degrees of freedom* of the system. These variables are referred to as *coordinates*. Quite often these coordinates represent physical quantities, such as translational or angular displacements, but at times they represent more abstract quantities, such as coefficients of a series. For this reason, they are referred to as *generalized coordinates*. Although it is possible to work with surplus coordinates, provided proper constraint equations are also included in the mathematical formulation, we shall assume for the most part that the number of coordinates coincides with the number of degrees of freedom of the system.

In this chapter, we begin with the formulation of the system Lagrange's differential equations of motion. Then, we linearize the equations about a given equilibrium by invoking the so-called 'small-motions assumption'. After discussing briefly certain energy considerations, the free vibration problem is introduced and the eigenvalue problem for a variety of cases is derived. This provides the motivation for much of the material in the next three chapters.

2.2 The system equations of motion

Let us consider an n-degree-of-freedom system and assume that its motion is fully described by the n coordinates $q_i(t)$ $(i = 1, 2, \ldots, n)$,

2   Free vibration of discrete systems

known as *generalized coordinates*. Under certain circumstances, the system kinetic energy can be written in the form

$$T = T_2 + T_1 + T_0 \tag{2.1}$$

where

$$T_2 = \tfrac{1}{2} \sum_{i=1}^{n} \sum_{j=1}^{n} m_{ij} \dot{q}_i \dot{q}_j \tag{2.2}$$

is a homogeneous quadratic function of the *generalized velocities* $\dot{q}_i(t)$,

$$T_1 = \sum_{j=1}^{n} f_j \dot{q}_j \tag{2.3}$$

is linear in the generalized velocities and T_0 contains no generalized velocities. In general, the coefficients m_{ij} and f_i and the function T_0 depend on the generalized coordinates q_i $(i = 1, 2, \ldots, n)$ and time. A system for which the kinetic energy can be written in the form (2.1) is known as *nonnatural* (M3, p. 77) and is most frequently encountered in the study of rotational motion of bodies. Whereas the term T_0 behaves like an apparent potential energy, giving rise to the so-called *centrifugal forces*, the term T_1 produces forces of the *Coriolis* type. Because the latter are related to gyroscopic phenomena most often (but not exclusively) associated with spinning bodies, the terms linear in velocities are referred to as gyroscopic terms. Note that such terms also appear in the vibration of pipes containing flowing liquids (P1).

   In addition to gyroscopic and centrifugal forces, other forces may act upon a system. One important class consists of forces derivable from the potential energy function V, which is a function of the generalized coordinates alone, $V = V(q_1, q_2, \ldots, q_n)$. Elastically restoring forces and gravitational forces fall in this general class. Another important class of forces is that consisting of viscous damping forces. The damping forces depend on the generalized velocities and are assumed to be derivable from the quadratic function

$$\mathscr{F} = \tfrac{1}{2} \sum_{i=1}^{n} \sum_{j=1}^{n} c_{ij} \dot{q}_i \dot{q}_j \tag{2.4}$$

which is known as *Rayleigh's dissipation function*. The coefficients c_{ij}, called *damping coefficients*, are generally constant and they are symmetric, $c_{ij} = c_{ji}$. We shall denote all the remaining forces, not falling into any of the above categories, by Q_i and assume that they can be obtained from the virtual work expression

$$\overline{\delta W} = \sum_{i=1}^{n} Q_i \, \delta q_i \qquad (2.5)$$

where δq_i are the *generalized virtual displacements*. The *generalized forces* Q_i generally depend on time but not on displacements or velocities. Introducing the system Lagrangian defined as

$$L = T - V \qquad (2.6)$$

the equations of motion take the form of Lagrange's equations (M3, Secs. 2.8 and 2.12)

$$\frac{d}{dt}\left(\frac{\partial L}{\partial \dot{q}_i}\right) - \frac{\partial L}{\partial q_i} + \frac{\partial \mathscr{F}}{\partial \dot{q}_i} = Q_i, \qquad i = 1, 2, \ldots, n \qquad (2.7)$$

In general, Eqs. (2.7) constitute a set of n nonhomogeneous nonlinear ordinary differential equations (ode's) of second order. General solutions of sets of nonlinear differential equations do not exist. Under given circumstances, however, it is possible to make certain simplifying assumptions permitting linearization of the equations and subsequent solution of the linearized equations.

2.3   Small motions about equilibrium points

⌐ne solution of Eqs. (2.7) consists of the n generalized coordinates $q_i(t)$ $(i = 1, 2, \ldots, n)$. This solution can be interpreted geometrically by conceiving of an n-dimensional Euclidean space with q_i as axes, where the space is known as the *configuration space*. At any time t, the solution represents a point in the configuration space defined by the tip of a vector $\mathbf{q}(t)$ whose components are the generalized coordinates $q_i(t)$. As time unfolds, the tip of the vector traces a curve in the configuration space known as the *dynamical path*, which describes the solution history in that space. Note that time plays the role of a parameter in this representation.

It turns out that the configuration space is not very convenient for a geometric representation of the motion. The main reason is that two dynamical paths can intersect, so that a given point in the configuration space does not define the state of the system uniquely. Indeed, the point can only specify the coordinates $q_i(t)$, and to define the state uniquely it is necessary to specify also the generalized velocities $\dot{q}_i(t)$ or, alternatively, the generalized momenta $p_i(t) = \partial L / \partial \dot{q}_i$. If the generalized velocities are used as a set of auxiliary variables, then the motion can be described in a $2n$-dimensional Euclidean space defined by q_i and $\dot{q}_i$ and known as the

31

state space. If the generalized momenta are used as auxiliary variables, then the motion can be described in a $2n$-dimensional Euclidean space defined by q_i and p_i and called the *phase space.* A point in either space defines the state of the system uniquely. The set of $2n$ variables q_i and $\dot{q}_i$ define a so-called *state vector* and the tip of this vector traces a *trajectory* in the state space, depicting the manner in which the solution unfolds with time. The advantage of the representation in the state space is that two trajectories never intersect, so that to a given point in the state space corresponds a unique trajectory. As a matter of interest, we observe that the configuration space is obtained through a projection of the state space parallel to the $\dot{q}_i$ axes. In the process, a trajectory is projected into a dynamical path in the configuration space.

A constant solution in the state space, $q_i(t) = q_{i0} = \text{const}$, $\dot{q}_i(t) = \dot{q}_{i0} = 0$, defines a so-called *equilibrium point,* so that all equilibrium points must lie in the configuration space. Note that at an equilibrium point all the generalized velocities and accelerations are zero, which explains the terminology. If the constants q_{i0} are different from zero, then the equilibrium point is said to be *nontrivial.* On the other hand, the equilibrium point defined by $q_{i0} = \dot{q}_{i0} = 0$ is said to be *trivial.* The point representing the trivial equilibrium can be easily identified as the origin of the state space.

We shall be concerned with the case in which Eqs. (2.7) admit the constant solutions $q_{i0} = \text{const}$, $\dot{q}_{i0} = 0$ $(i = 1, 2, \ldots, n)$. Such solutions carry the implication that $Q_i(t) = 0$ $(i = 1, 2, \ldots, n)$. Hence, assuming that all $Q_i(t)$ are zero, we conclude from Eqs. (2.1–7) that the constants q_{i0} must be solutions of the equations

$$\frac{\partial U}{\partial q_i} = 0, \qquad i = 1, 2, \ldots, n \tag{2.8}$$

where

$$U = V - T_0 \tag{2.9}$$

represents a modified potential energy referred to as the *dynamic potential.* Note that Eqs. (2.8) define the equilibrium points of the system. If U contains terms of degree higher than two in the generalized coordinates, then there can be more than one equilibrium point.

Our interest lies in *small motions about equilibrium points,* i.e., motions confined to small neighborhoods of the equilibrium points. A simple coordinate transformation, however, can translate the origin of the state space so as to make it coincide with an equilibrium point. Hence, without loss of generality, we shall consider the motion in the neighborhood of the trivial solution $q_{i0} = \dot{q}_{i0} = 0$ $(i = 1, 2, \ldots, n)$.

The assumption of small motions about an equilibrium point implies *linearization of the equations of motion*, which is tantamount to retaining in the Lagrangian only quadratic terms in the deviations of the generalized coordinates and velocities from equilibrium. As a result of linearization, the coefficients m_{ij} in T_2 become constant. These coefficients are symmetric, as can be concluded from

$$m_{ij} = m_{ji} = \frac{\partial^2 T_2}{\partial \dot{q}_i \, \partial \dot{q}_j}\bigg|_{\mathbf{q}=\dot{\mathbf{q}}=0} = \frac{\partial^2 T_2}{\partial \dot{q}_j \, \partial \dot{q}_i}\bigg|_{\mathbf{q}=\dot{\mathbf{q}}=0}, \qquad i, j = 1, 2, \ldots, n \quad (2.10)$$

and are generally known as *mass coefficients* or *inertia coefficients*. Clearly, T_2 remains in the form (2.2). On the other hand, the coefficients f_j in T_1 become linear in the generalized coordinates

$$f_j = \sum_{i=1}^{n} f_{ij} q_i, \qquad j = 1, 2, \ldots, n \tag{2.11}$$

where

$$f_{ij} = \frac{\partial f_j}{\partial q_i}\bigg|_{\mathbf{q}=0}, \qquad i, j = 1, 2, \ldots, n \tag{2.12}$$

are constant coefficients. Introducing Eq. (2.11) into Eq. (2.3), we can rewrite T_1 in the form

$$T_1 = \sum_{i=1}^{n} \sum_{j=1}^{n} f_{ij} q_i \dot{q}_j \tag{2.13}$$

In practice, it is often unnecessary to use Eqs. (2.10) and (2.12) to calculate the coefficients m_{ij} and f_{ij}, respectively, as the coefficients can be readily identified by expanding the kinetic energy T (see Example 2.1). There remains the question of reducing the dynamic potential U to quadratic form. Because our interest lies in the equilibrium point representing the trivial solution, let us expand the dynamic potential in a Taylor's series about the origin and write

$$U(q_1, q_2, \ldots, q_n) = U(0, 0, \ldots, 0) + \sum_{i=1}^{n} \frac{\partial U}{\partial q_i} q_i$$

$$+ \tfrac{1}{2} \sum_{i=1}^{n} \sum_{j=1}^{n} \frac{\partial^2 U}{\partial q_i \, \partial q_j} q_i q_j + \cdots \tag{2.14}$$

where the various partial derivatives are evaluated at $q_i = 0$. Recognizing that $U(0, 0, \ldots, 0)$ is merely a constant, which has no effect on the equations of motion, and that Eqs. (2.8) render the second term on the right side of Eq. (2.14) equal to zero, we conclude that the lowest degree terms in the dynamic potential are indeed quadratic. Hence, consistent with the linearization implied by the small motions assumption, we can ignore terms of degree higher than two and write

$$U(q_1, q_2, \ldots, q_n) \cong \tfrac{1}{2} \sum_{i=1}^{n} \sum_{j=1}^{n} k_{ij} q_i q_j \tag{2.15}$$

where

$$k_{ij} = k_{ji} = \frac{\partial^2 U}{\partial q_i\, \partial q_j}\bigg|_{\mathbf{q=0}} = \frac{\partial^2 U}{\partial q_j\, \partial q_i}\bigg|_{\mathbf{q=0}} \tag{2.16}$$

are constant symmetric coefficients known as *stiffness coefficients*. These coefficients can be regarded as consisting of several types, the most important being *elastic stiffness coefficients* and *geometric stiffness coefficients*. The first arise from the elastic potential energy, or strain energy, and the second from the centrifugal forces, where the latter can be traced to the term T_0 in the kinetic energy.

Now we are in a position of deriving explicit equations of motion. Indeed, introducing Eqs. (2.1), (2.2), (2.4), (2.13) and (2.15), in conjunction with Eq. (2.6), into Lagrange's equations, Eqs. (2.7), we obtain

$$\sum_{j=1}^{n} [m_{ij}\ddot{q}_j + (g_{ij} + c_{ij})\dot{q}_j + k_{ij}q_j] = Q_i, \qquad i = 1, 2, \ldots, n \tag{2.17}$$

where the skew symmetric coefficients

$$g_{ij} = f_{ji} - f_{ij} = -g_{ji}, \qquad i, j = 1, 2, \ldots, n \tag{2.18}$$

are referred to as *gyroscopic coefficients*.

There is another important class of forces, not accounted for in Eqs. (2.17). These forces also depend on the generalized coordinates alone, but they are not derivable from the potential energy V. These forces have the form $h_{ij}q_j$, where the coefficients h_{ij} are skew symmetric

$$h_{ji} = -h_{ij} \tag{2.19}$$

The terms $h_{ij}q_j$ arise in power-transmitting devices such as cranks, shafts, pulleys, etc., and represent so-called *circulatory forces* (Z1, p. 29; B5, p. 131; H9, p. 58). They can also occur in the case of dual-spin satellites with internal damping and have come to be known as *constraint damping forces* (L2, M15). If circulatory forces are included, then Eqs. (2.17) become simply

$$\sum_{j=1}^{n} [m_{ij}\ddot{q}_j + (g_{ij} + c_{ij})\dot{q}_j + (k_{ij} + h_{ij})q_j] = Q_i, \qquad i = 1, 2, \ldots, n \qquad (2.20)$$

Equations (2.20) constitute a set of linear ode's with constant coefficients describing the small motions of the system in the neighborhood of the origin.

Equations (2.20) can be conveniently written in matrix form. To this end, we recall the n-dimensional configuration vector $\mathbf{q}$, whose elements are the generalized coordinates q_i, and introduce the associated generalized force vector $\mathbf{Q}$. Moreover, introducing the $n \times n$ matrices

$$M = [m_{ij}], \qquad G = [g_{ij}], \qquad C = [c_{ij}],$$
$$K = [k_{ij}], \qquad H = [h_{ij}] \quad (2.21)$$

Eqs. (2.20) become simply

$$M\ddot{\mathbf{q}} + (G + C)\dot{\mathbf{q}} + (K + H)\mathbf{q} = \mathbf{Q} \qquad (2.22)$$

Note that the symmetry of the coefficients m_{ij}, c_{ij} and k_{ij} implies the symmetry of the corresponding matrices

$$M = M^T, \qquad C = C^T, \qquad K = K^T \qquad (2.23)$$

where M is known as the *mass matrix*, or *inertia matrix*, C is called the *damping matrix*, and K is the *stiffness matrix*. On the other hand, the skew symmetry of the coefficients g_{ij} and h_{ij} implies the skew symmetry of the matrices G and H, as expressed by

$$G = F^T - F = -G^T, \qquad H = -H^T \qquad (2.24)$$

where $F = [f_{ij}]$. The matrix G is referred to at times as the *gyroscopic matrix* and at other times as the *Coriolis matrix*. We shall refer to H as the *circulatory matrix*.

Equations (2.20) can be kept within the framework of Lagrange's equations by enlarging the scope of Rayleigh's dissipation function $\mathscr{F}$. Indeed, introducing the more general dissipation function

$$\mathscr{F}^* = \frac{1}{2}\sum_{i=1}^{n}\sum_{j=1}^{n} c_{ij}\dot{q}_i\dot{q}_j + \sum_{i=1}^{n}\sum_{j=1}^{n} h_{ij}\dot{q}_i q_j \qquad (2.25)$$

Eqs. (2.20) can be obtained from Lagrange's equations by simply replacing $\mathscr{F}$ by $\mathscr{F}^*$ in Eqs. (2.7).

Equation (2.22) represents the equations of motion of a damped linear gyroscopic system with circulatory forces. Mathematically, it constitutes a set of linear ode's with constant coefficients. Note that a system with constant coefficients is often referred to as *time-invariant*. The behavior of the system depends on the matrices M, G, C, K and H, as well as on the generalized force vector $\mathbf{Q}$. These matrices are known as soon as the quadratic forms T_2, T_1, $\mathscr{F}^*$ and U are known, at which point the task of deriving the equations of motion can be considered as completed (with the exception of the vector $\mathbf{Q}$ which can be obtained from the virtual work, Eq. (2.5)). Hence, a discussion of the nature of these quadratic forms is in order. It is not difficult to verify that the kinetic energy terms T_2 and T_1 can be written in the form of the triple matrix products

$$T_2 = \tfrac{1}{2}\dot{\mathbf{q}}^T M \dot{\mathbf{q}} \tag{2.26a}$$

and

$$T_1 = \mathbf{q}^T F \dot{\mathbf{q}} \tag{2.26b}$$

Rayleigh's dissipation function has the matrix form

$$\mathscr{F} = \tfrac{1}{2}\dot{\mathbf{q}}^T C \dot{\mathbf{q}} \tag{2.27}$$

and the more general dissipation function can be expressed as

$$\mathscr{F}^* = \mathscr{F} + \dot{\mathbf{q}}^T H \mathbf{q} \tag{2.28}$$

Moreover, the linearized dynamic potential, Eq. (2.15), has the expression

$$U = \tfrac{1}{2}\mathbf{q}^T K \mathbf{q} \tag{2.29}$$

Next, we wish to introduce several definitions concerning the sign properties of a real quadratic form and the matrix of the coefficients of the quadratic form. To this end, let us consider the typical quadratic form

$$f = \mathbf{x}^T A \mathbf{x} \tag{2.30}$$

where A is a real symmetric $n \times n$ matrix and $\mathbf{x}$ is a real n-vector. The real quadratic form f is said to be *positive definite* if its value is always positive for $\mathbf{x} \neq \mathbf{0}$ and is zero only for the trivial case $\mathbf{x} = \mathbf{0}$. If f is positive definite, then the matrix A is said to be positive definite. The real quadratic form f is said to be *positive semidefinite* if its value is nonnegative, i.e., f is generally positive but it can be zero without $\mathbf{x}$ being identically zero. If f is positive semidefinite, then the matrix A is said to be positive semidefinite. Analogous definitions exist for *negative definite* and *negative semidefinite* quadratic forms and associated matrices. If the quadratic form f can take either sign, then the quadratic form and the coefficient matrix are said to be *sign-variable*. In Sec. 4.6, we shall study methods for testing the sign properties of real symmetric matrices.

The kinetic energy term T_2 is by definition a positive definite function of the generalized velocities $\dot{q}_i$ ($i = 1, 2, \ldots, n$), so that the mass matrix M is always positive definite. On the other hand, the kinetic energy term T_1 leads to a skew symmetric matrix G, so that the question of its sign properties does not even arise. The same thing can be said about the skew symmetric matrix H. The Rayleigh dissipation function $\mathscr{F}$ is generally a nonnegative quantity, but not necessarily positive definite. Indeed, because not all the generalized velocities $\dot{q}_i$ need appear in $\mathscr{F}$, the Rayleigh dissipation function, and hence the damping matrix C, need not be positive definite but only positive semidefinite. The dynamic potential U is due to a variety of sources, including elastic restoring forces, gravitational forces, and centrifugal forces. The latter depend on the angular velocity Ω of the reference frame, so that U can be regarded as a function of the parameter Ω. Depending on Ω, U can be positive definite, positive semidefinite, sign-variable, negative semidefinite, or negative definite. This determines whether the system is stable or not, and hence whether the motion is consistent with the small motions assumption or it violates it. Later in this text, we shall study methods for determining the system stability.

Example 2.1

Derive the equations of motion for the system of Fig. 2.1. Then, linearize the equations and identify the coefficient matrices. Show how the same matrices can be obtained directly from the kinetic and potential energy and Rayleigh's dissipation function. The springs are nonlinear, with the

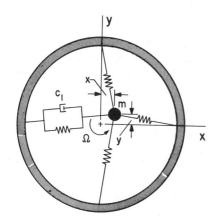

Figure 2.1

relations between the spring forces and the corresponding displacements having the form

$$F_{s1}(x) = -k_1(x + \varepsilon_1 x^3), \qquad F_{s2}(y) = -k_2(y + \varepsilon_2 y^3) \tag{a}$$

where the minus sign implies that the force is opposed to the displacement.

Attaching axes x and y to the frame rotating with the constant angular velocity Ω relative to the inertial space, the kinetic energy can be written in the form

$$T = \tfrac{1}{2}m[(\dot{x} - \Omega y)^2 + (\dot{y} + \Omega x)^2] \tag{b}$$

whereas Rayleigh's dissipation function is simply

$$\mathscr{F} = \tfrac{1}{2}c_1\dot{x}^2 \tag{c}$$

On the other hand, the potential energy is obtained by the following integrations (M3, p. 190)

$$V = \int_x^0 F_{s1}(\xi)\,d\xi + \int_y^0 F_{s2}(\eta)\,d\eta = -\int_x^0 k_1(\xi + \varepsilon_1\xi^3)\,d\xi$$

$$-\int_y^0 k_2(\eta + \varepsilon_2\eta^3)\,d\eta = \tfrac{1}{2}[k_1(x^2 + \tfrac{1}{2}\varepsilon_1 x^4) + k_2(y^2 + \tfrac{1}{2}\varepsilon_2 y^4)] \tag{d}$$

Recalling Eq. (2.6), introducing Eqs. (b)–(d) into Lagrange's equations, Eqs. (2.7), and performing the necessary differentiations, we obtain the equations of motion

$$m\ddot{x} - 2m\Omega\dot{y} + c_1\dot{x} - m\Omega^2 x + k_1(x + \varepsilon_1 x^3) = 0$$
$$m\ddot{y} + 2m\Omega\dot{x} - m\Omega^2 y + k_2(y + \varepsilon_2 y^3) = 0 \tag{e}$$

Ignoring the nonlinear terms $k_1\varepsilon_1 x^3$ and $k_2\varepsilon_2 y^3$, we can write Eqs. (e) in the matrix form (2.22), where the configuration vector is $\mathbf{q} = [x \ y]^T$, the matrix H and the force vector $\mathbf{Q}$ are zero, and the remaining coefficient matrices are

$$M = \begin{bmatrix} m & 0 \\ 0 & m \end{bmatrix}, \qquad G = \begin{bmatrix} 0 & -2m\Omega \\ 2m\Omega & 0 \end{bmatrix},$$

$$C = \begin{bmatrix} c_1 & 0 \\ 0 & 0 \end{bmatrix}, \qquad K = \begin{bmatrix} k_1 - m\Omega^2 & 0 \\ 0 & k_2 - m\Omega^2 \end{bmatrix} \tag{f}$$

Alternatively, expanding Eq. (b), we can write the kinetic energy in the matrix form

$$T = \tfrac{1}{2}m(\dot{x}^2 + \dot{y}^2) + m\Omega(x\dot{y} - \dot{x}y) + \tfrac{1}{2}m\Omega^2(x^2 + y^2) \tag{g}$$

where the first term on the right side is recognized as T_2, the second term as T_1, and the third term as T_0. In matrix form, we can write

$$T_2 = \tfrac{1}{2}\begin{bmatrix} \dot{x} \\ \dot{y} \end{bmatrix}^T \begin{bmatrix} m & 0 \\ 0 & m \end{bmatrix}\begin{bmatrix} \dot{x} \\ \dot{y} \end{bmatrix}, \qquad T_1 = \begin{bmatrix} x \\ y \end{bmatrix}^T \begin{bmatrix} 0 & m\Omega \\ -m\Omega & 0 \end{bmatrix}\begin{bmatrix} \dot{x} \\ \dot{y} \end{bmatrix}$$

$$T_0 = \tfrac{1}{2}\begin{bmatrix} x \\ y \end{bmatrix}^T \begin{bmatrix} m\Omega^2 & 0 \\ 0 & m\Omega^2 \end{bmatrix}\begin{bmatrix} x \\ y \end{bmatrix}$$

(h)

On the other hand, the linearized potential energy is

$$V = \tfrac{1}{2}\begin{bmatrix} x \\ y \end{bmatrix}^T \begin{bmatrix} k_1 & 0 \\ 0 & k_2 \end{bmatrix}\begin{bmatrix} x \\ y \end{bmatrix}$$

(i)

and Rayleigh's dissipation function is

$$\mathscr{F} = \tfrac{1}{2}\begin{bmatrix} \dot{x} \\ \dot{y} \end{bmatrix}^T \begin{bmatrix} c_1 & 0 \\ 0 & 0 \end{bmatrix}\begin{bmatrix} \dot{x} \\ \dot{y} \end{bmatrix}$$

(j)

Hence, from the first two of Eqs. (h), we obtain

$$M = \begin{bmatrix} m & 0 \\ 0 & m \end{bmatrix}, \qquad F = \begin{bmatrix} 0 & m\Omega \\ -m\Omega & 0 \end{bmatrix}$$

(k)

so that

$$G = F^T - F = \begin{bmatrix} 0 & -2m\Omega \\ 2m\Omega & 0 \end{bmatrix}$$

(l)

Moreover, subtracting the third of Eqs. (h) from Eq. (i), we can write

$$U = V - T_0 = \tfrac{1}{2}\begin{bmatrix} x \\ y \end{bmatrix}^T \begin{bmatrix} k_1 - m\Omega^2 & 0 \\ 0 & k_2 - m\Omega^2 \end{bmatrix}\begin{bmatrix} x \\ y \end{bmatrix}$$

(m)

so that the 2×2 matrix in Eq. (m) is recognized as K. Similarly, the 2×2 matrix in (j) can be identified as C.

It is obvious that M is positive definite and C is positive semidefinite. On the other hand, the sign properties of K depend on the parameter Ω. The matrix is positive definite for $k_1 > m\Omega^2$ and $k_2 > m\Omega^2$, positive semidefinite for $k_1 = m\Omega^2$ and $k_2 > m\Omega^2$ or $k_1 > m\Omega^2$ and $k_2 = m\Omega^2$, sign-variable for $k_1 > m\Omega^2$ and $k_2 < m\Omega^2$ or $k_1 < m\Omega^2$ and $k_2 > m\Omega^2$, etc.

2.4   Energy considerations

To gain some insight into the system behavior, let us probe into the nature of the system from the point of view of energy. To this end, let

2 Free vibration of discrete systems

$\mathbf{Q} = \mathbf{0}$ in Eq. (2.22) and consider the homogeneous system

$$M\ddot{\mathbf{q}} + (G+C)\dot{\mathbf{q}} + (K+H)\mathbf{q} = \mathbf{0} \tag{2.31}$$

Then, premultiplying Eq. (2.31) through by $\dot{\mathbf{q}}^T$ and recalling that M and K are symmetric and G is skew symmetric, we obtain

$$\dot{\mathbf{q}}^T M\ddot{\mathbf{q}} + \dot{\mathbf{q}}^T (G+C)\dot{\mathbf{q}} + \dot{\mathbf{q}}^T (K+H)\mathbf{q}$$

$$= \frac{\mathrm{d}}{\mathrm{d}t}(\tfrac{1}{2}\dot{\mathbf{q}}^T M\dot{\mathbf{q}} + \tfrac{1}{2}\mathbf{q}^T K\mathbf{q}) + \dot{\mathbf{q}}^T C\dot{\mathbf{q}} + \dot{\mathbf{q}}^T H\mathbf{q} = 0 \tag{2.32}$$

which, in view of Eqs. (2.26a) and (2.27–29), can be rewritten in the form

$$\frac{\mathrm{d}\mathscr{H}}{\mathrm{d}t} = -(\mathscr{F} + \mathscr{F}^*) \tag{2.33}$$

where

$$\mathscr{H} = T_2 + U \tag{2.34}$$

is recognized as the Hamiltonian for this linear gyroscopic system (M3, p. 96). Hence, the rate of change of the Hamiltonian is equal to the negative of the sum of Rayleigh's dissipation function $\mathscr{F}$ and the more general dissipation function $\mathscr{F}^*$.

A case of particular interest is that in which the viscous damping forces and the circulatory forces are zero. Because in this case the right side of Eq. (2.33) reduces to zero, we obtain

$$\mathscr{H} = \text{const} \tag{2.35}$$

The implication is that the Hamiltonian is conserved for an undamped gyroscopic system. Equation (2.35) represents a so-called Jacobi integral (M3, p. 96). Hence, an undamped gyroscopic system can be regarded as conservative in the sense that the Hamiltonian is conserved.

When the kinetic energy is only quadratic in the generalized velocities and there are no terms in the generalized coordinates, i.e., when $T_1 = T_0 = 0$, we obtain a *natural system* characterized by

$$T = T_2 = \tfrac{1}{2}\dot{\mathbf{q}}^T M\dot{\mathbf{q}} \tag{2.36}$$

In this case the dynamic potential reduces to the potential energy,

$$U = V \tag{2.37}$$

and the Hamiltonian reduces to

$$\mathscr{H} = T + V = E \tag{2.38}$$

where E is recognized as the system total energy. Hence, for an undamped natural system the Jacobi integral is identified as the total energy

$$E = \text{const} \qquad (2.39)$$

and the system is said to be conservative. Equation (2.39) is known as the *principle of conservation of energy*.

From the above discussion, we conclude that for the Hamiltonian to be conserved it is necessary that both $\mathbf{Q}$ and $\mathscr{F}^*$ be zero. The components Q_i of the vector $\mathbf{Q}$ can be identified as externally impressed forces and the quantities $-\partial \mathscr{F}^*/\partial \dot{q}_i$ as the sum of viscous damping forces and circulatory forces. Because they are the factors destroying the conservation of the Hamiltonian, Q_i and $-\partial \mathscr{F}^*/\partial \dot{q}_i$ are referred to as *nonconservative forces*. On the other hand, the forces $-\partial U/\partial q_i$, i.e., the forces derivable from the dynamic potential U, are said to be *conservative*. The nature of the solution is governed by the type of forces acting upon the system.

2.5 Free vibration and the eigenvalue problem

Equation (2.22) represents a set of linear simultaneous ode's with constant coefficients. Using linear system theory, a general closed-form solution of Eq. (2.22) can be shown to exist, and will be in fact derived in Ch. 6. However, an attempt to produce the actual numerical solution is likely to meet with serious computational difficulties, particularly for high-order systems. More success can be expected for certain special cases of the equation. A convenient way of deriving the solution is modal analysis, which requires the solution of the so-called eigenvalue problem for the system. In the following, we shall derive the eigenvalue problem for various cases, and in the next chapter we shall discuss properties of its solution.

i. *Conservative nongyroscopic system*

In the absence of gyroscopic, viscous damping, circulatory and impressed forces, $G = C = H = 0$ and $\mathbf{Q} = \mathbf{0}$, Eq. (2.22) reduces to

$$M\ddot{\mathbf{q}}(t) + K\mathbf{q}(t) = \mathbf{0} \qquad (2.40)$$

where M and K are real symmetric matrices of order n. Moreover, M is positive definite. Let us explore the possibility of *synchronous motion*, in which all coordinates execute the same motion in time. To this end, we assume that the solution of Eq. (2.40) has the exponential form

$$\mathbf{q}(t) = e^{st}\mathbf{u} \qquad (2.41)$$

where s is a constant scalar and $\mathbf{u}$ is a constant n-vector. Introducing Eq. (2.41) into Eq. (2.40) and dividing through by e^{st}, we obtain

$$K\mathbf{u} = \lambda M\mathbf{u}, \qquad \lambda = -s^2 \tag{2.42}$$

Equation (2.42) represents a set of simultaneous homogeneous algebraic equations in the unknowns u_i $(i = 1, 2, \ldots, n)$, with λ playing the role of a parameter. The problem of determining the constants λ for which Eq. (2.42) possesses nontrivial solutions $\mathbf{u}$ is known as the *characteristic value problem*, or more commonly the *eigenvalue problem*. From Sec. 1.12, however, we recall that the necessary and sufficient condition for a set of homogeneous algebraic equations to possess a nontrivial solution is that the determinant of the coefficients be zero. Hence, Eq. (2.42) has nontrivial solutions if

$$\det[K - \lambda M] = 0 \tag{2.43}$$

where $\det[K - \lambda M]$ is called the *characteristic determinant;* the equation itself is known as the *characteristic equation,* or the *frequency equation.* Because K and M are square matrices of order n, the characteristic determinant represents a polynomial of degree n in λ. In general, it has n distinct roots λ_r $(r = 1, 2, \ldots, n)$ called *characteristic values* or *eigenvalues.* To each of these eigenvalues there corresponds a vector $\mathbf{u}_r$ satisfying the equation

$$K\mathbf{u}_r = \lambda_r M\mathbf{u}_r, \qquad r = 1, 2, \ldots, n \tag{2.44}$$

where $\mathbf{u}_r$ is known as the *characteristic vector* or *eigenvector* belonging to λ_r.

The eigenvalue problem (2.42) is in terms of two real symmetric matrices. Many computational algorithms, however, are in terms of a single real symmetric matrix; such an eigenvalue problem is said to be in *standard form.* It turns out that a relatively simple coordinate transformation can reduce the eigenvalue problem (2.42) to one in terms of a single real symmetric matrix, so that it too can be regarded as having a standard form. Indeed, it is shown in Sec. 3.3 that, because M is a positive definite real symmetric matrix, it can be decomposed into

$$M = Q^T Q \tag{2.45}$$

where Q is a real nonsingular matrix. Inserting Eq. (2.45) into Eq. (2.42), we obtain

$$K\mathbf{u} = \lambda Q^T Q\mathbf{u} \tag{2.46}$$

Next, let us consider the linear transformation

$$\mathbf{u} = Q^{-1}\mathbf{v} \tag{2.47}$$

where Q^{-1} exists because Q is nonsingular. Introducing Eq. (2.47) into Eq. (2.46) and multiplying on the left by Q^{-T}, where we have adopted the notation $Q^{-T} = (Q^{-1})^T = (Q^T)^{-1}$, we obtain the eigenvalue problem in standard form

$$A\mathbf{v} = \lambda\mathbf{v} \tag{2.48}$$

where

$$A = Q^{-T}KQ^{-1} \tag{2.49}$$

is a real symmetric matrix. The matrix A is positive definite (semidefinite) if K is positive definite (semidefinite). Hence, the eigenvalue problems (2.42) and (2.48) can be regarded as being equivalent. Whereas both eigenvalue problems yield the same eigenvalues λ_r, the eigenvectors are not the same but related by Eq. (2.47).

The transformation matrix Q can be of different types, the most desirable type being that which is the easiest to invert. In Sec. 5.8, we shall discuss the Cholesky decomposition seeking a transformation matrix Q in the form of an upper triangular matrix U. Of course, when M is diagonal Q is simply equal to $M^{1/2}$, a diagonal matrix with the diagonal elements $\sqrt{m_{ii}}$ being real numbers, which are also taken as positive.

Returning to Eq. (2.41) and recognizing that to each eigenvalue λ_r correspond two exponents, $s_r = \pm\sqrt{-\lambda_r}$, we conclude that Eq. (2.40) admits solutions of the form

$$\mathbf{q}_r(t) = (a_r e^{\sqrt{-\lambda_r}\,t} + b_r e^{-\sqrt{-\lambda_r}\,t})\mathbf{u}_r, \qquad r = 1, 2, \ldots, n \tag{2.50}$$

which are often referred to as *eigensolutions*, where a_r and b_r are constants. Hence, synchronous motion is possible in any one of the forms given by Eqs. (2.50). Because the system under consideration is linear, the general solution of Eq. (2.40) is a linear combination of the eigensolutions, or

$$\mathbf{q}(t) = \sum_{r=1}^{n} (a_r e^{\sqrt{-\lambda_r}\,t} + b_r e^{-\sqrt{-\lambda_r}\,t})\mathbf{u}_r \tag{2.51}$$

where the constants a_r and b_r depend on the initial displacements $q_i(0)$ and initial velocities $\dot{q}_i(0)$. They can be obtained by solving the $2n$ equations

$$\mathbf{q}(0) = \sum_{r=1}^{n} (a_r + b_r)\mathbf{u}_r, \qquad \dot{\mathbf{q}}(0) = \sum_{r=1}^{n} \sqrt{-\lambda_r}\,(a_r - b_r)\mathbf{u}_r \tag{2.52}$$

The behavior of the system is controlled by the time-dependent part of the solution $a_r e^{\sqrt{-\lambda_r} t} + b_r e^{-\sqrt{-\lambda_r} t}$, which in turn is controlled by the exponents $s_r = \pm\sqrt{-\lambda_r}$. The values of these exponents depend on the nature of the matrix K. Indeed, because M is positive definite, if K is positive definite, then the system is positive definite. Clearly, in this case the matrix A is positive definite and all its eigenvalues are real and positive (see Sec. 3.3), so that $s_r = \pm i\omega_r$ $(r = 1, 2, \ldots, n)$, where $\omega_r = \sqrt{\lambda_r}$. When K is positive semidefinite, the system, and hence A, is positive semidefinite and all the eigenvalues λ_r are real and nonnegative, i.e., some of the eigenvalues are zero and the remaining ones are positive. Moreover, in both cases the eigenvectors $\mathbf{u}_r$ are real. Because the solution (2.51) must be real, it follows that for nonzero eigenvalues, $\omega_r \neq 0$, a_r and b_r must be complex conjugates, so the Eq. (2.51) reduces to

$$\mathbf{q}(t) = \sum_{r=1}^{n} (a_r e^{i\omega_r t} + \bar{a}_r e^{-i\omega_r t}) \mathbf{u}_r \tag{2.53}$$

where $\bar{a}_r$ is the complex conjugate of a_r.

The quantities ω_r are recognized as the *natural frequencies* of the system and the associated eigenvectors $\mathbf{u}_r$ are known as the *modal vectors*. The two quantities together constitute the *natural modes of vibration*. Equation (2.53) can be expressed in terms of real quantities alone. Indeed, it can be verified that solution (2.53) can be reduced to the form

$$\mathbf{q}(t) = \sum_{r=1}^{n} c_r \cos (\omega_r t - \phi_r) \mathbf{u}_r \tag{2.54}$$

where c_r is referred to as the *amplitude* and ϕ_r as the *phase angle* of the eigensolution $\mathbf{q}_r(t)$. Their values depend on the initial conditions and can be related to the constants a_r and b_r. Equation (2.54) implies that the free vibration consists of a superposition of harmonic terms, which is consistent with the concept of a conservative system.

When K is positive semidefinite, the eigenvalue problem (2.44) admits zero eigenvalues. Corresponding to a *zero eigenvalue*, say $\lambda_s = 0$, we have the eigensolution

$$\mathbf{q}_s(t) = (a_s + t b_s) \mathbf{u}_s \tag{2.55}$$

which is divergent. Zero eigenvalues occur when the *system is unrestrained*, in which case the associated modes can be identified as *rigid-body modes*. If the rigid-body modes are suppressed, then the motion about the rigid-body motion consists of a superposition of harmonic terms.

In the case in which K is sign-variable, the eigenvalue problem (2.44) admits negative eigenvalues. The exponents s_s corresponding to a negative eigenvalue λ_s are real, $s_s = \pm\sqrt{|-\lambda_s|}$, so that one part of the eigensolution approached zero exponentially, but, more importantly, the second part of the eigensolution diverges exponentially. Such a solution is clearly unstable and before long the small-motions assumption is violated. Similar statements can be made in connection with the cases in which K is negative semidefinite and negative definite. Note that unstable eigensolutions associated with negative eigenvalues λ_s are still consistent with conservative systems (M3, Sec. 5.3).

ii. *Conservative gyroscopic systems*

The equation describing the motion of a conservative gyroscopic system is obtained by removing from Eq. (2.22) the terms arising from nonconservative forces, namely, the damping matrix C, the circulatory matrix H and the impressed force vector $\mathbf{Q}$. The result is

$$M\ddot{\mathbf{q}}(t) + G\dot{\mathbf{q}}(t) + K\mathbf{q}(t) = \mathbf{0} \qquad (2.56)$$

where M and K are real symmetric matrices and G is a real skew symmetric matrix. All three matrices are of order n. We shall confine ourselves to the case in which both M and K are positive definite.

The eigenvalue problem associated with Eq. (2.56) can be reduced to a standard form, similar to that corresponding to the nongyroscopic system (2.40), by a method developed by Meirovitch (M4). Following is a brief description of the method.

Introducing the $2n$-dimensional state vector

$$\mathbf{x}(t) = [\dot{\mathbf{q}}(t)^T \vdots \mathbf{q}(t)^T]^T \qquad (2.57)$$

Eq. (2.56) can be written in the form

$$M^*\dot{\mathbf{x}}(t) + G^*\mathbf{x}(t) = \mathbf{0} \qquad (2.58)$$

where

$$M^* = \begin{bmatrix} M & \vdots & 0 \\ \cdots & \cdots & \cdots \\ 0 & \vdots & K \end{bmatrix}, \qquad G^* = \begin{bmatrix} G & \vdots & K \\ \cdots & \cdots & \cdots \\ -K & \vdots & 0 \end{bmatrix} \qquad (2.59)$$

are $2n \times 2n$ real nonsingular matrices, the first symmetric and the second skew symmetric. Moreover, M^* is positive definite. We shall seek a solution of Eq. (2.58) in the form

$$\mathbf{x}(t) = e^{st}\mathbf{x} \qquad (2.60)$$

where s is a constant complex scalar and $\mathbf{x}$ is a $2n$-dimensional constant complex vector. Introducing Eq. (2.60) into (2.58) and dividing through by e^{st}, we obtain the eigenvalue problem

$$sM^*\mathbf{x} + G^*\mathbf{x} = \mathbf{0} \tag{2.61}$$

It is well known (M4) that the eigenvalues of Eq. (2.61) consist of n pairs of pure imaginary complex conjugates, $s_r = \pm i\omega_r$ $(r = 1, 2, \ldots, n)$. Correspondingly, the eigenvectors also occur in pairs of complex conjugates, $\mathbf{x}_r = \mathbf{y}_r + i\mathbf{z}$, $\bar{\mathbf{x}}_r = \mathbf{y} - i\mathbf{z}_r$, where $\mathbf{y}_r$ is the real part and $\mathbf{z}_r$ the imaginary part of the eigenvector $\mathbf{x}_r$.

Eigenvalue problem (2.61) is in terms of a symmetric and a skew symmetric matrix and its solution is complex. Our object is to transform the eigenvalue problem into one in terms of symmetric matrices alone, where the latter is known to possess real solutions. Introducing $s = i\omega$ and $\mathbf{x} = \mathbf{y} + i\mathbf{z}$ into Eq. (2.61) and separating the real and the imaginary parts, we obtain two companion equations in terms of $\mathbf{y}$ and $\mathbf{z}$. Eliminating $\mathbf{z}$ from the first and $\mathbf{y}$ from the second, we obtain

$$K^*\mathbf{y} = \lambda M^*\mathbf{y}, \qquad K^*\mathbf{z} = \lambda M^*\mathbf{z}, \qquad \lambda = \omega^2 \tag{2.62}$$

where

$$K^* = G^{*T}M^{*-1}G^* \tag{2.63}$$

is a real symmetric positive definite matrix. Hence, the eigenvalue problem (2.61) has been reduced to one in terms of two real symmetric matrices, similar in form to the standard form given by Eq. (2.42). Moreover, because M^* is positive definite, the eigenvalue problem (2.62) can be further reduced to one in terms of a single real symmetric matrix, similar to the standard form (2.48).

To be sure, dissimilarities exist between the eigenvalue problems (2.42) and (2.62). In the first place, we note that the eigenvalue problem (2.62) is of order $2n$ instead of order n. Because both $\mathbf{y}$ and $\mathbf{z}$ satisfy the same eigenvalue problem, the solution of Eq. (2.62) consists of n pairs of repeated eigenvalues λ_r and n pairs of $2n$-dimensional eigenvectors $\mathbf{y}_r$ and $\mathbf{z}_r$ $(r = 1, 2, \ldots, n)$ belonging to λ_r. On the other hand, the similarities are more pervasive. In particular, both eigenvalue problems (2.42) and (2.62) are in terms of real symmetric matrices, so that their solutions exhibit the same characteristics. In fact, the same efficient computational algorithms can be used to obtain the solution of both. The fact that λ_r is a repeated eigenvalue turns out to be inconsequential in view of the fact that both M^* and K^* are real and symmetric (see Sec. 3.3). Moreover, because M^* and K^* are positive definite all λ_r are positive, so that all ω_r are real, as expected, where ω_r are recognized as the natural frequencies of oscillation.

Recalling that $\mathbf{x}(t)$ is real, the eigensolutions can be written in the form

$$\mathbf{x}_r(t) = a_r e^{i\omega_r t} \mathbf{x}_r + \bar{a}_r e^{-i\omega_r t} \bar{\mathbf{x}}_r$$
$$= c_r[\cos(\omega_r t - \phi_r)\mathbf{y}_r + \sin(\omega_r t - \phi_r)\mathbf{z}_r], \qquad r = 1, 2, \ldots, n \qquad (2.64)$$

where once again c_r is identified as the amplitude and ϕ_r as the phase angle, both determined by the initial conditions. Because the system (2.58) is linear, the solution $\mathbf{x}(t)$ consists of a superposition of the eigensolutions $\mathbf{x}_r(t)$. Note from Eq. (2.57) that the solution $\mathbf{x}(t)$ yields not only the displacement vector $\mathbf{q}(t)$ but also the velocity vector $\dot{\mathbf{q}}(t)$.

iii. *Damped nongyroscopic systems*

Letting G, H and $\mathbf{Q}$ be zero in Eq. (2.22), we obtain the equation

$$M\ddot{\mathbf{q}}(t) + C\dot{\mathbf{q}}(t) + K\mathbf{q}(t) = \mathbf{0} \qquad (2.65)$$

describing the free vibration of a damped nongyroscopic system, where all three matrices are real and symmetric. The solution of this problem has been discussed elsewhere (M2, Ch. 9), but here we wish to adopt a slightly different approach.

Equation (2.65) can be rewritten in the form

$$M^*\dot{\mathbf{x}}(t) + K^*\mathbf{x}(t) = \mathbf{0} \qquad (2.66)$$

where $\mathbf{x}(t)$ is the $2n$-dimensional state vector given by Eq. (2.57) and

$$M^* = \begin{bmatrix} M & 0 \\ \hline 0 & -K \end{bmatrix}, \qquad K^* = \begin{bmatrix} C & K \\ \hline K & 0 \end{bmatrix} \qquad (2.67)$$

are $2n \times 2n$ real symmetric matrices. However, neither matrix is positive definite. Hence, the decomposition (2.45) is no longer possible so that the problem cannot be reduced to one in terms of a single real symmetric matrix.

Letting a solution of Eq. (2.66) have the form

$$\mathbf{x}(t) = e^{\lambda t}\mathbf{x} \qquad (2.68)$$

the associated eigenvalue problem becomes

$$\lambda M^* \mathbf{x} + K^* \mathbf{x} = \mathbf{0} \tag{2.69}$$

which can be reduced to the form

$$A\mathbf{x} = \lambda \mathbf{x} \tag{2.70}$$

where, assuming that M^* is nonsingular,

$$A = -M^{*-1}K^* = \left[\begin{array}{c|c} -M^{-1}C & -M^{-1}K \\ \hline I & 0 \end{array}\right] \tag{2.71}$$

is an arbitrary real matrix.

Because A is real, if the eigenvalues λ are complex, then they must occur in pairs of complex conjugates, and so must the eigenvectors $\mathbf{x}$. The eigenvalue problem (2.70) is discussed in more detail in Ch. 3.

iv. *General dynamical systems*

The simultaneous presence of gyroscopic and viscous damping forces destroys the symmetry or skew symmetry of the coefficient matrix multiplying $\dot{\mathbf{q}}$. Similarly, the presence of circulatory forces destroys the symmetry of the coefficient matrix multiplying $\mathbf{q}$. Because in both these cases the corresponding coefficient matrices must be regarded as being arbitrary, no special cases other than Cases i–iii exist. In view of this, there is no reason to treat the various effects separately, so that we shall turn our attention to the general dynamical system.

Letting $\mathbf{Q} = \mathbf{0}$ in Eq. (2.22), we obtain the equation describing the free vibration of a general dynamical system, namely,

$$M\ddot{\mathbf{q}}(t) + (G + C)\dot{\mathbf{q}}(t) + (K + H)\mathbf{q}(t) = \mathbf{0} \tag{2.72}$$

where all the matrices are real, M, C and K being symmetric and G and H being skew symmetric. We shall assume that M is positive definite.

The matrix differential equation for the system can be written in the form (2.66), but now

$$M^* = \left[\begin{array}{c|c} M & 0 \\ \hline 0 & I \end{array}\right], \qquad K^* = \left[\begin{array}{c|c} G+C & K+H \\ \hline -I & 0 \end{array}\right] \tag{2.73}$$

where M^* is a positive definite real symmetric matrix and K^* is an arbitrary real matrix.

The eigenvalue problem can be expressed in the form (2.70), in which

$$A = -M^{*-1}K^* = \left[\begin{array}{c:c} -M^{-1}(G+C) & -M^{-1}(K+H) \\ \hdashline I & 0 \end{array}\right] \tag{2.74}$$

is an arbitrary real matrix. In general the solution of the eigenvalue problem is complex. The eigenvalue problem for arbitrary real matrices is discussed in Ch. 3.

Equation (2.72) can also be treated by a different approach. Indeed, because M is positive definite, it admits the decomposition

$$M = Q^T Q \tag{2.75}$$

where Q is a real nonsingular matrix. Then, introducing the linear transformation

$$\mathbf{q}(t) = Q^{-1}\mathbf{p}(t) \tag{2.76}$$

into Eq. (2.72) and premultiplying the result by $Q^{-T} = (Q^{-1})^T$, we obtain

$$\ddot{\mathbf{p}} + Q^{-T}(G+C)Q^{-1}\dot{\mathbf{p}} + Q^{-T}(K+H)Q^{-1}\mathbf{p} = \mathbf{0} \tag{2.77}$$

Defining the state vector

$$\mathbf{v} = [\dot{\mathbf{p}}^T \; \vdots \; \mathbf{p}^T]^T \tag{2.78}$$

Eq. (2.77) can be shown to lead to the eigenvalue problem

$$A\mathbf{v} = \lambda\mathbf{v} \tag{2.79}$$

where

$$A = \left[\begin{array}{c:c} -Q^{-T}(G+C)Q^{-1} & -Q^{-T}(K+H)Q^{-1} \\ \hdashline I & 0 \end{array}\right] \tag{2.80}$$

is a real arbitrary matrix.

The use of the decomposition (2.75) may seem a little contrived and difficult to justify at this point, as the resulting matrix A is not symmetric. We shall see in Ch. 5, however, that the object is to use a special matrix Q, namely, an upper triangular matrix U. The advantage of this procedure is that a triangular matrix is considerably easier to invert than a full square matrix, so that the decomposition of M into $U^T U$ and inversion of U can be more efficient and accurate than the direct inversion of M, particularly for high-order M. The decomposition into triangular matrices is known as the Cholesky decomposition (see Sec. 5.8).

3

The eigenvalue problem

3.1 General discussion

In Ch. 2 we derived the equations of motion and introduced the eigen-value problem for a variety of dynamical systems. It became immediately clear that the solution of the eigenvalue problem, and in particular the system eigenvalues, contains a great deal of information about the system dynamical characteristics. Moreover, the idea began to emerge that the eigenvalue problem plays a central role in the treatment of the response problem.

In this chapter, we delve a little deeper into the eigenvalue problem and identify various classes of eigenvalue problems by the structure of the matrices defining them. This represents a first attempt to extract informa-tion concerning the nature of the solution without actually carrying out the solution. This information involves questions as to whether the eigenvalues and eigenvectors are real or complex and whether the eigen-vectors are orthogonal in an ordinary sense or in some special sense. Finally, attention is focused on the so-called expansion theorem, which specifies the manner in which n-vectors representing either the system configuration or the system state can be constructed in the form of linear combinations of the system eigenvectors. The expansion theorem pro-vides the mathematical foundation for the system response by modal analysis.

3.2 The general eigenvalue problem

Let us consider any $n \times n$ matrix A with elements in the field F. Then, the *eigenvalues* of A are defined as those numbers λ for which the equation

$$A\mathbf{x} = \lambda \mathbf{x} \tag{3.1}$$

has a nontrivial solution $\mathbf{x} \neq \mathbf{0}$. The vector $\mathbf{x}$ itself is a vector in the n-dimensional space and is called an *eigenvector belonging to the eigen-value* λ. Although in Ch. 2 we used the symbol n to denote the number of

50

degrees of freedom of the system, in this chapter the integer n can be regarded as representing either the number of degrees of freedom of the system or the order of the system.

Because Eq. (3.1) is homogeneous, if **x** is an eigenvector corresponding to the eigenvalue λ, then α**x** is also an eigenvector corresponding to λ, where $\alpha \neq 0$ is an arbitrary scalar. Actually **x** and α**x** represent the same eigenvector, because normalization (Sec. 1.5) of **x** and α**x** yields the same vector.

From Sec. 1.12, we recall that a set of n homogeneous algebraic equations possesses a nontrivial solution if and only if the determinant of the coefficients vanishes. Hence, Eq. (3.1) has a nontrivial solution provided

$$\det [A - \lambda I] = 0 \tag{3.2}$$

Equations (3.2) can be interpreted as implying that the eigenvalue λ is a number which renders the matrix $A - \lambda I$ singular. Because when the matrix $A - \lambda I$ is singular the matrix $[A - \lambda I]^T = A^T - \lambda I$ is also singular, it follows that *the matrix A and the matrix A^T possess the same eigenvalues.*

Equation (3.2) is known as the *characteristic equation of A* and $\det [A - \lambda I]$ is called the *characteristic determinant of A*, where the latter represents a polynomial of order n in λ called the *characteristic polynomial of A*. Of course, the roots of the characteristic polynomial are the eigenvalues of A. Denoting the eigenvalues by $\lambda_1, \lambda_2, \ldots, \lambda_n$, the characteristic determinant can be written in the factored form

$$\det [A - \lambda I] = (\lambda_1 - \lambda)(\lambda_2 - \lambda) \cdots (\lambda_n - \lambda) = \prod_{r=1}^{n} (\lambda_r - \lambda) \tag{3.3}$$

and we note that the coefficient of λ^n is $(-1)^n$.

Equation (3.3) carries the tacit assumption that all the eigenvalues are distinct, which may not always be the case. If any of the roots, say λ_i, is repeated m_i times, where m_i is a positive integer, then the root is said to possess *multiplicity* m_i. If there are only k distinct roots, then the factorization has the form

$$\det [A - \lambda I] = (\lambda_1 - \lambda)^{m_1}(\lambda_2 - \lambda)^{m_2} \cdots (\lambda_k - \lambda)^{m_k} = \prod_{i=1}^{k} (\lambda_i - \lambda)^{m_i} \tag{3.4}$$

where $m_1 + m_2 + \cdots + m_k = n$.

An eigenvector can correspond to only one eigenvalue, but an eigenvalue can have many eigenvectors. Indeed, if $\mathbf{x}_1, \mathbf{x}_2, \ldots, \mathbf{x}_{m_i}$ are eigenvectors corresponding to the same eigenvalue λ_i, then

$$A(\alpha_1 \mathbf{x}_1 + \alpha_2 \mathbf{x}_2 + \cdots + \alpha_{m_i} \mathbf{x}_{m_i}) = \alpha_1 A\mathbf{x}_1 + \alpha_2 A\mathbf{x}_2 + \cdots + \alpha_{m_i} A\mathbf{x}_{m_i}$$

$$= \alpha_1 \lambda_i \mathbf{x}_1 + \alpha_2 \lambda_i \mathbf{x}_2 + \cdots + \alpha_{m_i} \lambda_i \mathbf{x}_{m_i}$$

$$= \lambda_i (\alpha_1 \mathbf{x}_1 + \alpha_2 \mathbf{x}_2 + \cdots + \alpha_{m_i} \mathbf{x}_{m_i}) \qquad (3.5)$$

It follows that *any linear combination of the eigenvectors* $\mathbf{x}_i, \mathbf{x}_2, \ldots, \mathbf{x}_{m_i}$ *corresponding to the same eigenvalue* λ_i *is also an eigenvector corresponding to* λ_i.

An important property of a set of vectors is mutual independence, because independent vectors can be used as a basis for a linear space (see Sec. 1.4). Hence, a pertinent question is whether the eigenvectors of a given matrix are linearly independent. The answer is provided by a theorem from linear algebra stating that if all n eigenvalues $\lambda_1, \lambda_2, \ldots, \lambda_n$ of the $n \times n$ matrix A are distinct, *then the n eigenvectors* $\mathbf{x}_1, \mathbf{x}_2, \ldots, \mathbf{x}_n$ *belonging to these eigenvalues are linearly independent* (F3, p. 73).

If A is a nonsingular matrix and λ and $\mathbf{x}$ are an eigenvalue and an eigenvector of A, respectively, then λ^{-1} and $\mathbf{x}$ are an eigenvalue and an eigenvector of the matrix A^{-1}. This can be easily shown by premultiplying both sides of Eq. (3.1) by $\lambda^{-1} A^{-1}$.

Next, let us investigate how the eigenvalues of a matrix change as we subtract from the diagonal elements of the matrix A a constant μ. Considering Eq. (3.1), we can write

$$[A - \mu I]\mathbf{x} = A\mathbf{x} - \mu\mathbf{x} = (\lambda - \mu)\mathbf{x} \qquad (3.6)$$

Hence, *if λ is an eigenvalue of the matrix A, then $\lambda - \mu$ is an eigenvalue of the matrix $A - \mu I$*. Note that the eigenvectors of $A - \mu I$ are the same as those of A. Clearly, subtraction of μ from the diagonal elements of A produces a *shift in the eigenvalues* by the same constant μ. This procedure can often be used to accelerate convergence in calculating the eigenvalues of A by an iterative process (Secs. 5.13 and 10.6).

The matrix A can be regarded as a *linear transformation* on the vector space L^n which maps the vector $\mathbf{x}$ into a vector $\mathbf{x}'$, or $\mathbf{x}' = A\mathbf{x}$. The components $x_1, x_2, \ldots, x_n$ of the vector $\mathbf{x}$ can be regarded as the *coordinates* of the vector $\mathbf{x}$ with respect to the basis $\mathbf{e}_1, \mathbf{e}_2, \ldots, \mathbf{e}_n$. Let $\mathbf{p}_1, \mathbf{p}_2, \ldots, \mathbf{p}_n$ be another basis for L^n and write $\mathbf{x}$ in the form

$$\mathbf{x} = y_1 \mathbf{p}_1 + y_2 \mathbf{p}_2 + \cdots + y_n \mathbf{p}_n \qquad (3.7)$$

where $y_1, y_2, \ldots, y_n$ are recognized as the coordinates of $\mathbf{x}$ with respect to the basis $\mathbf{p}_1, \mathbf{p}_2, \ldots, \mathbf{p}_n$. Figure 3.1 shows the decomposition of the three-dimensional vector $\mathbf{x}$ in terms of the basis $\mathbf{e}_1, \mathbf{e}_2, \mathbf{e}_3$ and the basis $\mathbf{p}_1$,

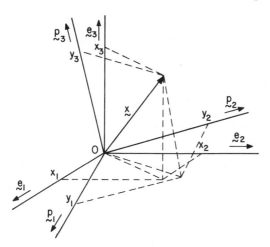

Figure 3.1

$\mathbf{p}_2$, $\mathbf{p}_3$. Equation (3.7) can be written in the compact form

$$\mathbf{x} = P\mathbf{y} \tag{3.8}$$

where P is an $n \times n$ matrix with its columns equal to the vectors $\mathbf{p}_1, \mathbf{p}_2, \ldots, \mathbf{p}_n$ and $\mathbf{y}$ is an n-vector with its elements equal to y_1, $y_2, \ldots, y_n$. By the definition of a basis, the vectors $\mathbf{p}_1, \mathbf{p}_2, \ldots, \mathbf{p}_n$ are linearly independent, so that the matrix P is nonsingular. Similarly, denoting by $y'_1, y'_2, \ldots, y'_n$ the coordinates of $\mathbf{x}'$ with respect to the basis $\mathbf{p}_1, \mathbf{p}_2, \ldots, \mathbf{p}_n$, we can write

$$\mathbf{x}' = P\mathbf{y}' \tag{3.9}$$

Hence, the transformation $\mathbf{x}' = A\mathbf{x}$ has the representation

$$P\mathbf{y}' = A(P\mathbf{y}) \tag{3.10}$$

Multiplying both sides of Eq. (3.10) on the left by P^{-1}, we obtain

$$\mathbf{y}' = B\mathbf{y} \tag{3.11}$$

where

$$B = P^{-1}AP \tag{3.12}$$

The matrix B represents the same linear transformation as A but in a different coordinate system. Two square matrices A and B related by an equation of the type (3.12) are said to be *similar* and the relationship (3.12) itself is called a *similarity transformation*.

53

3   The eigenvalue problem

We now propose to show that *similar matrices possess the same eigenvalues*. To this end, let us consider the characteristic determinant associated with B and write

$$\det[B - \lambda I] = \det[P^{-1}AP - \lambda I)$$
$$= \det[P^{-1}[A - \lambda I]P]$$
$$= \det P^{-1} \det[A - \lambda I] \det P \tag{3.13}$$

But

$$\det P^{-1} \det P = \det[P^{-1}P] = \det I = 1 \tag{3.14}$$

so that

$$\det[B - \lambda I] = \det[A - \lambda I] \tag{3.15}$$

Because matrices B and A possess the same characteristic determinant, they must possess the same characteristic equation, and hence the same eigenvalues.

A similarity transformation of particular interest is that in which the matrix A is similar to a diagonal matrix D. Letting $B = D$ in Eq. (3.12), we can write

$$AP = PD \tag{3.16}$$

which represents a set of n equations of the type

$$A\mathbf{p}_j = d_j\mathbf{p}_j, \qquad j = 1, 2, \ldots, n \tag{3.17}$$

where d_j are the diagonal elements of D. Comparing Eqs. (3.17) to Eq. (3.1) we conclude that *if A is similar to a diagonal matrix D, then the diagonal elements of D are the eigenvalues of A*. Moreover, *the independent vectors $\mathbf{p}_j$ $(j = 1, 2, \ldots, n)$ are the eigenvectors of A*. Hence, *a necessary and sufficient condition that an $n \times n$ matrix A be similar to a diagonal matrix is that A have n linearly independent eigenvectors*. In particular, *A is similar to a diagonal matrix if all its eigenvalues are distinct*.

The foregoing discussion suggests the possibility of computing the eigenvalues of a square matrix A by means of a series of similarity transformations reducing A to a simple form in which the eigenvalues are in evidence. Of course, the diagonal form is the most desirable one, but not every matrix can be diagonalized.

Although not every matrix can be reduced to a diagonal matrix by a similarity transformation, it can be shown that *every matrix is similar to an upper triangular matrix*. Indeed, *if A is an $n \times n$ matrix with eigenvalues*

$\lambda_1, \lambda_2, \ldots, \lambda_n$ *then there exists a nonsingular matrix P such that*

$$T = P^{-1}AP = \begin{bmatrix} \lambda_1 & t_{12} & \cdots & t_{1n} \\ 0 & \lambda_2 & \cdots & t_{2n} \\ \multicolumn{4}{c}{\dotfill} \\ 0 & 0 & \cdots & \lambda_n \end{bmatrix} \qquad (3.18)$$

The theorem can be proved by induction and such a proof can be found in the text by Murdoch (M16, Sec. 6.3). Note that if the interest is strictly in the eigenvalues of A, then triangularization of the matrix A is sufficient.

A theorem concerned with arbitrary matrices A whose eigenvalues are not necessarily distinct is known as *Jordan's theorem* and can be stated as follows: *Let A be an $n \times n$ matrix with eigenvalues $\lambda_1, \lambda_2, \ldots, \lambda_k$ with multiplicities $m_1, m_2, \ldots, m_k$, respectively, so that det $[A - \lambda I]$ has the form (3.4). Then, A is similar to a block-diagonal matrix of the form*

$$J = \begin{bmatrix} \Lambda_1 & 0 & \cdots & 0 \\ 0 & \Lambda_2 & \cdots & 0 \\ \multicolumn{4}{c}{\dotfill} \\ 0 & 0 & \cdots & \Lambda_k \end{bmatrix} \qquad (3.19)$$

where Λ_i are $m_i \times m_i$ matrices of the type

$$\Lambda_i = \begin{bmatrix} \lambda_i & x & \cdots & 0 & 0 \\ 0 & \lambda_i & \cdots & 0 & 0 \\ \multicolumn{5}{c}{\dotfill} \\ 0 & 0 & \cdots & \lambda_i & x \\ 0 & 0 & \cdots & & \lambda_i \end{bmatrix} \qquad (3.20)$$

in which x is equal to either 0 or 1. Proof of the theorem is given in the text by Franklin (F3, Sec. 5.3). The matrix J is called the *Jordan canonical form* of A. In the special case in which all the eigenvalues are distinct the Jordan form reduces to a diagonal form.

Earlier in the section we made the statement that the characteristic determinant does not change in similarity transformations. Indeed, expanding Eq. (3.3), we obtain the characteristic polynomial

$$\begin{aligned} \det[A - \lambda I] &= (-1)^n (\lambda - \lambda_1)(\lambda - \lambda_2) \cdots (\lambda - \lambda_n) \\ &= (-1)^n (\lambda^n + c_1 \lambda^{n-1} + \cdots + c_{n-1} \lambda + c_n) \end{aligned} \qquad (3.21)$$

The implication of the above statement is that *the coefficients c_1, $c_2, \ldots, c_n$ are invariant in similarity transformations.*

Two of the coefficients $c_1, c_2, \ldots, c_n$ have special meaning, namely, c_1 and c_n. To show this, let us first define the *trace* of the matrix A as the sum of its diagonal elements.

$$\text{tr } A = \sum_{i=1}^{n} a_{ii} \tag{3.22}$$

Then, using a Laplace expansion of $\det [A - \lambda I]$, it can be verified that the only term involving λ^{n-1} is in the product $(a_{11} - \lambda)$ $(a_{22} - \lambda) \cdots (a_{nn} - \lambda)$, so that the coefficient of λ^{n-1} is $c_1 = -\sum_{i=1}^{n} a_{ii} = -\text{tr } A$. On the other hand, from Eq. (3.21), we note that the coefficient of λ^{n-1} is $c_1 = -\sum_{i=1}^{n} \lambda_i$. Because the two expressions must yield the same coefficient c_1 and because this coefficient is constant, we conclude that

$$\text{tr } A = \sum_{i=1}^{n} \lambda_i \tag{3.23}$$

or *the sum of the diagonal elements of A is invariant in similarity transformations and is equal to the sum of the eigenvalues of A.*

Next, let us set $\lambda = 0$ in Eq. (3.21). On the one hand, we observe that $c_n = (-1)^n \det A$ and on the other hand that $c_n = \prod_{i=1}^{n} (-\lambda_i)$. Using the same argument as above, we conclude that

$$\det A = \prod_{i=1}^{n} \lambda_i \tag{3.24}$$

or *the determinant of A is invariant in similarity transformations and is equal to the product of the eigenvalues of A.*

The two invariants, namely, $\text{tr } A$ and $\det A$, can sometimes be used to check computational accuracy.

3.3 The eigenvalue problem for real symmetric matrices

The discussion of Sec. 3.2 was quite general in nature and applicable to arbitrary square matrices A. In the study of vibration of structures, however, there is a strong interest in real symmetric matrices, as can be concluded from Ch. 2. It turns out that the solution of the eigenvalue problem for real symmetric matrices represents an appreciably simpler problem than that for nonsymmetric matrices. It suffices to say that a real symmetric matrix is diagonalizable and its eigenvalues and eigenvectors possess many properties that are not only interesting but also extremely useful.

Let us consider a real symmetric matrix A of order n. There are several basic theorems peculiar to such matrices. One of the most important ones reads as follows: *The eigenvalues of a real symmetric matrix are real.* To prove the theorem, let us consider an eigenvalue λ and assume that it is complex, $\lambda = \alpha + i\beta$. Because all the elements of A are real, we conclude from Eq. (3.23) that the complex conjugate $\bar{\lambda} = \alpha - i\beta$ must also be an eigenvalue. Accordingly, if $\mathbf{x}$ is an eigenvector belonging to λ, then the complex conjugate $\bar{\mathbf{x}}$ must be an eigenvector belonging to $\bar{\lambda}$. Hence, we must have

$$A\mathbf{x} = \lambda\mathbf{x}, \qquad A\bar{\mathbf{x}} = \bar{\lambda}\bar{\mathbf{x}} \tag{3.25}$$

Next, premultiply the first of Eqs. (3.25) by $\bar{\mathbf{x}}^T$, postmultiply the transpose of the second of Eqs. (3.25) by $\mathbf{x}$, subtract one result from the other and obtain

$$\bar{\mathbf{x}}^T A\mathbf{x} - (A\bar{\mathbf{x}})^T\mathbf{x} = 0 = (\lambda - \bar{\lambda})\bar{\mathbf{x}}^T\mathbf{x} = 2i\beta\bar{\mathbf{x}}^T\mathbf{x} \tag{3.26}$$

where use has been made of the fact that A is symmetric. But the product $\bar{\mathbf{x}}^T\mathbf{x}$ is a positive quantity, so that β must be zero, which proves the theorem. As a corollary, it follows that *the eigenvectors of a real symmetric matrix are real.*

Another basic theorem can be stated in the form: *Two eigenvectors of a real symmetric matrix A corresponding to distinct eigenvalues are orthogonal.* Let us consider two eigenvectors $\mathbf{x}_i$ and $\mathbf{x}_j$ corresponding to the distinct eigenvalues λ_i and λ_j, so that

$$A\mathbf{x}_i = \lambda_i\mathbf{x}_i, \qquad A\mathbf{x}_j = \lambda_j\mathbf{x}_j \tag{3.27}$$

Premultiply the first of Eqs. (3.27) by $\mathbf{x}_j^T$, postmultiply the transpose of the second of Eqs. (3.27) by $\mathbf{x}_i$, subtract one result from the other, recall that $A^T = A$ and write

$$0 = (\lambda_i - \lambda_j)\mathbf{x}_j^T\mathbf{x}_i \tag{3.28}$$

But the eigenvalues are distinct, $\lambda_i \neq \lambda_j$, so that Eq. (3.28) is satisfied if and only if

$$\mathbf{x}_j^T\mathbf{x}_i = 0, \qquad \lambda_i \neq \lambda_j \tag{3.29}$$

Actually the eigenvectors of a real symmetric matrix can be regarded as being orthogonal regardless of whether the eigenvalues are distinct or not. Indeed, let us consider the theorem: *If an eigenvalue λ_i of a real* *symmetric matrix A has multiplicity* m_i, *then there exist exactly* m_i *linearly independent corresponding eigenvectors* (M16, p. 230). Because any linear combination of the m_i linearly independent eigenvectors is also an eigenvector belonging to λ_i, it is always possible to choose such linear combinations that *the* m_i *eigenvectors corresponding to the same eigenvalue* λ_i *are mutually orthogonal.* Of course, such eigenvectors are clearly orthogonal to eigenvectors belonging to the remaining eigenvalues.

Let us assume that the eigenvectors $\mathbf{x}_1, \mathbf{x}_2, \ldots, \mathbf{x}_n$ of the real symmetric matrix A have been normalized so as to satisfy $\mathbf{x}_i^T \mathbf{x}_i = 1$ $(i = 1, 2, \ldots, n)$. This permits the replacement of Eqs. (3.29) by

$$\mathbf{x}_j^T \mathbf{x}_i = \delta_{ij}, \quad i, j = 1, 2, \ldots, n \tag{3.30}$$

which clearly defines the set of eigenvectors as an *orthonormal set of vectors* (Sec. 1.5).

Next, let us arrange the eigenvalues $\lambda_1, \lambda_2, \ldots, \lambda_n$ in the diagonal matrix $\Lambda = \mathrm{diag}\,[\lambda_i]$ and the eigenvectors $\mathbf{x}_1, \mathbf{x}_2, \ldots, \mathbf{x}_n$ in the orthonormal matrix $X = [\mathbf{x}_1, \mathbf{x}_2, \ldots, \mathbf{x}_n]$. Then, in terms of this notation, Eqs. (3.27) and (3.30) can be combined into the compact forms

$$AX = X\Lambda, \qquad X^T X = I \tag{3.31}$$

Premultiplying the first of Eqs. (3.31) by X^T and considering the second of Eqs. (3.31), we obtain simply

$$X^T A X = \Lambda \tag{3.32}$$

where Λ is the diagonal matrix of the eigenvalues. It follows that *a real symmetric matrix A is diagonalizable by means of a similarity transformation such that the transformation matrix is the orthonormal matrix of the eigenvectors,* $P = X$ *and* $P^{-1} = X^T$, *and the diagonal matrix has as its diagonal elements the eigenvalues of A,* $D = \Lambda$. Many computational algorithms solve the eigenvalue problem for a real symmetric matrix A by reducing it to a diagonal form by means of similarity transformations. We shall study such methods later in this text.

The foregoing discussion has been concerned with a real symmetric matrix A without regard to whether A is positive definite or not. In the following, we wish to introduce certain theorems and results peculiar to positive definite matrices, as well as to positive semidefinite matrices. In the process, we shall make use of either quadratic forms associated with A or of the matrix A itself.

Let us consider the diagonal matrix Λ of order n with its diagonal elements equal to $\lambda_1, \lambda_2, \ldots, \lambda_n$. Then, *the diagonal matrix* Λ *is positive definite if and only if all* λ_i $(r = 1, 2, \ldots, n)$ *are positive.* To show this, we consider any arbitrary real n-vector $\mathbf{x}$ and write the quadratic form

$$\mathbf{x}^T \Lambda \mathbf{x} = \sum_{i=1}^{n} \lambda_i x_i^2 \tag{3.33}$$

The quadratic form $\sum_{i=1}^{n} \lambda_i x_i^2$ is positive for all real nontrivial vectors $\mathbf{x}$ only if all $\lambda_i > 0$ ($i = 1, 2, \ldots, n$). On the other hand, if some of the λ_i are zero and the remaining λ_i are positive, then the quadratic form $\sum_{i=1}^{n} \lambda_i x_i^2$ is generally positive, but for a certain choice of x_i the quadratic form can be zero. For example, if $\lambda_i > 0$ for $1 \leq i \leq r$ and $\lambda_i = 0$ for $r < i \leq n$, then the quadratic form is zero for $x_i = 0$ for $1 \leq i \leq r$ and $x_i \neq 0$ for $r < i \leq n$. It follows that *the diagonal matrix Λ is positive semidefinite if and only if all λ_i ($i = 1, 2, \ldots, n$) are nonnegative.*

A real symmetric matrix A is positive definite if and only if for every real nonsingular matrix P of the same order, the matrix $P^T A P$ is positive definite. By definition, the matrix A is positive definite if the quadratic form $\mathbf{x}^T A \mathbf{x}$ is positive for all real nontrivial vectors $\mathbf{x}$. Next, let us consider the linear transformation $\mathbf{x} = P\mathbf{y}$, where $\mathbf{y} \neq \mathbf{0}$, which implies that P is nonsingular. It follows that $\mathbf{x}^T A \mathbf{x} = (P\mathbf{y})^T A (P\mathbf{y}) = \mathbf{y}^T P^T A P \mathbf{y} > 0$ for all $\mathbf{y} \neq \mathbf{0}$, so that $P^T A P$ is also positive definite.

Next, let us consider Eq. (3.32). As a consequence of the preceding theorems, Eq. (3.32) permits us to conclude that *if the real symmetric matrix A is positive definite, then Λ must also be positive definite, so that every $\lambda_i > 0$ ($i = 1, 2, \ldots, n$). Conversely, a real symmetric matrix A is positive definite if and only if all its eigenvalues are positive.* Recognizing that for orthonormal matrices X we have $XX^T = I$, because $X^T = X^{-1}$, we can premultiply Eq. (3.32) through by X and postmultiply the result by X^T to obtain

$$A = X \Lambda X^T \tag{3.34}$$

so that if Λ is positive definite, then A is positive definite. The above theorems can be extended to the case in which A is only positive semidefinite.

Equation (3.34) can be used to produce some interesting results. Let p be an integer and write

$$A^p = (X \Lambda X^T)^p = \overbrace{(X \Lambda X^T)(X \Lambda X^T) \cdots (X \Lambda X^T)}^{p \text{ times}} = X \Lambda^p X^T \tag{3.35}$$

so that *if a real symmetric matrix A is positive definite, then A^p is also* *positive definite.* Moreover, the eigenvalues of A^p are $\lambda_1^p, \lambda_2^p, \ldots, \lambda_n^p$. This points to a method of raising a positive definite matrix to an integer power. Actually, *p can be any rational number.* Indeed, we observe that

$$A = (A^{1/p})^p = X\Lambda X^T = \overbrace{(X\Lambda^{1/p}X^T)(X\Lambda^{1/p}X^T)\cdots(X\Lambda^{1/p}X^T)}^{p \text{ times}} \quad (3.36)$$

from which we conclude that

$$A^{1/p} = X\Lambda^{1/p}X^T \quad (3.37)$$

Finally, a theorem with important computational implications reads as follows: *A real symmetric matrix A is positive definite if and only if there exists a nonsingular matrix Q such that $A = Q^TQ$.* If A is symmetric and positive definite, then there exists a nonsingular matrix P such that

$$P^{-1}AP = P^TAP = D \quad (3.38)$$

where D is a diagonal matrix with real and positive diagonal elements. Because $D = D^{1/2}D^{1/2}$, Eq. (3.38) yields

$$A = PDP^{-1} = PD^{1/2}D^{1/2}P^T = (PD^{1/2})(PD^{1/2})^T = Q^TQ \quad (3.39)$$

where

$$Q = (PD^{1/2})^T \quad (3.40)$$

is a nonsingular matrix because both P and $D^{1/2}$ are nonsingular. This theorem can be used to calculate the inverse of a positive definite real symmetric matrix by triangular decomposition, as we shall see later in this text. It should be recalled that the theorem was used in Sec. 2.5 to reduce an eigenvalue problem in terms of two real symmetric matrices to one in terms of a single symmetric matrix.

Because the eigenvectors $x_1, x_2, \ldots, x_n$ of the real symmetric matrix A constitute a set of orthonormal n-vectors, they can be used as a basis for a linear space L^n. This implies that any arbitrary nonzero vector x in that space can be written as the linear combination

$$x = \alpha_1 x_1 + \alpha_2 x_2 + \cdots + \alpha_n x_n = X\alpha \quad (3.41)$$

where X is the orthonormal matrix of the eigenvectors and α is an n-vector of coefficients, $\alpha = [\alpha_1 \ \alpha_2 \cdots \alpha_n]^T$. Multiplying Eq. (3.41) on the left by X^T and considering the matrix orthonormality equation, the second of Eqs. (3.31), we obtain

$$\alpha = X^T x \quad (3.42)$$

Equations (3.41) and (3.42) are referred to as the *expansion theorem.*

60

The above discussion is in terms of a single real symmetric matrix. As shown in Sec. 2.5, however, the eigenvalue problem for conservative nongyroscopic as well as gyroscopic systems can be expressed in terms of two real symmetric matrices. From Sec. 2.5, we can write the eigenvalue problem in the form

$$K\mathbf{u} = \lambda M\mathbf{u}, \qquad \lambda = \omega^2 \qquad (3.43)$$

where for nongyroscopic systems the dimension of the vector $\mathbf{u}$ is equal to the number of degrees of freedom of the system, whereas for gyroscopic systems it is equal to twice that number. It was also shown in Sec. 2.5 that the eigenvalue problem (3.43) can be reduced to one in terms of a single real symmetric matrix. To this end, we first decompose the matrix M into

$$M = Q^T Q \qquad (3.44)$$

where Q is a real nonsingular matrix, and note that the decomposition is always possible if M is positive definite. Then, considering the linear transformation

$$\mathbf{u} = Q^{-1}\mathbf{x} \qquad (3.45)$$

introducing Eqs. (3.44) and (3.45) into Eq. (3.43), and multiplying on the left by $Q^{-T} = (Q^T)^{-1}$, we obtain the eigenvalue problem

$$A\mathbf{x} = \lambda \mathbf{x} \qquad (3.46)$$

where

$$A = Q^{-T} K Q^{-1} \qquad (3.47)$$

is a real symmetric matrix.

The eigenvectors $\mathbf{x}_i$ $(i = 1, 2, \ldots, n)$ of the matrix A are mutually orthogonal, as indicated by Eq. (3.30). In fact, Eq. (3.30) indicates that the eigenvectors $\mathbf{x}_i$ are not merely orthogonal but orthonormal. The question arises naturally as to whether the eigenvectors $\mathbf{u}_i$ $(i = 1, 2, \ldots, n)$ associated with the eigenvalue problem (3.43) possess the same property. From Eq. (3.45) we can write

$$\mathbf{x}_i = Q\mathbf{u}_i, \qquad \mathbf{x}_j = Q\mathbf{u}_j \qquad (3.48)$$

so that, introducing Eqs. (3.48) into Eq. (3.30), we obtain

$$\mathbf{u}_j^T Q^T Q \mathbf{u}_i = \delta_{ij}, \qquad i, j = 1, 2, \ldots, n \qquad (3.49)$$

Recalling Eq. (3.44), we conclude that

$$\mathbf{u}_j^T M \mathbf{u}_i = \delta_{ij}, \qquad i, j = 1, 2, \ldots, n \qquad (3.50)$$

or, the eigenvectors $\mathbf{u}_i$ are *orthogonal with respect to the mass matrix* rather than orthogonal in an ordinary sense. Note that Eq. (3.50) represents not only orthogonality relations but also a normalization scheme.

Introducing the *modal matrix* $U = [\mathbf{u}_1, \mathbf{u}_2, \ldots, \mathbf{u}_n]$ and recalling that $\Lambda = \mathrm{diag}\,[\lambda_i]$ represents the diagonal matrix of the eigenvalues, the complete set of solutions of the eigenvalue problem (3.43) can be written in the compact form

$$KU = MU\Lambda \tag{3.51}$$

Moreover, the orthogonality relations (3.50) have the matrix counterpart

$$U^T M U = I \tag{3.52}$$

so that, multiplying Eq. (3.51) on the left by U^T, we conclude that

$$U^T K U = \Lambda \tag{3.53}$$

As one may expect, the expansion theorem can be formulated also in terms of the vectors $\mathbf{u}_i$. Indeed, the vectors constitute a basis for an n-dimensional linear space, so that an arbitrary vector $\mathbf{u}$ in that space can be expressed as

$$\mathbf{u} = \alpha_1 \mathbf{u}_1 + \alpha_2 \mathbf{u}_2 + \cdots + \alpha_n \mathbf{u}_n = U\boldsymbol{\alpha} \tag{3.54}$$

where $\boldsymbol{\alpha} = [\alpha_1\,\alpha_2 \cdots \alpha_n]^T$ is the n-vector of coefficients. Multiplying Eq. (3.54) on the left by $U^T M$ and considering Eq. (3.52), we obtain

$$\boldsymbol{\alpha} = U^T M \mathbf{u} \tag{3.55}$$

Equations (3.54) and (3.55) constitute the expansion theorem in terms of the eigenvectors satisfying Eq. (3.43), where the equation is defined by two real symmetric matrices. The theorem plays a crucial role in the evaluation of the response of vibrating linear systems by modal analysis and we shall make extensive use of it in later chapters. Because the expansion theorem just presented is in terms of real vectors, however, its applicability is limited only to the undamped nongyroscopic and gyroscopic systems discussed in Sec. 2.5, as well as to some special type of damped nongyroscopic systems, namely, systems possessing proportional damping.

3.4   Geometric interpretation of the eigenvalue problem

Let us consider a real symmetric matrix A of order n and a real nonzero n-vector $\mathbf{x}$ and write the associated quadratic form

$$f = \mathbf{x}^T A \mathbf{x} = \sum_{i=1}^{n} \sum_{j=1}^{n} a_{ij} x_i x_j \tag{3.56}$$

The equation

$$f = \mathbf{x}^T A \mathbf{x} = \sum_{i=1}^{n} \sum_{j=1}^{n} a_{ij} x_i x_j = 1 \qquad (3.57)$$

represents a surface in an n-dimensional Euclidean space E^n. If A is positive definite, then the surface is closed and can be regarded as an n-dimensional ellipsoid centered at the origin of the Euclidean space. Figure 3.2 shows the surface for $n = 3$. We propose to show that *the principal axes of the ellipsoid are the eigenvectors of the matrix A.*

The gradient of f represents a vector in the direction normal to the ellipsoid surface. In n dimensions, the gradient of f is a vector ∇f having the components $\partial f/\partial x_1, \partial f/\partial x_2, \ldots, \partial f/\partial x_n$. Considering Eq. (3.57), we can write a typical component $\partial f/\partial x_k$ in the form

$$\frac{\partial f}{\partial x_k} = \sum_{i=1}^{n} \sum_{j=1}^{n} a_{ij}\left(\frac{\partial x_i}{\partial x_k} x_j + x_i \frac{\partial x_j}{\partial x_k}\right) = \sum_{i=1}^{n} \sum_{j=1}^{n} a_{ij}(x_j \delta_{ik} + x_i \delta_{jk})$$

$$= \sum_{j=1}^{n} a_{kj} x_j + \sum_{i=1}^{n} a_{ik} x_i = 2 \sum_{j=1}^{n} a_{kj} x_j \qquad (3.58)$$

because the coefficients are symmetric, $a_{ik} = a_{ki}$, and i and j are dummy indices. Hence, the gradient can be written in the form

$$\nabla f = 2\left[\sum_{j=1}^{n} a_{1j} x_j \ \sum_{j=1}^{n} a_{2j} x_j \ \cdots \ \sum_{j=1}^{n} a_{nj} x_j\right]^T = 2A\mathbf{x} \qquad (3.59)$$

From analytic geometry, however, we recall that the principal axes of the ellipsoid are normal to the surface. Hence, denoting by $\mathbf{x}$ a vector from the origin of the Euclidean space to a point on the surface, we conclude that for $\mathbf{x}$ to be aligned with a principal axis it must be proportional to ∇f. Denoting the constant of proportionality by 2λ, the

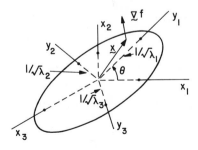

Figure 3.2

condition that $\mathbf{x}$ be aligned with a principal axis is

$$\nabla f = 2\lambda\mathbf{x} \tag{3.60}$$

Comparing Eqs. (3.59) and (3.60), we can write the condition in the form

$$A\mathbf{x} = \lambda\mathbf{x} \tag{3.61}$$

which represents the eigenvalue problem for the real symmetric positive definite matrix A. Hence, the eigenvectors of the matrix A can be interpreted geometrically as the principal axes of the ellipsoid corresponding to the quadratic form associated with A. Because the principal axes of an ellipsoid are orthogonal, we conclude that the eigenvectors of A are orthogonal, a fact established earlier algebraically.

From the above discussion, we conclude that the solution of the eigenvalue problem associated with A can be obtained by simply determining the principal axes of the ellipsoid

$$\mathbf{x}^T A\mathbf{x} = \lambda\mathbf{x}^T\mathbf{x} = 1 \tag{3.62}$$

In analytic geometry, this task is accomplished for $n = 2$ by a coordinate transformation representing a rotation of axes. For $n > 2$, the coordinate transformation can be regarded as a series of rotations. By analogy, we shall also consider a coordinate transformation, in particular a linear transformation of the type introduced in Sec. 3.2. To this end, let us consider a set of mutually orthogonal unit vectors $\hat{\mathbf{x}}_1, \hat{\mathbf{x}}_2, \ldots, \hat{\mathbf{x}}_n$ obtained by dividing the eigenvectors $\mathbf{x}_1, \mathbf{x}_2, \ldots, \mathbf{x}_n$ by their respective norms. Using the set $\hat{\mathbf{x}}_1, \hat{\mathbf{x}}_2, \ldots, \hat{\mathbf{x}}_n$ as an orthonormal basis in the Euclidean n-space, we can write

$$\mathbf{x} = y_1\hat{\mathbf{x}}_1 + y_2\hat{\mathbf{x}}_2 + \cdots + y_n\hat{\mathbf{x}}_n \tag{3.63}$$

where $y_1, y_2, \ldots, y_n$ are the coordinates of $\mathbf{x}$ with respect to the orthonormal basis. Note from Eq. (3.62) that the unit vectors $\hat{\mathbf{x}}_r$ are such that

$$\hat{\mathbf{x}}_r^T A\hat{\mathbf{x}}_r = \lambda_r\hat{\mathbf{x}}_r^T\hat{\mathbf{x}}_r = \lambda_r, \qquad r = 1, 2, \ldots, n \tag{3.64}$$

Equation (3.63) has the matrix form

$$\mathbf{x} = \hat{X}\mathbf{y} \tag{3.65}$$

where $\hat{X} = [\hat{\mathbf{x}}_1 \, \hat{\mathbf{x}}_2 \cdots \hat{\mathbf{x}}_n]$ is the orthonormal matrix of the eigenvectors. In view of this, Eq. (3.64) becomes

$$\hat{X}^T A\hat{X} = \Lambda \tag{3.66}$$

where Λ is the diagonal matrix of the eigenvalues.

Introducing Eq. (3.65) into Eq. (3.62), we can write

$$\mathbf{x}^T A \mathbf{x} = \mathbf{y}^T \hat{X}^T A \hat{X} \mathbf{y} = 1 \tag{3.67}$$

so that, considering Eq. (3.66), we obtain

$$\mathbf{y}^T \Lambda \mathbf{y} = 1 \tag{3.68}$$

which is *the equation of the ellipsoid in its canonical form.*
Letting $\mathbf{x} = \mathbf{x}_i$ and $\lambda = \lambda_i$ in Eq. (3.62), we obtain

$$\lambda_i \mathbf{x}_i^T \mathbf{x}_i = 1 \tag{3.69}$$

so that

$$\lambda_i = \frac{1}{\|\mathbf{x}_i\|^2} \tag{3.70}$$

where $\|\mathbf{x}_i\|$ is the Euclidean length of the eigenvector $\mathbf{x}_i$. When two eigenvalues are equal, the corresponding eigenvectors, although independent, are equal in length. This can be interpreted geometrically as the statement that the ellipsoid is an ellipsoid of revolution. Hence, any two orthogonal axes in the plane normal to the axis of revolution can be taken as principal axes.

The geometric interpretation can also be used for the case in which A is only positive semidefinite. In this case, the ellipsoid becomes an infinitely-long cylinder with the infinite axis corresponding to the zero eigenvalue, as can be concluded from Eq. (3.70).

The above geometric discussion can be regarded as a procedure for finding the Jordan form associated with A, which in this case is simply the matrix Λ. The Jacobi method for solving the eigenvalue problem of a real symmetric positive definite matrix is based on this procedure.

3.5   Hermitian matrices

Real symmetric matrices are not the only ones possessing interesting and useful properties. Indeed, many of the properties of real symmetric matrices are shared also by Hermitian matrices. We recall that a matrix A is Hermitian if $A = \bar{A}^T = A^H$.

Following the analogy with real symmetric matrices, we can state the theorem: *The eigenvalues of a Hermitian matrix are real.* As in Sec. 3.3, let λ and $\mathbf{x}$ be a solution of the eigenvalue problem

$$A \mathbf{x} = \lambda \mathbf{x} \tag{3.71}$$

where A is a Hermitian matrix of order n, $\lambda = \alpha + i\beta$ is a complex eigenvalue, and **x** is a complex eigenvector. It follows that the complex conjugates $\bar{\lambda}$ and $\bar{x}$ satisfy the eigenvalue problem

$$\bar{A}\bar{x} = \bar{\lambda}\bar{x} \tag{3.72}$$

Multiplying Eq. (3.71) on the left by $\bar{\mathbf{x}}^T$, postmultiplying the transpose of Eq. (3.72) on the right by **x**, and subtracting one result from the other, we obtain

$$\bar{\mathbf{x}}^T A \mathbf{x} - (\bar{A}\bar{\mathbf{x}})^T \mathbf{x} = 0 = (\lambda - \bar{\lambda})\bar{\mathbf{x}}^T \mathbf{x} = 2i\beta\bar{\mathbf{x}}^T \mathbf{x} \tag{3.73}$$

Because $\bar{\mathbf{x}}^T \mathbf{x}$ is positive and cannot be zero, it follows that $\beta = 0$, which proves the theorem.

Next, let $\mathbf{x}_i$ and $\mathbf{x}_j$ be nonzero eigenvectors belonging to the distinct real eigenvalues λ_i and λ_j, $\lambda_i \neq \lambda_j$, so that

$$A\mathbf{x}_i = \lambda_i \mathbf{x}_i, \qquad \bar{A}\bar{\mathbf{x}}_j = \lambda_j \bar{\mathbf{x}}_j \tag{3.74}$$

Then, we can write

$$\bar{\mathbf{x}}_j^T A \mathbf{x}_i = \lambda_i \bar{\mathbf{x}}_j^T \mathbf{x}_i, \qquad (\bar{A}\bar{\mathbf{x}}_j)^T \mathbf{x}_i = \lambda_j \bar{\mathbf{x}}_j^T \mathbf{x}_i \tag{3.75}$$

Subtracting the second of Eqs. (3.75) from the first, and recognizing that $\bar{\mathbf{x}}_j^T A \mathbf{x}_i = (\bar{A}\bar{\mathbf{x}}_j)^T \mathbf{x}_i$, we have

$$0 = (\lambda_i - \lambda_j)\bar{\mathbf{x}}_j^T \mathbf{x}_i \tag{3.76}$$

and, because the eigenvalues are distinct, it follows that

$$\bar{\mathbf{x}}_j^T \mathbf{x}_i = 0 \quad \text{for} \quad \lambda_i \neq \lambda_j \tag{3.77}$$

or, *two eigenvectors of a Hermitian matrix belonging to different eigenvalues are orthogonal.*

The eigenvectors can be normalized, so that

$$\bar{\mathbf{x}}_j^T \mathbf{x}_i = \delta_{ij}, \qquad i, j = 1, 2, \ldots, n \tag{3.78}$$

Introducing the matrix

$$X = [\mathbf{x}_1 \, \mathbf{x}_2 \cdots \mathbf{x}_n] \tag{3.79}$$

Eqs. (3.78) can be written in the compact form

$$X^H X = X X^H = I \tag{3.80}$$

where $X^H = \bar{X}^T$. It follows immediately from Eq. (3.80) that

$$X^{-1} = X^H \tag{3.81}$$

Square matrices X which satisfy Eq. (3.81) are called *unitary*. Matrices which have orthogonal eigenvectors are sometimes referred to as *normal*. Clearly, Hermitian matrices are normal, and so are symmetric and skew symmetric matrices, where the latter are two special cases of Hermitian matrices.

As a consequence of the orthogonality of the eigenvectors, if the eigenvalues of A are distinct, then we can use Eqs. (3.75) and write

$$X^{-1}AX = X^H AX = \Lambda \tag{3.82}$$

where Λ is the diagonal matrix of the eigenvalues. It follows that *a Hermitian matrix with distinct eigenvalues can be diagonalized by means of a unitary similarity transformation.*

The restriction that the eigenvalues be distinct is not really necessary. Indeed, the above theorem can be extended by stating that *any Hermitian matrix can be diagonalized by a unitary similarity transformation.* Proof of the theorem can be found in the text by Franklin (F3, p. 100).

Next, let us write the quadratic form associated with the Hermitian matrix A as

$$f = \bar{\mathbf{x}}^T A \mathbf{x} \tag{3.83}$$

and consider the linear transformation

$$\mathbf{x} = X\mathbf{y} \tag{3.84}$$

where X is defined by Eq. (3.79). Introducing Eq. (3.84) into Eq. (3.83) and considering Eq. (3.82), we obtain

$$\bar{\mathbf{x}}^T A \mathbf{x} = \bar{\mathbf{y}}^T X^H A X \mathbf{y} = \bar{\mathbf{y}}^T \Lambda \mathbf{y} = \sum_{i=1}^{n} \lambda_i |y_i|^2 \tag{3.85}$$

If all the eigenvalues λ_i of A are positive, then

$$\bar{\mathbf{x}}^T A \mathbf{x} = \sum_{i=1}^{n} \lambda_i |y_i|^2 > 0 \tag{3.86}$$

so that the quadratic form is positive definite. It follows that: *A Hermitian matrix is positive definite if and only if all its eigenvalues are positive.* Using the analogy with real symmetric matrices, the above theorem can be extended to the case in which all the eigenvalues of A are only nonnegative.

Before leaving the subject of Hermitian matrices, let us reconsider the expansion theorem introduced in Sec. 3.3 in terms of real vectors. As pointed out earlier, the eigenvectors of a Hermitian matrix A constitute a set of n orthogonal n-vectors $\mathbf{x}_i$ $(i = 1, 2, \ldots, , n)$. Hence, they can be used as a basis for a linear space L^n. This implies that any arbitrary vector $\mathbf{x}$ in that space can be represented by the linear combination

$$\mathbf{x} = c_1 \mathbf{x}_1 + c_2 \mathbf{x}_2 + \cdots + c_n \mathbf{x}_n = X\mathbf{c} \tag{3.87}$$

where X is the matrix of the eigenvectors, Eq. (3.79), and $\mathbf{c}$ is an n-vector, $\mathbf{c} = [c_1, c_2, \ldots, c_n]^T$. Now let us assume that the eigenvectors $\mathbf{x}_i$ have been normalized, so that X is an orthonormal matrix, i.e., it satisfies Eq. (3.80). Multiplying Eq. (3.87) on the left by $X^H = \bar{X}^T$ and considering Eq. (3.80), we obtain simply

$$\mathbf{c} = X^H \mathbf{x} \tag{3.88}$$

Equations (3.87) and (3.88) represent an *expansion theorem for a complex linear space*. Of course, when the matrix A is real and symmetric, Eqs. (3.87) and (3.88) reduce to Eqs. (3.41) and (3.42).

It is perhaps worth mentioning that the eigenvalue problem for the undamped gyroscopic system of Sec. 2.5 can be reduced to one in terms of a single skew symmetric matrix. Because a skew symmetric matrix is a special case of a Hermitian matrix, the approach of this section would be applicable. This would require the use of complex algebra, however, so that the approach presented in Sec. 2.5 appears superior.

3.6   The eigenvalue problem for two nonpositive definite real symmetric matrices

We have shown in Sec. 2.5 that the eigenvalue problem for damped nongyroscopic systems can be written in the form

$$\lambda M^* \mathbf{x} + K^* \mathbf{x} = \mathbf{0} \tag{3.89}$$

in which $\mathbf{x}$ is an n-vector, where n represents twice the number of degrees of freedom of the system and

$$M^* = \begin{bmatrix} M & 0 \\ 0 & -K \end{bmatrix}, \qquad K^* = \begin{bmatrix} C & K \\ K & 0 \end{bmatrix} \tag{3.90}$$

are real symmetric $n \times n$ matrices which are clearly not positive definite. This fact precludes a decomposition of the type (2.45), so that the problem cannot be reduced to one in terms of a single real symmetric matrix. This is to be expected, because the eigenvalues of a real symmetric matrix are real, whereas the eigenvalue problem (3.89) is known in general to possess pairs of complex conjugate eigenvalues.

A solution of the eigenvalue problem (3.89) can be obtained by reducing it to one in terms of a single nonsymmetric real matrix of the type (2.71). However, because the original eigenvalue problem, Eq. (3.89), is in terms of symmetric matrices, it possesses certain properties not shared by eigenvalue problems in terms of general nonsymmetric matrices. In particular, the eigenvectors corresponding to the eigenvalue problem (3.89) are orthogonal with respect to both matrices M^* and K^*. To prove this statement, let us assume that all the eigenvalues are distinct and consider two solutions of the eigenvalue problem (3.89), namely,

$$\lambda_i M^* \mathbf{x}_i + K^* \mathbf{x}_i = \mathbf{0}, \qquad \lambda_j M^* \mathbf{x}_j + K^* \mathbf{x}_j = \mathbf{0} \tag{3.91}$$

Next, premultiply the first of Eqs. (3.91) by $\mathbf{x}_j^T$ and the second by $\mathbf{x}_i^T$, so that

$$\lambda_i \mathbf{x}_j^T M^* \mathbf{x}_i + \mathbf{x}_j^T K^* \mathbf{x}_i = 0, \qquad \lambda_j \mathbf{x}_i^T M^* \mathbf{x}_j + \mathbf{x}_i^T K^* \mathbf{x}_j = \mathbf{0} \tag{3.92}$$

In view of the fact that M^* and K^* are symmetric, if we subtract the transpose of the second of Eqs. (3.92) from the first of Eqs. (3.92), then we obtain

$$(\lambda_i - \lambda_j) \mathbf{x}_j^T M^* \mathbf{x}_i = 0 \tag{3.93}$$

Because the eigenvalues are distinct, $\lambda_i \neq \lambda_j$, Eq. (3.93) yields the *orthogonality relations*

$$\mathbf{x}_j^T M^* \mathbf{x}_i = 0, \qquad \lambda_i \neq \lambda_j, \qquad i, j = 1, 2, \ldots, n \tag{3.94}$$

and it follows automatically from either one of Eqs. (3.92) that

$$\mathbf{x}_j^T K^* \mathbf{x}_i = 0, \qquad \lambda_i \neq \lambda_j, \qquad i, j = 1, 2, \ldots, n \tag{3.95}$$

The fact that the eigenvectors are orthogonal with respect to M^* (or K^*) can be used to obtain a solution of the eigenvalue problem by the power method using matrix deflation (Secs. 5.5 and 5.6). It can also be used to develop an expansion theorem, as shown in the sequel.

The eigenvectors $\mathbf{x}_i$ can be normalized by setting

$$\mathbf{x}_i^T M^* \mathbf{x}_i = 1, \qquad i = 1, 2, \ldots, n \tag{3.96}$$

so that, if $X = [\mathbf{x}_1 \, \mathbf{x}_2 \cdots \mathbf{x}_n]$ represents the square matrix of the normalized eigenvectors, then Eqs. (3.94) and (3.96) can be combined into

$$X^T M^* X = I \tag{3.97}$$

from which it follows automatically that

$$-X^T K^* X = \Lambda \tag{3.98}$$

where Λ is the diagonal matrix of the eigenvalues. Because the eigenvalues λ_i are distinct, the eigenvectors $\mathbf{x}_i$ $(i = 1, 2, \ldots, n)$ are independent. Hence, they constitute a basis for a linear space L^n, so that any arbitrary n-vector in that space can be expressed in the form

$$\mathbf{x} = X\mathbf{c} \tag{3.99}$$

where $\mathbf{c} = [c_1, c_2, \ldots, c_n]^T$ is the vector of the coefficients. Multiplying Eq. (3.99) on the left by $X^T M^*$ and considering Eq. (3.97), we conclude that

$$\mathbf{c} = X^T M^* \mathbf{x} \qquad (3.100)$$

Equations (3.99) and (3.100) represent the *expansion theorem for damped nongyroscopic systems*. Note that both X and $\mathbf{c}$ are in general complex.

3.7   The eigenvalue problem for real nonsymmetric matrices

Let us consider the general eigenvalue problem

$$A\mathbf{x} = \lambda \mathbf{x} \qquad (3.101)$$

where A is a real nonsymmetric $n \times n$ matrix. An example of such a matrix A is provided by Eq. (2.74) of Sec. 2.5. It should be pointed out, however, that our present discussion is more general and the only restriction placed on A is that it be real. Various properties of the eigenvalue problem (3.101) were discussed in Sec. 3.2. In this section, we wish to augment the discussion by exploring the existence of eigenvectors orthogonality as well as the existence of an expansion theorem. To this end, let us consider the eigenvalue problem associated with A^T and write it in the form

$$A^T \mathbf{y} = \lambda \mathbf{y} \qquad (3.102)$$

where we recognize that the eigenvalues of A^T are the same as those of A, as shown in Sec. 3.2. On the other hand, the eigenvectors of A^T are different from those of A. We shall confine ourselves to the case in which the eigenvalues of A are all distinct, so that the eigenvectors are independent. It follows that the eigenvectors of A^T are also independent.

Next, let us consider two distinct solutions of Eqs. (3.101) and (3.102). These solutions satisfy the equations

$$A\mathbf{x}_i = \lambda_i \mathbf{x}_i, \qquad i = 1, 2, \ldots, n \qquad (3.103)$$

and

$$A^T \mathbf{y}_j = \lambda_j \mathbf{y}_j, \qquad j = 1, 2, \ldots, n \qquad (3.104)$$

Equations (3.104) can also be written in the form

$$\mathbf{y}_j^T A = \lambda_j \mathbf{y}_j^T, \qquad j = 1, 2, \ldots, n \qquad (3.105)$$

Because of their position on the left side of the matrix A, the eigenvectors $\mathbf{y}_j$ are commonly referred to as *left eigenvectors of A*. Consistent with this

3.7   The eigenvalue problem for real nonsymmetric matrices

terminology, $\mathbf{x}_i$ are called *right eigenvectors of A*. Multiplying Eq. (3.103) on the left by $\mathbf{y}_j^T$ and Eq. (3.105) on the right by $\mathbf{x}_i$ and subtracting one result from the other, we obtain

$$(\lambda_i - \lambda_j)\mathbf{y}_j^T\mathbf{x}_i = 0 \tag{3.106}$$

so that for distinct eigenvalues we must have

$$\mathbf{y}_j^T\mathbf{x}_i = 0, \qquad \lambda_i \neq \lambda_j, \qquad i, j = 1, 2, \ldots, n \tag{3.107}$$

or the *left eigenvectors and right eigenvectors of A corresponding to distinct eigenvalues are orthogonal*. It should be stressed, however, that the eigenvectors are *not mutually orthogonal* in the same ordinary sense as those associated with a Hermitian matrix A. Indeed, the two sets of eigenvectors $\mathbf{x}_i$ and $\mathbf{y}_j$ are *biorthogonal*. The eigenvectors can be normalized by letting

$$\mathbf{y}_i^T\mathbf{x}_i = 1, \qquad i = 1, 2, \ldots, n \tag{3.108}$$

Then, introducing the matrices

$$X = [\mathbf{x}_1\,\mathbf{x}_2\cdots\mathbf{x}_n], \qquad Y = [\mathbf{x}_1\,\mathbf{x}_2\cdots\mathbf{x}_n] \tag{3.109}$$

Eqs. (3.107) and (3.108) can be combined into

$$Y^TX = I \tag{3.110}$$

from which we conclude that

$$Y^T = X^{-1} \tag{3.111}$$

and we note that relations (3.110) and (3.111) also imply the reciprocal relations

$$X^TY = I, \qquad X^T = Y^{-1} \tag{3.112}$$

Recalling that all the eigenvalues are distinct and denoting the diagonal matrix of the eigenvalues by Λ, Eqs. (3.103) can be written in the compact form

$$AX = X\Lambda \tag{3.113}$$

Multiplying Eq. (3.113) on the left by Y^T and considering Eqs. (3.110) and (3.111), we conclude that

$$Y^TAX = X^{-1}AX = \Lambda \tag{3.114}$$

so that the matrix A can be diagonalized by a similarity transformation, as already shown in Sec. 3.2. Equation (3.114), however, contains some information not available in Sec. 3.2 in that it identifies the matrix X^{-1} as

Y^T. The case in which the eigenvalues are not distinct was discussed to some extent in Sec. 3.2 and will not be pursued here.

The fact that the eigenvectors $\mathbf{x}_i$ and $\mathbf{y}_j$ are biorthogonal permits us to formulate an expansion theorem also for this more general case. In fact, we have the choice of expanding any arbitrary n-vector $\mathbf{x}$ in L^n in terms of the eigenvectors $\mathbf{x}_i$ or $\mathbf{y}_j$. An expansion in terms of $\mathbf{x}_i$ has the form

$$\mathbf{x} = X\boldsymbol{\alpha} \tag{3.115}$$

where $\boldsymbol{\alpha} = [\alpha_1 \, \alpha_2 \cdots \alpha_n]^T$ is the vector of associated coefficients. Multiplying Eq. (3.115) on the left by Y^T and considering Eq. (3.110), we obtain

$$\boldsymbol{\alpha} = Y^T\mathbf{x} \tag{3.116}$$

Similarly, an expansion in terms of the eigenvectors $\mathbf{y}_j$ has the form

$$\mathbf{x} = Y\boldsymbol{\beta}, \qquad \boldsymbol{\beta} = X^T\mathbf{x} \tag{3.117}$$

where $\boldsymbol{\beta} = [\beta_1 \, \beta_2 \cdots \beta_n]^T$ is the vector of coefficients associated with $\mathbf{y}_j$. Equations (3.115–117) are sometimes referred to as a *dual expansion theorem*. Note that the space L^n is generally complex.

The eigenvalue problem (3.102) is called the *adjoint eigenvalue problem*. Similarly, the set of vectors $\mathbf{y}_j$ $(j = 1, 2, \ldots, n)$ is known as the adjoint of the set of vectors $\mathbf{x}_i$ $(i = 1, 2, \ldots, n)$. When the matrix A is symmetric the two sets coincide and the eigenvalue problem is said to be *self-adjoint*. Of course, in this case the eigenvectors are real.

4

Qualitative behavior of the eigensolution

4.1 Introduction

A good understanding of the eigenvalue problem is essential to the understanding of the dynamic behavior of vibrating systems. Moreover, the solution of the eigenvalue problem is vital to the evaluation of the system response by analytic means. Much information concerning the nature of the eigensolution can be obtained by merely investigating the structure of the matrices defining the algebraic eigenvalue problem. A step in this direction was already made in Ch. 3, in which questions as to when the eigenvalues and eigenvectors are real and when they are complex were discussed, and the circumstances in which the system eigenvectors are orthogonal were examined.

By its very nature, the algebraic eigenvalue problem implies numerical computations. Indeed, there exists a large variety of computational algorithms for its solution. To enhance the understanding of these algorithms, it is advisable to study first the qualitative behavior of the eigensolution. This is the object of this chapter.

As demonstrated in Chs. 2 and 3, some very important classes of vibrating systems lead to eigenvalue problems in terms of real symmetric matrices, which are special cases of Hermitian matrices. It turns out that many principles concerned with the behavior of the eigensolution can be proved for general Hermitian matrices with no more effort than for real symmetric matrices, so that we shall choose the more general route. In this chapter, we present two principles of particular importance to the area of structural dynamics, namely, Rayleigh's principle and the inclusion principle. Also in this chapter, we examine how changes in the elements of the matrix defining the eigenvalue problem affect the eigenvalues.

4.2   The Rayleigh principle

Let A be a positive definite Hermitian matrix of order n, so that it has n real and positive eigenvalues λ_i and n mutually orthogonal eigenvectors $\mathbf{x}_i$ $(i = 1, 2, \ldots, n)$ satisfying the eigenvalue problem

$$A\mathbf{x}_i = \lambda_i \mathbf{x}_i, \qquad i = 1, 2, \ldots, n \tag{4.1}$$

For convenience, let us assume that $\lambda_1 \leq \lambda_2 \leq \cdots \leq \lambda_n$. Next, multiply Eq. (4.1) on the left by $\mathbf{x}_i^H$, recall that $\mathbf{x}^H = \bar{\mathbf{x}}^T$ for any complex vector $\mathbf{x}$, and divide the result by $\mathbf{x}_i^H \mathbf{x}_i = \|\mathbf{x}_i\|^2$ to obtain

$$\lambda_i = \frac{\mathbf{x}_i^H A \mathbf{x}_i}{\|\mathbf{x}_i\|^2}, \qquad i = 1, 2, \ldots, n \tag{4.2}$$

so that every eigenvalue λ_i can be obtained as the ratio of a quadratic form divided by the norm of the eigenvector squared. It will prove convenient to assume that the vectors have been normalized so as to satisfy $\|\mathbf{x}_i\| = 1$ $(i = 1, 2, \ldots, n)$, in which case the set of eigenvectors is orthonormal, so that

$$\mathbf{x}_i^H \mathbf{x}_j = \delta_{ij}, \qquad i, j = 1, 2, \ldots, n \tag{4.3}$$

Because $\|\mathbf{x}_i\| = 1$, Eqs. (4.2) reduce to

$$\lambda_i = \mathbf{x}_i^H A \mathbf{x}_i, \qquad \|\mathbf{x}_i\| = 1, \qquad i = 1, 2, \ldots, n \tag{4.4}$$

Equation (4.2) produces the eigenvalue λ_i only when $\mathbf{x}_i$ is the eigenvector belonging to λ_i. The question can be raised as to what happens to the ratio if, instead of using the eigenvector $\mathbf{x}_i$, we use an arbitrary vector $\mathbf{x}$ and let $\mathbf{x}$ range over the space L^n. To answer this question, let us form

$$\lambda(\mathbf{x}) = \frac{\mathbf{x}^H A \mathbf{x}}{\|\mathbf{x}\|^2} \tag{4.5}$$

which is known as *Rayleigh's quotient*, where the quotient is clearly a function of $\mathbf{x}$. If the vector $\mathbf{x}$ is a unit vector, then $\|\mathbf{x}\| = 1$ and Eq. (4.5) reduces to

$$\lambda(\mathbf{x}) = \mathbf{x}^H A \mathbf{x}, \qquad \|\mathbf{x}\| = 1 \tag{4.6}$$

Of course, in this case λ no longer represents a quotient, but we shall refer to Eq. (4.6) as Rayleigh's quotient as well, because of the special meaning of the equation.

We have shown in Sec. 3.5 that the eigenvectors of a Hermitian matrix A of order n constitute a set of n orthonormal vectors, and hence they can be used as a basis for an n-dimensional vector space L^n. Indeed, by the expansion theorem. Eq. (3.74), any arbitrary unit vector in L^n can be written as

$$\mathbf{x} = \sum_{i=1}^{n} c_i \mathbf{x}_i = X\mathbf{c} \tag{4.7}$$

where X is the unitary matrix of the eigenvectors and $\mathbf{c} = [c_1 \, c_2 \cdots c_n]^T$ is a complex vector. The matrix X satisfies

$$X^H A X = \Lambda \tag{4.8}$$

in which $X^H = \bar{X}^T$ and Λ is the diagonal matrix of the eigenvalues. On the other hand, the fact that $\mathbf{x}$ is a unit vector implies that

$$\mathbf{x}^H \mathbf{x} = (\overline{X\mathbf{c}})^T X\mathbf{c} = \mathbf{c}^H X^H X\mathbf{c} = 1 \tag{4.9}$$

so that, considering Eq. (3.80), we conclude that

$$\mathbf{c}^H \mathbf{c} = \|\mathbf{c}\|^2 = 1 \tag{4.10}$$

Hence, if $\mathbf{x}$ is a unit vector, then $\mathbf{c}$ is also a unit vector. Introducing Eq. (4.7) into Eq. (4.6) and considering Eq. (4.8), we obtain

$$\lambda = \mathbf{c}^H X^H A X\mathbf{c} = \mathbf{c}^H \Lambda \mathbf{c} = \sum_{i=1}^{n} \lambda_i \, |c_i|^2 \tag{4.11}$$

Next, let us assume that the vector $\mathbf{x}$ resembles closely one of the eigenvectors, say $\mathbf{x}_r$. In terms of the expansion (4.7), this implies that $|c_r|^2 \gg |c_i|^2$ $(i = 1, 2, \ldots, n; i \neq r)$. Introducing the notation

$$\frac{|c_i|^2}{|c_r|^2} = \varepsilon_i^2, \qquad i = 1, 2, \ldots, n; \qquad i \neq r \tag{4.12}$$

where ε_i is real, we conclude that Eq. (4.11) can be written in the form

$$\lambda = \lambda_r + |c_r|^2 \sum_{i=1}^{n} (\lambda_i - \lambda_r)\varepsilon_i^2 \tag{4.13}$$

Hence, if $\mathbf{x}$ differs from $\mathbf{x}_r$ by a small quantity of first order in ε_i, $\mathbf{x} = \mathbf{x}_r + 0(\varepsilon)$, then Rayleigh's quotient λ differs from λ_r by a small quantity of second order in ε_i, $\lambda = \lambda_r + 0(\varepsilon^2)$. This is equivalent to the statement that *Rayleigh's quotient has a stationary value in the neighborhood of an eigenvector*.

A case of particular significance is that corresponding to $r = 1$. In this case Eq. (4.13) becomes

$$\lambda = \lambda_1 + |c_1|^2 \sum_{i=2}^{n} (\lambda_i - \lambda_1)\varepsilon_i^2 \geq \lambda_1 \tag{4.14}$$

so that *Rayleigh's quotient is never lower than the lowest eigenvalue* λ_1. It is generally higher than λ_1 except when $\mathbf{x}$ is identically equal to $\mathbf{x}_1$, in which case *Rayleigh's quotient has a minimum at* $\mathbf{x} = \mathbf{x}_1$. This can be stated mathematically in the form

$$\lambda_1 = \min \mathbf{x}^H A \mathbf{x}, \qquad \|\mathbf{x}\| = 1 \tag{4.15}$$

This statement is often referred to as *Rayleigh's principle*. On the other hand, when $r = n$ Eq. (4.13) yields

$$\lambda = \lambda_n - |c_n|^2 \sum_{i=1}^{n-1} (\lambda_n - \lambda_i) \varepsilon_i^2 \le \lambda_n \tag{4.16}$$

or *Rayleigh's quotient is never higher than the highest eigenvalue* λ_n. It is generally lower than λ_n except when $\mathbf{x}$ is identically equal to $\mathbf{x}_n$, in which case *Rayleigh's quotient has a maximum at* $\mathbf{x} = \mathbf{x}_n$. This latter statement can be written as

$$\lambda_n = \max \mathbf{x}^H A \mathbf{x}, \qquad \|\mathbf{x}\| = 1 \tag{4.17}$$

Rayleigh's principle can be used to estimate eigenvalues, particularly the lowest one. To this end, we must make a fair guess of the first eigenvector and use it as a trial vector $\mathbf{x}$ in Eq. (4.6). Clearly, the estimated eigenvalue will generally be larger than the lowest eigenvalue λ_1. Similarly, it is possible to obtain an estimate for λ_n, although this may prove more difficult because of the difficulty in producing a good guess for the highest eigenvector. Because of the stationarity property of Rayleigh's quotient, remarkably good estimates of the eigenvalues can be obtained with only fair guesses of the eigenvectors.

The question remains as to how to estimate intermediate eigenvalues. First, let us consider a way of estimating λ_2. If the vector $\mathbf{c}$ in Eq. (4.7) has the form $\mathbf{c} = [0 \; c_2 \; c_3 \cdots c_n]^T$, then the eigenvector $\mathbf{x}_1$ is absent from the expansion for the trial vector $\mathbf{x}$. Now if we choose a trial vector $\mathbf{x}$ resembling $\mathbf{x}_2$ closely, by analogy with Eq. (4.14), we can write

$$\lambda = \lambda_2 + |c_2|^2 \sum_{i=3}^{n} (\lambda_i - \lambda_2) \varepsilon_i^2 \ge \lambda_2 \tag{4.18}$$

so that Rayleigh's quotient is never lower than λ_2 provided the trial vector $\mathbf{x}$ does not contain $\mathbf{x}_1$. Hence, we can write

$$\lambda_2 = \min \mathbf{x}^H A \mathbf{x}, \qquad \|\mathbf{x}\| = 1, \qquad \mathbf{x}^H \mathbf{x}_1 = 0 \tag{4.19}$$

or, *Rayleigh's quotient has a minimum of* λ_2 *for all trial vectors* $\mathbf{x}$ *which are orthogonal to the first eigenvector*. The minimum is reached for $\mathbf{x} = \mathbf{x}_2$. This statement can be generalized to higher eigenvalues by constraining the trial vectors to be orthogonal to an appropriate number of lower eigenvectors. For example, λ_r can be obtained from

$$\lambda_r = \min \mathbf{x}^H A \mathbf{x}, \qquad \|\mathbf{x}\| = 1, \qquad \mathbf{x}^H \mathbf{x}_i = 0, \qquad i = 1, 2, \ldots, r-1$$

$$(4.20)$$

The difficulty in using Rayleigh's quotient to estimate higher eigenvalues lies in the requirement that lower eigenvectors be known, because this requirement can seldom be met. Ways of circumventing this difficulty will be discussed later in this chapter.

Rayleigh's principle was derived for a positive definite Hermitian matrix and uses complex vectors as trial vectors. The widest application of the principle is for a special type of Hermitian matrix, namely, the real symmetric matrix. Hence, the above discussion is valid for the very important case in which A is a positive definite (semidefinite) real symmetric matrix. Of course, in this case the trial vector $\mathbf{x}$ is also real and Eq. (4.5) reduces to

$$\lambda(\mathbf{x}) = \frac{\mathbf{x}^T A \mathbf{x}}{\|\mathbf{x}\|^2} \tag{4.21}$$

and if the trial vector is assumed to be a unit vector, $\|\mathbf{x}\| = 1$, Eq. (4.21) reduces to

$$\lambda(\mathbf{x}) = \mathbf{x}^T A \mathbf{x}, \qquad \|\mathbf{x}\| = 1 \tag{4.22}$$

The real form of Rayleigh's quotient, Eq. (4.21), can be interpreted geometrically by an approach similar to that of Sec. 3.4. To this end, we recognize once again that the equation

$$\mathbf{x}^T A \mathbf{x} = 1 \tag{4.23}$$

defines a surface in an n-dimensional Euclidean space E^n. When A is positive definite, the surface represents an n-dimensional ellipsoid with the center at the origin of the Euclidean space. For obvious reasons, we shall confine ourselves to the three-dimensional case, $n = 3$. The three-dimensional ellipsoid is shown in Fig. 4.1. As demonstrated in Sec. 3.4, the principal axes of the ellipsoid are the three eigenvectors of the matrix A. Inserting Eq. (4.23) into Eq. (4.21), we obtain

$$\lambda(\mathbf{x}) = \frac{1}{\|\mathbf{x}\|^2} \tag{4.24}$$

so that Rayleigh's quotient can be interpreted geometrically as the reciprocal of the magnitude squared of a vector $\mathbf{x}$ from the origin of the

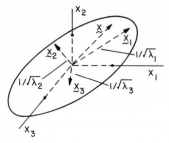

Figure 4.1

Euclidean space E^3 to a point on the surface of the ellipsoid. Hence, the eigenvalues and the eigenvectors of A are related by

$$\lambda_i = \lambda(\mathbf{x}_i) = \frac{1}{\|\mathbf{x}_i\|^2}, \qquad i = 1, 2, 3 \tag{4.25}$$

so that the semi-axes of the ellipsoid are inversely proportional to the square root of the eigenvalues, a fact established in Sec. 3.4.

As the trial vector $\mathbf{x}$ changes, the tip of the vector $\mathbf{x}$ slides across the surface of the ellipsoid. Because the eigenvectors represent the semi-axes of the ellipsoid, and hence they are normal to the ellipsoid surface, the rate of change of $\|\mathbf{x}\|$ is zero when $\mathbf{x}$ coincides with an eigenvector. It follows immediately that the rate of change of $\lambda(\mathbf{x})$ is also zero when $\mathbf{x}$ coincides with an eigenvector, which establishes the stationarity of $\lambda(\mathbf{x})$ in the neighborhood of an eigenvector geometrically. It is also obvious from Fig. 4.1 that

$$\lambda_1 \leq \lambda(\mathbf{x}) = \frac{1}{\|\mathbf{x}\|^2} \leq \lambda_3 \tag{4.26}$$

because $\|\mathbf{x}\|$ cannot exceed $\|\mathbf{x}_1\|$ and it cannot be smaller than $\|\mathbf{x}_3\|$.

The form (4.22) of Rayleigh's quotient can also be interpreted geometrically, but in this case we must confine ourselves to a two-dimensional system. Letting $\mathbf{x} = \mathbf{x}(x_1, x_2)$, we shall regard the quadratic form $\lambda(\mathbf{x}) = \mathbf{x}^T A \mathbf{x}$, representing Rayleigh's quotient, as a surface in a three-dimensional Euclidean space defined by x_1, x_2 and λ. For a positive definite matrix A, and hence a positive definite quadratic form $\mathbf{x}^T A \mathbf{x}$, this surface has the form of a cup of elliptic cross section with λ as the symmetry axis, as shown in Fig. 4.2a. The cylinder $\|\mathbf{x}\| = 1$ intersects this cup along a continuous curve C possessing two minimum and two maximum points. The minimum points are clearly

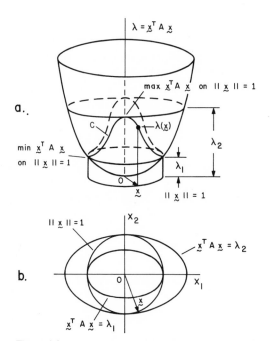

Figure 4.2

$$\lambda_1 = \min \mathbf{x}^T A \mathbf{x} \quad \text{on} \quad \|\mathbf{x}\| = 1 \tag{4.27a}$$

whereas the maximum points represent

$$\lambda_2 = \max \mathbf{x}^T A \mathbf{x} \quad \text{on} \quad \|\mathbf{x}\| = 1 \tag{4.27b}$$

The intersections of the planes $\lambda(\mathbf{x}) = \lambda_1 = \text{const}$ and $\lambda(\mathbf{x}) = \lambda_2 = \text{const}$ and the cup define two ellipses the first inscribed into the cylinder $\|\mathbf{x}\| = 1$ and the second circumscribing the cylinder. The projections of the ellipses and the cylinder onto the plane x_1, x_2 are shown in Fig. 4.2b. As the tip of the trial vector $\mathbf{x}$ moves around the circle $\|\mathbf{x}\| = 1$ in Fig. 4.2b, $\lambda(\mathbf{x})$ moves along the closed curve C in Fig. 4.2a, reaching the minimum value λ_1 and the maximum value λ_2 alternately. The extremum nature of $\lambda(\mathbf{x})$ is self-evident from this geometric construction. Note that in the two-dimensional case the stationary value is either a minimum or a maximum.

At this point, let us return to the problem of vibrating systems. We recall from Sec. 2.5 that the eigenvalue problem for a *conservative nongyroscopic system* is given by

$$K\mathbf{u} = \lambda M\mathbf{u}, \quad \lambda = \omega^2 \tag{4.28}$$

where K and M are $n \times n$ real symmetric stiffness and mass matrices, respectively, λ is a real nonnegative scalar, and $\mathbf{u}$ is a real vector in the L^n space, which in this case corresponds to the configuration space. The matrix M is positive definite and the matrix K is assumed to be either positive definite or positive semidefinite. Multiplying both sides of Eq. (4.28) on the left by $\mathbf{u}^T$ and dividing through by $\mathbf{u}^T M \mathbf{u}$, we obtain

$$\lambda = \omega^2 = \frac{\mathbf{u}^T K \mathbf{u}}{\mathbf{u}^T M \mathbf{u}} \tag{4.29}$$

which is the form of Rayleigh's quotient for conservative nongyroscopic systems.

Equation (4.29) can be brought into the form (4.21). Indeed, because M is a positive definite real symmetric matrix, we conclude from Sec. 3.3 that it can be decomposed into

$$M = Q^T Q \tag{4.30}$$

where Q is a real nonsingular matrix. Then, considering the linear transformation

$$\mathbf{u} = Q^{-1} \mathbf{x} \tag{4.31}$$

where $\mathbf{x}$ is a nonzero vector, and inserting Eqs. (4.30) and (4.31) into Eq. (4.29), we obtain Eq. (4.21) in which

$$A = Q^{-T} K Q^{-1} \tag{4.32}$$

is a real symmetric matrix, where $Q^{-T} = (Q^{-1})^T$. As pointed out in Sec. 2.5, the matrix A is positive definite (semidefinite) if K is positive definite (semidefinite).

The fact that Eq. (4.29) reduces to the form (4.21) carries the inference that all the properties of Rayleigh's quotient deduced from the form (4.5), or (4.21), hold also when the quotient is expressed in the form (4.29). For computational purposes, however, the form (4.29) of Rayleigh's quotient has many advantages. In particular, it is easier to select a trial vector $\mathbf{u}$ than a trial vector $\mathbf{x}$, because $\mathbf{u}$ can be identified with the mode of deformation of a structure.

As a matter of interest and for future reference, we would like to assess the manner in which the system natural frequencies change as the system properties change. To this end, Rayleigh's quotient can be particularly useful. Indeed, from Eq. (4.29), we observe that if the system is made stiffer, so that the elements of the matrix K increase, then the numerator of Rayleigh's quotient increases relative to the denominator. This permits us to conclude that *an increase in the system stiffness brings about an increase in the system natural frequencies.* On the other hand, using a parallel argument, we conclude that *an increase in the system mass causes a decrease in the system natural frequencies.*

In Sec. 2.5, we have shown that the eigenvalue problem for *conservative gyroscopic systems* has the form

$$sM^*\mathbf{x} + G^*\mathbf{x} = \mathbf{0} \tag{4.33}$$

where M^* and G^* are $2n \times 2n$ real nonsingular matrices, the first symmetric and the second skew symmetric, as can be seen from Eqs. (2.59), s is a complex scalar, and $\mathbf{x}$ is a complex vector in the L^{2n} space, which in this case corresponds to the state space. Moreover, M^* is positive definite. Multiplying Eq. (4.33) on the left by $\mathbf{x}^H$, we obtain Rayleigh's quotient

$$s = -\frac{\mathbf{x}^H G^* \mathbf{x}}{\mathbf{x}^H M^* \mathbf{x}} \tag{4.34}$$

Equation (4.34) can be reduced to the form (4.5), in which the equivalent matrix A would be skew symmetric, and hence a special case of a Hermitian matrix. Instead of working with complex quantities, as required by the form (4.5), it is perhaps more advantageous to consider the real counterpart of the eigenvalue problem (4.33), namely,

$$K^*\mathbf{y} = \lambda M^*\mathbf{y}, \qquad K^*\mathbf{z} = \lambda M^*\mathbf{z}, \qquad \lambda = \omega^2 \tag{4.35}$$

where $K^* = G^{*T} M^{*-1} G^*$ is a real symmetric positive definite matrix and $\mathbf{y}$ and $\mathbf{z}$ are real vectors. Letting the vector $\mathbf{v}$ represent either $\mathbf{y}$ or $\mathbf{z}$, Eqs. (4.35) yield Rayleigh's quotient (M7)

$$\lambda(\mathbf{v}) = \omega^2 = \frac{\mathbf{v}^T K^* \mathbf{v}}{\mathbf{v}^T M^* \mathbf{v}} \tag{4.36}$$

which is precisely of the form (4.29), except that here the real matrices M^* and K^* are of order $2n$ and the real vector $\mathbf{v}$ is $2n$-dimensional. Owing to the analogy with the form (4.29), we conclude that a stationarity principle exists for conservative gyroscopic systems in the same way as for conservative nongyroscopic systems. Because the eigenvalues ω_r^2 associated with Eqs. (4.35) have multiplicity two, however, Rayleigh's quotient in the form (4.36) has the same stationary value in the neighborhood of two eigenvectors, $\mathbf{y}_r$ and $\mathbf{z}_r$. Whereas the existence of a stationarity principle carries certain implications concerning the qualitative behavior of the eigenvalues, the Rayleigh quotient for gyroscopic systems is less of a computational tool than for nongyroscopic systems. The reasons is that for gyroscopic systems it is considerably more difficult to select trial vectors resembling the system eigenvectors.

A Rayleigh quotient can be defined for nonsymmetric real matrices A, such as those associated with the damped nongyroscopic and gyroscopic systems discussed in Sec. 2.5, and a stationarity principle can be shown to exist also for these cases (L1, Sec. 4.7; H9). However, this would involve the use of both right and left eigenvectors, which are generally complex. In addition, there is no rational way of selecting such vectors. In view of this, the usefulness of such a principle must be regarded as limited at best, and will not be pursued here.

Example 4.1

Consider the conservative nongyroscopic system of Fig. 4.3 and obtain an estimate of λ_1 by means of Rayleigh's principle.

The mass and stiffness matrices for the system are

$$M = m \begin{bmatrix} 1 & 0 & 0 \\ 0 & 1 & 0 \\ 0 & 0 & 2 \end{bmatrix}, \qquad K = k \begin{bmatrix} 2 & -1 & 0 \\ -1 & 3 & -2 \\ 0 & -2 & 2 \end{bmatrix} \tag{a}$$

Because M is diagonal, Eq. (4.30) yields

$$Q = M^{1/2} = m^{1/2} \begin{bmatrix} 1 & 0 & 0 \\ 0 & 1 & 0 \\ 0 & 0 & \sqrt{2} \end{bmatrix} \tag{b}$$

so that

$$Q^{-1} = m^{-1/2} \begin{bmatrix} 1 & 0 & 0 \\ 0 & 1 & 0 \\ 0 & 0 & 1/\sqrt{2} \end{bmatrix} \tag{c}$$

Introducing Eq. (c) into Eq. (4.32), we obtain

$$A = \frac{k}{m} \begin{bmatrix} 1 & 0 & 0 \\ 0 & 1 & 0 \\ 0 & 0 & 1/\sqrt{2} \end{bmatrix} \begin{bmatrix} 2 & -1 & 0 \\ -1 & 3 & -2 \\ 0 & -2 & 2 \end{bmatrix} \begin{bmatrix} 1 & 0 & 0 \\ 0 & 1 & 0 \\ 0 & 0 & 1/\sqrt{2} \end{bmatrix}$$

$$= \frac{k}{m} \begin{bmatrix} 2 & -1 & 0 \\ -1 & 3 & -\sqrt{2} \\ 0 & -\sqrt{2} & 1 \end{bmatrix} \tag{d}$$

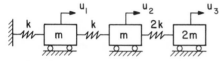

Figure 4.3

Next, let us use as a unit trial vector

$$\mathbf{x} = \frac{1}{\sqrt{14}}[1 \quad 2 \quad 3]^T \tag{e}$$

so that, inserting Eqs. (d) and (e) into Eq. (4.21), we obtain

$$\lambda = \frac{1}{14}\frac{k}{m}\begin{bmatrix}1\\2\\3\end{bmatrix}^T\begin{bmatrix}2 & -1 & 0\\-1 & 3 & -\sqrt{2}\\0 & -\sqrt{2} & 1\end{bmatrix}\begin{bmatrix}1\\2\\3\end{bmatrix}$$

$$= \frac{19-12\sqrt{2}}{14}\frac{k}{m} = 0.144960\,\frac{k}{m} \tag{f}$$

This compares with the exact value of the lowest eigenvalue $\lambda_1 = 0.139194\,k/m$. The estimate can be improved by selecting a trial vector resembling the first eigenvector more closely.

Example 4.2

Consider the conservative gyroscopic system of Example 2.1, let $k_1 = 2m\Omega^2$ and $k_2 = 3m\Omega^2$ and obtain an estimate for λ_1.

From Example 2.1, we can write

$$M = \begin{bmatrix} m & 0 \\ 0 & m \end{bmatrix}, \qquad G = \begin{bmatrix} 0 & -2m\Omega \\ 2m\Omega & 0 \end{bmatrix}, \qquad K = \begin{bmatrix} m\Omega^2 & 0 \\ 0 & 2m\Omega^2 \end{bmatrix} \tag{a}$$

so that, using Eqs. (2.59), we obtain

$$M^* = \begin{bmatrix} m & 0 & 0 & 0 \\ 0 & m & 0 & 0 \\ 0 & 0 & m\Omega^2 & 0 \\ 0 & 0 & 0 & 2m\Omega^2 \end{bmatrix},$$

$$G^* = \begin{bmatrix} 0 & -2m\Omega & m\Omega^2 & 0 \\ 2m\Omega & 0 & 0 & 2m\Omega^2 \\ -m\Omega^2 & 0 & 0 & 0 \\ 0 & -2m\Omega^2 & 0 & 0 \end{bmatrix} \tag{b}$$

Moreover, using Eq. (2.63), we have

$$K^* = G^{*T}M^{*-1}G^* = \begin{bmatrix} 5m\Omega^2 & 0 & 0 & 4m\Omega^3 \\ 0 & 6m\Omega^3 & -2m\Omega^3 & 0 \\ 0 & -2m\Omega^3 & m\Omega^4 & 0 \\ 4m\Omega^3 & 0 & 0 & 4m\Omega^4 \end{bmatrix} \qquad \text{(c)}$$

Using as a trial vector

$$\mathbf{v} = [0 \quad \Omega \quad 3 \quad 0]^T \qquad \text{(d)}$$

Rayleigh's quotient, Eq. (4.36), yields

$$\lambda = \Omega^2 \frac{\begin{bmatrix} 0 \\ \Omega \\ 3 \\ 0 \end{bmatrix}^T \begin{bmatrix} 5 & 0 & 0 & 4\Omega \\ 0 & 6 & -2\Omega & 0 \\ 0 & -2\Omega & \Omega^2 & 0 \\ 4\Omega & 0 & 0 & 4\Omega^2 \end{bmatrix} \begin{bmatrix} 0 \\ \Omega \\ 3 \\ 0 \end{bmatrix}}{\begin{bmatrix} 0 \\ \Omega \\ 3 \\ 0 \end{bmatrix}^T \begin{bmatrix} 1 & 0 & 0 & 0 \\ 0 & 1 & 0 & 0 \\ 0 & 0 & \Omega^2 & 0 \\ 0 & 0 & 0 & 2\Omega^2 \end{bmatrix} \begin{bmatrix} 0 \\ \Omega \\ 3 \\ 0 \end{bmatrix}}$$

$$= \frac{3\Omega^4}{10\Omega^2} = 0.3\Omega^2 \qquad \text{(e)}$$

This represents an estimate of λ_1, and a very good one. Indeed, it can be easily verified (by solving the eigenvalue problem) that the lowest eigenvalue is $\lambda_1 = 0.298437\Omega^2$ and, as expected, the estimated eigenvalue is higher than the computed eigenvalue. To explain the reason as to why the estimate is so good, we note that the eigenvectors belonging to λ_1 are

$$\mathbf{y}_1 = \begin{bmatrix} 0 \\ \Omega \\ 2.850781 \\ 0 \end{bmatrix}, \quad \mathbf{z}_1 = \begin{bmatrix} \Omega \\ 0 \\ 0 \\ -1.175391 \end{bmatrix} \qquad \text{(f)}$$

Clearly, the two eigenvectors are orthogonal. From Eqs. (f), we conclude that the estimate of the lowest eigenvalue λ_1 is so good because the trial vector $\mathbf{v}$ resembles $\mathbf{y}_1$ very closely. It can be easily verified that a trial vector resembling $\mathbf{z}_1$ closely will yield a good estimate of the same eigenvalue λ_1.

4.3   Rayleigh's theorem for systems with constraints

In Sec. 4.2, we have shown how Rayleigh's quotient $\lambda(\mathbf{x})$ can be used to study the nature of the eigenvalues of a given vibrating system. In particular, considering an n-degree-of-freedom system and combining Eqs. (4.6), (4.15) and (4.17), we can write

$$\lambda_1 \leq \lambda(\mathbf{x}) \leq \lambda_n \tag{4.37}$$

where λ_1 and λ_n are the lowest and highest eigenvalues, respectively. Moreover, if the first $r-1$ eigenvectors are known, then Eq. (4.20) can be used to construct a lower bound for the rth eigenvalue λ_r. This construction involves constraining the trial vector $\mathbf{x}$ to a subspace of lower dimension. More specifically, by requiring the trial vector $\mathbf{x}$ to be orthogonal to the first $r-1$ eigenvectors, the vector $\mathbf{x}$ is constrained to a subspace L^{n-r+1} which is the intersection of the space L^n and the space L^{n-r+1} orthogonal to the eigenvectors $\mathbf{x}_1, \mathbf{x}_2, \ldots, \mathbf{x}_{r-1}$. As pointed out in Sec. 4.2, however, an approach necessitating the eigenvectors $\mathbf{x}_1, \mathbf{x}_2, \ldots, \mathbf{x}_{r-1}$ is not practical because these eigenvectors are generally not available. Hence, a procedure capable of a characterization of the intermediate eigenvalues λ_r $(r = 2, 3, \ldots, n-1)$ that is independent of the eigenvectors $\mathbf{x}_1, \mathbf{x}_2, \ldots, \mathbf{x}_{r-1}$ is highly desirable.

Let us consider an n-degree-of-freedom vibrating system and assume that the system is subject to a constraint in the form of the equation

$$\mathbf{v}^H \mathbf{x} = \mathbf{x}^H \mathbf{v} = 0 \tag{4.38}$$

where $\mathbf{v}$ is a given n-vector. Geometrically, Eq. (4.38) implies that the vector $\mathbf{x}$ is constrained to a space orthogonal to the vector $\mathbf{v}$. Hence, if the arbitrary vector $\mathbf{x}$ is defined in the space L^n, then the *constrained vector* $\mathbf{x}$ is defined in the space L^{n-1}, which is the intersection of the space L^n and the hyperplane (plane of higher dimension) orthogonal to $\mathbf{v}$. The space L^{n-1} is known as a *subspace of constraint of L^n*. Physically, the constraint equation (4.38) implies that if the original unconstrained system has n degrees of freedom, then the constrained system has only $n-1$ degrees of freedom. For example, if $n = 3$ and the space L^3 is the Euclidean space E^3, then the subspace of constraint E^2 is simply the plane through the origin of E^3 and normal to the vector $\mathbf{v}$. This idea can be best illustrated by means of the geometric construction of Sec. 4.2. Assuming that A is a 3×3 positive definite real symmetric matrix, the equation $\lambda(\mathbf{x}) = x^T A x = 1$ describes the three-dimensional ellipsoid shown in Fig. 4.1. The imposition of the constraint (4.38) simply restricts the vector $\mathbf{x}$ to the ellipse defined by the intersection of the ellipsoid and the plane normal to $\mathbf{v}$.

Next, let us assume that an n-degree-of-freedom system with eigenvalues $\lambda_1, \lambda_2, \ldots, \lambda_n$ is subjected to a constraint of the type (4.38), and denote the eigenvalues of the constrained system by $\tilde{\lambda}_1, \tilde{\lambda}_2, \ldots, \tilde{\lambda}_{n-1}$. The question we propose to study is how the eigenvalues $\tilde{\lambda}_1, \tilde{\lambda}_2, \ldots, \tilde{\lambda}_{n-1}$ of the constrained system relate to the eigenvalues $\lambda_1, \lambda_2, \ldots, \lambda_n$ of the unconstrained system.

By analogy with Rayleigh's principle for unconstrained systems, Eq. (4.15), we can write for the constrained system

$$\tilde{\lambda}_1(\mathbf{v}) = \min \mathbf{x}^H A \mathbf{x}, \qquad \|\mathbf{x}\| = 1, \qquad \mathbf{x}^H \mathbf{v} = 0 \tag{4.39}$$

where $\tilde{\lambda}_1(\mathbf{v})$ is a function of the vector $\mathbf{v}$. As $\mathbf{v}$ ranges over the space L^n, the function $\tilde{\lambda}_1(\mathbf{v})$ varies. *Rayleigh's theorem for one constraint* states that: *The first eigenvalue of a system subjected to one constraint lies between the first and the second eigenvalue of the orginal unconstrained system*

$$\lambda_1 \leq \tilde{\lambda}_1 \leq \lambda_2 \tag{4.40}$$

We shall prove the theorem in two stages, first that $\lambda_1 \leq \tilde{\lambda}_1$ and then that $\tilde{\lambda}_1 \leq \lambda_2$.

To prove that $\lambda_1 \leq \tilde{\lambda}_1$ is relatively easy. Constraints have a tendency to increase the system stiffness, so that the eigenvalue $\tilde{\lambda}_1$ must be at least equal to λ_1 and in general it is larger. The only way $\tilde{\lambda}_1$ can be equal to λ_1 is for the constraint vector $\mathbf{v}$ to coincide with one of the eigenvectors belonging to a higher eigenvalue and for the trial vector $\mathbf{x}$ to coincide with the first eigenvector $\mathbf{x}_1$. Hence, in the case in which $\mathbf{x} = \mathbf{x}_1$, we have $\tilde{\lambda}_1 = \lambda_1$. This can be summarized in the form: *The lowest eigenvalue of the constrained system is not lower than the lowest eigenvalue of the unconstrained system, and it may be higher.*

Next, we want to show that $\tilde{\lambda}_1$ can be as high as λ_2, but not higher. For the particular choice $\mathbf{v} = \mathbf{x}_1$, the plane of constraint is orthogonal to the first eigenvector $\mathbf{x}_1$, so that, by the definition of the second eigenvalue presented in Sec. 4.2, we must have $\tilde{\lambda}_1(\mathbf{x}_1) = \lambda_2$. Hence, there is at least one choice of constraint vector, namely $\mathbf{v} = \mathbf{x}_1$, which raises the lowest eigenvalue of the constrained system as high as the second lowest eigenvalue of the original unconstrained system. Referring once again to Fig. 4.1, we see that for $\mathbf{v} = \mathbf{x}_1$, the vector $\mathbf{x}$ is constrained to the ellipse with semi-axes $\mathbf{x}_2$ and $\mathbf{x}_3$. The smallest value of Rayleigh's quotient for this constrained case is λ_2 and it occurs when $\mathbf{x}$ coincides with the semi-major axis $\mathbf{x}_2$.

We now propose to show that $\tilde{\lambda}_1$ cannot be higher than λ_2. To this end, let us assume that the vector $\mathbf{v}$ is given. The object is to find an admissible vector $\mathbf{x}$ such that $\mathbf{x}^H A \mathbf{x} \leq \lambda_2$. Hence, let us consider a trial vector $\mathbf{x}$ in the form

$$\mathbf{x} = c_1 \mathbf{x}_1 + c_2 \mathbf{x}_2 \tag{4.41}$$

There is no loss of generality in using $\mathbf{x}$ in the form (4.41), and this is always possible. Indeed, the orthogonality of the vectors $\mathbf{x}$ and $\mathbf{v}$ is expressed by

$$\mathbf{x}^H \mathbf{v} = \bar{c}_1 \mathbf{x}_1^H \mathbf{v} + \bar{c}_2 \mathbf{x}_2^H \mathbf{v} = 0 \qquad (4.42)$$

which represents an equation in the two unknowns c_1 and c_2. Hence, a solution of Eq. (4.42) exists. By the expansion theorem, Eq. (3.88), the components c_1 and c_2 can be written in the form

$$c_1 = \mathbf{x}_1^H \mathbf{x}, \qquad c_2 = \mathbf{x}_2^H \mathbf{x} \qquad (4.43)$$

where we note that $|c_1|^2 + |c_2|^2 = 1$ because $\|\mathbf{x}\| = 1$. Hence, the trial vector $\mathbf{x}$ in the form (4.41) satisfies all the constraints of the minimum problem, Eq. (4.39). The motivation for choosing the vector $\mathbf{x}$ in this particular form lies in our interest in a minimum for $\mathbf{x}^H A \mathbf{x}$.

Next, let us write

$$\begin{aligned}\mathbf{x}^H A \mathbf{x} &= (\bar{c}_1 \mathbf{x}_1^H + \bar{c}_2 \mathbf{x}_2^H) A (c_1 \mathbf{x}_1 + c_2 \mathbf{x}_2) \\ &= |c_1|^2 \lambda_1 + |c_2|^2 \lambda_2 \le \lambda_2 (|c_1|^2 + |c_2|^2) = \lambda_2 \end{aligned} \qquad (4.44)$$

But the minimum value of $\mathbf{x}^H A \mathbf{x}$ satisfies the inequality

$$\tilde{\lambda}_1(\mathbf{v}) \le \mathbf{x}^H A \mathbf{x} \qquad (4.45)$$

so that

$$\tilde{\lambda}_1(\mathbf{v}) \le \lambda_2 \qquad (4.46)$$

which proves the second half of the theorem.

Inequality (4.46) can also be explained by means of Fig. 4.1. If $\mathbf{v}$ does not coincide with $\mathbf{x}_1$, then the trial vector $\mathbf{x}$ must contain an $\mathbf{x}_1$-component. Geometrically, this implies that the intersection of the plane normal to $\mathbf{v}$ and the ellipsoid represents an ellipse with one axis at least as large as $\mathbf{x}_2$. Hence, $\tilde{\lambda}_1(\mathbf{v}) = \min (1/\|\mathbf{x}\|^2)$ subject to $\mathbf{x}^T \mathbf{v} = 0$ is such that inequality (4.46) is satisfied. Note that $\min (1/\|\mathbf{x}\|^2)$ on $\mathbf{x}^T A \mathbf{x} = 1$ is equivalent to $\min \mathbf{x}^T A \mathbf{x}$ on $\|\mathbf{x}\| = 1$.

We shall discuss the eigenvalues $\tilde{\lambda}_2, \tilde{\lambda}_3, \ldots, \tilde{\lambda}_{n-1}$ in Secs. 4.4–4.5.

As with Rayleigh's principle, Rayleigh's theorem for systems with constraints finds its widest application in vibrating structures for which A is a real symmetric matrix.

Example 4.3

Consider the system of Example 4.1, use the constraint vector

$$\mathbf{v} = [0 \quad 1 \quad -1]^T \qquad (a)$$

and show that inequality (4.40) is satisfied.

The constraint equation $\mathbf{x}^H \mathbf{v} = \mathbf{x}^T \mathbf{v} = 0$ yields

$$x_2 - x_3 = 0 \tag{b}$$

so that the constrained vector has the form

$$\mathbf{x} = [x_1 \quad x_2 \quad x_2]^T \tag{c}$$

Next, let us write

$$f = \mathbf{x}^T A \mathbf{x} = \frac{k}{m} [2x_1^2 + 4x_2^2 - 2(1+\sqrt{2})x_1 x_2] \tag{d}$$

The condition that $\mathbf{x}$ be a unit vector yields

$$x_1^2 + 2x_2^2 = 1 \tag{e}$$

so that Eq. (d) reduces to

$$f = 2\frac{k}{m}[1 - (1+\sqrt{2})x_2\sqrt{1 - 2x_2^2}] \tag{f}$$

The function f has a minimum for $x_2 = \frac{1}{2}$, which is

$$\tilde{\lambda}_1 = \min \mathbf{x}^T A \mathbf{x} = \left(1 - \frac{1}{\sqrt{2}}\right)\frac{k}{m} = 0.292893\,\frac{k}{m} \tag{g}$$

The two lowest eigenvalues for the system are

$$\lambda_1 = 0.139194\,\frac{k}{m}, \qquad \lambda_2 = 1.745810\,\frac{k}{m} \tag{h}$$

so that inequality (4.40) is satisfied.

4.4 Maximum-minimum characterization of eigenvalues

In Sec. 4.3 we have considered the characterization of the lowest eigenvalue $\tilde{\lambda}_1$ of a modified system, obtained by imposing a constraint on the original system. In particular, we have shown that, according to Rayleigh's theorem for systems with one constraint, $\tilde{\lambda}_1$ satisfies the inequality $\lambda_1 \le \tilde{\lambda}_1 \le \lambda_2$.

The same inequality, however, can be regarded as a characterization of the second eigenvalue λ_2 of the original unconstrained system. Moreover, unlike Eq. (4.19) of Sec. 4.2, *this characterization is independent of the first eigenvector* $\mathbf{x}_1$. This is very important, because the vector $\mathbf{x}_1$ is not generally known exactly. The new interpretation of Rayleigh's theorem for systems with one constraint can be stated in the form of a

4.4 Maximum-minimum characterization of eigenvalues

theorem attributed to Courant and Fischer, which reads as follows: *The second eigenvalue λ_2 of a Hermitian matrix A is the maximum value which can be given to $\min \mathbf{x}^H A \mathbf{x}$ by the imposition of the single constraint $\mathbf{x}^H \mathbf{v} = 0$,* or

$$\lambda_2 = \max \tilde{\lambda}_1(\mathbf{v}) = \max (\min \mathbf{x}^H A \mathbf{x}), \qquad \|\mathbf{x}\| = 1, \qquad \mathbf{x}^H \mathbf{v} = 0 \qquad (4.47)$$

The theorem can be extended to any number of constraints $r < n$, thus providing a characterization of the eigenvalues λ_{r+1} of a Hermitian matrix A which is independent of the eigenvectors of A. The mathematical statement of the theorem is

$$\lambda_{r+1} = \max (\min \mathbf{x}^H A \mathbf{x}), \qquad \|\mathbf{x}\| = 1, \qquad \mathbf{x}^H \mathbf{v}_i = 0,$$
$$i = 1, 2, \ldots, r \qquad (4.48)$$

and the proof follows the lines of that for one constraint. Hence, by analogy with Eq. (4.39), let us define the function

$$\tilde{\lambda}_r(\mathbf{v}_1, \mathbf{v}_2, \ldots, \mathbf{v}_r) = \min \mathbf{x}^H A \mathbf{x}, \qquad \|\mathbf{x}\| = 1, \qquad \mathbf{x}^H \mathbf{v}_i = 0,$$
$$i = 1, 2, \ldots, r \qquad (4.49)$$

where $\tilde{\lambda}_r$ is a continuous function of $\mathbf{v}_1, \mathbf{v}_2, \ldots, \mathbf{v}_r$. According to Rayleigh's principle of Sec. 4.2, if $\mathbf{v}_1, \mathbf{v}_2, \ldots, \mathbf{v}_r$ coincide with the eigenvectors $\mathbf{x}_1, \mathbf{x}_2, \ldots, \mathbf{x}_r$ belonging to the r lowest eigenvalues $\lambda_1, \lambda_2, \ldots, \lambda_r$, then

$$\tilde{\lambda}_r(\mathbf{x}_1, \mathbf{x}_2, \ldots, \mathbf{x}_r) = \lambda_{r+1} \qquad (4.50)$$

Next, let us assume that the vectors $\mathbf{v}_1, \mathbf{v}_2, \ldots, \mathbf{v}_r$ are given and consider a trial vector in the form

$$\mathbf{x} = c_1 \mathbf{x}_1 + c_2 \mathbf{x}_2 + \cdots + c_{r+1} \mathbf{x}_{r+1} = \sum_{i=1}^{r+1} c_i \mathbf{x}_i \qquad (4.51)$$

Using the same argument as for one constraint, it can be shown that such a vector satisfies all the constraints of the minimum problem (4.49). Hence, let us write

$$\mathbf{x}^H A \mathbf{x} = \left(\sum_{i=1}^{r+1} \bar{c}_i \mathbf{x}_i^H \right) A \left(\sum_{j=1}^{r+1} c_j \mathbf{x}_j \right)$$
$$= \sum_{i=1}^{r+1} |c_i|^2 \lambda_i \leq \lambda_{r+1} \sum_{i=1}^{r+1} |c_i|^2 = \lambda_{r+1} \qquad (4.52)$$

4   *Qualitative behavior of the eigensolution*

Because the minimum value of $\mathbf{x}^H A \mathbf{x}$ satisfies the inequality

$$\tilde{\lambda}_r(\mathbf{v}_1, \mathbf{v}_2, \ldots, \mathbf{v}_r) \leq \mathbf{x}^H A \mathbf{x} \tag{4.53}$$

it follows that

$$\tilde{\lambda}_r(\mathbf{v}_1, \mathbf{v}_2, \ldots, \mathbf{v}_r) \leq \lambda_{r+1} \tag{4.54}$$

where the equality sign holds when $\mathbf{v}_i = \mathbf{x}_i$ $(i = 1, 2, \ldots, r)$. This proves the theorem as stated by Eq. (4.48). In words, the theorem reads: *The eigenvalue λ_{r+1} of a Hermitian matrix A is the maximum value which can be given to $\min \mathbf{x}^H A \mathbf{x}$ by the imposition of the r constraints $\mathbf{x}^H \mathbf{v}_i = 0$ $(i = 1, 2, \ldots, r)$.*

As with Rayleigh's principle and Rayleigh's theorem, the maximum-minimum theorem is most useful for real symmetric matrices.

Sometimes in the technical literature the eigenvalues are assumed to be such that $\lambda_1 \geq \lambda_2 \geq \cdots \geq \lambda_n$. In this case the words maximum and minimum are interchanged and the theorem is referred to as the 'minimax' theorem.

Although the maximum-minimum theorem can be used to estimate higher eigenvalues of a system without prior knowledge of eigenvectors belonging to lower eigenvalues, the real use of the theorem is in proving other important results concerning the system eigenvalues.

4.5 The inclusion principle

Quite often, in the study of vibrations one is faced with the problem of truncation. In particular, the question arises as to the effect on the system eigenvalues of representing the system by a mathematical model with fewer degrees of freedom. To answer this question, let us consider a system described by a Hermitian matrix A of order n and with eigenvalues $\lambda_1, \lambda_2, \ldots, \lambda_n$ and then consider another system described by a Hermitian matrix B of order $n-1$ and with eigenvalues $\gamma_1, \gamma_2, \ldots, \gamma_{n-1}$, where B is obtained from A by deleting the last row and the last column. The object is to determine how the eigenvalues $\gamma_1, \gamma_2, \ldots, \gamma_{n-1}$ relate to the eigenvalues $\lambda_1, \lambda_2, \ldots, \lambda_n$.

The solution to the problem can be obtained by using the maximum-minimum theorem. In the first place, we observe that the quadratic form $\mathbf{y}^H B \mathbf{y}$ is identical to the quadratic form $\mathbf{x}^H A \mathbf{x}$ if $x_i = y_i$ $(i = 1, 2, \ldots, n-1)$ and $x_n = 0$. This can be easily verified by writing

$$\mathbf{y}^H B \mathbf{y} = \sum_{i=1}^{n-1}\sum_{j=1}^{n-1} b_{ij}\bar{y}_i y_j = \sum_{i=1}^{n-1}\sum_{j=1}^{n-1} a_{ij}\bar{x}_i x_j = \sum_{i=1}^{n}\sum_{j=1}^{n} a_{ij}\bar{x}_i x_j = \mathbf{x}^H A \mathbf{x} \tag{4.55}$$

The equation $x_n = 0$ can be regarded as a constraint equation $\mathbf{x}^H \mathbf{e}_n = 0$ imposed on the vector $\mathbf{x}$, where $\mathbf{e}_n$ is the unit vector $[0\ 0\ \cdots\ 0\ 1]^T$.

Regarding the system associated with B as unconstrained, Rayleigh's principle, Eq. (4.15) permits us to write

$$\gamma_1 = \min \mathbf{y}^H B \mathbf{y}, \qquad \|\mathbf{y}\| = 1 \tag{4.56}$$

But the problem associated with B is equivalent to the problem associated with A subject to the constraint $\mathbf{x}^H \mathbf{e}_n = 0$, so that

$$\gamma_1 = \tilde{\lambda}_1(\mathbf{e}_n) = \min \mathbf{x}^H A \mathbf{x}, \qquad \|\mathbf{x}\| = 1, \qquad \mathbf{x}^H \mathbf{e}_n = 0 \tag{4.57}$$

It follows from Rayleigh's theorem for one constraint, inequality (4.40), that

$$\lambda_1 \le \gamma_1 \le \lambda_2 \tag{4.58}$$

Next, let us assume that the vector $\mathbf{x}$ satisfies the constraints $\mathbf{x}^H \mathbf{v}_i = 0$ $(i = 1, 2, \ldots, r-1;\ r < n)$ and introduce the notation

$$\tilde{\lambda}_{r-1}(\mathbf{v}_1, \mathbf{v}_2, \ldots, \mathbf{v}_{r-1}) = \min \mathbf{x}^H A \mathbf{x}, \qquad \|\mathbf{x}\| = 1, \qquad \mathbf{x}^H \mathbf{v}_i = 0$$
$$i = 1, 2, \ldots, r-1;\ r < n \tag{4.59}$$

Moreover, let us assume that, in addition to the constraints $\mathbf{x}^H \mathbf{v}_i = 0$, the vector $\mathbf{x}$ is also subject to the constraint $\mathbf{x}^H \mathbf{e}_n = 0$, and introduce the notation

$$\tilde{\tilde{\lambda}}_{r-1}(\mathbf{v}_1, \mathbf{v}_2, \ldots, \mathbf{v}_{r-1}) = \min \mathbf{x}^H A \mathbf{x}, \qquad \|\mathbf{x}\| = 1, \qquad \mathbf{x}^H \mathbf{v}_i = 0, \qquad \mathbf{x}^H \mathbf{e}_n = 0$$
$$i = 1, 2, \ldots, r-1;\ r < n \tag{4.60}$$

Because in the latter case there is an additional constraint, we can write

$$\tilde{\lambda}_{r-1} \le \tilde{\tilde{\lambda}}_{r-1}, \qquad r = 2, 3, \ldots, n-1 \tag{4.61}$$

But a unit vector $\mathbf{x}$ subject to the constraints $\mathbf{x}^H \mathbf{e}_n = 0$ and $\mathbf{x}^H \mathbf{v}_i = 0$ $(i = 1, 2, \ldots, r-1)$ corresponds to a unit vector $\mathbf{y}$ subject to the constraints $\mathbf{y}^H \mathbf{w}_i = 0$ $(i = 1, 2, \ldots, r-1)$, where the $(n-1)$-dimensional vectors $\mathbf{w}_i$ are obtained from the n-dimensional vectors $\mathbf{v}_i$ by deleting the last component. Hence, if we denote

$$\tilde{\gamma}_{r-1}(\mathbf{w}_1, \mathbf{w}_2, \ldots, \mathbf{w}_{r-1}) = \min \mathbf{y}^H B \mathbf{y}, \qquad \|\mathbf{y}\| = 1, \qquad \mathbf{y}^H \mathbf{w}_i = 0$$
$$i = 1, 2, \ldots, r-1, r < n \tag{4.62}$$

then we obtain

$$\tilde{\gamma}_{r-1} = \tilde{\tilde{\lambda}}_{r-1}, \qquad r = 2, 3, \ldots, n-1 \tag{4.63}$$

Using the maximum-minimum theorem, we can write

$$\lambda_r = \max \tilde{\lambda}_{r-1} \tag{4.64}$$

Moreover, identical functions have identical maximum values, so that

$$\gamma_r = \max \tilde{\gamma}_{r-1} = \max \tilde{\tilde{\lambda}}_{r-1} \tag{4.65}$$

Introducing Eqs. (4.64) and (4.65) into inequality (4.61), it follows that

$$\lambda_r \le \gamma_r, \qquad r = 2, 3, \ldots, n-1 \tag{4.66}$$

Equation (4.65), in conjunction with Eq. (4.60), implies that

$$\gamma_r = \max (\min \mathbf{x}^H A \mathbf{x}), \qquad \|\mathbf{x}\| = 1, \qquad \mathbf{x}^H \mathbf{v}_i = 0, \qquad \mathbf{x}^H \mathbf{e}_n = 0$$

$$i = 1, 2, \ldots, r-1; r < n \tag{4.67}$$

On the other hand, Eq. (4.59) yields

$$\lambda_{r+1} = \max (\min \mathbf{x}^H A \mathbf{x}), \qquad \|\mathbf{x}\| = 1, \qquad \mathbf{x}^H \mathbf{v}_i = 0$$

$$i = 1, 2, \ldots, r; r < n \tag{4.68}$$

Comparing Eqs. (4.67) and (4.68), we observe that in Eq. (4.67) the rth constraint vector is $\mathbf{e}_n$ whereas in Eq. (4.68) the rth constraint vector is $\mathbf{v}_r$. Therefore, there are more admissible candidates to maximize λ_r than there are to maximize $\tilde{\gamma}_{r-1}$. It follows that

$$\lambda_{r+1} \ge \gamma_r, \qquad r = 1, 2, \ldots, n-1 \tag{4.69}$$

Combining Eqs. (4.58), (4.66) and (4.69), we conclude that

$$\lambda_1 \le \gamma_1 \le \lambda_2 \le \gamma_2 \le \cdots \le \lambda_{n-1} \le \gamma_{n-1} \le \lambda_n \tag{4.70}$$

Inequalities (4.70) constitute the so-called *inclusion principle.*

In certain approximate methods converting a continuous conservative nongyroscopic system into a discrete one, the effect of including an additional degree of freedom in the mathematical model representing the same physical system results in adding an extra row and column to the matrices M and K, and hence to the matrix A. This can actually be interpreted as relaxing one constraint. Hence, the inclusion principle permits us to conclude that *the estimated natural frequencies tend to decrease with each additional degree of freedom.* At the same time there is a new frequency added which is higher than any of the previous ones. As the number of degrees of freedom is increased, the estimated frequencies decrease monotonically and approach the actual natural frequencies of the system asymptotically. This is the mathematical basis for the Rayleigh–Ritz method to be discussed later in this text.

The same approach can be used also in the case of continuous conservative gyroscopic systems. Care must be exercised, however, because in this case Rayleigh's quotient is in terms of $2n \times 2n$ matrices M^* and K^* and $2n$-dimensional vectors $\mathbf{v}$, as shown in Sec. 4.2, and not in terms of $n \times n$ matrices M and K and n-dimensional vectors $\mathbf{x}$. Indeed, it is important that we recognize that the addition of one degree of freedom to the mathematical model adds one row and one column to the matrices M, G and K, and at the same time we observe from Eqs. (2.59) and (2.63) that it adds two rows and two columns to the matrices M^*, G^* and K^*. Recalling, however, that the eigenvalues associated with the eigenvalue problem defined by M^* and K^*, Eqs. (4.29), have multiplicity two, we conclude that the addition of one degree of freedom to the mathematical model results in only one additional natural frequency, although it results in two additional eigenvectors. Hence, the inclusion principle, inequalities (4.70), and any conclusions reached by means of the principle are equally valid for conservative gyroscopic systems. We shall return to this subject later in this text, when we discuss the Rayleigh–Ritz method.

Example 4.4

Consider the matrices

$$A = \begin{bmatrix} 2.5 & -1 & 0 \\ -1 & 5 & -\sqrt{2} \\ 0 & -\sqrt{2} & 10 \end{bmatrix}, \qquad B = \begin{bmatrix} 2.5 & -1 \\ -1 & 5 \end{bmatrix} \tag{a}$$

and verify that the eigenvalues of A and those of B conform to the inclusion principle.

The eigenvalues of A satisfy the characteristic equation

$$|A - \lambda I| = \begin{vmatrix} 2.5 - \lambda & -1 & 0 \\ -1 & 5 - \lambda & -\sqrt{2} \\ 0 & -\sqrt{2} & 10 - \lambda \end{vmatrix}$$

$$= (5 - \lambda)(\lambda^2 - 12.5\lambda + 22) = 0 \tag{b}$$

which has the roots

$$\lambda_1 = 2.119322, \qquad \lambda_2 = 5, \qquad \lambda_3 = 10.380678 \tag{c}$$

The eigenvalues of B satisfy the characteristic equation

$$|B - \gamma I| = \begin{vmatrix} 2.5 - \gamma & -1 \\ -1 & 5 - \gamma \end{vmatrix} = \gamma^2 - 7.5\gamma + 11.5 = 0 \tag{d}$$

which has the roots

$$\gamma_1 = 2.149219, \qquad \gamma_2 = 5.350781 \tag{e}$$

Clearly,

$$\lambda_1 < \gamma_1 < \lambda_2 < \gamma_2 < \lambda_3 \tag{f}$$

so that the eigenvalues of A and those of B do indeed conform to the inclusion principle.

4.6   A criterion for the positive definiteness of a Hermitian matrix

The inclusion principle is qualitative in nature and not very suitable for estimating eigenvalues. The principle can be used, however, to develop a criterion for the positive definiteness of a Hermitian matrix.

Let us consider a Hermitian matrix A of order n. If A is positive definite, then all its eigenvalues are positive. Because det A is equal to the product of the eigenvalues of A, it follows that if A is positive definite, then det A is positive. Next, let us consider the quadratic form $\mathbf{x}^H A \mathbf{x}$, which is positive definite if A is positive definite, and introduce a k-vector $\mathbf{y}$ obtained from $\mathbf{x}$ by retaining the first k components. Moreover, let A_k be the $k \times k$ matrix obtained from A by retaining the first k rows and k columns. Then, the quadratic form $\mathbf{y}^H A_k \mathbf{y}$ is also positive definite, because it is equal to the quadratic form $\mathbf{x}^H A \mathbf{x}$ when $x_{k+1} = x_{k+2} = \cdots = x_n = 0$. It follows that if A is positive definite, then A_k is also positive definite, so that det A_k is also positive. Because k can take any integer value from 1 to n, we conclude that *if the Hermitian matrix A is positive definite, then all its principal minor determinants are positive.*

The converse can also be shown to be true. Let us assume that all the principal minor determinants of the matrix A are positive, det $A_i > 0$, $i \leq n$. We propose to prove that if A_k is positive definite for some $k < n$, then A_{k+1} is also positive definite. Let us denote the eigenvalues of the matrix A_k by $\beta_1, \beta_2, \ldots, \beta_k$. Assuming that A_k is positive definite, all the eigenvalues $\beta_1, \beta_2, \ldots, \beta_k$ are positive, where $\beta_1 \leq \beta_2 \leq \cdots \leq \beta_k$. Next, let us consider the matrix A_{k+1} and denote its eigenvalues by $\alpha_1, \alpha_2, \ldots, \alpha_{k+1}$, where $\alpha_1 \leq \alpha_2 \leq \cdots \leq \alpha_{k+1}$. By the inclusion principle, the two sets of eigenvalues are related by

$$\alpha_1 \leq \beta_1 \leq \alpha_2 \leq \beta_2 \leq \cdots \leq \alpha_k \leq \beta_k \leq \alpha_{k+1} \tag{4.71}$$

Because the eigenvalues $\beta_1, \beta_2, \ldots, \beta_k$ are positive, the eigenvalues $\alpha_2, \alpha_3, \ldots, \alpha_{k+1}$ are also positive. There remains the question of α_1, which could be negative. But the determinant of a matrix equals the product of its eigenvalues. Hence, because det A_{k+1} is positive, we can write

$$\alpha_1\alpha_2\cdots\alpha_{k+1}=\det A_{k+1}>0 \qquad (4.72)$$

and, since $\alpha_2, \alpha_3, \ldots, \alpha_{k+1}$ are positive, we must conclude that α_1 is also positive. It follows that if A_k is positive definite, then A_{k+1} is also positive definite. But, $A_1 = a_{11}>0$, so that letting $k = 1, 2, \ldots, n-1$, we conclude that $A_2, A_3, \ldots, A_n$ are all positive definite. This permits us to state the theorem: *A Hermitian matrix A of order n is positive definite if and only if all its principal minor determinants are positive*, $\det A_i >0$ $(i = 1, 2, \ldots, n)$. The theorem is sometimes referred to as *Sylvester's theorem*, or *Sylvester's criterion*.

Sylvester's criterion requires the evaluation of the principal minor determinants of the Hermitian matrix A, and in general it may not be the most expedient method for establishing the positive definiteness of A. At times, however, it can help rule out the positive definiteness of a matrix. Indeed, an implication of the criterion is the requirement that all the elements on the main diagonal be positive. This is so because the principal minor determinant of lowest order is the upper left corner element of the matrix. But the character of the problem does not change if the rth row and rth column trade positions with the sth row and sth column, respectively. In this manner, any element on the main diagonal can move to the upper left corner. Hence, if at least one of the elements on the main diagonal is not positive, the matrix cannot be positive definite.

Of course, the positive definiteness of a Hermitian matrix can always be checked by examining the eigenvalues of the matrix. To this end, it is not really necessary to solve the eigenvalue problem and it is sufficient to reduce the matrix to triangular form. Note that the eigenvalues of a matrix are simply the diagonal elements of the reduced triangular form.

Example 4.5

Use Sylvester's criterion and check whether the matrix A of Example 4.4 is positive definite.

The principal minor determinants of A have the values

$$\det A_1 = \det 2.5 = 2.5$$

$$\det A_2 = \begin{vmatrix} 2.5 & -1 \\ -1 & 5 \end{vmatrix} = 11.5 \qquad (a)$$

$$\det A_3 = \det A = \begin{vmatrix} 2.5 & -1 & 0 \\ -1 & 5 & -\sqrt{2} \\ 0 & -\sqrt{2} & 10 \end{vmatrix} = 110$$

All the principal minor determinants of A are positive, so that by Sylvester's criterion the matrix A is positive definite. The same conclusion could have been reached in Example 4.4 by noticing that all the eigenvalues of A are positive.

4.7   Eigenvalues of the sum of two Hermitian matrices

Let the properties of a vibrating system be described by a Hermitian matrix A of order n. For various reasons, however, the system properties are not known exactly, so that the matrix A contains some error. Denoting by B the known matrix and by E the error matrix, we can write

$$B = A + E \tag{4.73}$$

Next, let us denote the eigenvalues of B by $\beta_1, \beta_2, \ldots, \beta_n$, the eigenvalues of A by $\lambda_1, \lambda_2, \ldots, \lambda_n$, and the eigenvalues of E by $\rho_1, \rho_2, \ldots, \rho_n$, where the various eigenvalues are arranged in order of increasing algebraic values. Of course, only the eigenvalues $\beta_1, \beta_2, \ldots, \beta_n$ can be regarded as known. The object is to obtain an estimate of the eigenvalues $\lambda_1, \lambda_2, \ldots, \lambda_n$, having only a sketchy knowledge of the matrix E.

By the maximum-minimum theorem, Eq. (4.48), we have

$$\beta_r = \max (\min \mathbf{x}^H B \mathbf{x}), \qquad \|\mathbf{x}\| = 1, \qquad \mathbf{x}^H \mathbf{v}_i = 0,$$
$$i = 1, 2, \ldots, r-1; \qquad r < n \tag{4.74}$$

where $\mathbf{v}_i$ are constraint vectors. On the other hand, using Eqs. (4.14) and (4.16), we can write

$$\rho_1 \leq \mathbf{x}^H E \mathbf{x} \leq \rho_n, \qquad \|\mathbf{x}\| = 1 \tag{4.75}$$

it follows that

$$\mathbf{x}^H A \mathbf{x} + \rho_1 \leq \mathbf{x}^H B \mathbf{x} \leq \mathbf{x}^H A \mathbf{x} + \rho_n, \qquad \|\mathbf{x}\| = 1 \tag{4.76}$$

If the vector $\mathbf{x}$ is subject to the constraints $\mathbf{x}^H \mathbf{v}_i = 0$ $(i = 1, 2, \ldots, r-1)$, then

$$\min \mathbf{x}^H A \mathbf{x} + \rho_1 \leq \min \mathbf{x}^H B \mathbf{x} \leq \min \mathbf{x}^H A \mathbf{x} + \rho_n, \qquad \|\mathbf{x}\| = 1, \qquad \mathbf{x}^H \mathbf{v}_i = 0,$$
$$i = 1, 2, \ldots, r-1; \qquad r < n \tag{4.77}$$

Hence, using the maximum-minimum theorem for the eigenvalues λ_r of A and the eigenvalues β_r of B, inequality (4.77) can be replaced by

$$\lambda_r + \rho_1 \leq \beta_r \leq \lambda_r + \rho_n \tag{4.78}$$

Inequality (4.78) implies that when the error matrix E is added to A all of the eigenvalues of A are changed by an amount which lies between the lowest and highest eigenvalue of E. Clearly, if E is positive definite, then $\beta_r > \lambda_r$ $(r = 1, 2, \ldots, n)$.

Inequality (4.78) is valid without regard to the magnitude of the elements of the error matrix E. The question can be raised as to whether the eigenvalues of B are near the eigenvalues of A if the elements of E are small. This is the same as asking whether the eigenvalues of E are small if the elements of E are small. To answer this question, let us recall the definition of the matrix norm (1.67), and write

$$\|E\| = \max \frac{\|E\mathbf{x}\|}{\|\mathbf{x}\|}, \qquad \|\mathbf{x}\| \neq 0 \tag{4.79}$$

Letting $\mathbf{x}$ be the unit eigenvectors $\mathbf{x}_1$ and $\mathbf{x}_n$ belonging to ρ_1 and ρ_n, respectively, and using inequality (1.68), we conclude that

$$\begin{aligned}
\|E\mathbf{x}_1\| &= \|\rho_1\mathbf{x}_1\| = |\rho_1| \leq \|E\| \\
\|E\mathbf{x}_n\| &= \|\rho_n\mathbf{x}_n\| = |\rho_n| \leq \|E\|
\end{aligned} \tag{4.80}$$

Next, let us look upon the matrix E as the difference between B and A, $E = B - A$, so that from inequality (4.78) we can write

$$\rho_1 \leq \beta_r - \lambda_r \leq \rho_n \tag{4.81}$$

But inequalities (4.80) imply that

$$-\rho_1 \leq \|E\|, \qquad \rho_n \leq \|E\| \tag{4.82}$$

where we note that ρ_1 can be negative if E is not positive definite. Inequalities (4.81) and (4.82) yield

$$\|E\| \geq -\rho_1 \geq -(\beta_r - \lambda_r), \qquad \|E\| \geq \rho_n \geq \beta_r - \lambda_r \tag{4.83}$$

which can be combined into

$$|\beta_r - \lambda_r| \leq \|B - A\| \tag{4.84}$$

where

$$\|B - A\| = \|E\| \tag{4.85}$$

Inequality (4.84) is known as *Weyl's theorem* and it states that the difference between the corresponding eigenvalues of B and A are smaller than the norm of the error matrix E.

It remains for us to establish ways of estimating the norm of E. Letting $\mathbf{x}$ be a unit vector, Eq. (4.79) can be rewritten in the form

$$\|E\| = \max \|E\mathbf{x}\|, \qquad \|\mathbf{x}\| = 1 \tag{4.86}$$

97

Then, recalling the quadratic norm of a vector, Eq. (1.10), and the assumption that E is Hermitian, we can write

$$\|E\|^2 = \max \|E\mathbf{x}\|^2 = \max (E\mathbf{x}, E\mathbf{x})$$
$$= \max \mathbf{x}^H E^H E \mathbf{x} = \max \mathbf{x}^H E^2 \mathbf{x}, \qquad \|\mathbf{x}\| = 1 \qquad (4.87)$$

But the eigenvalues of E^2 are $\rho_1^2, \rho_2^2, \ldots, \rho_n^2$, so that, by Rayleigh's principle, $\|E\|^2$ is the largest eigenvalue of E^2, or

$$\|E\|^2 = \max (\rho_1^2, \rho_n^2) \qquad (4.88)$$

and because $\rho_1 \leq \rho_2 \leq \cdots \leq \rho_n$, we obtain

$$\|E\| = \max (-\rho_1, \rho_n) \qquad (4.89)$$

Equation (4.89) can be explained by the fact that, although E was assumed to be Hermitian, it need not be positive definite.

Next, we would like to obtain some estimates which do not depend explicitly on the eigenvalues of E, but only on the elements of E. From Eq. (4.88), we conclude that

$$\|E\|^2 \leq \sum_{r=1}^{n} \rho_r^2 = \mathrm{tr}\, E^2 = \sum_{i=1}^{n} \left(\sum_{j=1}^{n} e_{ij} e_{ji} \right)$$
$$= \sum_{i=1}^{n} \left(\sum_{j=1}^{n} \bar{e}_{ij} e_{ij} \right) = \sum_{i=1}^{n} \sum_{j=1}^{n} |e_{ij}|^2 \qquad (4.90)$$

so that

$$\|E\| \leq \left(\sum_{i=1}^{n} \sum_{j=1}^{n} |e_{ij}|^2 \right)^{1/2} \qquad (4.91)$$

Another estimate can be obtained by considering the eigenvalue problem associated with E. Let $\mathbf{x}$ be an eigenvector of E belonging to the eigenvalue ρ, where ρ is either ρ_1 or ρ_n. The eigenvalue problem can be written in the form

$$\sum_{j=1}^{n} e_{kj} x_j = \rho x_k, \qquad k = 1, 2, \ldots, n \qquad (4.92)$$

Now let x_m be the component of $\mathbf{x}$ with the largest modulus, $|x_m| = \max |x_j| > 0$ $(j = 1, 2, \ldots, n)$, so that letting $k = m$ in Eqs. (4.92), we can write

$$|\rho| \cdot |x_m| \leq \sum_{j=1}^{n} |e_{mj}| \cdot |x_j| \leq |x_m| \sum_{j=1}^{n} |e_{mj}| \qquad (4.93)$$

which yields

$$|\rho| \le \sum_{j=1}^{n} |e_{mj}| \le \max_{i} \sum_{j=1}^{n} |e_{ij}| \tag{4.94}$$

where the maximum is with respect to any row i of E. Introducing inequality (4.94) into Eq. (4.89), we obtain

$$\|E\| \le \max_{i} \sum_{j=1}^{n} |e_{ij}| \tag{4.95}$$

As an illustration, let us assume that all we know about E is that it is a small Hermitian matrix and that the magnitudes of its elements do not exceed a given number ε. Then,

$$|e_{ij}| \le \varepsilon \tag{4.96}$$

so that, from inequality (4.95), it follows that

$$\|E\| \le n\varepsilon \tag{4.97}$$

Hence, using Weyl's theorem, we obtain

$$|\beta_r - \lambda_r| \le n\varepsilon \tag{4.98}$$

4.8   Gerschgorin's theorems

Let us write the eigenvalue problem associated with the $n \times n$ matrix A in the form

$$\sum_{j=1}^{n} a_{kj} x_j = \lambda x_k, \qquad k = 1, 2, \ldots, n \tag{4.99}$$

Next, let us assume that x_m is the component of the vector $\mathbf{x}$ with the largest modulus, $|x_m| = \max |x_j| \ (j = 1, 2, \ldots, n)$, so that letting $k = m$ in Eqs. (4.99), we can write

$$(\lambda - a_{mm}) x_m = \sum_{\substack{j=1 \\ j \neq m}}^{n} a_{mj} x_j \tag{4.100}$$

Hence,

$$|\lambda - a_{mm}| \cdot |x_m| \le \sum_{\substack{j=1 \\ j \neq m}}^{n} |a_{mj}| \cdot |x_j| \le |x_m| \sum_{\substack{j=1 \\ j \neq m}}^{n} |a_{mj}| \tag{4.101}$$

Dividing inequality (4.101) through by $|x_m|$, we obtain

$$|\lambda - a_{mm}| \leq \sum_{\substack{j=1 \\ j \neq m}}^{n} |a_{mj}| \qquad (4.102)$$

Inequality (4.102) is known as the *first theorem of Gerschgorin* and can be stated as follows: *Every eigenvalue of the matrix A lies in at least one of the circular disks with centers at a_{mm} and radii $r_m = \sum_{\substack{j=1 \\ j \neq m}}^{n} |a_{mj}|$.* The disks are sometimes referred to as *Gerschgorin's disks.*

Of course, for the theorem to be of much use in estimating eigenvalues, it is important that the off-diagonal elements of A be small relative to its diagonal elements.

A second theorem by Gerschgorin furnishes more detailed information concerning the distribution of the eigenvalues among the disks. Proof of the theorem is based on the concept of continuity. In particular, *The eigenvalues of a matrix A are continuous functions of the elements of A.* Proof of the theorem on continuity can be found in the text by Franklin (F3, Sec. 6.13).

Let us write the matrix A in the form

$$A = D + C \qquad (4.103)$$

where $D = \text{diag}(a_{ii})$ is the diagonal matrix obtained from A by omitting its off-diagonal elements and C is the matrix of these off-diagonal elements. Next, let us introduce the matrix

$$A(\varepsilon) = D + \varepsilon C \qquad (4.104)$$

where $0 \leq \varepsilon \leq 1$. Clearly, for $\varepsilon = 0$, we have $A(0) = \text{diag}(a_{ii})$ and for $\varepsilon = 1$, we have $A(1) = A$. The coefficients of the characteristic polynomial of $A(\varepsilon)$ are polynomials in ε and, by continuity, they are continuous functions of ε. But, by the first theorem of Gerschgorin, for any value of ε the eigenvalues of $A(\varepsilon)$ lie in circular disks centered at a_{ii} and with radii $\varepsilon r_i = \sum_{\substack{j=1 \\ j \neq i}}^{n} \varepsilon |a_{ij}|$. As ε varies from 1 to 0, the n eigenvalues of $A(\varepsilon)$ move continuously to $a_{11}, a_{22}, \ldots, a_{nn}$.

Next, let us assume that the first k of the disks corresponding to A are disjoint and that the remaining $n - k$ disks are connected. Because the first k disks of A with radii $r_1, r_2, \ldots, r_k$ are disjoint, certainly the first k disks of $A(\varepsilon)$ with radii $\varepsilon r_1, \varepsilon r_2, \ldots, \varepsilon r_k$ are disjoint. As ε decreases from 1 to 0, the first k Gerschgorin disks shrink to the points $\lambda = a_{ii}$

$(i = 1, 2, \ldots, k)$. In the process, the eigenvalues contained in these disks must remain inside them. Assuming that the first k eigenvalues of A are distinct, it follows that each disjoint disk contains exactly one eigenvalue. It also follows that the balance of the eigenvalues must be contained in the connected domain which is the union of the remaining $n - k$ disks. This permits us to state the *second Gerschgorin theorem*: *If k of the Gerschgorin disks of inequality (4.102) are disjoint and the remaining $n - k$ form a connected domain, which is isolated from the first k disks, then there are exactly $n - k$ eigenvalues of A contained in the connected domain.*

Example 4.6

Consider the matrix

$$A = \begin{bmatrix} 2.5 & -1 & 0 \\ -1 & 5 & -\sqrt{2} \\ 0 & -\sqrt{2} & 10 \end{bmatrix} \tag{a}$$

and verify Gerschgorin's theorems.

The centers of the Gerschgorin's disks are at

$$a_{11} = 2.5, \qquad a_{22} = 5, \qquad a_{33} = 10 \tag{b}$$

and the corresponding radii are

$$\begin{aligned}
r_1 &= |a_{12}| + |a_{13}| = |-1| + 0 = 1 \\
r_2 &= |a_{21}| + |a_{23}| = |-1| + |-\sqrt{2}| = 1 + \sqrt{2} \\
r_3 &= |a_{31}| + |a_{32}| = 0 + |-\sqrt{2}| = \sqrt{2}
\end{aligned} \tag{c}$$

The disks are shown in Fig. 4.4. The eigenvalues of the matrix A are

$$\lambda_1 = 2.119322, \qquad \lambda_2 = 5, \qquad \lambda_3 = 10.380678 \tag{d}$$

and it can be easily seen that they are consistent with both theorems.

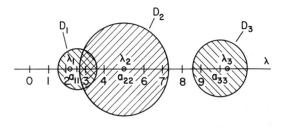

Figure 4.4

4.9 First-order perturbation of the eigenvalue problem

A frequently encountered problem in structural dynamics is how to take into account in analysis design changes introduced after the structural dynamic analysis has been completed and the natural frequencies and modes have been computed. If the new design is drastically different from the old one, then a completely new analysis and computational cycle is necessary. But if the new design varies only slightly from the old one, then the question is whether the information from the old design can be used to extract information concerning the new design. In particular, the question of interest here is whether the eigensolution already available can be used to derive the eigensolution corresponding to the new data, without extensive additional computations. In the sequel, we propose to study this problem by first considering an arbitrary real matrix and then specializing the results to real symmetric matrices.

Let us consider the $n \times n$ matrix A_0 and denote its eigenvalues by λ_{0i}, its right eigenvectors by $\mathbf{x}_{0i}$ and its left eigenvectors by $\mathbf{y}_{0i}$, where the eigenvalues and eigenvectors satisfy

$$A_0 \mathbf{x}_{0i} = \lambda_{0i} \mathbf{x}_{0i}, \qquad i = 1, 2, \ldots, n \tag{4.105a}$$

$$\mathbf{y}_{0i}^T A_0 = \lambda_{0i} \mathbf{y}_{0i}^T, \qquad i = 1, 2, \ldots, n \tag{4.105b}$$

For simplicity, A_0 was assumed to be real, but otherwise arbitrary. Moreover, we assume that the eigenvalues λ_{0i} are all distinct. Of course, the eigenvalues and eigenvectors are in general complex. The eigenvalue problem associated with a real arbitrary matrix was discussed in Sec. 3.7, where it was shown that the right and left eigenvectors corresponding to distinct eigenvalues are orthogonal. For convenience, we shall assume that the eigenvectors have been normalized so as to satisfy

$$\mathbf{y}_{0j}^T \mathbf{x}_{0i} = \delta_{ij}, \qquad \mathbf{y}_{0j}^T A_0 \mathbf{x}_{0i} = \lambda_{0i} \delta_{ij}, \qquad i, j = 1, 2, \ldots, n \tag{4.106}$$

Next, let us consider the eigenvalue problem associated with the real $n \times n$ matrix

$$A = A_0 + A_1 \tag{4.107}$$

where A_0 is the original matrix discussed above and A_1 is an $n \times n$ matrix representing a small change from A_0. We shall refer to A as the perturbed matrix and to A_1 as the first-order perturbation matrix. By contrast, A_0 represents the unperturbed matrix. The perturbed eigenvalue problem can be written in the form

$$A \mathbf{x}_i = \lambda_i \mathbf{x}_i, \qquad i = 1, 2, \ldots, n \tag{4.108a}$$

$$\mathbf{y}_i^T A = \lambda_i \mathbf{y}_i^T, \qquad i = 1, 2, \ldots, n \tag{4.108b}$$

where λ_i are the perturbed eigenvalues, x_i the perturbed right eigenvectors and y_i the perturbed left eigenvectors. In this context, Eq. (4.105) represents the unperturbed, or the zero-order eigenvalue problem. As with that problem, the eigenvalues are assumed to be distinct and the eigenvectors biorthonormal, i.e.,

$$y_j^T x_i = \delta_{ij} \qquad y_j^T A x_i = \lambda_i \delta_{ij}, \qquad i, j = 1, 2, \ldots, n \qquad (4.109)$$

Our interest lies in a first-order perturbation eigensolution. Because A is obtained from A_0 through a small perturbation, it is natural to assume that the perturbed eigenvalues and eigenvectors can be written in the form

$$\lambda_i = \lambda_{0i} + \lambda_{1i}, \qquad x_i = x_{0i} + x_{1i}, \qquad y_i = y_{0i} + y_{1i}, \qquad i = 1, 2, \ldots, n \qquad (4.110)$$

where λ_{1i}, x_{1i} and y_{1i} are first-order perturbations. Introducing Eqs. (4.107) and (4.110) into Eqs. (4.108), we can write

$$(A_0 + A_1)(x_{0i} + x_{1i}) = (\lambda_{0i} + \lambda_{1i})(x_{0i} + x_{1i}),$$
$$i = 1, 2, \ldots, n \qquad (4.111a)$$
$$(y_{0i}^T + y_{1i}^T)(A_0 + A_1) = (\lambda_{0i} + \lambda_{1i})(y_{0i}^T + y_{1i}^T),$$
$$i = 1, 2, \ldots, n \qquad (4.111b)$$

The problem before us is the determination of the perturbations λ_{1i}, x_{1i} and y_{1i} on the assumption that A_0, A_1, λ_{0i}, x_{0i} and y_{0i} are known. We shall concentrate on the determination of λ_{1i} and x_{1i}, as y_{1i} can be determined by analogy.

Care must be exercised in defining x_{1i}. We observe that if $A_1 = 0$, then $\lambda_{1i} = 0$ $(i = 1, 2, \ldots, n)$, in which case Eq. (4.111a) is satisfied by any x_{1i} which is a scalar multiple of x_{0i}. Because the vectors $x_{01}, x_{02}, \ldots, x_{0n}$ are linearly independent, they can be regarded as a basis in an n-space, so that the vector x_{1i} can be expressed as a linear combination of these vectors. To guarantee that $x_{1i} = 0$ when $A_1 = 0$ and that the coefficient of x_{0i} remains equal to 1 when A_0 is replaced by $A_0 + A_1$, we shall assume that the perturbed eigenvector x_i has the expansion

$$x_i = x_{0i} + \sum_{k=1}^{n} \varepsilon_{ik} x_{0k}, \qquad \varepsilon_{ii} = 0, \qquad i = 1, 2, \ldots, n \qquad (4.112)$$

where ε_{ik} $(i \neq k)$ are small coefficients. Hence, the eigenvector perturbation is simply

$$x_{1i} = \sum_{k=1}^{n} \varepsilon_{ik} x_{0k}, \qquad \varepsilon_{ii} = 0, \qquad i = 1, 2, \ldots, n \qquad (4.113)$$

so that the problem has been reduced to the determination of λ_{1i} and ε_{ik} $(i \neq k)$.

We shall seek a solution accurate to the first order, so that second-order terms in Eq. (4.111a) will be ignored. In view of this, if we consider the zero-order eigenvalue problem (4.105a), Eq. (4.111a) reduces to

$$A_0 \mathbf{x}_{1i} + A_1 \mathbf{x}_{0i} = \lambda_{0i} \mathbf{x}_{1i} + \lambda_{1i} \mathbf{x}_{0i}, \qquad i = 1, 2, \ldots, n \qquad (4.114)$$

Premultiplying Eq. (4.114) by $\mathbf{y}_{0j}^T$, we obtain

$$\mathbf{y}_{0j}^T A_0 \mathbf{x}_{1i} + \mathbf{y}_{0j}^T A_1 \mathbf{x}_{0i} = \lambda_{0i} \mathbf{y}_{0j}^T \mathbf{x}_{1i} + \lambda_{1i} \mathbf{y}_{0j}^T \mathbf{x}_{0i},$$
$$i = 1, 2, \ldots, n \qquad (4.115)$$

Recalling expansion (4.113), however, we can write

$$
\begin{aligned}
\mathbf{y}_{0j}^T A_0 \mathbf{x}_{1i} &= \mathbf{y}_{0j}^T A_0 \sum_{k=1}^{n} \varepsilon_{ik} \mathbf{x}_{0k} = \sum_{k=1}^{n} \varepsilon_{ik} \mathbf{y}_{0j}^T A_0 \mathbf{x}_{0k} \\
&= \sum_{k=1}^{n} \varepsilon_{ik} \lambda_{0j} \delta_{jk} = \varepsilon_{ij} \lambda_{0j} \\
\mathbf{y}_{0j}^T \mathbf{x}_{1i} &= \mathbf{y}_{0j}^T \sum_{k=1}^{n} \varepsilon_{ik} \mathbf{x}_{0k} = \sum_{k=1}^{n} \varepsilon_{ik} \mathbf{y}_{0j}^T \mathbf{x}_{0k} \\
&= \sum_{k=1}^{n} \varepsilon_{ik} \delta_{jk} = \varepsilon_{ij}
\end{aligned}
\qquad (4.116)
$$

so that, considering the first of Eqs. (4.106), Eq. (4.115) becomes

$$\varepsilon_{ij}(\lambda_{0j} - \lambda_{0i}) + \mathbf{y}_{0j}^T A_1 \mathbf{x}_{0i} = \lambda_{1i} \delta_{ij} \qquad (4.117)$$

But when $i = j$, $\varepsilon_{ij} = 0$, so that Eq. (4.117) yields the eigenvalue perturbations

$$\lambda_{1i} = \mathbf{y}_{0i}^T A_1 \mathbf{x}_{0i}, \qquad i = 1, 2, \ldots, n \qquad (4.118)$$

On the other hand, when $i \neq j$, $\delta_{ij} = 0$ and Eq. (4.117) reduces to

$$\varepsilon_{ik} = \frac{\mathbf{y}_{0k}^T A_1 \mathbf{x}_{0i}}{\lambda_{0i} - \lambda_{0k}}, \qquad i, k = 1, 2, \ldots, n; \qquad i \neq k \qquad (4.119)$$

where the subscript j has been replaced by k. The formal determination of the eigenvectors perturbations is completed by inserting Eqs. (4.119) into Eqs. (4.113).

By analogy, the adjoint perturbation vectors can be written in the form

$$\mathbf{y}_{ij} = \sum_{k=1}^{n} \gamma_{jk} \mathbf{y}_{0k}, \qquad \gamma_{jk} = 0, \qquad j = 1, 2, \ldots, n \qquad (4.120)$$

where the coefficients γ_{jk} can be shown to have the expressions

$$\gamma_{jk} = \frac{\mathbf{x}_{0k}^T A_1 \mathbf{y}_{0j}}{\lambda_{0j} - \lambda_{0k}}, \qquad j, k = 1, 2, \ldots, n; \qquad j \neq k \qquad (4.121)$$

A case of particular interest is that in which the matrices A_0 and A_1 are real and symmetric. Clearly, if the properties of the original system are such that the eigenvalue problem can be described by a real symmetric matrix A_0, then any design will yield a real symmetric matrix A, from which it follows that A_1 is also real and symmetric. This implies that the system is self-adjoint and the right and left eigenvectors coincide. Of course, the eigenvalues and eigenvectors are all real. Hence, the eigenvalue perturbations become

$$\lambda_{1i} = \mathbf{x}_{0i}^T A_1 \mathbf{x}_{0i}, \qquad i = 1, 2, \ldots, n \qquad (4.122)$$

and it is obvious that they are real. Moreover, the coefficients of the eigenvector perturbation expansions have the expressions

$$\varepsilon_{ik} = \frac{\mathbf{x}_{0k}^T A_1 \mathbf{x}_{0i}}{\lambda_{0i} - \lambda_{0k}}, \qquad i, k = 1, 2, \ldots, n; \qquad i \neq k \qquad (4.123)$$

and they are all real, so that the eigenvector perturbations are real, as expected.

Higher-order perturbations than first-order can be derived by expressing the matrix A in the form

$$A = A_0 + A_1 + A_2 + \cdots \qquad (4.124)$$

where the subscripts indicates the order of magnitude of the various matrices. In a likewise manner, the eigenvalues and eigenvectors have the expansions

$$\lambda_i = \lambda_{0i} + \lambda_{1i} + \lambda_{2i} + \cdots$$

$$\mathbf{x}_i = \mathbf{x}_{0i} + \mathbf{x}_{1i} + \mathbf{x}_{2i} + \cdots \qquad (4.125)$$

$$\mathbf{y}_i = \mathbf{y}_{0i} + \mathbf{y}_{1i} + \mathbf{y}_{2i} + \cdots$$

where the notation is obvious. Inserting Eqs. (4.124) and (4.125) into Eq. (4.108a) and following a procedure similar to that for the first-order perturbation, higher-order perturbations can be obtained. Note that higher-order perturbation eigensolutions can be obtained also when A remains in the form (4.107).

Example 4.7

Obtain a first-order perturbation eigensolution for the matrix

$$A = \begin{bmatrix} 2.6 & -1.1 & 0 \\ -1.1 & 5.2 & -\sqrt{2} \\ 0 & -\sqrt{2} & 10 \end{bmatrix} \tag{a}$$

based on the eigensolution of

$$A_0 = \begin{bmatrix} 2.5 & -1 & 0 \\ -1 & 5 & -\sqrt{2} \\ 0 & -\sqrt{2} & 10 \end{bmatrix} \tag{b}$$

The eigensolution of A_0 can be shown to be

$$\lambda_{01} = 2.119322, \qquad \lambda_{02} = 5, \qquad \lambda_{03} = 10.380678 \tag{c}$$

$$\mathbf{x}_{01} = \begin{bmatrix} 0.932674 \\ 0.355049 \\ 0.063715 \end{bmatrix}, \qquad \mathbf{x}_{02} = \begin{bmatrix} 0.359211 \\ -0.898027 \\ -0.254000 \end{bmatrix},$$

$$\mathbf{x}_{03} = \begin{bmatrix} 0.032965 \\ -0.259786 \\ 0.965103 \end{bmatrix} \tag{d}$$

where the eigenvectors $\mathbf{x}_{0i}$ ($i = 1, 2, 3$) have been normalized.
First we observe that

$$A_1 = A - A_0 = \begin{bmatrix} 0.1 & -0.1 & 0 \\ -0.1 & 0.2 & 0 \\ 0 & 0 & 0 \end{bmatrix} \tag{e}$$

which must be considered as 'small' relative to A_0.

The corrections to the eigenvalues can be obtained from Eqs. (4.122) in the form

$$\lambda_{11} = \mathbf{x}_{01}^T A_1 \mathbf{x}_{01} = \begin{bmatrix} 0.932674 \\ 0.355049 \\ 0.063715 \end{bmatrix}^T \begin{bmatrix} 0.1 & -0.1 & 0 \\ -0.1 & 0.2 & 0 \\ 0 & 0 & 0 \end{bmatrix} \begin{bmatrix} 0.932674 \\ 0.355049 \\ 0.063715 \end{bmatrix}$$

$$= 0.045971$$

$$\lambda_{12} = \mathbf{x}_{02}^T A_1 \mathbf{x}_{02} = \begin{bmatrix} 0.359211 \\ -0.898027 \\ -0.254000 \end{bmatrix}^T \begin{bmatrix} 0.1 & -0.1 & 0 \\ -0.1 & 0.2 & 0 \\ 0 & 0 & 0 \end{bmatrix} \begin{bmatrix} 0.359211 \\ -0.898027 \\ -0.254000 \end{bmatrix}$$

$$= 0.238710$$

$$\lambda_{13} = \mathbf{x}_{03}^T A_1 \mathbf{x}_{03} = \begin{bmatrix} 0.032965 \\ -0.259786 \\ 0.965103 \end{bmatrix}^T \begin{bmatrix} 0.1 & -0.1 & 0 \\ -0.1 & 0.2 & 0 \\ 0 & 0 & 0 \end{bmatrix} \begin{bmatrix} 0.032965 \\ -0.259786 \\ 0.965103 \end{bmatrix}$$

$$= 0.015319 \tag{f}$$

The eigenvector perturbations are given by Eqs. (4.113) in conjunction with Eqs. (4.123). Hence,

$$\varepsilon_{12} = -\varepsilon_{21} = \frac{\mathbf{x}_{02}^T A_1 \mathbf{x}_{01}}{\lambda_{01} - \lambda_{02}}$$

$$= \frac{\begin{bmatrix} 0.359211 \\ -0.898027 \\ -0.254000 \end{bmatrix}^T \begin{bmatrix} 0.1 & -0.1 & 0 \\ -0.1 & 0.2 & 0 \\ 0 & 0 & 0 \end{bmatrix} \begin{bmatrix} 0.932674 \\ 0.355049 \\ 0.063715 \end{bmatrix}}{2.119322 - 5}$$

$$= -0.014141$$

$$\varepsilon_{13} = -\varepsilon_{31} = \frac{\mathbf{x}_{03}^T A_1 \mathbf{x}_{01}}{\lambda_{01} - \lambda_{03}}$$

$$= \frac{\begin{bmatrix} 0.032965 \\ -0.259786 \\ 0.965103 \end{bmatrix}^T \begin{bmatrix} 0.1 & -0.1 & 0 \\ -0.1 & 0.2 & 0 \\ 0 & 0 & 0 \end{bmatrix} \begin{bmatrix} 0.932674 \\ 0.355049 \\ 0.063715 \end{bmatrix}}{2.119322 - 10.380678}$$

$$= -0.000930$$

$$\varepsilon_{23} = -\varepsilon_{32} = \frac{\mathbf{x}_{03}^T A_1 \mathbf{x}_{02}}{\lambda_{02} - \lambda_{03}}$$

$$= \frac{\begin{bmatrix} 0.032965 \\ -0.259786 \\ 0.965103 \end{bmatrix}^T \begin{bmatrix} 0.1 & -0.1 & 0 \\ -0.1 & 0.2 & 0 \\ 0 & 0 & 0 \end{bmatrix} \begin{bmatrix} 0.359211 \\ -0.898027 \\ -0.254000 \end{bmatrix}}{5 - 10.380678}$$

$$= -0.011176 \tag{g}$$

Using Eqs. (4.113), we obtain

$$\mathbf{x}_{11} = \varepsilon_{12}\mathbf{x}_{02} + \varepsilon_{13}\mathbf{x}_{03}$$

$$= -0.014141 \begin{bmatrix} 0.359211 \\ -0.898027 \\ -0.254000 \end{bmatrix} - 0.000930 \begin{bmatrix} 0.032965 \\ -0.259786 \\ 0.965103 \end{bmatrix}$$

$$= \begin{bmatrix} -0.005110 \\ 0.012941 \\ 0.002694 \end{bmatrix}$$

$$\mathbf{x}_{12} = \varepsilon_{21}\mathbf{x}_{01} + \varepsilon_{23}\mathbf{x}_{03}$$

$$= 0.014141 \begin{bmatrix} 0.932674 \\ 0.355049 \\ 0.063715 \end{bmatrix} - 0.011176 \begin{bmatrix} 0.032965 \\ -0.259786 \\ 0.965103 \end{bmatrix}$$

$$= \begin{bmatrix} 0.012821 \\ 0.007924 \\ -0.009885 \end{bmatrix}$$

$$\mathbf{x}_{13} = \varepsilon_{31}\mathbf{x}_{01} + \varepsilon_{32}\mathbf{x}_{02}$$

$$= 0.000930 \begin{bmatrix} 0.932674 \\ 0.355049 \\ 0.063715 \end{bmatrix} + 0.011176 \begin{bmatrix} 0.359211 \\ -0.898027 \\ -0.254000 \end{bmatrix}$$

$$= \begin{bmatrix} 0.004882 \\ -0.009706 \\ -0.002779 \end{bmatrix} \tag{h}$$

The first-approximation eigenvalues are obtained from

$$\lambda_i \cong \lambda_{0i} + \lambda_{1i}, \qquad i = 1, 2, 3 \tag{i}$$

Adding the respective values in Eqs. (c) and (f), we have

$$\lambda_1 = 2.165293, \qquad \lambda_2 = 5.238710, \qquad \lambda_3 = 10.395997 \tag{j}$$

The first-approximation eigenvectors are obtained from

$$\mathbf{x}_i \cong \mathbf{x}_{0i} + \mathbf{x}_{1i}, \qquad i = 1, 2, 3 \tag{k}$$

Combining (d) and (h) and normalizing, we obtain

$$\mathbf{x}_1 = \begin{bmatrix} 0.927470 \\ 0.367953 \\ 0.066402 \end{bmatrix}, \qquad \mathbf{x}_2 = \begin{bmatrix} 0.371971 \\ -0.889958 \\ -0.263842 \end{bmatrix},$$

$$\mathbf{x}_3 = \begin{bmatrix} 0.037845 \\ -0.269475 \\ 0.962263 \end{bmatrix} \tag{1}$$

For comparison purposes, the solution of the eigenvalue problem for the matrix A, Eq. (a), was obtained exactly, i.e., without any perturbation scheme. The results are

$$\lambda_1 = 2.164748, \qquad \lambda_2 = 5.238546, \qquad \lambda_3 = 10.396796 \tag{m}$$

$$\mathbf{x}_1 = \begin{bmatrix} 0.927810 \\ 0.367120 \\ 0.066263 \end{bmatrix}, \qquad \mathbf{x}_2 = \begin{bmatrix} 0.371103 \\ -0.890158 \\ -0.264388 \end{bmatrix},$$

$$\mathbf{x}_3 = \begin{bmatrix} 0.038078 \\ -0.269893 \\ 0.962137 \end{bmatrix} \tag{n}$$

Comparing Eqs. (j) and (m) on the one hand and Eqs. (l) and (n) on the other hand, we conclude that the first-order perturbation solution produced reasonable results in this particular case.

Computational methods for the eigensolution

5.1 General discussion

In Ch. 4, we discussed the qualitative behavior of the eigensolution. In this chapter, we shift our attention to the quantitative aspects of the solution. The algebraic eigenvalue problem has enjoyed a great deal of popularity in recent years, as witnessed by the wealth of papers and books on the subject. The interest in the eigenvalue problem is not new, and can be traced to the enhanced understanding of vibratory systems that the eigensolution provides. The renewed interest in the eigenvalue problem, however, can be attributed to the increasingly powerful digital computer as it permitted the numerical solution of problems of magnitude and complexity undreamed of before. The intent of this chapter is to present a selection of modern computational algorithms of particular importance to the field of vibrations.

The computational algorithms for the eigensolution fall in different classes. Some of them solve the complete eigenvalue problem, others produce only the eigenvalues, and yet others produce the eigenvectors belonging to given eigenvalues. In one form or another, the latter class of algorithms requires the solution of sets of simultaneous algebraic equations. Yet another class of algorithms is not concerned with the actual solution of the eigenvalue problem but with its reduction to a simpler form, one that lends itself to an efficient solution. Hence, quite often a combination of several algorithms proves to be the most effective way of carrying out solutions of relatively complex eigenvalue problems.

In this chapter, we begin by discussing the solution of sets of simultaneous algebraic equations, with special emphasis on the Gaussian elimination. Then, several iteration techniques for the eigensolution, such as the power method and the Jacobi method, are presented. Among techniques for reducing matrices to simpler form, we single out Givens' method and Householder's method, which can be used to reduce matrices to tridiagonal form. Several techniques for determining eigenvalues and eigenvectors of a matrix that are especially efficient when used in conjunction with this simpler form are discussed. In particular, we present a method based on Sturm's theorem and the QR method for the determination of the eigenvalues and inverse iteration for the eigenvectors.

5.2 Gaussian elimination

Let us consider the system of linear algebraic equations

$$
\begin{aligned}
a_{11}x_1 + a_{12}x_2 + \cdots + a_{1n}x_n &= c_1 \\
a_{21}x_1 + a_{22}x_2 + \cdots + a_{2n}x_n &= c_2 \\
&\cdots \cdots \cdots \cdots \cdots \cdots \\
a_{n1}x_1 + a_{n2}x_2 + \cdots + a_{nn}x_n &= c_n
\end{aligned}
\tag{5.1}
$$

Equations (5.1) can be written in the compact form

$$
A\mathbf{x} = \mathbf{c} \tag{5.2}
$$

where the matrix A of the coefficients a_{ij} $(i, j = 1, 2, \ldots, n)$ is assumed to be nonsingular. The solution of Eq. (5.2) can be written symbolically in the form

$$
\mathbf{x} = A^{-1}\mathbf{c} \tag{5.3}
$$

so that a solution of the system of equations can be obtained by calculating the inverse of A. Because A is nonsingular, this can always be done. To this end, use can be made of Eq. (1.60). The question is whether such a course of action is advisable. To calculate the inverse of A by means of Eq. (1.60) it is necessary to calculate a variety of determinants, one of the determinants being det A. We have shown in Sec. 1.10 that the evaluation of the determinant of a matrix of order n involves $n!$ multiplications. For large n, this is a prohibitive proposition because of the large computer time required. Hence, we wish to explore ways of obtaining the solution $\mathbf{x}$ without the need to calculate A^{-1} explicitly.

Let us premultiply both sides of Eq. (5.2) by the matrix P, where P is a nonsingular $n \times n$ matrix, and obtain

$$
PA\mathbf{x} = P\mathbf{c} \tag{5.4}
$$

If $\mathbf{x}$ is a solution of Eq. (5.2), then it is also a solution of Eq. (5.4). The matrix P can be regarded as representing a linear transformation in the space L^n. The object of the transformation is to permit an accurate and efficient solution of the system of equations. In particular, we seek a matrix P which renders the matrix of coefficients triangular. Hence, let us introduce the notation

$$PA = U, \qquad Pc = b. \tag{5.5}$$

where U is an $n \times n$ upper triangular matrix and b is an n-vector, so that Eq. (5.4) reduces to

$$Ux = b \tag{5.6}$$

Now the solution of Eq. (5.6) is relatively easy to obtain. We observe that the nth equation involves only the unknown x_n, so that solving this last equation we obtain

$$x_n = b_n / u_{nn} \tag{5.7}$$

Next, we turn our attention to the $(n-1)$th equation. The equations involve only x_{n-1} and x_n. Because x_n is known already, we can solve for x_{n-1} and obtain

$$x_{n-1} = \frac{b_{n-1}}{u_{n-1,n-1}} - \frac{u_{n-1,n}}{u_{n-1,n-1}} x_n \tag{5.8}$$

Having x_n and x_{n-1} permits us to solve the $(n-2)$th equation for x_{n-2}. Continuing the procedure, we can solve for $x_{n-3}, x_{n-4}, \ldots, x_1$, in sequence. The method of solving a set of simultaneous algebraic equations with an upper triangular matrix of coefficients, whereby the unknowns are determined in the sequence $x_n, x_{n-1}, \ldots, x_1$, is known as *back substitution*.

There remains the problem of reducing Eq. (5.1) to Eq. (5.6), which amounts to the problem of determining the matrix U and the vector b. A procedure for such a reduction is discussed in the sequel. The procedure represents a formalization of the method of elimination learned in high-school algebra and is known as the *Gaussian elimination*.

Let us return to Eqs. (5.1) and introduce the notation

$$c_j = a_{j,n+1} \tag{5.9}$$

Then, assuming that $a_{11} \neq 0$ and subtracting a_{i1}/a_{11} times the first equation from the ith equation ($i = 2, 3, \ldots, n$), we obtain

$$\begin{aligned}
a_{11}x_1 + a_{12}x_2 + \cdots + a_{1n}x_n &= a_{1,n+1} \\
a_{22}^{(1)}x_2 + \cdots + a_{2n}^{(1)}x_n &= a_{2,n+1}^{(1)} \\
\cdots \cdots \cdots \cdots \cdots \cdots \\
a_{n2}^{(1)}x_2 + \cdots + a_{nn}^{(1)}x_n &= a_{n,n+1}^{(1)}
\end{aligned} \tag{5.10}$$

where the new coefficients have the expressions

$$a_{ij}^{(1)} = a_{ij} - \frac{a_{i1}}{a_{11}} a_{1j}, \qquad i = 2, 3, \ldots, n; \qquad j = 2, 3, \ldots, n+1 \tag{5.11}$$

The question remains as to what happens if $a_{11} = 0$. In this case, we simply interchange two rows or two columns, so that the element in the upper left corner is different from zero, and proceed with the solution. This is always possible because the above interchanges do not affect the results. The interchange of two rows corresponds simply to a reordering of the equations and the interchange of two columns to a reordering of the unknowns.

Next, let us assume that $a_{22}^{(1)} \neq 0$ and subtract $a_{i2}^{(1)}/a_{22}^{(1)}$ times the second equation from the ith equation ($i = 3, 4, \ldots, n$), to obtain

$$
\begin{aligned}
a_{11}x_1 + a_{12}x_2 + a_{13}x_3 + \cdots + a_{1n}x_n &= a_{1,n+1} \\
a_{22}^{(1)}x_2 + a_{23}^{(1)}x_3 + \cdots + a_{2n}^{(1)}x_n &= a_{2,n+1}^{(1)} \\
a_{33}^{(2)}x_3 + \cdots + a_{3n}^{(2)}x_n &= a_{3,n+1}^{(2)} \\
\cdots \cdots \cdots \cdots \cdots \cdots \\
a_{n3}^{(2)}x_3 + \cdots + a_{nn}^{(2)}x_n &= a_{n,n+1}^{(2)}
\end{aligned}
\tag{5.12}
$$

where

$$
a_{ij}^{(2)} = a_{ij}^{(1)} - \frac{a_{i2}^{(1)}}{a_{22}^{(1)}} a_{2j}^{(1)}, \qquad i = 3, 4, \ldots, n; \qquad j = 3, 4, \ldots, n+1 \tag{5.13}
$$

After $n - 1$ steps, the procedure yields finally

$$
\begin{aligned}
a_{11}x_1 + a_{12}x_2 + a_{13}x_3 + \cdots + a_{1n}x_n &= a_{1,n+1} \\
a_{22}^{(1)}x_2 + a_{23}^{(1)}x_3 + \cdots + a_{2n}^{(1)}x_n &= a_{2,n+1}^{(1)} \\
a_{33}^{(2)}x_3 + \cdots + a_{3n}^{(2)}x_n &= a_{3,n+1}^{(2)} \\
\cdots \cdots \cdots \cdots \cdots \cdots \\
a_{nn}^{(n-1)}x_n &= a_{n,n+1}^{(n-1)}
\end{aligned}
\tag{5.14}
$$

Comparing Eqs. (5.6) and (5.14), we conclude that

$$
U = \begin{bmatrix}
a_{11} & a_{12} & a_{13} & \cdots & a_{1n} \\
0 & a_{22}^{(1)} & a_{23}^{(1)} & \cdots & a_{2n}^{(1)} \\
0 & 0 & a_{33}^{(2)} & \cdots & a_{3n}^{(2)} \\
\cdots & \cdots & \cdots & \cdots & \cdots \\
0 & 0 & 0 & \cdots & a_{nn}^{(n-1)}
\end{bmatrix}, \qquad
b = \begin{bmatrix}
a_{1,n+1} \\
a_{2,n+1}^{(1)} \\
a_{3,n+1}^{(2)} \\
\cdots \\
a_{n,n+1}^{(n-1)}
\end{bmatrix}
\tag{5.15}
$$

so that the matrix of the coefficients has been reduced to an upper triangular form, as desired. The solution of Eqs. (5.14) can be obtained by back substitution, as indicated earlier in this section.

Instead of reducing the matrix of the coefficients to triangular form, we could have gone one step farther and reduce it to diagonal form. Indeed, if we subtract $a_{12}/a_{22}^{(1)}$ times the second of Eqs. (5.12) from the first, we obtain

$$
\begin{aligned}
a_{11}x_1 + \qquad\qquad a_{13}^{(2)}x_3 + \cdots + a_{1n}^{(2)}x_n &= a_{1,n+1}^{(2)} \\
a_{22}^{(1)}x_2 + a_{23}^{(1)}x_3 + \cdots + a_{2n}^{(1)}x_n &= a_{2,n+1}^{(1)} \\
a_{33}^{(2)}x_3 + \cdots + a_{3n}^{(2)}x_n &= a_{3,n+1}^{(2)} \\
\cdots\cdots\cdots\cdots\cdots \\
a_{n3}^{(2)}x_3 + \cdots + a_{nn}^{(2)}x_n &= a_{n,n+1}^{(2)}
\end{aligned}
\tag{5.16}
$$

so that now the second column also has only one element. Hence, the procedure involves subtraction not only from the equations below the subtracted equation but also from the equations above it. It is clear that the procedure leads ultimately to a diagonal matrix of coefficients. This variation on the Gaussian elimination is called the *Gauss–Jordan reduction*. The question arises naturally as to why should one triangularize a matrix when one can diagonalize it, and thus obtain the desired solution. As it turns out, the Gaussian elimination is more efficient because it involves only $n^3/3 + 0(n^2)$ multiplications and divisions, as opposed to the Gauss–Jordan reduction which involves $n^3/2 + 0(n^2)$. Hence, for large n, the Gaussian elimination is to be preferred because it requires fewer operations.

The rth step in the Gaussian elimination involved division by the element $a_{rr}^{(r-1)}$. This element is known as the rth *pivotal element* or the rth *pivot*. In the procedure described, the pivots are taken in order down the main diagonal of the matrix of coefficients. Of course, the procedure fails when a pivot is zero. To circumvent that problem, it becomes necessary to interchange rows or columns, as explained earlier. Even when a pivot is not zero but very small, large errors can occur, so that such an interchange of rows or columns is desirable also in this case. The procedure can be enhanced by choosing as pivot for the rth step the element defined by

$$
a_{rr}^{(r-1)} = \max_{i=r,\dots,n} \left| a_{ir}^{(r-1)} \right|
\tag{5.17}
$$

The algorithm thus altered is called *Gaussian elimination with interchanges* or *with pivoting for size*.

Example 5.1

Solve the set of equations

$$
\begin{aligned}
2x_1 + 2x_2 - x_3 &= 5 \\
x_1 + 1.5x_2 + 2x_3 &= 13.5 \\
x_1 + 6x_2 - 2x_3 &= 6.5
\end{aligned}
\tag{a}
$$

by the Gaussian elimination.

First, subtract $\frac{1}{2}$ of the first of Eqs. (a) from the second and the third. The result is

$$
\begin{aligned}
2x_1 + 2x_2 - \quad x_3 &= 5 \\
0.5x_2 + 2.5x_3 &= 11 \\
5x_2 - 1.5x_3 &= 4
\end{aligned}
\tag{b}
$$

The coefficient of the second is relatively small. Although it is not sufficiently small to cause instability, for the purpose of illustration, let us interchange the second and third rows, so that

$$
\begin{aligned}
2x + 2x_2 - \quad x_3 &= 5 \\
5x_2 - 1.5x_3 &= 4 \\
0.5x_2 + 2.5x_3 &= 11
\end{aligned}
\tag{c}
$$

Next, let us subtract $\frac{1}{10}$ of the second of Eqs. (c) from the third and write

$$
\begin{aligned}
2x_1 + 2x_2 - \quad x_3 &= 5 \\
5x_2 - 1.5x_3 &= 4 \\
2.65x_3 &= 10.6
\end{aligned}
\tag{d}
$$

Equations (d) have the desired triangular form, so that we shall obtain their solution by back substitution. The third of Eqs. (d) yields directly

$$
x_3 = \frac{10.6}{2.65} = 4
\tag{e}
$$

Inserting $x_3 = 4$ into the second of Eqs. (d), we obtain

$$
x_2 = \tfrac{1}{5}(4+6) = 2
\tag{f}
$$

Finally, inserting $x_2 = 2$ and $x_3 = 4$ into the first of Eqs. (d), we obtain

$$
x_1 = \tfrac{1}{2}(5+4-4) = 2.5
\tag{g}
$$

so that the solution is

$$
x_1 = 2.5, \qquad x_2 = 2, \qquad x_3 = 4
\tag{h}
$$

5.3   Reduction to triangular form by elementary row operations

In the solution of sets of algebraic equations there are three basic operations:

1. Multiply equation p by a nonzero constant k and add the result to equation q.

115

2. Interchange equations p and q.
3. Multiply equation p by a nonzero constant α.

These are called *elementary operations*. If we regard the set of equations as being represented by the matrix of the coefficients, then the description of the above operations remains the same except that we should replace the word 'equation' by the word 'row', so that we shall refer to them as *elementary row operations*. If we regard Eqs. (e) of Example 5.1 as being the result of multiplying the third of Eqs. (d) by 1/2.65, etc., then we conclude that all three operations listed above were used in producing the solution (h). The three elementary operations can be represented in matrix form by premultiplying the matrix A of the coefficients by the corresponding matrices

$$
\begin{bmatrix}
 & p & q & \\
1 \ \ldots \ 0 & \ldots & 0 & \ldots \ 0 \\
0 \ \ldots \ 1 & \ldots & 0 & \ldots \ 0 \\
0 \ \ldots \ k & \ldots & 1 & \ldots \ 0 \\
0 \ \ldots \ 0 & \ldots & 0 & \ldots \ 1
\end{bmatrix}
\begin{matrix} \\ p \\ q \\ \end{matrix}
,\qquad
\begin{bmatrix}
 & p & q & \\
1 \ \ldots \ 0 & \ldots & 0 & \ldots \ 0 \\
0 \ \ldots \ 0 & \ldots & 1 & \ldots \ 0 \\
0 \ \ldots \ 1 & \ldots & 0 & \ldots \ 0 \\
0 \ \ldots \ 0 & \ldots & 0 & \ldots \ 1
\end{bmatrix}
\begin{matrix} \\ p \\ q \\ \end{matrix}
,
$$

$$
\begin{bmatrix}
 & p & & \\
1 \ \cdots \ 0 & \cdots & 0 & \cdots \ 0 \\
0 \ \cdots \ \alpha & \cdots & 0 & \cdots \ 0 \\
0 \ \cdots \ 0 & \cdots & 1 & \cdots \ 0 \\
0 \ \cdots \ 0 & \cdots & 0 & \cdots \ 1
\end{bmatrix}
\begin{matrix} \\ p \\ \\ \end{matrix}
\tag{5.18}
$$

where the second matrix is often called a *permutation matrix*.

Next, let us formalize the Gaussian elimination by expressing the various steps in matrix form. To this end, let us denote the original matrix A of the coefficients by A_0 and the matrix of coefficients corresponding to Eqs. (5.10) by A_1. Then, we can say that A_1 is obtained from A_0 by means of the linear transformation

$$
A_1 = P_1 A_0 \tag{5.19}
$$

where the transformation matrix P_1 can be verified to be the unit lower

116

triangular matrix

$$P_1 = \begin{bmatrix} 1 & 0 & 0 & \cdots & 0 \\ -p_{21} & 1 & 0 & \cdots & 0 \\ -p_{31} & 0 & 1 & \cdots & 0 \\ \cdots & \cdots & \cdots & \cdots & \cdots \\ -p_{n1} & 0 & 0 & \cdots & 1 \end{bmatrix} \tag{5.20}$$

in which

$$p_{j1} = \frac{a_{j1}}{a_{11}}, \qquad j = 2, 3, \ldots, n \tag{5.21}$$

Moreover, denoting the matrix of coefficients corresponding to Eqs. (5.12) by A_2, the matrix can be shown to have the form

$$A_2 = P_2 A_1 = P_2 P_1 A_0 \tag{5.22}$$

where

$$P_2 = \begin{bmatrix} 1 & 0 & 0 & \cdots & 0 \\ 0 & 1 & 0 & \cdots & 0 \\ 0 & -p_{32} & 1 & \cdots & 0 \\ \cdots & \cdots & \cdots & \cdots & \cdots \\ 0 & -p_{n2} & 0 & \cdots & 1 \end{bmatrix} \tag{5.23}$$

in which

$$p_{j2} = \frac{a_{j2}^{(1)}}{a_{22}^{(1)}}, \qquad j = 3, 4, \ldots, n \tag{5.24}$$

By now, a pattern is beginning to emerge. Indeed, the rth step involves premultiplication of the matrix A_{r-1} by the matrix P_r, where P_r is the unit matrix of order n with the exception of the elements in the rth column below the main diagonal, which have the values

$$-p_{jr} = -\frac{a_{jr}^{(r-1)}}{a_{rr}^{(r-1)}}, \qquad j = r+1, r+2, n \tag{5.25}$$

At the end of $n-1$ steps, we obtain the upper triangular matrix

$$U = P_{n-1} A_{n-2} = P_{n-1} P_{n-2} \cdots P_1 A_0 = P A_0 \tag{5.26}$$

where

$$P = P_{n-1} P_{n-2} \cdots P_1 \tag{5.27}$$

The above result can be presented in a slightly different manner. To this end, let us introduce the vector

$$\mathbf{p}_r = [0 \quad 0 \quad \cdots \quad 0 \quad p_{r+1,r} \quad p_{r+2,r} \quad \cdots \quad p_{nr}]^T \tag{5.28}$$

Then, the matrix P_r can be written in the compact form

$$P_r = I - \mathbf{p}_r \mathbf{e}_r^T, \qquad r = 1, 2, \ldots, n-1 \tag{5.29}$$

where I is the unit matrix of order n and $\mathbf{e}_r$ is the rth standard unit vector. The inverse of P_r has the simple form

$$P_r^{-1} = I + \mathbf{p}_r \mathbf{e}_r^T \tag{5.30}$$

Indeed, we can write

$$
\begin{aligned}
P_r^{-1} P_r &= (I + \mathbf{p}_r \mathbf{e}_r^T)(I - \mathbf{p}_r \mathbf{e}_r^T) \\
&= I + \mathbf{p}_r \mathbf{e}_r^T - \mathbf{p}_r \mathbf{e}_r^T - \mathbf{p}_r (\mathbf{e}_r^T \mathbf{p}_r) \mathbf{e}_r^T = I
\end{aligned}
\tag{5.31}
$$

because $\mathbf{e}_r^T \mathbf{p}_r = 0$.

Recalling that $A_0 = A$, Eq. (5.26) can be written in the form

$$A = P^{-1} U \tag{5.32}$$

Using Eqs. (5.27) and (5.30), we obtain

$$
\begin{aligned}
P^{-1} &= P_1^{-1} \cdots P_{n-2}^{-1} P_{n-1}^{-1} \\
&= (I + \mathbf{p}_1 \mathbf{e}_1^T) \cdots (I + \mathbf{p}_{n-2} \mathbf{e}_{n-2}^T)(I + \mathbf{p}_{n-1} \mathbf{e}_{n-1}^T)
\end{aligned}
\tag{5.33}
$$

Recognizing that $\mathbf{e}_i^T \mathbf{p}_j = 0$ for $i < j$, Eq. (5.33) yields the unit lower triangular matrix

$$L = P^{-1} = I + \sum_{r=1}^{n-1} \mathbf{p}_r \mathbf{e}_r^T \tag{5.34}$$

so that Eq. (5.32) becomes

$$A = LU \tag{5.35}$$

Hence, the Gaussian elimination method is tantamount to the decomposition of the matrix A of the coefficients into a unit lower triangular matrix L (Sec. 1.7) and an upper triangular matrix U. If the *triangular decomposition* (5.35) is possible, then it is unique.

Note that in the actual solution, it is often advisable to interchange various rows, which can be done by premultiplication of a given matrix A_r by a permutation matrix. This process is known as *partial pivoting*. Interchange of various columns can be achieved by postmultiplication of a given matrix by a permutation matrix. A process in which both rows and columns are interchanged is known as *complete pivoting*.

5.3 Reduction to triangular form by elementary row operations

Example 5.2

Perform triangular decomposition on the matrix of coefficients of Example 5.1.

The matrix of coefficients is

$$A = \begin{bmatrix} 2 & 2 & -1 \\ 1 & 1.5 & 2 \\ 1 & 6 & -2 \end{bmatrix} \tag{a}$$

From Eqs. (5.21), we obtain

$$p_{21} = \frac{a_{21}}{a_{11}} = 0.5, \qquad p_{31} = \frac{a_{31}}{a_{11}} = 0.5 \tag{b}$$

so that the first transformation matrix is

$$P_1 = \begin{bmatrix} 1 & 0 & 0 \\ -0.5 & 1 & 0 \\ -0.5 & 0 & 1 \end{bmatrix} \tag{c}$$

Hence, using Eq. (5.19), we obtain

$$A_1 = P_1 A = \begin{bmatrix} 1 & 0 & 0 \\ -0.5 & 1 & 0 \\ -0.5 & 0 & 1 \end{bmatrix} \begin{bmatrix} 2 & 2 & -1 \\ 1 & 1.5 & 2 \\ 1 & 6 & -2 \end{bmatrix} = \begin{bmatrix} 2 & 2 & -1 \\ 0 & 0.5 & 2.5 \\ 0 & 5 & -1.5 \end{bmatrix} \tag{d}$$

At this point, let us introduce the permutation matrix

$$I_{23} = \begin{bmatrix} 1 & 0 & 0 \\ 0 & 0 & 1 \\ 0 & 1 & 0 \end{bmatrix} \tag{e}$$

and write

$$I_{23} A_1 = \begin{bmatrix} 2 & 2 & -1 \\ 0 & 5 & -1.5 \\ 0 & 0.5 & 2.5 \end{bmatrix} \tag{f}$$

Clearly, the purpose of the premultiplication of A_1 by I_{23} is to use the largest number as the next pivot. Using Eqs. (5.24), we obtain

$$p_{32} = \frac{a_{32}^{(1)}}{a_{22}^{(1)}} = 0.1 \tag{g}$$

119

so that

$$P_2 = \begin{bmatrix} 1 & 0 & 0 \\ 0 & 1 & 0 \\ 0 & -0.1 & 1 \end{bmatrix} \tag{h}$$

it follows that

$$U = A_2 = P_2 I_{23} P_1 A = \begin{bmatrix} 2 & 2 & -1 \\ 0 & 5 & -1.5 \\ 0 & 0 & 2.65 \end{bmatrix} \tag{i}$$

Using Eq. (5.32), we conclude that

$$P^{-1} = P_1^{-1} I_{23}^{-1} P_2^{-1} = \begin{bmatrix} 1 & 0 & 0 \\ 0.5 & 0.1 & 1 \\ 0.5 & 1 & 0 \end{bmatrix} \tag{j}$$

which is a *triangular matrix with the rows permuted.*
    Next, let us write Eq. (i) in the form

$$U = P_2 I_{23} P_1 I_{23} I_{23} A \tag{k}$$

and observe that

$$L = I_{23}^{-1} P_1^{-1} I_{23}^{-1} P_2^{-1} = P_1^{-1} P_2^{-1} = \begin{bmatrix} 1 & 0 & 0 \\ 0.5 & 1 & 0 \\ 0.5 & 0.1 & 1 \end{bmatrix} \tag{l}$$

so that in this case the decomposition is

$$LU = I_{23} A = \begin{bmatrix} 2 & 2 & -1 \\ 1 & 6 & -2 \\ 1 & 1.5 & 2 \end{bmatrix} \tag{m}$$

which is equivalent to making the row interchange first and then proceeding with the ordinary Gaussian elimination with no pivoting. In practice, however, we do not know whether a pivot is zero (or close to zero) until it has been computed. But, assuming that all the interchanges have been performed in advance, we can regard the decomposition of A as one of the type (5.35).

5.4   Computation of eigenvectors belonging to known eigenvalues

Let us consider a vibrating system of the type discussed in Sec. 2.5 and assume that the eigenvalue problem can be written in the form

$$A\mathbf{x} = \lambda \mathbf{x} \tag{5.36}$$

where A is a real $n \times n$ matrix, not necessarily symmetric, $\mathbf{x}$ is an n-vector, and λ a scalar. Let us assume for the moment that λ is a known simple eigenvalue of A and consider the problem of computing the eigenvector $\mathbf{x}$ belonging to λ. This eigenvector must satisfy the equation

$$(A - \lambda I)\mathbf{x} = \mathbf{0} \tag{5.37}$$

which represents a set of n homogeneous algebraic equations in the unknowns x_i ($i = 1, 2, \ldots, n$), where x_i are the components of the vector $\mathbf{x}$.

Next, let $\Delta(\lambda) = \det(A - \lambda I)$ represent the determinant of the coefficients in Eq. (5.37). Then, by the definition of an eigenvalue, this determinant must be zero.

$$\Delta(\lambda) = 0 \tag{5.38}$$

and, because by assumption λ is a simple eigenvalue, we must also have

$$\Delta'(\lambda) \neq 0 \tag{5.39}$$

where prime denotes the derivative with respect to λ. It can be shown, however, that

$$\Delta'(\lambda) = \frac{d\Delta(\lambda)}{d\lambda} = \sum_{k=1}^{n} \Delta_k(\lambda) \tag{5.40}$$

where $\Delta_k(\lambda)$ is the determinant obtained from $\Delta(\lambda)$ by replacing the kth row of $\Delta(\lambda)$ by its derivative with respect to λ (see, for example, F3, p. 23). Hence, recalling Eq. (5.39), we can write

$$\frac{d}{d\lambda} \begin{vmatrix} a_{11} - \lambda & a_{12} \cdots a_{1n} \\ a_{21} & a_{22} - \lambda \cdots a_{2n} \\ \cdot \cdot \cdot \cdot \cdot \cdot \cdot \cdot \cdot \cdot \cdot \cdot \\ a_{n1} & a_{n2} \cdots a_{nn} - \lambda \end{vmatrix} = -\sum_{k=1}^{n} |M_{kk}(\lambda)| \neq 0 \tag{5.41}$$

where $|M_{kk}(\lambda)|$ is the minor determinant of order $n-1$ obtained by omitting the kth row and column from $\Delta(\lambda)$. The conclusion is that $\Delta(\lambda)$ possesses at least one minor determinant of order $n-1$ that is different from zero, so that the matrix $A - \lambda I$ has rank $n-1$. The implication is that there is an eigenvector $\mathbf{x}$ belonging to λ which is unique except for its magnitude.

To compute the eigenvector $\mathbf{x}$ belonging to λ, we shall set arbitrarily $x_r = 1$ and obtain the set of equations

$$\sum_{\substack{j=1 \\ j \neq r}}^{n} (a_{ij} - \lambda \delta_{ij})x_j = -a_{ir}, \qquad i = 1, 2, \ldots, n \tag{5.42}$$

121

which represents a set of n equations and only $n-1$ unknowns. It follows that one of the above equations is redundant and can be omitted. We shall omit the rth equation, so that

$$\sum_{\substack{j=1 \\ j \neq r}}^{n} (a_{ij} - \lambda\delta_{ij})x_j = -a_{ir}, \qquad i = 1, 2, \ldots, r-1, r+1, \ldots, n \qquad (5.43)$$

For Eqs. (5.43) to have a unique solution, the determinant $|M_{rr}(\lambda)|$ must be different from zero. By (5.41), there is at least one such determinant. If by accident the subset of $n-1$ equations chosen has the determinant of the coefficients equal to zero, then by trial and error a certain r can be found such that $|M_{rr}(\lambda)| \neq 0$.

Example 5.3

The eigenvalue problem for the conservative nongyroscopic system of Example 4.1 has the form

$$Kx = \omega^2 Mx \qquad (a)$$

where

$$M = m \begin{bmatrix} 1 & 0 & 0 \\ 0 & 1 & 0 \\ 0 & 0 & 2 \end{bmatrix}, \qquad K = k \begin{bmatrix} 2 & -1 & 0 \\ -1 & 3 & -2 \\ 0 & -2 & 2 \end{bmatrix} \qquad (b)$$

Multiplying Eq. (a) on the left by M^{-1}/k, the eigenvalue problem reduces to the form (5.36), where

$$A = \begin{bmatrix} 2 & -1 & 0 \\ -1 & 3 & -2 \\ 0 & -1 & 1 \end{bmatrix}, \qquad \lambda = \omega^2 m/k \qquad (c)$$

Knowing that the lowest eigenvalue of A is $\lambda_1 = 0.139194$, determine the corresponding eigenvector.

Introducing the above values of A and λ_1 into Eq. (5.37), we obtain the set of homogeneous algebraic equations

$$\begin{aligned} 1.860806x_1 - x_2 \quad &= 0 \\ -x_1 + 2.860806x_2 - 2x_3 &= 0 \\ -x_2 + 0.860806x_3 &= 0 \end{aligned} \qquad (d)$$

Letting $x_3 = 1$ and omitting the last equation, we can write

$$\begin{aligned} 1.860806x_1 - x_2 &= 0 \\ -x_1 + 2.860806x_2 &= 2 \end{aligned} \qquad (e)$$

122

We shall obtain the solution of Eqs. (e) by Gaussian elimination. To this end, we add the first equation divided by 1.860806 to the second, so that

$$1.860806x_1 - x_2 \qquad = 0$$
$$\left(2.860806 - \frac{1}{1.860806}\right)x_2 = 2 \qquad \text{(f)}$$

The second of Eqs. (f) yields

$$x_2 = \frac{2}{2.860806 - 1/1.860806} = 0.860806 \qquad \text{(g)}$$

Inserting this result into the first of Eqs. (f), we obtain

$$1.860806x_1 - 0.860806 = 0 \qquad \text{(h)}$$

which has the solution

$$x_1 = 0.462598 \qquad \text{(i)}$$

Hence, the eigenvector belonging to λ_1 is

$$\mathbf{x}_1 = [0.462598 \quad 0.860806 \quad 1]^T \qquad \text{(j)}$$

As explained earlier, the magnitude of the vector $\mathbf{x}_1$ is arbitrary, so that if the above vector is multiplied by any scalar, the resulting vector still represents the eigenvector belonging to λ_1.

5.5 Matrix iteration by the power method

The solution of the eigenvalue problem presented in Sec. 5.4 consists of first obtaining the eigenvalues and then determining the eigenvectors by solving sets of algebraic equations. The determination of the eigenvalues can present a serious problem, especially for high-order systems, and we shall consider this problem later in this chapter. At this point, we wish to turn our attention to a different class of methods for the eigensolution, namely, iteration methods. Many of the iteration methods yield both eigenvalues and eigenvectors simultaneously.

Let us consider a real square matrix A of order n. The eigenvalue problem associated with A possesses n solutions satisfying the equation

$$A\mathbf{x}_r = \lambda_r\mathbf{x}_r, \qquad r = 1, 2, \ldots, n \qquad (5.44)$$

where the eigenvalues are ordered so that $|\lambda_1| \geq |\lambda_2| \geq \cdots \geq |\lambda_n|$. We shall consider only two cases, depending on the nature of λ_1. If λ_1 is complex, then λ_2 is its complex conjugate, and λ_1 and λ_2 are the eigenvalues with maximum magnitudes, $|\lambda_1| = |\lambda_2| > |\lambda_3|$. In both cases we shall refer to λ_1 as the *dominant eigenvalue*. We further assume that the n eigenvectors $\mathbf{x}_r$ are linearly independent and that they span the space L^n. Hence, any vector $\mathbf{v}_0$ in L^n can be expressed as the linear combination

$$\mathbf{v}_0 = \sum_{r=1}^{n} \alpha_r \mathbf{x}_r \qquad (5.45)$$

Next, let us consider the iterative scheme defined by

$$\mathbf{v}_p = A\mathbf{v}_{p-1}, \qquad p = 1, 2, \ldots . \qquad (5.46)$$

If $\mathbf{v}_0$ is expressed in the form (5.45), then the pth *iterated vector* is given by

$$\mathbf{v}_p = A\mathbf{v}_{p-1} = A^2\mathbf{v}_{p-2} = \cdots = A^p\mathbf{v}_0 = \sum_{r=1}^{n} \alpha_r \lambda_r^p \mathbf{x}_r \qquad (5.47)$$

Factoring out λ_1^p, we obtain

$$\mathbf{v}_p = \lambda_1^p \left[\alpha_1 \mathbf{x}_1 + \sum_{r=2}^{n} \alpha_r \left(\frac{\lambda_r}{\lambda_1} \right)^p \mathbf{x}_r \right] \qquad (5.48)$$

Assuming that $\alpha_1 \neq 0$, for a sufficiently large p, Eq. (5.48) can be written in the form

$$\mathbf{v}_p = \lambda_1^p (\alpha_1 \mathbf{x}_1 + \boldsymbol{\varepsilon}_p) \qquad (5.49)$$

where $\boldsymbol{\varepsilon}_p$ is a vector with very small components. In fact, $\boldsymbol{\varepsilon}_p$ approaches the null vector as $p \to \infty$, so that

$$\lim_{p \to \infty} \mathbf{v}_p = \lambda_1^p \alpha_1 \mathbf{x}_1 \qquad (5.50)$$

Hence, the iteration process produces the eigenvector $\mathbf{x}_1$ belonging to the dominant eigenvalue λ_1. One more iteration yields

$$\lim_{p \to \infty} \mathbf{v}_{p+1} = \lambda_1^{p+1} \alpha_1 \mathbf{x}_1 \qquad (5.51)$$

Equations (5.50) and (5.51) are vector equations, which implies that they are satisfied by every component of the vectors. It follows that for the ith components

$$\lim_{p \to \infty} v_{p,i} = \lambda_1^p \alpha_1 x_{1,i}, \qquad \lim_{p \to \infty} v_{p+1,i} = \lambda_1^{p+1} \alpha_1 x_{1,i} \qquad (5.52)$$

Dividing one by the other, we can write

$$\lim_{p \to \infty} \frac{v_{p+1,i}}{v_{p,i}} = \lambda_1 \tag{5.53}$$

so that, after achieving convergence, the dominant eigenvalue can be obtained as the ratio of homologous (having the same relative position) elements of two consecutive iterated vectors.

Implicit in Eq. (5.45) is the assumption that α_1 is different from zero. If $\alpha_1 = 0$, then the trial vector v_0 does not contain the eigenvector x_1. Hence, every subsequent iterated vector will be free of x_1, so that the process cannot converge to x_1. In practice, however, a computational process works with a fixed number of figures, so that rounding is likely to introduce a component in the iterated vector which is a nonzero multiple of x_1. This vector grows with every iteration and ultimately it becomes the dominant factor in v_p. This fact is reassuring if one desires to obtain the dominant eigenvalue and the corresponding eigenvector, but it also points out why the process is not suitable for determining eigenvalues other than the dominant one. However, the process can be modified to permit iteration to subdominant eigenvalues.

The speed of convergence depends on two factors. The most important one is how large the magnitude of the dominant eigenvalue is compared to that of the first subdominant eigenvalue. Clearly, the larger the ratio of the two the faster the process converges. The second factor relates to the magnitude of α_1 compared to α_r $(r = 2, 3, \ldots, n)$. If α_1 is large compared to $\alpha_2, \alpha_3, \ldots, \alpha_n$, then the process is likely to converge faster. In physical terms, if v_0 is a good estimate of the first eigenvector, the process will converge faster.

The iteration process is self-correcting. Indeed, any error in computing an iterated vector only delays convergence, but does not destroy it. This is so because an erroneous iterated vector can be regarded as the first trial vector in a new iteration sequence.

In the above discussion, we assumed tacitly that the iteration process works with real vectors. Of course, this assumption is fully justified when λ_1 is real. When λ_1 is complex and $\lambda_2 = \bar{\lambda}_1$ is its complex conjugate, convergence is not recognizable in the form (5.50), so that further elaboration is necessary. However, we shall continue to insist on working with real iterated vectors, in spite of the fact that the eigenvectors are known to be complex. To demonstrate the procedure, let us introduce the notation

$$\lambda_1 = \rho e^{i\theta}, \qquad \lambda_2 = \rho e^{-i\theta} \tag{5.54}$$

125

where the magnitude ρ is such that $\rho > |\lambda_r|$, $r = 3, 4, \ldots, n$. Then, expansion (5.45) can be written in the form

$$\mathbf{v}_0 = \alpha_1 \mathbf{x}_1 + \alpha_2 \mathbf{x}_2 + \sum_{r=3}^{n} \alpha_r \mathbf{x}_r \tag{5.55}$$

where $\alpha_2 = \bar{\alpha}_1$ and $\mathbf{x}_2 = \bar{\mathbf{x}}_1$ are complex conjugates, so that the sum $\alpha_1 \mathbf{x}_1 + \alpha_2 \mathbf{x}_2$ represents a real vector. It follows that the pth iterated vector can be written as

$$\mathbf{v}_p = \rho^p (\alpha_1 e^{ip\theta} \mathbf{x}_1 + \alpha_2 e^{-ip\theta} \mathbf{x}_2 + \boldsymbol{\varepsilon}_p) \tag{5.56}$$

so that

$$\lim_{p \to \infty} \mathbf{v}_p = \lambda_1^p \alpha_1 \mathbf{x}_1 + \lambda_2^p \alpha_2 \mathbf{x}_2 \tag{5.57}$$

Next, let us write

$$\mathbf{v}_{p+2} + \xi \mathbf{v}_{p+1} + \eta \mathbf{v}_p \cong (\lambda_1^2 + \xi \lambda_1 + \eta) \lambda_1^p \alpha_1 \mathbf{x}_1 + (\lambda_2^2 + \xi \lambda_2 + \eta) \lambda_2^p \alpha_2 \mathbf{x}_2 \tag{5.58}$$

If λ_1 and λ_2 are solutions of the quadratic equation

$$\lambda^2 + \xi \lambda + \eta = 0 \tag{5.59}$$

i.e., if they are such that

$$\begin{matrix} \lambda_1 \\ \lambda_2 \end{matrix} = \tfrac{1}{2}(-\xi \pm i\sqrt{4\eta^2 - \xi^2}) \tag{5.60}$$

then

$$\mathbf{v}_{p+2} + \xi \mathbf{v}_{p+1} + \eta \mathbf{v}_p \cong \mathbf{0} \tag{5.61}$$

so that three successive iterated vectors are almost dependent.

Equation (5.61) can be written in the form

$$A^{p+2} \mathbf{v}_0 + \xi A^{p+1} \mathbf{v}_0 + \eta A^p \mathbf{v}_0 \cong \mathbf{0} \tag{5.62}$$

Denoting by a_{p+2}, a_{p+1} and a_p homologous elements of the matrices A^{p+2}, A^{p+1} and A^p, respectively, we conclude that, upon reaching convergence, these elements must satisfy

$$a_{p+2} + \xi a_{p+1} + \eta a_p = 0 \tag{5.63}$$

By analogy, we can also write

$$a_{p+3} + \xi a_{p+2} + \eta a_{p+1} = 0 \tag{5.64}$$

where the notation is obvious. Solving the two equations for ξ and η, we

126

obtain

$$\xi = \frac{a_p a_{p+3} - a_{p+1} a_{p+2}}{a_{p+1}^2 - a_p a_{p+2}}, \qquad \eta = \frac{a_{p+2}^2 - a_{p+2} a_{p+3}}{a_{p+1}^2 - a_p a_{p+2}} \qquad (5.65)$$

When the homologous elements a_p, a_{p+1}, a_{p+2} and a_{p+3} are such that ξ and η, as calculated by means of Eqs. (5.65), reach constant values for any p, convergence has been achieved. Then these values of ξ and η can be introduced into Eqs. (5.60) to calculate the eigenvalues λ_1 and λ_2.

To obtain the eigenvectors belonging to λ_1 and λ_2, we can write without loss of generality

$$\mathbf{v}_p = \mathrm{Re}\,\mathbf{x}_1 \qquad (5.66)$$

Then, from the equation

$$\mathbf{v}_{p+1} = \lambda_1 (\mathrm{Re}\,\mathbf{x}_1 + i\,\mathrm{Im}\,\mathbf{x}_1) + \lambda_2 (\mathrm{Re}\,\mathbf{x}_1 - i\,\mathrm{Im}\,\mathbf{x}_1) \qquad (5.67)$$

we conclude that

$$\mathrm{Im}\,\mathbf{x}_1 = -\frac{\xi}{\sqrt{4\eta^2 - \xi^2}}\mathbf{v}_p - \frac{1}{\sqrt{4\eta^2 - \xi^2}}\mathbf{v}_{p+1} \qquad (5.68)$$

As in the real case, the matrix iteration process just described is error proof.

The question of subdominant eigenvalues will be examined in Sec. 5.6.

Example 5.4

Consider the eigenvalue problem (a) of Example 5.3, multiply the equation on the left by $K^{-1}k/m\omega^2$ and obtain the eigenvalue problem in the form (5.36), where

$$A = \begin{bmatrix} 1 & 1 & 2 \\ 1 & 2 & 4 \\ 1 & 2 & 5 \end{bmatrix}, \qquad \lambda = \frac{k}{m\omega^2} \qquad (a)$$

Obtain the first eigenvalue and eigenvector by matrix iteration. Note that the matrix A here is the inverse of that in Example 5.3, so that the lowest eigenvalue there is the reciprocal of the highest eigenvalue here.

As a first trial vector, let us use

$$\mathbf{v}_0 = [0.5 \quad 0.9 \quad 1.0]^T \qquad (b)$$

which is reasonably close to x_1. The first iteration yields

$$\begin{bmatrix} 1 & 1 & 2 \\ 1 & 2 & 4 \\ 1 & 2 & 5 \end{bmatrix} \begin{bmatrix} 0.5 \\ 0.9 \\ 1.0 \end{bmatrix} = \begin{bmatrix} 0.5+0.9+2.0 \\ 0.5+1.8+4.0 \\ 0.5+1.8+5.0 \end{bmatrix} = \begin{bmatrix} 3.4 \\ 6.3 \\ 7.3 \end{bmatrix} = 7.3 \begin{bmatrix} 0.465753 \\ 0.863014 \\ 1.000000 \end{bmatrix} \qquad (c)$$

so that the second iteration is formed as follows

$$\begin{bmatrix} 1 & 1 & 2 \\ 1 & 2 & 4 \\ 1 & 2 & 5 \end{bmatrix} \begin{bmatrix} 0.465753 \\ 0.863014 \\ 1.000000 \end{bmatrix} = \begin{bmatrix} 0.465753+0.863014+2.000000 \\ 0.465753+1.726028+4.000000 \\ 0.465753+1.726028+5.000000 \end{bmatrix}$$

$$= \begin{bmatrix} 3.328767 \\ 6.191781 \\ 7.191781 \end{bmatrix} = 7.191781 \begin{bmatrix} 0.462857 \\ 0.860952 \\ 1.000000 \end{bmatrix} \qquad (d)$$

which is already very close to the desired value. The seventh iteration yields

$$\begin{bmatrix} 1 & 1 & 2 \\ 1 & 2 & 4 \\ 1 & 2 & 5 \end{bmatrix} \begin{bmatrix} 0.462598 \\ 0.860806 \\ 1.000000 \end{bmatrix} = \begin{bmatrix} 0.462598+0.860806+2.000000 \\ 0.462598+1.721612+4.000000 \\ 0.462598+1.721612+5.000000 \end{bmatrix}$$

$$= \begin{bmatrix} 3.323404 \\ 6.184210 \\ 7.184210 \end{bmatrix} = 7.184210 \begin{bmatrix} 0.462598 \\ 0.860806 \\ 1.000000 \end{bmatrix} \qquad (e)$$

at which point convergence has clearly been achieved, because premultiplication of the trial vector by A reproduced the trial vector. Hence, the vector on the right side of Eq. (e) represents the first modal vector x_1; it is identical to that obtained in Example 5.3. Moreover, the scaling factor $\lambda_1 = 7.184210$ can be verified as being the reciprocal of the lowest eigenvalue 0.139194 of Example 5.3. Using the second of Eqs. (a), we obtain the lowest natural frequency

$$\omega_1 = \frac{1}{\sqrt{\lambda_1}} \sqrt{\frac{k}{m}} = \frac{1}{\sqrt{7.184210}} \sqrt{\frac{k}{m}} = 0.373087 \sqrt{\frac{k}{m}} \qquad (f)$$

5.6   Hotelling's deflation

The matrix iteration presented in Sec. 5.5 can yield only the dominant eigenvalue and the corresponding eigenvector. Any attempt to obtain the first subdominant eigenvalue is likely to fail as long as the same matrix A is used for the iteration process. Hence, we wish to look into the possibility of modifying A in a way which will permit iteration to the first subdominant eigenvalue.

An iteration procedure for the first subdominant eigenvalue for arbitrary matrices possessing complex eigenvalues exists (F3, Sec. 7.10), but it is very complicated. We shall present here another technique known as *matrix deflation*. The method is due to Hotelling and is restricted to real symmetric matrices.

Let us assume that A is a real symmetric matrix, so that it possesses orthogonal eigenvectors. We shall further assume that the eigenvectors have been normalized, so that they satisfy

$$\mathbf{x}_r^T \mathbf{x}_s = \delta_{rs}, \qquad r, s = 1, 2, \ldots, n \tag{5.69}$$

Then, if λ_1 and $\mathbf{x}_1$ are the first eigenvalue and eigenvector of A, the matrix

$$A_2 = A - \lambda_1 \mathbf{x}_1 \mathbf{x}_1^T \tag{5.70}$$

has the eigenvalues $0, \lambda_2, \lambda_3, \ldots, \lambda_n$ and the eigenvectors $\mathbf{x}_1, \mathbf{x}_2, \mathbf{x}_3, \ldots, \mathbf{x}_n$, respectively. Indeed, postmultiplying A_2 by the eigenvector $\mathbf{x}_i$ and using Eq. (5.69), we can write

$$A_2\mathbf{x}_i = A\mathbf{x}_i - \lambda_1\mathbf{x}_1\mathbf{x}_1^T\mathbf{x}_i = \lambda_i\mathbf{x}_i - \delta_{1i}\lambda_1\mathbf{x}_1 = \begin{cases} \mathbf{0} & \text{if } i = 1 \\ \lambda_i\mathbf{x}_i & \text{if } i \neq 1 \end{cases} \tag{5.71}$$

Hence, using the trial vector $\mathbf{v}_0$ in the form (5.45), in conjunction with the matrix A_2, we obtain the first iterated vector

$$\begin{aligned} \mathbf{v}_1 = A_2\mathbf{v}_0 &= \sum_{r=1}^{n} \alpha_r A_2\mathbf{x}_r = \sum_{r=1}^{n} \alpha_r A\mathbf{x}_r - \lambda_1\mathbf{x}_1 \sum_{r=1}^{n} \alpha_r \mathbf{x}_1^T\mathbf{x}_r \\ &= \sum_{r=1}^{n} \alpha_r\lambda_r\mathbf{x}_r - \lambda_1\mathbf{x}_1 \sum_{r=1}^{n} \alpha_r\delta_{1r} = \sum_{r=2}^{n} \alpha_r\lambda_r\mathbf{x}_r \end{aligned} \tag{5.72}$$

which is entirely free of $\mathbf{x}_1$. It follows that, if we use the iteration process

$$\mathbf{v}_p = A_2\mathbf{v}_{p-1}, \qquad p = 1, 2, \ldots \tag{5.73}$$

then the pth iterated vector can be written in the form

$$\mathbf{v}_p = \lambda_2^p \left[\alpha_2\mathbf{x}_2 + \sum_{r=3}^{n} \alpha_r \left(\frac{\lambda_r}{\lambda_2}\right)^p \mathbf{x}_r \right] \tag{5.74}$$

Clearly, the iteration process converges to $\mathbf{x}_2$. The matrix A_2 is known as a *deflated matrix*.

Matrix deflation can be used to obtain the remaining subdominant

129

eigenvalues. Indeed, it can be easily verified that the deflated matrix

$$A_k = A_{k-1} - \lambda_{k-1} \mathbf{x}_{k-1} \mathbf{x}_{k-1}^T = A - \sum_{i=1}^{k-1} \lambda_i \mathbf{x}_i \mathbf{x}_i^T \qquad (5.75)$$

has the eigenvalues $0, 0, \ldots, \lambda_k, \lambda_{k+1}, \ldots, \lambda_n$ and the eigenvectors $\mathbf{x}_1, \mathbf{x}_2, \ldots, \mathbf{x}_k, \mathbf{x}_{k+1}, \ldots, \mathbf{x}_n$, respectively, so that it can be used to iterate to the eigenvalue λ_k and the eigenvector $\mathbf{x}_k$.

The above procedure is tailored to real symmetric matrices. A similar procedure, also due to Hotelling, exists for an arbitrary real matrix A, not necessarily symmetric. The procedure requires the eigenvectors of both A and A^T. The eigenvalue problem for real nonsymmetric matrices was discussed in Sec. 3.7. Denoting the eigenvectors of A by $\mathbf{x}_r$ ($r = 1, 2, \ldots, n$) and the eigenvectors of A^T by $\mathbf{y}_s$ ($s = 1, 2, \ldots, n$), it is shown in Sec. 3.7 that, for distinct eigenvalues λ_r and λ_s, the eigenvectors satisfy

$$\mathbf{y}_s^T \mathbf{x}_r = 0, \qquad \lambda_r \neq \lambda_s, \qquad r, s = 1, 2, \ldots, n \qquad (5.76)$$

or, the set of eigenvectors $\mathbf{x}_r$ is orthogonal to the set of eigenvectors $\mathbf{y}_s$, a property known as biorthogonality. The two sets of eigenvectors can be normalized by writing

$$\mathbf{y}_s^T \mathbf{x}_r = \delta_{rs}, \qquad r, s = 1, 2, \ldots, n \qquad (5.77)$$

Then, by analogy with Eq. (5.70), we can define the deflated matrix

$$A_2 = A - \lambda_1 \mathbf{x}_1 \mathbf{y}_1^T \qquad (5.78)$$

A postmultiplication of A_2 by $\mathbf{x}_i$ yields

$$A_2 \mathbf{x}_i = A\mathbf{x}_i - \lambda_1 \mathbf{x}_1 \mathbf{y}_1^T \mathbf{x}_i = \lambda_i \mathbf{x}_i - \delta_{1i} \lambda_1 \mathbf{x}_1 = \begin{cases} \mathbf{0} & \text{if } i = 1 \\ \lambda_i \mathbf{x}_i & \text{if } i \neq 1 \end{cases} \qquad (5.79)$$

so that A_2, as defined by Eq. (5.78), has the eigenvalues $0, \lambda_2, \lambda_3, \ldots, \lambda_n$ and the eigenvectors $\mathbf{x}_1, \mathbf{x}_2, \mathbf{x}_3, \ldots, \mathbf{x}_n$. Hence, an iteration process using A_2 in conjunction with an initial trial vector in the form (5.45) yields the first subdominant eigenvalue λ_2 and the corresponding eigenvector $\mathbf{x}_2$. Of course, one must first solve the eigenvalue problem for A^T to obtain the eigenvector $\mathbf{y}_1$. Similarily, to compute $\mathbf{y}_2$ one must solve the eigenvalue problem associated with A_2^T.

In general, the eigenvalue λ_k and eigenvector $\mathbf{x}_k$ can be obtained by an iteration process using the deflated matrix

$$A_k = A_{k-1} - \lambda_{k-1} \mathbf{x}_{k-1} \mathbf{y}_{k-1}^T, \qquad k = 2, 3, \ldots, n \qquad (5.80)$$

and to obtain the eigenvector $\mathbf{y}_k$ one iterates with A_k^T. It is clear that the iteration process for arbitrary matrices requires twice the work needed for symmetric matrices.

The question remains as to how to obtain higher eigenvalues and eigenvectors by deflation when the eigenvalue problem is given in terms of two matrices instead of one, as expressed by Eq. (3.43), for example. Of course, following the procedure of Sec. 3.3, one could reduce the eigenvalue problem to one in terms of a single real symmetric matrix, which would require first a decomposition of the type (3.44). Such a decomposition will be discussed in Sec. 5.8. At this point, however, the interest lies in showing how Hotelling's deflation can be modified so as to permit working with eigenvectors satisfying orthogonality relations such as expressed by Eqs. (3.50).

It is not difficult to see that Eq. (3.43) is equivalent to

$$A\mathbf{u} = \lambda\mathbf{u}, \qquad \lambda = 1/\omega^2 \tag{5.81}$$

where

$$A = K^{-1}M \tag{5.82}$$

is clearly a nonsymmetric matrix. Note that Eq. (5.82) implies the nonsingularity of K. The iteration process for the dominant eigenvalue and eigenvector of the matrix A remains as described in Sec. 5.5. Of course, one could obtain the subdominant eigenvalues and eigenvectors of A by the deflation procedure for real nonsymmetric matrices just described. This would require solving both for the right and left eigenvectors of A. However, because the eigenvectors of A are known to satisfy the orthogonality relations

$$\mathbf{u}_j^T M \mathbf{u}_i = \delta_{ij}, \qquad i, j = 1, 2, \ldots, n \tag{5.83}$$

this is not really necessary. Indeed, it is not difficult to verify that an iteration using the deflated matrix

$$A_2 = A - \lambda_1 \mathbf{u}_1 \mathbf{u}_1^T M \tag{5.84}$$

will yield the first subdominant eigenvalue and eigenvector. It should be pointed out that Eqs. (5.83) imply that the eigenvectors have actually been normalized, so that $\mathbf{u}_1^T M \mathbf{u}_1 = 1$. The procedure can be generalized by observing that an iteration process using the deflated matrix

$$A_k = A_{k-1} - \lambda_{k-1} \mathbf{u}_{k-1} \mathbf{u}_{k-1}^T M, \qquad k = 2, 3, \ldots, n \tag{5.85}$$

will produce the eigenvalue λ_k and the eigenvector $\mathbf{u}_k$.

The main drawback of Hotelling's deflation is that the deflated matrices are based on the previously computed eigenvectors, so that accuracy suffers with each additional subdominant eigenvalue sought.

131

Example 5.5

Consider the eigenvalue problem of Example 5.4 and obtain the subdominant eigenvalues and eigenvectors by Hotelling's method
From Examples 5.3 and 5.4 we recall that

$$
A = \begin{bmatrix} 1 & 1 & 1 \\ 1 & 2 & 4 \\ 1 & 2 & 5 \end{bmatrix}, M = m \begin{bmatrix} 1 & 0 & 0 \\ 0 & 1 & 0 \\ 0 & 0 & 2 \end{bmatrix} \tag{a}
$$

Moreover, the first eigenvalue is $\lambda_1 = 7.184210$ and the first eigenvector is $\mathbf{u}_1 = [0.462598 \quad 0.860806 \quad 1.000000]^T$.

To compute the second eigenvalue and eigenvector, we must first produce the first deflated matrix according to Eq. (5.84). This equation, however, calls for a normalized vector $\mathbf{u}_1$ satisfying $\mathbf{u}_1^T M \mathbf{u}_1 = 1$. Let the normalized vector be $\mathbf{u}_1 = \alpha[0.462598 \quad 0.860806 \quad 1.000000]^T$, where α is a constant scalar whose value must be such that

$$
\mathbf{u}_1^T M \mathbf{u}_1 = \alpha^2 m \begin{bmatrix} 0.462598 \\ 0.860806 \\ 1.000000 \end{bmatrix}^T \begin{bmatrix} 1 & 0 & 0 \\ 0 & 1 & 0 \\ 0 & 0 & 2 \end{bmatrix} \begin{bmatrix} 0.462598 \\ 0.860806 \\ 1.000000 \end{bmatrix}
$$

$$
= 2.954984 \alpha^2 m = 1 \tag{b}
$$

from which we obtain

$$
\alpha = 0.581731 m^{-1/2} \tag{c}
$$

Hence, the first normalized eigenvector is

$$
\mathbf{u}_1 = m^{-1/2}[0.269108 \quad 0.500758 \quad 0.581731]^T \tag{d}
$$

Next, we use Eq. (5.84) and form the first deflated matrix

$$
A_2 = A - \lambda_1 \mathbf{u}_1 \mathbf{u}_1^T M = \begin{bmatrix} 1 & 1 & 2 \\ 1 & 2 & 4 \\ 1 & 2 & 5 \end{bmatrix}
$$

$$
- 7.184210 \begin{bmatrix} 0.269108 \\ 0.500758 \\ 0.581731 \end{bmatrix} \begin{bmatrix} 0.269108 \\ 0.500758 \\ 0.581731 \end{bmatrix}^T \begin{bmatrix} 1 & 0 & 0 \\ 0 & 1 & 0 \\ 0 & 0 & 2 \end{bmatrix}
$$

$$
= \begin{bmatrix} 0.479727 & 0.031870 & -0.249355 \\ 0.031870 & 0.198495 & -0.185614 \\ -0.124674 & -0.092803 & 0.137569 \end{bmatrix} \tag{e}
$$

Because the second mode is known to contain a node, we shall use the first trial vector for the second eigenvector

$$\mathbf{v}_0 = [1 \quad 1 \quad -1]^T \tag{f}$$

Hence, the first iteration to the second eigenvector is

$$\begin{bmatrix} 0.479727 & 0.031870 & -0.249355 \\ 0.031870 & 0.198495 & -0.185614 \\ -0.124674 & -0.092803 & 0.137569 \end{bmatrix} \begin{bmatrix} 1 \\ 1 \\ -1 \end{bmatrix}$$

$$= 0.760952 \begin{bmatrix} 1.000000 \\ 0.546656 \\ -0.466581 \end{bmatrix} \tag{g}$$

and the second iteration is

$$\begin{bmatrix} 0.479727 & 0.031870 & -0.249355 \\ 0.031870 & 0.198495 & -0.185614 \\ -0.124674 & -0.092803 & 0.137569 \end{bmatrix} \begin{bmatrix} 1.000000 \\ 0.546656 \\ -0.466581 \end{bmatrix}$$

$$= 0.613493 \begin{bmatrix} 1.000000 \\ 0.369983 \\ -0.390537 \end{bmatrix} \tag{h}$$

Finally, the sixteenth iteration yields

$$\begin{bmatrix} 0.479727 & 0.031870 & -0.249355 \\ 0.031870 & 0.198495 & -0.185614 \\ -0.124674 & -0.092803 & 0.137569 \end{bmatrix} \begin{bmatrix} 1.000000 \\ 0.254095 \\ -0.340659 \end{bmatrix}$$

$$= 0.572770 \begin{bmatrix} 1.000000 \\ 0.254095 \\ -0.340659 \end{bmatrix} \tag{i}$$

so that convergence has been achieved. From Eq. (i), we conclude that the second eigenvalue is $\lambda_2 = 0.572770$ and the second eigenvector is

$$\mathbf{u}_2 = [1.000000 \quad 0.254095 \quad -0.340659]^T \tag{j}$$

Using the second of Eqs. (a) once again, we obtain the second natural frequency

$$\omega_2 = \frac{1}{\sqrt{\lambda_2}} \sqrt{\frac{k}{m}} = \frac{1}{\sqrt{0.572770}} \sqrt{\frac{k}{m}} = 1.32136 \sqrt{\frac{k}{m}} \tag{k}$$

The determination of the third eigenvalue and eigenvector is left as an exercise to the reader. Note that before computing the second deflated matrix according to Eq. (5.85), one must normalize the vector $\mathbf{u}_2$ so as to satisfy $\mathbf{u}_2^T M \mathbf{u}_2 = 1$.

5.7   Wielandt's deflation

Wielandt's method can be regarded as a generalization of Hotelling's method. Let us consider an arbitrary real matrix A and assume that its dominant eigenvalue λ_1 and the corresponding eigenvector $\mathbf{x}_1$ have been calculated. Then, to obtain the first subdominant eigenvalue λ_2, we consider the deflated matrix

$$A_2 = A - \mathbf{x}_1 \mathbf{u}_1^T \tag{5.86}$$

where $\mathbf{u}_1$ is any vector such that

$$\mathbf{u}_1^T \mathbf{x}_1 = \lambda_1 \tag{5.87}$$

Postmultiplying Eq. (5.86) by $\mathbf{x}_1$, we obtain

$$A_2 \mathbf{x}_1 = A \mathbf{x}_1 - \mathbf{x}_1 \mathbf{u}_1^T \mathbf{x}_1 = \lambda_1 \mathbf{x}_1 - \lambda_1 \mathbf{x}_1 = \mathbf{0} \tag{5.88}$$

so that $\mathbf{x}_1$ is an eigenvector of A_2 belonging to the null eigenvalue.
Next, let us consider

$$A_2(\mathbf{x}_i - \gamma_i \mathbf{x}_1) = A\mathbf{x}_i - \gamma_i A\mathbf{x}_1 - \mathbf{x}_1 \mathbf{u}_1^T \mathbf{x}_i + \gamma_i \mathbf{x}_1 \mathbf{u}_1^T \mathbf{x}_1$$

$$= \lambda_i \mathbf{x}_i - \mathbf{x}_1 \mathbf{u}_1^T \mathbf{x}_i$$

$$= \lambda_i \left[\mathbf{x}_i - \left(\frac{1}{\lambda_i} \mathbf{u}_1^T \mathbf{x}_i \right) \mathbf{x}_1 \right], \qquad i = 2, 3, \ldots, n \tag{5.89}$$

Choosing $\gamma_i = \mathbf{u}_1^T \mathbf{x}_i / \lambda_i$ and introducing the notation

$$\mathbf{x}_i^{(2)} = \mathbf{x}_i - \left(\frac{1}{\lambda_i} \mathbf{u}_1^T \mathbf{x}_i \right) \mathbf{x}_1, \qquad i = 2, 3, \ldots, n \tag{5.90}$$

we can write Eq. (5.89) in the form

$$A_2 \mathbf{x}_i^{(2)} = \lambda_i \mathbf{x}_i^{(2)}, \qquad i = 2, 3, \ldots, n \tag{5.91}$$

from which we conclude that the nonzero eigenvalues of A_2 are $\lambda_2, \lambda_3, \ldots, \lambda_n$, and hence the same as the subdominant eigenvalues of A. On the other hand, the corresponding eigenvectors of A_2 are $\mathbf{x}_2^{(2)}, \mathbf{x}_3^{(2)}, \ldots, \mathbf{x}_n^{(2)}$, as opposed to the eigenvectors $\mathbf{x}_2, \mathbf{x}_3, \ldots, \mathbf{x}_n$ of A. The eigenvectors $\mathbf{x}_1, \mathbf{x}_2^{(2)}, \ldots, \mathbf{x}_n^{(2)}$ are linearly independent, as no relation of the type

$$\alpha_1 \mathbf{x}_1 + \sum_{i=2}^{n} \alpha_i \mathbf{x}_i^{(2)} = \mathbf{0} \tag{5.92}$$

can be satisfied without all α_i $(i = 1, 2, \ldots, n)$ being identically zero.

Note that by taking $\mathbf{u}_1 = \lambda_1 \mathbf{y}_1$, where $\mathbf{y}_1$ is the first normalized eigenvector of A^T, we obtain Hotelling's deflation.

Now let us consider any vector $\mathbf{w}_1$ such that

$$\mathbf{w}_1^T \mathbf{x}_1 = 1 \tag{5.93}$$

from which it follows that

$$\mathbf{w}_1^T A \mathbf{x}_1 = \mathbf{w}_1^T \lambda_1 \mathbf{x}_1 = \lambda_1 \tag{5.94}$$

Comparing Eqs. (5.87) and (5.94), we conclude that we can take the vector $\mathbf{u}_1^T$ in the form $\mathbf{w}_1^T A$, so that Eq. (5.86) becomes

$$A_2 = A - \mathbf{x}_1 \mathbf{w}_1^T A \tag{5.95}$$

The matrix A_2, as given by Eq. (5.95), has the eigenvalues $0, \lambda_2, \lambda_3, \ldots, \lambda_n$ and the eigenvectors $\mathbf{x}_1, \mathbf{x}_2^{(2)}, \mathbf{x}_3^{(2)}, \ldots, \mathbf{x}_n^{(2)}$, where

$$\mathbf{x}_i^{(2)} = \mathbf{x}_i - \left(\frac{1}{\lambda_i} \mathbf{w}_1^T A \mathbf{x}_i \right) \mathbf{x}_1, \qquad i = 2, 3, \ldots, n \tag{5.96}$$

If the eigenvector $\mathbf{x}_1$ is normalized so that its component with the largest magnitude is in the jth row and that this magnitude is equal unity, then the unit vector $\mathbf{e}_j$ is clearly a suitable choice for $\mathbf{w}_1$. In this case, the product $\mathbf{w}_1^T A = \mathbf{e}_j^T A$ is simply the jth row of the matrix A, so that the matrix A_2 has the form

$$A_2 = A - \mathbf{x}_1 \mathbf{e}_j^T A = (I - \mathbf{x}_1^T \mathbf{e}_j) A \tag{5.97}$$

which has the jth row equal to zero. Indeed, because $x_{1,j} = 1$, where $x_{1,j}$ is the jth component of $\mathbf{x}_1$, it follows that the jth row of A_2 is null and every other row l of A_2 is obtained by subtracting $x_{1,l}$ times the jth row of A from the lth row of A. Hence, all the eigenvectors $\mathbf{x}_i^{(2)}$ have the jth component equal to zero. In view of this, we can omit from the matrix A_2 the jth row and column and from the eigenvectors $\mathbf{x}_i^{(2)}$ the jth component, so that the iteration process can be performed with a matrix of order $n - 1$ instead of a matrix of order n.

5.8   The Cholesky decomposition

For computational reasons, it is often desirable to decompose a matrix into a product of simpler matrices. In particular, Eq. (3.44) indicates that a positive definite real symmetric matrix A can be decomposed into the product $Q^T Q$, where Q is a real nonsingular matrix. In this section, we shall study the case in which Q has a triangular form. Before discussing the decomposition of real symmetric matrices, however, we shall first discuss the decomposition of arbitrary real square matrices.

Let us consider the arbitrary upper triangular matrix

$$
T = \begin{bmatrix} t_{11} & t_{12} & \cdots & t_{1n} \\ 0 & t_{22} & \cdots & t_{2n} \\ \cdot & \cdot & \cdot \cdot \cdot \cdot & \cdot \\ 0 & 0 & \cdots & t_{nn} \end{bmatrix} \tag{5.98}
$$

The matrix T can be written as the product of a diagonal matrix D and a unit upper triangular matrix U in the form

$$
T = DU \tag{5.99}
$$

where

$$
D = \begin{bmatrix} d_{11} & 0 & \cdots & 0 \\ 0 & d_{22} & \cdots & 0 \\ \cdot & \cdot & \cdot \cdot \cdot \cdot \cdot & \cdot \\ 0 & 0 & \cdots & d_{nn} \end{bmatrix}, \qquad U = \begin{bmatrix} 1 & u_{12} & \cdots & u_{1n} \\ 0 & 1 & \cdots & u_{2n} \\ \cdot & \cdot & \cdot \cdot \cdot \cdot \cdot & \cdot \\ 0 & 0 & \cdots & 1 \end{bmatrix} \tag{5.100}
$$

in which

$$
\begin{aligned}
d_{ii} &= t_{ii}, & i = 1, 2, \ldots, n \\
u_{ij} &= t_{ij}/t_{ii}, & i = 1, 2, \ldots, n; \qquad j = 2, 3, \ldots, n
\end{aligned} \tag{5.101}
$$

Returning to Eq. (5.35), we observe that only the matrix L is a unit triangular matrix. In view of Eq. (5.99), we can write the triangular decomposition (5.35) in the form

$$
A = LDU \tag{5.102}
$$

in which L and U are unit lower and upper triangular matrices and D is a diagonal matrix.

If A is symmetric and has a unique decomposition of the type (5.102), then it must have the form

$$
A = LDL^T \tag{5.103}
$$

In general, the decomposition need not exist. When A is positive definite, however, the decomposition must exist. Indeed, writing Eq. (5.103) in the form

$$
A = (LD^{1/2})(LD^{1/2})^T \tag{5.104}
$$

we conclude that the decomposition (5.106) is equivalent to the decomposition (3.39) in which $Q = D^{1/2}L^T$. If we regard L as being merely lower triangular instead of unit lower triangular, we conclude that *if A is a positive definite symmetric matrix, then it can be decomposed into*

$$A = LL^T \qquad (5.105)$$

where L is a unique lower triangular matrix with positive diagonal elements. Equation (5.105) is known as the *Cholesky decomposition* of A.

There remains the question of producing the matrix L for any positive definite real symmetric matrix A. To this end, let us write the matrix L in the form

$$L = \begin{bmatrix} l_{11} & 0 & 0 & \cdots & 0 \\ l_{21} & l_{22} & 0 & \cdots & 0 \\ l_{31} & l_{32} & l_{33} & \cdots & 0 \\ \cdots & \cdots & \cdots & \cdots & \cdots \\ l_{n1} & l_{n2} & l_{n3} & \cdots & l_{nn} \end{bmatrix} \qquad (5.106)$$

Then, by induction, it can be shown that the elements of L are given by the expressions

$$l_{ii} = \left(a_{ii} - \sum_{j=1}^{i-1} l_{ij}^2 \right)^{1/2}, \qquad i = 1, 2, \ldots, n \qquad (5.107a)$$

$$l_{ki} = \frac{1}{l_{ii}} \left(a_{ik} - \sum_{j=1}^{i-1} l_{ij}l_{kj} \right),$$

$$k = i+1, i+2, \ldots, n; \qquad i = 1, 2, \ldots, n \qquad (5.107b)$$

Equations (5.107a) and (5.107b) must be used alternately, as shown in Example 5.6.

The Cholesky decomposition requires only $n^3/6$ multiplications, which is about one half of the multiplications required by the Gaussian elimination. This is to be expected, as the Cholesky decomposition takes full advantage of the symmetry of the matrix A.

Returning to the reduction of an eigenvalue problem in terms of two real symmetric matrices to one in terms of a single real symmetric matrix discussed in Sec. 3.3, it should become immediately obvious that the matrix Q in Eq. (3.44) can be taken simply as L^T, where L is a lower triangular matrix obtainable by the Cholesky decomposition.

Example 5.6

Consider the matrix

$$A = \begin{bmatrix} 16 & -20 & -24 \\ -20 & 89 & -50 \\ -24 & -50 & 280 \end{bmatrix} \tag{a}$$

and use the Cholesky decomposition to express it as the product of two triangular matrices. The matrix A is known to be positive definite.

To obtain the matrix L, we use Eqs. (5.107). First, letting $i = 1$ in Eqs. (5.107a), we obtain

$$l_{11} = \sqrt{a_{11}} = \sqrt{16} = 4 \tag{b}$$

so that, from Eqs. (5.107b), we can write for $i = 1$ and $k = 2$

$$l_{21} = \frac{1}{l_{11}} a_{12} = \frac{-20}{4} = -5 \tag{c}$$

Returning to Eqs. (5.107a), we obtain for $i = 2$

$$l_{22} = (a_{22} - l_{21}^2)^{1/2} = [89 - (-5)^2]^{1/2} = 8 \tag{d}$$

so that, letting $i = 1, 2$ and $k = 3$ in Eqs. (5.107b), we have

$$l_{31} = \frac{1}{l_{11}} a_{13} = \frac{-24}{4} = -6 \tag{e}$$

$$l_{32} = \frac{1}{l_{22}} (a_{23} - l_{21}l_{31}) = \tfrac{1}{8}[-50 - (-5)(-6)] = -10$$

Finally, letting $i = 3$ in Eqs. (5.107a), we obtain

$$l_{33} = (a_{33} - l_{31}^2 - l_{32}^2)^{1/2} = [280 - (-6)^2 - (-10)^2]^{1/2} = 12 \tag{f}$$

Hence, the matrix A can be expressed in the form (5.105), in which

$$L = \begin{bmatrix} 4 & 0 & 0 \\ -5 & 8 & 0 \\ -6 & -10 & 12 \end{bmatrix} \tag{g}$$

5.9   The Jacobi method

The matrix iteration using the power method, in conjunction with one of the deflation procedures described in Secs. 5.6–7, produces the eigen-values and eigenvectors of a given matrix A one at a time. The Jacobi

138

method is also an iterative method, but, by contrast, it produces all the eigenvalues and eigenvectors of A simultaneously, where A is a real symmetric matrix. The method uses similarity transformations to diagonalize the matrix A, where the diagonal matrix is simply the matrix of the eigenvalues of A.

We have shown in Sec. 3.4 that the solution of the eigenvalue problem for a real positive definite symmetric matrix A is equivalent to finding the principal axes of the ellipsoid representing the quadratic form associated with A. For the two-dimensional case shown in Fig. 3.2, this amounts to a transformation of coordinates representing a rotation of axes in the plane by an angle θ. The angle θ must be such that axes y_1 and y_2 are the principal axes of the ellipse. The relation between axes x_1, x_2 and y_1, y_2 is

$$x_1 = y_1 \cos \theta - y_2 \sin \theta$$
$$x_2 = y_1 \sin \theta + y_2 \cos \theta$$
(5.108)

which can be written in the matrix form

$$\mathbf{x} = R\mathbf{y}$$
(5.109)

where R has the form

$$R = \begin{bmatrix} \cos \theta & -\sin \theta \\ \sin \theta & \cos \theta \end{bmatrix}$$
(5.110)

and is known as a *rotation matrix*. It should be pointed out that R is an orthonormal matrix, as it satisfies

$$R^T R = RR^T = I$$
(5.111)

which is typical of any matrix representing a coordinate transformation between two cartesian sets of axes. Recalling that the ellipse of Fig. 3.2 is given by

$$\mathbf{x}^T A \mathbf{x} = 1$$
(5.112)

and using the transformation (5.109), we obtain

$$\mathbf{y}^T R^T A R \mathbf{y} = 1$$
(5.113)

Comparing Eqs. (5.113) and (3.68), we conclude that the diagonal matrix of the eigenvalues of A has the form

$$\Lambda = R^T A R$$
(5.114)

Moreover, from Eq. (3.66), we conclude that R is simply the orthonormal matrix $\hat{X}$ of the eigenvectors of A.

Equation (5.114) has the explicit form

$$\begin{bmatrix} \lambda_1 & 0 \\ 0 & \lambda_2 \end{bmatrix} = \begin{bmatrix} \cos\theta & \sin\theta \\ -\sin\theta & \cos\theta \end{bmatrix} \begin{bmatrix} a_{11} & a_{12} \\ a_{12} & a_{22} \end{bmatrix} \begin{bmatrix} \cos\theta & -\sin\theta \\ \sin\theta & \cos\theta \end{bmatrix} \tag{5.115}$$

so that, equating homologous elements on both sides, we can write

$$\lambda_1 = a_{11}\cos^2\theta + 2a_{12}\sin\theta\cos\theta + a_{22}\sin^2\theta$$

$$\lambda_2 = a_{11}\sin^2\theta - 2a_{12}\sin\theta\cos\theta + a_{22}\cos^2\theta \tag{5.116}$$

$$0 = -(a_{11} - a_{22})\sin\theta\cos\theta + a_{12}(\cos^2\theta - \sin^2\theta)$$

From the last of Eqs. (5.116), we conclude that, for y_1 and y_2 to be principal axes, the angle θ must satisfy

$$\tan 2\theta = \frac{2a_{12}}{a_{11} - a_{22}} \tag{5.117}$$

On the other hand, the first two of Eqs. (5.116) yield the eigenvalues λ_1 and λ_2. The eigenvectors belonging to these eigenvalues are simply the columns of R

$$\mathbf{x}_1 = \begin{bmatrix} \cos\theta \\ \sin\theta \end{bmatrix}, \qquad \mathbf{x}_2 = \begin{bmatrix} -\sin\theta \\ \cos\theta \end{bmatrix} \tag{5.118}$$

For the two-dimensional case, i.e., when the matrix A is of order 2, the solution of the eigenvalue problem is obtained in a single step. In the general case, in which the matrix A is of order n, this is not possible and the principal axes of the n-dimensional ellipsoid can be obtained only iteratively, where each iteration step represents a planar rotation. This is the essence of the *Jacobi method*, for which reason the Jacobi method is also referred to as *diagonalization by successive rotations*. Because the process is iterative, the number of steps cannot be determined in advance and, in fact, it depends on how close the matrix A is to a diagonal matrix and on the desired accuracy.

The iteration sequence is given by the equation

$$A_k = R_k^T A_{k-1} R_k, \qquad k = 1, 2, 3, \ldots \tag{5.119}$$

where $A_0 = A$ and R_k is a rotation matrix. Assuming that the off-diagonal element of A_{k-1} of maximum modulus is in the (p, q) position, the matrix R_k is taken to represent a rotation in the (p, q) plane, so that the matrix

R_k can be exhibited in the form

$$
R_k = \begin{bmatrix}
1 & 0 & \cdots & 0 & \cdots & 0 & \cdots & 0 \\
0 & 1 & \cdots & 0 & \cdots & 0 & \cdots & 0 \\
\cdot & \cdot & \cdot & \cdot & \cdot & \cdot & \cdot & \cdot \\
0 & 0 & \cdots & \cos\theta_k & \cdots & -\sin\theta_k & \cdots & 0 \\
\cdot & \cdot & \cdot & \cdot & \cdot & \cdot & \cdot & \cdot \\
0 & 0 & \cdots & \sin\theta_k & \cdots & \cos\theta_k & \cdots & 0 \\
\cdot & \cdot & \cdot & \cdot & \cdot & \cdot & \cdot & \cdot \\
0 & 0 & \cdots & 0 & \cdots & 0 & \cdots & 1
\end{bmatrix}
\begin{matrix} \\ \\ \\ p \\ \\ q \\ \\ \\ \end{matrix} \tag{5.120}
$$

The angle θ_k is then chosen so as to reduce the (p, q) element of A_k to zero. Denoting the elements of the matrix A_k by $a_{ij}^{(k)}$ and introducing Eq. (5.120) into Eq. (5.119), we obtain the scalar equations

$$
a_{pp}^{(k)} = a_{pp}^{(k-1)} \cos^2\theta_k + 2a_{pq}^{(k-1)} \sin\theta_k \cos\theta_k + a_{qq}^{(k-1)} \sin^2\theta_k
$$
$$
a_{qq}^{(k)} = a_{pp}^{(k-1)} \sin^2\theta_k - 2a_{pq}^{(k-1)} \sin\theta_k \cos\theta_k + a_{qq}^{(k-1)} \cos^2\theta_k \tag{5.121a}
$$
$$
a_{pq}^{(k)} = -(a_{pp}^{(k-1)} - a_{qq}^{(k-1)}) \sin\theta_k \cos\theta_k + a_{pq}^{(k-1)}(\cos^2\theta_k - \sin^2\theta_k)
$$

$$
\left.\begin{aligned}
a_{ip}^{(k)} &= a_{ip}^{(k-1)} \cos\theta_k + a_{iq}^{(k-1)} \sin\theta_k \\
a_{iq}^{(k)} &= -a_{ip}^{(k-1)} \sin\theta_k + a_{iq}^{(k-1)} \cos\theta_k
\end{aligned}\right\} \; i \neq p, q
$$
$$
\left.\begin{aligned}
a_{pj}^{(k)} &= a_{pj}^{(k-1)} \cos\theta_k + a_{qj}^{(k-1)} \sin\theta_k \\
a_{qj}^{(k)} &= -a_{pj}^{(k-1)} \sin\theta_k + a_{qj}^{(k-1)} \cos\theta_k
\end{aligned}\right\} \; j \neq p, q \tag{5.121b}
$$

$$
a_{ij}^{(k)} = a_{ij}^{(k-1)}, \qquad i, j \neq p, q \tag{5.121c}
$$

so that only the elements in the p and q row and column of A_k differ from those in the p and q row and column of A_{k-1}. To annihilate the element $a_{pq}^{(k)}$, we must choose θ_k so as to satisfy

$$
\tan 2\theta_k = \frac{2a_{pq}^{(k-1)}}{a_{pp}^{(k-1)} - a_{qq}^{(k-1)}} \tag{5.122}
$$

Without loss of generality, we can take θ_k to lie in the range

$$
|\theta_k| \leq \tfrac{1}{4}\pi \tag{5.123}
$$

and when $a_{pp}^{(k-1)} = a_{qq}^{(k-1)}$ we can take $\theta_k = \pm\tfrac{1}{4}\pi$ depending on the sign of $a_{pq}^{(k-1)}$. Because in general $a_{pq}^{(k+1)} \neq 0$, the process is iterative. It remains to prove that the process is convergent, i.e., that the off-diagonal elements of A_k reduce to zero as $k \to \infty$, so that

$$
\lim_{k \to \infty} A_k = \Lambda \tag{5.124}
$$

5 Computational methods for the eigensolution

To prove Eq. (5.124), let us write the matrix A_k in the form

$$A_k = D_k + L_k + U_k \qquad (5.125)$$

where $D_k = \text{diag}\,(a_{ii}^{(k)})$, L_k is a lower triangular matrix with zero diagonal elements, and $U_k = L_k^T$ is an upper triangular matrix. Then, the Euclidean norm squared of A_k is simply

$$\|A_k\|_E^2 = \|D_k\|_E^2 + 2\,\|U_k\|_E^2 \qquad (5.126)$$

Using Eqs. (5.121a) and (5.121b) it can be shown that, independently of θ_k,

$$(a_{pp}^{(k)})^2 + (a_{qq}^{(k)})^2 + 2(a_{pq}^{(k)})^2 = (a_{pp}^{(k-1)})^2 + (a_{qq}^{(k-1)})^2 + 2(a_{pq}^{(k-1)})^2 \qquad (5.127)$$

and

$$
\begin{aligned}
(a_{ip}^{(k)})^2 + (a_{iq}^{(k)})^2 &= (a_{ip}^{(k-1)})^2 + (a_{iq}^{(k-1)})^2, \qquad i \ne p, q \\
(a_{pj}^{(k)})^2 + (a_{qj}^{(k)})^2 &= (a_{pj}^{(k-1)})^2 + (a_{qj}^{(k-1)})^2, \qquad j \ne p, q
\end{aligned}
\qquad (5.128)
$$

Equations (5.127) and (5.128) together with Eqs. (5.121c) indicate that

$$\|A_k\|_E^2 = \|A_{k-1}\|_E^2 \qquad (5.129)$$

confirming that the Euclidean norm is invariant under an orthonormal transformation. Moreover, Eqs. (5.121c) and (5.127) yield

$$\|U_k\|_E^2 - (a_{pq}^{(k)})^2 = \|U_{k-1}\|_E^2 - (a_{pq}^{(k-1)})^2 \qquad (5.130)$$

But, if θ_k is chosen so as to satisfy Eq. (5.122), then $a_{pq}^{(k)} = 0$, so that

$$\|U_k\|_E^2 = \|U_{k-1}\|_E^2 - (a_{pq}^{(k-1)})^2 \qquad (5.131)$$

it follows from Eqs. (5.126) and (5.129) that

$$\|D_k\|_E^2 = \|D_{k-1}\|_E^2 + 2(a_{pq}^{(k-1)})^2 \qquad (5.132)$$

or, the sum of the squares of the diagonal elements of A_k is larger than the sum of the squares of the diagonal elements of A_{k-1} by the amount $2(a_{pq}^{(k-1)})^2$. It is easy to see that the process is convergent, so that Eq. (5.124) holds. This is true whether the eigenvalues λ_r of A are all distinct or A has multiple eigenvalues (W1, Secs. 5.5–5.6).

From Eqs. (5.119), we can write

$$A_k = R_k^T R_{k-1}^T \cdots R_2^T R_1^T A R_1 R_2 \cdots R_{k-1} R_k \qquad (5.133)$$

Hence, considering Eqs. (3.66) and (5.124), we conclude that

$$\lim_{k \to \infty} R_1 R_2 \cdots R_{k-1} R_k = \hat{X} \qquad (5.134)$$

142

or, the product of the rotation matrices is equal to the matrix of the eigenvectors. Due to the fact that the matrices R_k are orthonormal, the matrix $\hat{X}$ is orthonormal, so that the Jacobi method produces automatically a set of orthonormal eigenvectors.

The fact that the element $a_{pq}^{(k-1)}$ to be annihilated is chosen as the off-diagonal element of A_{k-1} of largest modulus ensures the fastest convergence. On a digital computer, this implies a search through all the elements of U_{k-1}, so that a procedure avoiding this search is sometimes desirable. The simplest such procedure is to take the rotations sequentially in the order $(1, 2)$, $(1, 3), \ldots, (1, n)$, $(2, 3)$, $(2, 4), \ldots,$ $(2, n), \ldots, (n-1, n)$, where the sequence of $N = n(n-1)/2$ rotations is called a *sweep*. This procedure is referred to as the *serial Jacobi method*. Note that each rotation requires approximately $4n$ multiplications and a complete sweep approximately $2n^3$ multiplications.

If the element $a_{pq}^{(k-1)}$ is much smaller than the general level of the elements of U_{k-1}, then the annihilation of $a_{pq}^{(k-1)}$ serves little purpose. To render the annihilation of $a_{pq}^{(k-1)}$ purposeful, a search through the elements of U_{k-1} can be undertaken until an element of magnitude greater than some given *threshold value* is found. The first element of U_{k-1} exceeding this threshold value is then annihilated. This variant of the Jacobi method is known as the *threshold Jacobi method*. The question remains as to the numerical value of the threshold. Generally a threshold value is associated with each sweep and any rotation involving an off-diagonal element whose magnitude lies below the threshold value is simply omitted. The threshold value for a complete sweep can be taken as the average value of the off-diagonal elements squared, namely,

$$\sigma_l = \frac{1}{N} \|U_{l-1}\|_E, \qquad l = 1, 2, \ldots \tag{5.135}$$

where l is an integer designating the iteration step at the beginning of the sweep in question. This threshold value is lowered with each sweep until it reaches an overall threshold value, below which an element is regarded as zero. The process is terminated when N consecutive rotations are skipped.

The accuracy of the Jacobi method depends on how accurately $\sin \theta_k$ and $\cos \theta_k$ are calculated. If $\sin \theta_k$ and $\cos \theta_k$ are calculated with reasonable accuracy, then no significant growth of error occurs because of roundoff. The accuracy of the eigenvectors depends on the separation of the eigenvalues. Even when some eigenvalues are very close, the eigenvectors are almost exactly orthonormal. This is one of the most significant features of the Jacobi method. Indeed, if the interest lies not only in the eigenvalues but also in a set of orthonormal eigenvectors, then the Jacobi method may be more desirable than faster methods that do not produce orthonormal eigenvectors. Of course, when the diagonal elements of the matrix A are dominant the Jacobi method should be given serious consideration.

Example 5.7

Use the Jacobi method to solve the eigenvalue problem for the system of Fig. 4.3. The eigenvalue problem has the form

$$Ku = \omega^2 Mu \tag{a}$$

where, from Example 4.1, we have

$$M = m\begin{bmatrix} 1 & 0 & 0 \\ 0 & 1 & 0 \\ 0 & 0 & 2 \end{bmatrix}, \quad K = k\begin{bmatrix} 2 & -1 & 0 \\ -1 & 3 & -2 \\ 0 & -2 & 2 \end{bmatrix} \tag{b}$$

First, let us transform the eigenvalue problem (a) into one in terms of a single symmetric matrix. To this end, let us introduce the transformation

$$u = M^{-1/2}x \tag{c}$$

into Eq. (a), premultiply the result by $M^{-1/2}$ and obtain

$$M^{-1/2}KM^{-1/2}x = \omega^2 x \tag{d}$$

Hence, the eigenvalue problem reduces to the form

$$Ax = \lambda x \tag{e}$$

where

$$A = M^{-1/2}KM^{-1/2} = \begin{bmatrix} 2 & -1 & 0 \\ -1 & 3 & -\sqrt{2} \\ 0 & -\sqrt{2} & 1 \end{bmatrix}, \quad \lambda = \omega^2 m/k \tag{f}$$

To annihilate the element $a_{12}^{(0)} = a_{12}$, let $p = 1$, $q = 2$ and $k = 1$ in Eq. (5.122), so that

$$\tan 2\theta_1 = \frac{2a_{12}^{(0)}}{a_{11}^{(0)} - a_{22}^{(0)}} = \frac{2a_{12}}{a_{11} - a_{22}} = \frac{-2}{2-3} = 2$$

$$\therefore \ \theta_1 = 31°43'02.9''$$

Hence, from Eq. (5.120) with $k = 1$, the first rotation matrix is

$$R_1 = \begin{bmatrix} \cos\theta & -\sin\theta_1 & 0 \\ \sin\theta_1 & \cos\theta_1 & 0 \\ 0 & 0 & 1 \end{bmatrix} = \begin{bmatrix} 0.850651 & -0.525731 & 0 \\ 0.525731 & 0.850651 & 0 \\ 0 & 0 & 1 \end{bmatrix} \tag{g}$$

Introducing Eqs. (f) and (g) into Eq. (5.119) with $k = 1$, we obtain

$$A_1 = R_1^T A_0 R_1 = R_1^T A R_1 = \begin{bmatrix} 1.381966 & 0 & -0.743496 \\ 0 & 3.618034 & -1.203002 \\ -0.743496 & -1.203002 & 1.000000 \end{bmatrix}$$

(h)

Next, we wish to annihilate the element $a_{13}^{(1)}$. Letting $p = 1$, $q = 3$, and $k = 2$ in Eq. (5.122), we can write

$$\tan 2\theta_2 = \frac{2a_{13}^{(1)}}{a_{11}^{(1)} - a_{33}^{(1)}} = \frac{-2 \times 0.743496}{1.381966 - 1.000000} = -3.892996$$

$$\therefore \ \theta_2 = -37°47'48.9''$$

so that the second rotation matrix is

$$R_2 = \begin{bmatrix} \cos \theta_2 & 0 & -\sin \theta_2 \\ 0 & 1 & 0 \\ \sin \theta_2 & 0 & \cos \theta_2 \end{bmatrix} = \begin{bmatrix} 0.790188 & 0 & 0.612864 \\ 0 & 1 & 0 \\ -0.612864 & 0 & 0.790188 \end{bmatrix}$$

(i)

Hence, Eq. (5.119) with $k = 2$ yields

$$A_2 = R_2^T A_1 R_2 = \begin{bmatrix} 1.958616 & 0.737277 & 0 \\ 0.737277 & 3.618034 & -0.950598 \\ 0 & -0.950598 & 0.423350 \end{bmatrix}$$

(j)

At this point let us pause and examine the matrices $A_0 = A$, A_1 and A_2. We notice that the sum of the squares of the diagonal elements of these matrices increases at the expense of the sum of the squares of the off-diagonal elements as the iteration number increases. The increase is such that Eq. (5.132) is satisfied. In annihilating the element $a_{13}^{(1)}$, the element in the position $(1, 2)$, which was annihilated in the preceding iteration, acquires some nonzero value, $a_{12}^{(2)} = 0.737277$. This value, however, is smaller in magnitude than the original value, $a_{12}^{(0)} = -1$. Note also that the trace of the matrices A_0, A_1 and A_2 remains the same.

After nine iterations, we achieve convergence. According to Eq. (5.124), the resulting diagonal matrix is the matrix of the eigenvalues

$$\Lambda = A_9 = \begin{bmatrix} 1.745898 & 0 & 0 \\ 0 & 4.114908 & 0 \\ 0 & 0 & 0.139194 \end{bmatrix}$$

(k)

Moreover, according to Eq. (5.134), the product of the rotation matrices is the modal matrix $\hat{X}$. This must be premultiplied by $M^{-1/2}$ to obtain the

145

modal matrix U. The result is

$$U = M^{-1/2}\hat{X} = M^{-1/2}R_1 R_2 \cdots R_9$$

$$= m^{-1/2}\begin{bmatrix} 0.878182 & -0.395445 & 0.269108 \\ 0.223148 & 0.836329 & 0.500758 \\ -0.299166 & -0.268493 & 0.581731 \end{bmatrix} \tag{1}$$

and we observe that the modal matrix U is orthonormal with respect to M, i.e.,

$$U^T M U = I \tag{m}$$

The eigenvalues and eigenvectors are not arranged in increasing order. For example, the first eigenvalue, λ_1, is in the bottom right corner of Λ and the eigenvector $\mathbf{u}_1$ is the third column of U, or

$$\lambda_1 = 0.139194$$
$$\mathbf{u}_1 = m^{-1/2}[0.269108 \quad 0.500758 \quad 0.581731]^T \tag{n}$$

The eigenvector can be verified as being the same as that obtained in Example 5.3. The natural frequencies are related to the eigenvalues by the second of Eqs. (f). This permits us to calculate the lowest natural frequency

$$\omega_1 = \sqrt{\lambda_1 k/m} = \sqrt{0.139194}\,\sqrt{\frac{k}{m}} = 0.373087\,\sqrt{\frac{k}{m}} \tag{o}$$

Finally, we note from the matrix U that the first column, representing the second mode, has one sign change, the second column, representing the third mode, has two sign changes, and the third column has no sign change. This is typical of vibrating systems of the type shown in Fig. 4.3 (see M6, p. 160).

5.10   Givens' method

The Jacobi method can be used to produce a solution of the eigenvalue problem associated with a real symmetric A by diagonalizing the matrix. It is an iterative method based on the concept of coordinate rotations and is quite efficient when the diagonal elements of A are dominant. Another method based on the coordinate rotations concept is the *Givens method*. By contrast, however, Givens' method does not really produce a solution of the eigenvalue problem, but only reduces a real symmetric matrix A to tridiagonal form. The solution of the eigenvalue problem associated with the tridiagonal matrix must be obtained by some other method. The important feature of Givens' method is that the reduction to tridiagonal form is not iterative, as it involves only a finite number of steps.

What makes the Jacobi method an iterative one is the fact that an element (p, q) annihilated in the kth step does not generally remain zero in the $(k + 1)$st step. The question is whether it is possible to arrange the order of the computations so that previously annihilated elements remain zero, so that the procedure ceases to be iterative. It turns out that this is only partially possible, in the sense that this approach can be used to reduce a real symmetric matrix to tridiagonal form but not to diagonal form. This is the essence of Givens' method. But this is not the only difference between the Jacobi method and Givens' method. Instead of annihilating the element $a_{pq}^{(k)}$, as in the Jacobi method, in the Givens' method one seeks to annihilate the element $a_{iq}^{(k)}$, $i \neq p, q$. Of course, because of symmetry, the element $a_{qi}^{(k)}$ is annihilated at the same time. From Eqs. (5.121b), we conclude that

$$a_{iq}^{(k)} = -a_{ip}^{(k-1)} \sin \theta_k + a_{iq}^{(k-1)} \cos \theta_k, \qquad i \neq p, q \qquad (5.136)$$

so that, to reduce $a_{iq}^{(k)}$ to zero, we write

$$\sin \theta_k = \alpha_k a_{iq}^{(k-1)}, \qquad \cos \theta_k = \alpha_k a_{ip}^{(k-1)} \qquad (5.137)$$

where

$$\alpha_k = \{[a_{ip}^{(k-1)}]^2 + [a_{iq}^{(k-1)}]^2\}^{-1/2} \qquad (5.138)$$

Note also that the calculation of the angles of rotation is much simpler for Givens' method than for the Jacobi method.

The procedure can be carried out in a series of steps designed to reduce to zero all the elements in a given row, with the exception of the tridiagonal elements, which automatically annihilates the corresponding elements in the associated column. For example, a rotation in the $(2, 3)$ plane can be used to reduce $a_{13}^{(1)}$ and $a_{31}^{(1)}$ to zero. Indeed, letting $i = 1$, $p = 2$, $q = 3$ in Eq. (5.136), we obtain

$$-a_{12}^{(0)} \sin \theta_1 + a_{13}^{(0)} \cos \theta_1 = 0 \qquad (5.139)$$

where $a_{12}^{(0)} = a_{12}$ and $a_{13}^{(0)} = a_{13}$, so that

$$\sin \theta_1 = \alpha_1 a_{13}^{(0)}, \qquad \cos \theta_1 = \alpha_1 a_{12}^{(0)} \qquad (5.140)$$

in which

$$\alpha_1 = \{[a_{12}^{(0)}]^2 + [a_{13}^{(0)}]^2\}^{-1/2} \qquad (5.141)$$

Similarly, a rotation by an angle θ_2 in the $(2, 4)$ plane can be used to annihilate the elements $a_{14}^{(2)}$ and $a_{41}^{(2)}$ without affecting $a_{13}^{(1)} = a_{31}^{(1)} = 0$.

Following the same pattern, rotations in the $(2, 5)$, $(2, 6), \ldots, (2, n)$ planes can be used to annihilate the elements $a_{15}^{(3)}$, $a_{51}^{(3)}$, $a_{16}^{(4)}, \ldots, a_{1n}^{(n-2)}$, $a_{n1}^{(n-2)}$, in sequence, so that the elements reduced to zero previously remain zero. Hence, using $n - 2$ rotations, one can reduce the matrix A to a matrix A_{n-2} with all the elements in the first row and column equal to zero, with the exception of the first two elements.

The above procedure can be repeated to reduce all the elements in the second row and column to zero, with the exception of the first three elements. This requires $n - 3$ rotations in the planes $(3, 4)$, $(3, 5), \ldots, (3, n)$, in sequence. In general, $(n - 1)(n - 2)/2$ rotations yield the tridiagonal matrix

$$A_k = \begin{bmatrix} \alpha_1 & \beta_2 & 0 & \cdots & 0 & 0 \\ \beta_2 & \alpha_2 & \beta_3 & \cdots & 0 & 0 \\ 0 & \beta_3 & \alpha_3 & \cdots & 0 & 0 \\ \multicolumn{6}{c}{\cdots\cdots\cdots\cdots\cdots\cdots} \\ 0 & 0 & 0 & \cdots & \alpha_{n-1} & \beta_n \\ 0 & 0 & 0 & \cdots & \beta_n & \alpha_n \end{bmatrix}, \qquad k = \tfrac{1}{2}(n-1)(n-2) \quad (5.142)$$

The complete reduction to tridiagonal form requires approximately $\tfrac{4}{3}n^3$ multiplications, as opposed to $2n^3$ for just one sweep in the Jacobi method, so that Givens' method is considerably faster. Of course, the task of solving the eigenvalue problem for the tridiagonal matrix still remains.

We shall consider the solution of the eigenvalue problem associated with the matrix A_k, Eq. (5.142), in Secs. 5.12–15. At this point, a discussion of the manner in which the solution of the eigenvalue problem associated with A_k relates to that associated with A is in order.

Because A_k is obtained from A by means of a similarity transformation, the eigenvalues of A_k are the same as those of A. On the other hand, the eigenvectors of A_k are not the same as the eigenvectors of A. Denoting the eigenvectors of A by $\mathbf{u}$ and those of A_k by $\mathbf{x}$, the two eigenvalue problems can be written in the form

$$A\mathbf{u} = \lambda\mathbf{u} \qquad (5.143)$$

and

$$A_k\mathbf{x} = \lambda\mathbf{x} \qquad (5.144)$$

The question remains as to how the eigenvectors $\mathbf{u}$ relate to the eigenvectors $\mathbf{x}$. To answer this question, we recognize first that the matrix A_k is obtained from A by means of the recursive formula

$$A_k = R_k^T A_{k-1} R_k = R_k^T R_{k-1}^T A_{k-2} R_{k-1} R_k = \cdots = R^T A R \qquad (5.145)$$

where

$$R = R_1 R_2 \cdots R_{k-1} R_k = \prod_{i=1}^{k} R_i \qquad (5.146)$$

is an orthonormal matrix. But, unlike in the Jacobi method, here R does not represent the matrix of the eigenvectors of A because A_k is merely a tridiagonal matrix and not the diagonal matrix of the eigenvalues. Inserting Eq. (5.145) into Eq. (5.144), we obtain

$$R^T A R \mathbf{x} = \lambda \mathbf{x} \qquad (5.147)$$

Multiplying both sides of Eq. (5.147) on the left by R and recalling that for an orthonormal matrix $RR^T = I$, we can write

$$A R \mathbf{x} = \lambda R \mathbf{x} \qquad (5.148)$$

so that comparing Eqs. (5.143) and (5.148), we conclude that the eigenvectors of A and A_k are related by

$$\mathbf{u} = R \mathbf{x} \qquad (5.149)$$

where R is given by Eq. (5.146).

With proper modifications, Givens' method can also be applied to arbitrary matrices, in which case the reduced matrix has an upper Hessenberg form (W1, Sec. 6.2).

Example 5.8

The eigenvalue problem for the system shown in Fig. 5.1 can be written in the form

$$A\mathbf{u} = \lambda \mathbf{u} \qquad (a)$$

where $A = K^{-1}M$. Write down the matrix A and tridiagonalize the matrix by Givens' method.

The matrix A can be shown to have the explicit form

$$A = \begin{bmatrix} 1 & 1 & 1 & 1 \\ 1 & 2 & 2 & 2 \\ 1 & 2 & 3 & 3 \\ 1 & 2 & 3 & 4 \end{bmatrix} \qquad (b)$$

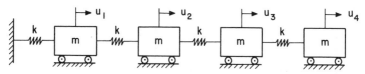

Figure 5.1

5   Computational methods for the eigensolution

and note that in this case $\lambda = k/m\omega^2$. To annihilate the element $a_{13}^{(0)}$ we must use a rotation θ_1 in the $(2, 3)$ plane. Hence, using Eqs. (5.140) and (5.141), we can write

$$\sin \theta_1 = \alpha_1 a_{13}^{(0)} = \frac{a_{13}^{(0)}}{\{[a_{12}^{(0)}]^2 + [a_{13}^{(0)}]^2\}^{1/2}} = \frac{1}{(1^2 + 1^2)^{1/2}} = 0.707107$$

$$\cos \theta_1 = \alpha_1 a_{12}^{(0)} = \frac{a_{12}^{(0)}}{\{[a_{12}^{(0)}]^2 + [a_{13}^{(0)}]^2\}^{1/2}} = \frac{1}{(1^2 + 1^2)^{1/2}} = 0.707107$$

(c)

so that the first rotation matrix is

$$R_1 = \begin{bmatrix} 1 & 0 & 0 & 0 \\ 0 & \cos\theta_1 & -\sin\theta_1 & 0 \\ 0 & \sin\theta_1 & \cos\theta_1 & 0 \\ 0 & 0 & 0 & 1 \end{bmatrix} = \begin{bmatrix} 1 & 0 & 0 & 0 \\ 0 & 0.707107 & -0.707107 & 0 \\ 0 & 0.707107 & 0.707107 & 0 \\ 0 & 0 & 0 & 1 \end{bmatrix}$$

(d)

and the first transformed matrix is

$$A_1 = R_1^T A_0 R_1 = R_1^T A R_1$$

$$= \begin{bmatrix} 1 & 1.414214 & 0 & 1 \\ 1.414214 & 4.5 & 0.5 & 3.535534 \\ 0 & 0.5 & 0.5 & 0.707107 \\ 1 & 3.535534 & 0.707107 & 4 \end{bmatrix}$$

(e)

Next, we use a rotation θ_2 in the $(2, 4)$ plane to annihilate $a_{14}^{(1)}$. From Eqs. (5.137) and (5.138), with $k = 2$, $i = 1$, $p = 2$ and $q = 4$, we can write

$$\sin \theta_2 = \frac{a_{14}^{(1)}}{\{[a_{12}^{(1)}]^2 + [a_{14}^{(1)}]^2\}^{1/2}} = \frac{1}{(1.414214^2 + 1^2)^{1/2}} = 0.577350$$

$$\cos \theta_2 = \frac{a_{12}^{(1)}}{\{[a_{12}^{(1)}]^2 + [a_{14}^{(1)}]^2\}^{1/2}} = \frac{1.414214}{(1.414214^2 + 1^2)^{1/2}} = 0.816497$$

(f)

so that

$$R_2 = \begin{bmatrix} 1 & 0 & 0 & 0 \\ 0 & 0.816497 & 0 & -0.577350 \\ 0 & 0 & 1 & 0 \\ 0 & 0.577350 & 0 & 0.816497 \end{bmatrix}$$

(g)

150

Hence,

$$A_2 = R_2^T A_1 R_2 = \begin{bmatrix} 1 & 1.732051 & 0 & 0 \\ 1.732051 & 7.666667 & 0.816497 & 0.942809 \\ 0 & 0.816497 & 0.5 & 0.288675 \\ 0 & 0.942809 & 0.288675 & 0.833333 \end{bmatrix}$$ (h)

Finally, to annihilate the element $a_{24}^{(2)}$ we use a rotation θ_3 in the $(3, 4)$ plane. Letting $k = 3$, $i = 2$, $p = 3$ and $q = 4$ in Eqs. (5.137) and (5.138), we obtain

$$\sin \theta_3 = \frac{a_{24}^{(2)}}{\{[a_{23}^{(2)}]^2 + [a_{24}^{(2)}]^2\}^{1/2}} = \frac{0.942809}{(0.816497^2 + 0.942809^2)^{1/2}} = 0.755929$$

$$\cos \theta_3 = \frac{a_{23}^{(2)}}{\{[a_{23}^{(2)}]^2 + [a_{24}^{(2)}]^2\}^{1/2}} = \frac{0.816497}{(0.816497^2 + 0.942809^2)^{1/2}} = 0.654654$$ (i)

Hence,

$$R_3 = \begin{bmatrix} 1 & 0 & 0 & 0 \\ 0 & 1 & 0 & 0 \\ 0 & 0 & 0.654654 & -0.755929 \\ 0 & 0 & 0.755929 & 0.654654 \end{bmatrix}$$ (j)

and

$$A_3 = R_3^T A_2 R_3 = \begin{bmatrix} 1 & 1.732051 & 0 & 0 \\ 1.732051 & 7.666667 & 1.247219 & 0 \\ 0 & 1.247219 & 0.976190 & 0.123718 \\ 0 & 0 & 0.123718 & 0.357143 \end{bmatrix}$$ (k)

which is the desired tridiagonal matrix. Note that the trace of A_3 is equal to 10, and so is the trace of A_1 and A_2.

For future reference, let us compute

$$R = R_1 R_2 R_3 = \begin{bmatrix} 1 & 0 & 0 & 0 \\ 0 & 0.577350 & -0.771517 & 0.267261 \\ 0 & 0.577350 & 0.154303 & -0.801784 \\ 0 & 0.577350 & 0.617213 & 0.534522 \end{bmatrix}$$ (l)

5.11   Householder's method

Householder's method also reduces a real symmetric matrix A to tridiagonal form, but it does it more efficiently than Givens' method, as it requires only half as many multiplications (W1, p. 293). Instead of annihilating one element at a time, Householder's method annihilates a whole row and column at a time, with the exception, of course, of the tridiagonal elements in that row and column. Like Givens' method, no successive transformation affects previous rows and columns already reduced to zero, so that a total of $n-2$ transformations are necessary. These are orthonormal similarity transformations, but they do not represent coordinate rotations as in the Jacobi and Givens' methods.

Let us consider the transformation

$$A_k = P_k A_{k-1} P_k, \qquad A_0 = A \tag{5.150}$$

where

$$P_k = I - 2\mathbf{v}_k \mathbf{v}_k^T, \qquad \mathbf{v}_k^T \mathbf{v}_k = 1 \tag{5.151}$$

in which the matrix P_k is not only symmetric but also orthonormal. Indeed, we have

$$P_k P_k^T = (I - 2\mathbf{v}_k \mathbf{v}_k^T)(I - 2\mathbf{v}_k \mathbf{v}_k^T) = I - 4\mathbf{v}_k \mathbf{v}_k^T + 4\mathbf{v}_k (\mathbf{v}_k^T \mathbf{v}_k)\mathbf{v}_k^T = I \tag{5.152}$$

The matrix P_k defined by Eq. (5.151) can be regarded as representing a linear transformation in a real Euclidean space transforming an n-vector into another n-vector. The transformation can be interpreted geometrically as a *reflection* through a given plane and is referred to as a *Householder transformation* (N1, p. 286).

The matrix A_1, corresponding to $k=1$ in Eq. (5.150), must have the form

$$A_1 = \begin{bmatrix} a_{11}^{(1)} & a_{12}^{(1)} & 0 & \dots & 0 \\ a_{12}^{(1)} & a_{22}^{(1)} & a_{23}^{(1)} & \dots & a_{2n}^{(1)} \\ 0 & a_{23}^{(1)} & a_{33}^{(1)} & \dots & a_{3n}^{(1)} \\ \cdot & \cdot & \cdot & \cdot & \cdot \\ 0 & a_{2n}^{(1)} & a_{3n}^{(1)} & \dots & a_{nn}^{(1)} \end{bmatrix} \tag{5.153}$$

The requirements

$$a_{13}^{(1)} = a_{14}^{(1)} = \cdots = a_{1n}^{(1)} = 0, \qquad \mathbf{v}_1^T \mathbf{v}_1 = 1 \tag{5.154}$$

can be regarded as $n-1$ constraints imposed on the n components $v_{1,j}$ of the vector $\mathbf{v}_1$. Hence, one of the components of $\mathbf{v}_1$ can be chosen arbitrarily. We shall choose $v_{1,1} = 0$, so that

$$
P_1 = \begin{bmatrix}
1 & 0 & 0 & \cdots & 0 \\
0 & 1 - 2v_{1,2}^2 & -2v_{1,2}v_{1,3} & \cdots & -2v_{1,2}v_{1,n} \\
0 & -2v_{1,2}v_{1,3} & 1 - 2v_{1,3}^2 & \cdots & -2v_{1,3}v_{1,n} \\
\cdots & \cdots & \cdots & \cdots & \cdots \\
0 & -2v_{1,2}v_{1,n} & -2v_{1,3}v_{1,n} & \cdots & 1 - 2v_{1,n}^2
\end{bmatrix} \tag{5.155}
$$

Next, let us consider the matrix product $A_1 = P_1 A_0 P_1 = P_1 A P_1$. First, we notice that the first row of $P_1 A$ is simply the first row of A. Denoting the first row of A_k by $\mathbf{a}_1^{(k)^T}$ ($k = 0, 1, 2, \ldots$), it follows that the first row of A_1 is

$$
\mathbf{a}_1^{(1)^T} = \mathbf{a}_1^T P_1 = \mathbf{a}_1^T - 2(\mathbf{a}_1^T \mathbf{v}_1)\mathbf{v}_1^T \tag{5.156}
$$

where $\mathbf{a}_1^T = \mathbf{a}_1^{(0)^T}$ is the first row of A. Equation (5.156) implies that

$$
a_{11}^{(1)} = a_{11}, \qquad a_{1j}^{(1)} = a_{1j} - 2(\mathbf{a}_1^T \mathbf{v}_1)v_{1,j} \qquad j = 2, 3, \ldots, n \tag{5.157}
$$

Considering Eqs. (5.154), we conclude that the nonzero components $v_{1,2}, v_{1,3}, \ldots, v_{1,n}$ of the vector $\mathbf{v}_1$ must solve the $n - 1$ equations

$$
a_{1j} - 2(\mathbf{a}_1^T \mathbf{v}_1)v_{1,j} = 0, \qquad j = 3, 4, \ldots, n \tag{5.158}
$$

and

$$
v_{1,2}^2 + v_{1,3}^2 + \cdots + v_{1,n}^2 = 1 \tag{5.159}
$$

To solve Eqs. (5.158) and (5.159), we first observe that the vectors $\mathbf{a}_1^{(1)}$ and $\mathbf{a}_1$ have the same length. Indeed, $\mathbf{a}_1^{(1)^T}\mathbf{a}_1^{(1)} = \mathbf{a}_1^T P_1 P_1 \mathbf{a}_1 = \mathbf{a}_1^T \mathbf{a}_1$. Because $a_{11}^{(1)} = a_{11}$, we must have

$$
(a_{12}^{(1)})^2 = \sum_{j=2}^n a_{1j}^2 = \alpha_1^2 \tag{5.160}
$$

where $\alpha_1 = \left(\sum_{j=2}^n a_{1j}^2 \right)^{1/2}$ is regarded as a positive constant. From Eqs. (5.157) and (5.160), we conclude that

$$
a_{12}^{(1)} = a_{12} - 2(\mathbf{a}_1^T \mathbf{v}_1)v_{1,2} = \pm\alpha_1 \tag{5.161}
$$

Now, let us multiply Eq. (5.160) by $v_{1,2}$ and Eqs. (5.158) by $v_{1,j}$ ($j = 3, 4, \ldots, n$) correspondingly and add the results to obtain

$$
\sum_{j=1}^n [a_{1j} - 2(\mathbf{a}_1^T \mathbf{v}_1)v_{1,j}]v_{1,j} = \pm\alpha_1 v_{1,2} \tag{5.162}
$$

which can be rewritten in the more compact form

$$
\mathbf{a}_1^T \mathbf{v}_1 - 2(\mathbf{a}_1^T \mathbf{v}_1)(\mathbf{v}_1^T \mathbf{v}_1) = \pm\alpha_1 v_{1,2} \tag{5.163}
$$

153

Because the vector $\mathbf{v}_1$ has unit length, Eq. (5.163) reduces to

$$\mathbf{a}_1^T \mathbf{v}_1 = \mp \alpha_1 v_{1,2} \tag{5.164}$$

Introducing Eq. (5.164) into Eq. (5.161), we obtain

$$v_{1,2} = \left[\frac{1}{2} \left(1 \mp \frac{a_{12}}{\alpha_1} \right) \right]^{1/2} \tag{5.165}$$

which permits us to calculate the components $v_{1,j}$ $(j = 3, 4, \ldots, n)$ of $\mathbf{v}_1$. Indeed, introducing Eq. (5.164) into Eqs. (5.158), we can write

$$v_{1,j} = \mp \frac{a_{1j}}{2\alpha_1 v_{1,2}}, \qquad j = 3, 4, \ldots, n \tag{5.166}$$

where $v_{1,2}$ is given by Eq. (5.165). If the sign in Eq. (5.165) is chosen to be that of a_{12}, then $v_{1,2}$ is as large as possible, which avoids the possibility of dividing by a small number in Eq. (5.166). Note that the choice of signs in Eqs. (5.165) and (5.166) must be the same. Because $v_{1,1} = 0$, Eqs. (5.165) and (5.166) define the vector $\mathbf{v}_1$ completely.

The procedure can be generalized by writing the vector $\mathbf{v}_k$ in Eq. (5.151) in the form

$$\mathbf{v}_k = [0 \quad 0 \quad \cdots \quad 0 \quad v_{k,k+1} v_{k,k+2} \quad \cdots \quad v_{k,n}]^T \tag{5.167}$$

where

$$v_{k,k+1} = \left[\frac{1}{2} \left(1 \mp \frac{a_{k,k+1}^{(k-1)}}{\alpha_k} \right) \right]^{1/2}$$

$$v_{k,j} = \mp \frac{a_{kj}^{(k-1)}}{2\alpha_k v_{k,k+1}}, \qquad j = k+2, k+3, \ldots, n \tag{5.168}$$

in which

$$\alpha_k = \left[\sum_{j=k+1}^{n} (a_{kj}^{(k-1)})^2 \right]^{1/2}, \qquad k = 1, 2, \ldots, n-2 \tag{5.169}$$

The reduction to tridiagonal form by Householder's method requires $\frac{2}{3}n^3$ multiplications, compared to $\frac{4}{3}n^3$ multiplications required by the Givens' method.

Upon completion of $k = n-2$ transformations, the matrix A is reduced to a tridiagonal matrix A_k of the form (5.142). As with Givens' method, the question arises also here as to how the solution of the eigenvalue problem associated with A_k relates to that associated with A. Clearly, the eigenvalues of the two matrices are the same. Moreover, denoting the eigenvectors of A by $\mathbf{u}$ and the eigenvectors of A_k by $\mathbf{x}$, the relation between the two can be shown with ease to be

$$\mathbf{u} = P\mathbf{x} \tag{5.170}$$

where P is the orthonormal matrix

$$P = \prod_{i=1}^{k} P_i \tag{5.171}$$

in which P_i is a matrix of the type (5.151).

In spite of algebraic operations differing in appearance, the methods of Givens and Householder are essentially equivalent (W1, Sec. 6.7). In fact, the matrices A_k produced by the two methods should be identical, with the possible difference in the signs of various off-diagonal entries, and a similar statement can be made about the transformation matrices R and P. In practice, rounding errors may cause them to differ.

Householder's method, like Givens' method, can be modified so as to permit the reduction of an arbitrary matrix to upper Hessenberg form (W1, Sec. 6.3).

Example 5.9

Tridiagonalize the matrix A of Example 5.8 by Householder's method.

The procedure consists of first computing the transformation matrices P_k, as given by Eq. (5.151), and then transforming the matrix $A = A_0$ according to Eq. (5.150). To this end, we use Eqs. (5.168) and (5.169) to determine the components of the vectors $\mathbf{v}_k$. Letting $k = 1$ in Eq. (5.169) and referring to the matrix $A = A_0$ in Example 5.8, we can write

$$\alpha_1 = \left\{ \sum_{j=2}^{4} [a_{1j}^{(0)}]^2 \right\}^{1/2} = (1^2 + 1^2 + 1^2)^{1/2} = \sqrt{3} \tag{a}$$

Then, from Eqs. (5.168) with $k = 1$, we obtain

$$v_{1,2} = \left[\frac{1}{2} \left(1 \mp \frac{a_{12}^{(0)}}{\alpha_1} \right) \right]^{1/2} = \left[\frac{1}{2} \left(1 + \frac{1}{\sqrt{3}} \right) \right]^{1/2} = 0.888074$$

$$v_{1,3} = \mp \frac{a_{13}^{(0)}}{2\alpha_1 v_{1,2}} = \frac{1}{2\sqrt{3} \times 0.888074} = 0.325058 \tag{b}$$

$$v_{1,4} = \mp \frac{a_{14}^{(0)}}{2\alpha_1 v_{1,2}} = \frac{1}{2\sqrt{3} \times 0.888074} = 0.325058$$

so that

$$\mathbf{v}_1 = [0 \quad 0.888074 \quad 0.325058 \quad 0.325058]^T \tag{c}$$

155

5 Computational methods for the eigensolution

Letting $k = 1$ in Eq. (5.151) and using Eq. (c), we can write

$$P_1 = I - 2\mathbf{v}_1\mathbf{v}_1^T = \begin{bmatrix} 1 & 0 & 0 & 0 \\ 0 & -0.577350 & -0.577350 & -0.577350 \\ 0 & -0.577350 & 0.788675 & -0.211325 \\ 0 & -0.577350 & -0.211325 & 0.788675 \end{bmatrix} \tag{d}$$

so that, letting $k = 1$ in Eq. (5.150) and using Eq. (d) in conjunction with the matrix $A = A_0$ of Example 5.8, we have

$$A_1 = P_1 A_0 P_1 = P_1 A P_1$$

$$= \begin{bmatrix} 1 & -1.732051 & 0 & 0 \\ -1.732051 & 7.666667 & -0.544658 & -1.122085 \\ 0 & -0.544658 & 0.377991 & 0.166667 \\ 0 & -1.122085 & 0.166667 & 0.955342 \end{bmatrix} \tag{e}$$

Next, letting $k = 2$ in Eq. (5.169), we obtain

$$\alpha_2 = \left\{ \sum_{j=3}^{4} [a_{2j}^{(1)}]^2 \right\}^{1/2} = [(-0.544658)^2 + (-1.122085)^2]^{1/2} = 1.247219 \tag{f}$$

so that, from Eqs. (5.168) with $k = 2$, we can write

$$v_{2,3} = \left[\frac{1}{2}\left(1 \mp \frac{a_{23}^{(1)}}{\alpha_2} \right) \right]^{1/2} = \left[\frac{1}{2}\left(1 + \frac{0.544658}{1.247219} \right) \right]^{1/2} = 0.847555$$

$$v_{2,4} = \mp \frac{a_{24}^{(1)}}{2\alpha_2 v_{2,3}} = \frac{1.122085}{2 \times 1.247219 \times 0.847555} = 0.530708 \tag{g}$$

Hence, the vector $\mathbf{v}_2$ is

$$\mathbf{v}_2 = [0 \quad 0 \quad 0.847555 \quad 0.530708]^T \tag{h}$$

and the matrix P_2 can be obtained from Eqs. (5.151) in the form

$$P_2 = I - 2\mathbf{v}_2\mathbf{v}_2^T = \begin{bmatrix} 1 & 0 & 0 & 0 \\ 0 & 1 & 0 & 0 \\ 0 & 0 & -0.436698 & -0.899608 \\ 0 & 0 & -0.899608 & 0.436698 \end{bmatrix} \tag{i}$$

Moreover, letting $k = 2$ in Eq. (5.150), we obtain

$$A_2 = P_2 A_1 P_2 = \begin{bmatrix} 1 & -1.732051 & 0 & 0 \\ -1.732051 & 7.666667 & 1.247219 & 0 \\ 0 & 1.247219 & 0.976190 & -0.123718 \\ 0 & 0 & -0.123718 & 0.357143 \end{bmatrix} \tag{j}$$

which is identical to the matrix A_3 of Example 5.8, except for the signs of the entries $(1, 2)$ and $(3, 4)$ and their symmetric counterparts. Finally, using Eq. (5.171), we can write

$$P = P_1 P_2 = \begin{bmatrix} 1 & 0 & 0 & 0 \\ 0 & -0.577350 & 0.771517 & 0.267261 \\ 0 & -0.577350 & -0.154303 & -0.801784 \\ 0 & -0.577350 & -0.617213 & 0.534522 \end{bmatrix} \tag{k}$$

which is identical to the matrix R of Example 5.8, with the exception of the signs in the second and third columns. Note, however, that the signs of the entries of A_2 and those in the columns of P in this example are consistent, as $PAP = A_2$, and a similar statement can be made concerning A_3 and R in Example 5.8.

5.12   Eigenvalues of a tridiagonal symmetric matrix. Sturm's theorem

As pointed out earlier, the methods of Givens and Householder only reduce a real symmetric matrix to tridiagonal form, so that the task of solving the eigenvalue problem still remains. In this section, we shall consider the problem of finding the eigenvalues of a tridiagonal matrix working with the characteristic polynomial. In particular, our interest is in a method not requiring the coefficients of the characteristic polynomial.

Let us consider the characteristic determinant associated with the matrix A_k, Eq. (5.144), and write it in the form

$$\det (A_k - \lambda I) = \begin{bmatrix} \alpha_1 - \lambda & \beta_2 & 0 & \cdots & 0 & 0 \\ \beta_2 & \alpha_2 - \lambda & \beta_3 & \cdots & 0 & 0 \\ 0 & \beta_3 & \alpha_3 - \lambda & \cdots & 0 & 0 \\ \cdots & \cdots & \cdots & \cdots & \cdots & \cdots \\ 0 & 0 & 0 & \cdots & \alpha_{n-1} - \lambda & \beta_n \\ 0 & 0 & 0 & \cdots & \beta_n & \alpha_n - \lambda \end{bmatrix}$$

$$\tag{5.172}$$

Denoting by $p_i(\lambda)$ the principal minor determinant of order i of the matrix $A_k - \lambda I$, one can show by induction that

$$p_1(\lambda) = \alpha_1 - \lambda$$
$$p_i(\lambda) = (\alpha_i - \lambda) p_{i-1}(\lambda) - \beta_i^2 p_{i-2}(\lambda), \qquad i = 2, 3, \ldots, n \tag{5.173}$$

where $p_0(\lambda) \equiv 1$. The characteristic equation is, of course,

$$p_n(\lambda) = 0 \qquad (5.174)$$

We shall assume that no β_i is zero. If any β_i is zero, then the determinant (5.172) reduces to the product of two determinants of lower order and the discussion to follow applies to each of these determinants.

Next, let us consider a given interval $a < \lambda < b$, where neither a nor b is a root of any of the polynomials (5.173). We shall be interested in the number of roots of $p_n(\lambda) = 0$ in the above interval, where the interval is denoted by (a, b). To this end, we propose to show that the sequence of polynomials $p_0(\lambda), p_1(\lambda), \ldots, p_n(\lambda)$ possesses the following properties:

i. $p_0(\lambda) \neq 0$.
ii. If $p_{i-1}(\mu) = 0$, then $p_i(\mu)$ and $p_{i-2}(\mu)$ are nonzero and of opposite signs.
iii. As λ passes through a zero of $p_n(\lambda)$, the quotient $p_n(\lambda)/p_{n-1}(\lambda)$ changes from positive to negative.

A sequence of polynomials possessing the above properties is known as a *Sturm sequence*.

The first property is obvious. In fact, $p_0(\lambda)$ is identically equal to unity. To show the second property, we consider Eqs. (5.173). Letting $\lambda = \mu$ be a zero of $p_{i-1}(\lambda)$, we obtain

$$p_i(\mu) = -\beta_i^2 p_{i-2}(\mu) \qquad (5.175)$$

Let us assume that $p_i(\lambda)$ has a zero at $\lambda = \mu$. Then, $p_{i-2}(\lambda)$ must also have a zero at $\lambda = \mu$. Using the recursion formulas (5.173), we conclude that $p_{i-3}(\lambda) = p_{i-4}(\lambda) = \cdots = p_0(\lambda) = 0$. The last equality, however, contradicts the fact that $p_0(\lambda) \equiv 1$. It follows that no two adjacent polynomials can have the same root, so that $p_i(\mu) \neq 0$ and $p_{i-2}(\mu) \neq 0$. In view of this, if we consider Eq. (5.175), then we conclude immediately that $p_i(\mu)$ and $p_{i-2}(\mu)$ must have opposite signs.

Let us assume that the roots $\lambda_1, \lambda_2, \ldots, \lambda_n$ of $p_n(\lambda)$ are ordered so that they satisfy $\lambda_1 > \lambda_2 > \cdots > \lambda_n$. Moreover, let us denote the roots of $p_{n-1}(\lambda)$ by $\gamma_1, \gamma_2, \ldots, \gamma_{n-1}$ and assume that they satisfy $\gamma_1 > \gamma_2 > \cdots > \gamma_{n-1}$. Then, by the inclusion principle (Sec. 4.5), we can write

$$\lambda_1 > \gamma_1 > \lambda_2 > \gamma_2 > \cdots > \gamma_{n-1} > \lambda_n \qquad (5.176)$$

The polynomials $p_n(\lambda)$ and $p_{n-1}(\lambda)$ can be plotted as functions of λ. Typical plots are shown in Fig. 5.2. The vertical lines through $\lambda_n, \gamma_{n-1}, \lambda_{n-1}, \ldots, \gamma_1, \lambda_1$ separate regions of opposite signs of $p_n(\lambda)/p_{n-1}(\lambda)$. The plots correspond to n even, but this fact has no

5.12 Eigenvalues of a tridiagonal symmetric matrix. Sturm's theorem

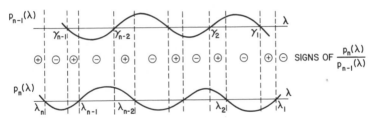

Figure 5.2

bearing on the signs of the ratio $p_n(\lambda)/p_{n-1}(\lambda)$. It is clear from the figure that, as λ crosses the roots $\lambda_n, \lambda_{n-1}, \ldots, \lambda_1$, the sign of $p_n(\lambda)/p_{n-1}(\lambda)$ changes from plus to minus.

Our interest is in locating the roots of $p_n(\lambda)$. This task is made easier by *Sturm's theorem* which can be stated as follows: *Let $p_0(\lambda), p_1(\lambda), \ldots, p_n(\lambda)$ be a Sturm sequence on the interval (a, b) and denote by $s(\mu)$ the number of sign changes in the consecutive sequence of numbers $p_0(\mu), p_1(\mu), \ldots, p_n(\mu)$. Then, the number of zeros of the polynomial $p_n(\lambda)$ in the interval (a, b) is equal to $s(b) - s(a)$.* If $p_i(\mu) = 0$ for some μ, then the sign of $p_i(\mu)$ is taken as the opposite of that of $p_{i-1}(\mu)$. This does not affect the count of sign changes in any way because the sign of $p_{i+1}(\mu)$ must be the opposite of that of $p_{i-1}(\mu)$, so that we get exactly one sign change from $p_{i-1}(\mu)$ to $p_{i+1}(\mu)$.

Sturm's theorem can be proved by induction. Let us assume that the number of sign changes $s(\mu)$ in the sequence $p_0(\mu), p_1(\mu), \ldots, p_n(\mu)$ is equal to the number of roots of $p_n(\lambda)$ corresponding to $\lambda < \mu$. As an example, Fig. 5.3 shows that there are three sign changes in the sequence $p_0(\mu), p_1(\mu), \ldots, p_6(\mu)$ and there are exactly three roots λ of $p_6(\lambda)$ lower in value than μ. As μ increases, $s(\mu)$ remains the same until μ crosses the next root of $p_n(\lambda)$, at which point $s(\mu)$ increases by one. Indeed, by the second property of the Sturm sequence, the net sign change remains the same as μ crosses a root of $p_{i-1}(\lambda)$ ($i = 2, 3, \ldots, n$). On the other hand, by the third property, there is one net sign change as μ crosses a root of $p_n(\mu)$. Hence, $s(\mu)$ increases by one each time μ crosses a root of $p_n(\lambda)$, which proves the theorem.

Sturm's theorem can be used to locate the roots of $p_n(\lambda)$. To this end, one selects an interval (a, b) and determines the number of zeros of $p_n(\lambda)$ in the interval by calculating $s(b) - s(a)$. To evaluate the numbers $p_0(\lambda), p_1(\lambda), \ldots, p_n(\lambda)$ for any given value of λ, it is not actually necessary to obtain the polynomials $p_i(\lambda)$ explicitly. Indeed, one can merely make use of the recursion formulas (5.173) to calculate these numbers. There remains the question as to how to select the interval (a, b). If the

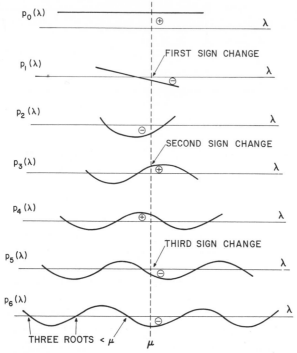

Figure 5.3

matrix A_k is positive definite, then it is known that all its eigenvalues are positive. In this case, one can simply take $a = 0$. To actually isolate a root, one can use the *method of bisection*. If $s(b) \neq s(a)$, we know that there is at least one root in the interval (a, b). Then, the next step is to evaluate $s(a + b/2)$ and check $s(a + b/2) - s(a)$ and $s(b) - s(a + b/2)$, which permits a further narrowing of the interval with one or more roots.

Sturm's theorem is generally used to locate roots only approximately. When the interval containing the root of interest has been narrowed sufficiently, one can use some other method, such as the Newton–Raphson method, to calculate the root to any desired accuracy.

A method for determining the eigenvectors of a matrix belonging to known eigenvalues was discussed in Sec. 5.4. In Sec. 5.16 we shall present a method which is particularly effective when the matrix is tridiagonal.

Example 5.10

Use Sturm's theorem to find the approximate location of the eigenvalues of the matrix A of Example 5.8. Then, calculate the eigenvalues by the Newton–Raphson method.

5.12 Eigenvalues of a tridiagonal symmetric matrix. Sturm's theorem

Before we can use Sturm's theorem, it is necessary to reduce the matrix A to tridiagonal form. This was done in Example 5.8, however, so that from Eq. (j) of Example 5.8 we can write

$$A_3 = \begin{bmatrix} 1 & 1.732051 & 0 & 0 \\ 1.732051 & 7.666667 & 1.247219 & 0 \\ 0 & 1.247219 & 0.976190 & 0.123718 \\ 0 & 0 & 0.123718 & 0.357143 \end{bmatrix} \quad \text{(a)}$$

The principal minor determinants of $A_3 - \lambda I$ are

$$\begin{aligned} p_1 &= 1 - \lambda \\ p_2 &= (7.666667 - \lambda)p_1 - 1.732051^2 \\ p_3 &= (0.976190 - \lambda)p_2 - 1.247219^2 p_1 \\ p_4 &= (0.357143 - \lambda)p_3 - 0.123718^2 p_2 \end{aligned} \quad \text{(b)}$$

Because the matrix A is positive definite, all its eigenvalues are positive. In addition, $\sum_{i=1}^{4} \lambda_i = \text{tr } A = 10$, so that we shall begin with $a = 0$ and $b = 10$. Hence let us construct the table:

Table 5.1

λ	p_1	p_2	p_3	p_4	$s(\lambda)$
10	−9	18	−148.428571	1431	4
5	−4	−13.666667	61.214291	−284.000016	3
7.5	−6.5	−4.083333	36.750000	−262.437495	3
8.75	−7.75	5.395833	−29.890622	250.785131	4
8.125	−7.125	0.265625	9.184430	−71.347406	3
8.4375	−7.4375	2.733073	−8.822860	71.287677	4
8.28125	−7.28125	1.474934	0.551908	−4.395954	3

At this point, we conclude that the open interval $8.28125 < \lambda < 8.4375$, containing the highest eigenvalue λ_4, has been narrowed sufficiently. Because an eigenvalue must render p_4 equal to zero, we further conclude that $\lambda = 8.28125$ is reasonably close to λ_4, so that an iteration beginning with this value will converge rapidly.

Before proceeding with the iteration, a few words about the bisection process involved in Table 5.1 appears in order. For $\lambda = 10$, we obtain four sign changes, which indicates that all four eigenvalues lie in the open interval $0 < \lambda < 10$, as expected. Bisecting that interval and observing that for $\lambda = 5$, there are only three sign changes, we conclude that one

161

eigenvalue must lie in the interval $5 < \lambda < 10$. Bisecting the interval $5 < \lambda < 10$ and noting that for $\lambda = 7.5$ there are still only three sign changes, we conclude that the eigenvalue must lie in the interval $7.5 < \lambda < 10$ and not in the interval $5 < \lambda < 7.5$. Hence, the procedure is continued by next bisecting the interval $7.5 < \lambda < 10$, etc.

The eigenvalue λ_4 is calculated by using the Newton–Raphson method. The iteration sequence is given by the formula (see, for example, R1, p. 332):

$$\lambda^{(i+1)} = \lambda^{(i)} - \frac{p_4(\lambda)}{p_4'(\lambda)}\bigg|_{\lambda = \lambda^{(i)}}, \qquad i = 0, 1, \ldots \qquad \text{(c)}$$

where

$$p_4'(\lambda) = -p_3(\lambda) - (0.357143 - \lambda)p_2(\lambda) - [(0.357143 - \lambda)(0.976190 - \lambda)$$
$$- 0.015306] \times [p_1(\lambda) + (7.666667 - \lambda)] + 1.555555(0.357143 - \lambda)$$
$$\text{(d)}$$

is the derivative of $p_4(\lambda)$ with respect to λ. Beginning with the value $\lambda^{(0)} = 8.28125$, we obtain

$$\lambda^{(1)} = 8.28125 - \frac{-4.395954}{449.933521} = 8.28125 + 0.009770 = 8.291020$$

$$\lambda^{(2)} = 8.291020 - \frac{0.073646}{459.232639} = 8.291020 - 0.000160 = 8.290860$$

$$\text{(e)}$$

The trend established by the correction $p_4(\lambda)/p_4'(\lambda)$ causes us to conclude that no additional iteration is necessary, so that the highest eigenvalue is

$$\lambda_4 = 8.290860 \qquad \text{(f)}$$

Because λ_4 is the only eigenvalue in the interval $5 < \lambda < 10$, the search for the three remaining eigenvalues is narrowed to the interval $0 < \lambda < 5$. The calculation of the remaining eigenvalues is left as an exercise to the reader.

5.13 The QR method

A class of methods for the computation of the eigenvalues and eigenvectors of a general matrix (not necessarily real and symmetric) consists of reducing the given matrix to upper triangular form by similarity transformations. We recall from Sec. 3.2 that every matrix A is similar to an upper triangular matrix T, with the diagonal elements of T being equal to the eigenvalues of A. It can be shown (W1, Sec. 1.47) that the transformation can be taken as unitary. Moreover, if the matrix A is real and its eigenvalues are real, then the transformation can be carried out using orthonormal matrices. If A is Hermitian, then the matrix T is actually diagonal.

One of the most attractive methods for the reduction of a general matrix A to a triangular form is an iteration technique known as the *QR algorithm*, developed independently by J. G. F. Francis (F2) and V. N. Kublanovskaya (K1). The QR method is closely related to the LR algorithm developed earlier by H. Rutishauser (R5). The method yields only the eigenvalues, so that the eigenvectors must be obtained by a different method. The QR algorithm has many desirable characteristics, one of the most important ones being stability. If the matrix A to be reduced to triangular form is full, then each iteration is extremely laborious. For this reason, the QR method becomes feasible only when used in conjunction with a simpler form for A, such as an upper Hessenberg form or a tridiagonal form. We recall from Secs. 5.10 and 5.11 that reduction to Hessenberg form or tridiagonal form can be effected by either Givens' method or Householder's method.

The iteration process is defined by the relations

$$A_s = Q_s R_s, \qquad A_{s+1} = R_s Q_s, \qquad s = 1, 2, \ldots \tag{5.177}$$

where $A_1 = A$ is the given matrix, Q_s is a unitary matrix, and R_s is an upper triangular matrix. Premultiplying the first of Eqs. (5.177) by Q_s^H and recalling that for unitary matrices $Q_s^H Q_s = I$, we obtain

$$R_s = Q_s^H A_s \tag{5.178}$$

so that, introducing Eq. (5.178) into the second of Eqs. (5.177), we can write

$$A_{s+1} = Q_s^H A_s Q_s, \qquad s = 1, 2, \ldots \tag{5.179}$$

from which we conclude that Eqs. (5.177) represent a unitary similarity transformation. The factorization is unique if the diagonal elements of R_s are taken to be real and positive (W1, p. 516). If A_s is real, then both Q_s and R_s are real, so that the transformation is actually orthonormal.

Equation (5.179) can be expanded in the form

$$A_{s+1} = Q_s^H A_s Q_s = Q_s^H Q_{s-1}^H A_{s-1} Q_{s-1} Q_s = \cdots$$
$$= Q_s^H \cdots Q_2^H Q_1^H A_1 Q_1 Q_2 \cdots Q_s \tag{5.180}$$

so that all A_s are unitarily similar to $A_1 = A$. Premultiplying Eq. (5.180)

5 Computational methods for the eigensolution

by $Q_1 Q_2 \cdots Q_s$, we obtain

$$Q_1 Q_2 \cdots Q_s A_{s+1} = A_1 Q_1 Q_2 \cdots Q_s \tag{5.181}$$

Next, let us introduce the notation

$$Q_1 Q_2 \cdots Q_s = P_s, \qquad R_s R_{s-1} \cdots R_1 = U_s \tag{5.182}$$

so that, using Eqs. (5.177) and (5.181), we can write

$$\begin{aligned} P_s U_s &= Q_1 Q_2 \cdots Q_{s-1} Q_s R_s R_{s-1} \cdots R_2 R_1 \\ &= Q_1 Q_2 \cdots Q_{s-1} A_s R_{s-1} \cdots R_2 R_1 \\ &= A_1 Q_1 Q_2 \cdots Q_{s-1} R_{s-1} \cdots R_2 R_1 = A_1 P_{s-1} U_{s-1} \end{aligned} \tag{5.183}$$

from which we conclude that

$$P_s U_s = A_1^s \tag{5.184}$$

or, $P_s U_s$ is the QR decomposition of A_1^s. This suggests the possibility that the QR method is related to the power method. Indeed, because U_s is an upper triangular matrix, if we denote its upper left corner element by $\rho_{11}^{(s)}$, then it is clear that

$$U_s \mathbf{e}_1 = \rho_{11}^{(s)} \mathbf{e}_1 \tag{5.185}$$

where $\mathbf{e}_1 = [1 \quad 0 \cdots 0]^T$ is a standard unit vector. Hence, postmultiplying Eq. (5.184) by $\mathbf{e}_1$ and considering Eq. (5.185), we obtain

$$\rho_{11}^{(s)} \mathbf{p}_1^{(s)} = A_1^s \mathbf{e}_1 \tag{5.186}$$

where $\mathbf{p}_1^{(s)}$ is the first column of P_s. If A_1 has an eigenvalue of largest modulus λ_n and $\mathbf{e}_1$ is not deficient in the corresponding eigenvector, then $\mathbf{p}_1^{(s)}$ approaches that eigenvector.

Under given conditions, the matrix A_s tends to an upper triangular form. The iteration always converges if the matrix A_1 is positive definite. Proof of convergence can be found in the treatise by Wilkinson (W1, Secs. 8.29–30). The iteration not only converges but the eigenvalues λ_i arrange themselves in proper order on the diagonal. For example, if $|\lambda_1| < |\lambda_2| < \cdots < |\lambda_n|$, then λ_1 will appear in the bottom right corner of A_s upon convergence. Convergence is not simultaneous but to one eigenvalue at a time.

Next, we shall discuss the actual computational process. To this end, we shall sacrifice some generality by assuming that A_1 is real. The algorithm requires the factorization of A_s into $Q_s R_s$, according to the first of Eqs. (5.177). An orthonormal matrix Q_s can be constructed by means of an orthonormalization of the columns of A_s by the Gram–Schmidt

164

process (Sec. 1.6). In practice, however, the process can yield a matrix Q_s that may differ from an orthonormal matrix to an arbitrary extent (W1, Sec. 4.55). A more satisfactory approach is to determine the matrix Q_s in the form of the product of $n-1$ rotation matrices. To this end, we write the first of Eqs. (5.177) in the form

$$Q_s^T A_s = R_s \tag{5.187}$$

If A_s has an upper Hessenberg form, then the $n-1$ rotations are used to annihilate its subdiagonal elements. This can be done either by Givens' method or by Householder's method. Because there is only one sub-diagonal element to be reduced to zero in each of the first $n-1$ columns of A_s, there is no advantage in using Householder's method and Givens' method is quite adequate. The rotations are in the planes $(1, 2)$, $(2, 3), \ldots, (n-1, n)$, in sequence, and the rotation matrices have the form

$$
\Theta_k =
\begin{matrix}
 & & & k & k+1 & & \\
\end{matrix}
\begin{bmatrix}
1 & 0 & \cdots & 0 & 0 & \cdots & 0 \\
0 & 1 & \cdots & 0 & 0 & \cdots & 0 \\
\cdots & \cdots & \cdots & \cdots & \cdots & \cdots & \cdots \\
0 & 0 & \cdots & \cos\theta_k & \sin\theta_k & \cdots & 0 \\
0 & 0 & \cdots & -\sin\theta_k & \cos\theta_k & \cdots & 0 \\
\cdots & \cdots & \cdots & \cdots & \cdots & \cdots & \cdots \\
0 & 0 & \cdots & 0 & 0 & \cdots & 1
\end{bmatrix}
\begin{matrix}
 \\ \\ \\ k, \\ k+1 \\ \\ \\
\end{matrix}
$$

$$k = 1, 2, \ldots, n-1 \tag{5.188}$$

where

$$\sin\theta_k = \frac{a_{k+1,k}^{(k-1)}}{\{[a_{k,k}^{(k-1)}]^2 + [a_{k+1,k}^{(k-1)}]^2\}^{1/2}}, \qquad \cos\theta_k = \frac{a_{k,k}^{(k-1)}}{\{[a_{k,k}^{(k-1)}]^2 + [a_{k+1,k}^{(k-1)}]^2\}^{1/2}}$$

$$k = 1, 2, \ldots, n-1 \tag{5.189}$$

in which θ_k denotes the rotation angle and $a_{k+1,k}^{(k-1)}$ denotes the element in the $k+1$ row and k column of the matrix $A_s^{(k-1)}$, where the matrix is obtained from A_s by means of the recursive formula

$$
\begin{aligned}
A_s^{(k)} &= \Theta_k A_s^{(k-1)}, \qquad k = 1, 2, \ldots, n-1 \\
A_s^{(0)} &= A_s, \qquad A_s^{(n-1)} = R_s
\end{aligned}
\tag{5.190}
$$

Moreover, the matrix Q_s^T can be written in the form

$$Q_s^T = \Theta_{n-1} \cdots \Theta_2 \Theta_1 \tag{5.191}$$

and note that Q_s^T is actually never calculated explicitly, although Q_s is calculated explicitly because it is needed to compute A_{s+1}.

There are $4n^2$ multiplications in one complete step of the QR method, where a complete step means the computation of A_{s+1} for a given matrix A_s, so that if convergence is achieved after s steps, then $4sn^2$ multiplications are required to compute all the eigenvalues of $A_1 = A$. In practice, however, fewer multiplications may be necessary, as demonstrated below.

Convergence can be accelerated greatly by using so-called *shifts in origin*. The QR algorithm incorporating shifts is defined by the relations

$$A_s - \mu_s I = Q_s R_s, \qquad A_{s+1} = R_s Q_s + \mu_s I, \qquad s = 1, 2, \ldots \qquad (5.192)$$

where the shifts μ_s are generally different for each iteration step. Then, by analogy with Eqs. (5.180) and (5.184), it can be shown that

$$A_{s+1} = (Q_1 Q_2 \cdots Q_s)^T A_1 Q_1 Q_2 \cdots Q_s \qquad (5.193)$$

and

$$(Q_1 Q_2 \cdots Q_s)(R_s \cdots R_2 R_1) = (A_1 - \mu_1 I)(A_1 - \mu_2 I) \cdots (A_1 - \mu_s I) \qquad (5.194)$$

Because convergence of the QR iteration without shifts is relatively slow, shifts in origin are essential in practice. The main question is as to the strategy to be employed in deciding on the values of the shifts.

For real matrices with real eigenvalues a good strategy is to take the shift μ_s equal to the eigenvalue of the matrix

$$\begin{bmatrix} a_{n-1,n-1}^{(s)} & a_{n-1,n}^{(s)} \\ a_{n,n-1}^{(s)} & a_{nn}^{(s)} \end{bmatrix} \qquad (5.195)$$

that is closest to $a_{nn}^{(s)}$ if the two eigenvalues are real, and equal to the real part of the eigenvalues if they are complex. Note that the matrix (5.195) is the 2×2 matrix in the bottom right corner of A_s. Complex shifts are to be avoided. The question remains as to the timing of the shift. Wilkinson (W1, Sec. 8.24) recommends using the shift when the shift gives some indication of convergence. He suggests accepting μ_s as a shift as soon as the criterion

$$|\mu_s/\mu_{s-1} - 1| < \tfrac{1}{2} \qquad (5.196)$$

is satisfied.

When one of the subdiagonal elements of A_s becomes negligibly small the problem can be deflated by reducing A_s to two Hessenberg matrices of smaller order. In particular, if convergence to the eigenvalue λ_1 of lowest modulus has been achieved, then the bottom row of A_s contains λ_1 as the only element. In this case the deflated matrix is of order $n - 1$ and is obtained by simply removing the last row and column from A_s. While the bottom right corner element of A_s converges to λ_1, the remaining diagonal elements approach the other eigenvalues of A_s. Hence, the new iteration cycle not only starts with a matrix of lower order but also with one that is much closer to the desired triangular form than the original matrix. Because of this, the effort necessary to iterate to a higher eigenvalue decreases rapidly with each computed eigenvalue.

A variant of the *QR* method uses lower triangular matrices instead of upper triangular matrices and is known as the *QL method* (W2, p. 227). When A_1 is real and symmetric, the Hessenberg form is actually tridiagonal and the *QR* method is particularly simple, requiring considerably fewer multiplications (W2).

Example 5.11

Use the QR method to obtain the eigenvalues of the tridiagonal matrix computed in Example 5.9.

The tridiagonal matrix of Example 5.9 is

$$A = A_1 = \begin{bmatrix} 1.000000 & -1.732051 & 0 & 0 \\ -1.732051 & 7.666667 & 1.247219 & 0 \\ 0 & 1.247219 & 0.976190 & -0.123718 \\ 0 & 0 & -0.123718 & 0.357143 \end{bmatrix} \quad \text{(a)}$$

Because the matrix (a) is symmetric, the QR method will actually diagonalize the matrix. Moreover, because the matrix is known to be positive definite, the eigenvalues are real and positive.

We shall annihilate the subdiagonal elements of A_1, as well as their symmetric counterparts, by Givens' method. To this end, we use Eqs. (5.188) and (5.189) to calculate the rotation matrices Θ_k. Letting $k = 1$ in Eqs. (5.189) and using the elements $a_{11}^{(0)}$ and $a_{21}^{(0)}$ of the matrix A_1, Eq. (a), we obtain

$$\sin \theta_1 = \frac{-1.732051}{\sqrt{1.000000^2 + (-1.732051)^2}} = -0.866025$$

$$\cos \theta_1 = \frac{1.000000}{\sqrt{1.000000^2 + (-1.732051)^2}} = 0.500000$$

so that, from Eq. (5.188), we obtain the first rotation matrix

$$\Theta_1 = \begin{bmatrix} 0.500000 & -0.866025 & 0 & 0 \\ 0.866025 & 0.500000 & 0 & 0 \\ 0 & 0 & 1 & 0 \\ 0 & 0 & 0 & 1 \end{bmatrix} \qquad \text{(b)}$$

Hence, Eq. (5.190) with $k = 1$ and $s = 1$ yields

$$A_1^{(1)} = \Theta_1 A_1^{(0)} = \begin{bmatrix} 2.000000 & -7.505553 & -1.080123 & 0 \\ 0 & 2.333333 & 0.623610 & 0 \\ 0 & 1.247219 & 0.976190 & -0.123718 \\ 0 & 0 & -0.123718 & 0.357143 \end{bmatrix} \qquad \text{(c)}$$

where $A_1^{(0)} = A_1$. Letting $k = 2$ in Eqs. (5.189) and using the elements $a_{22}^{(1)}$ and $a_{32}^{(1)}$ of the matrix $A_1^{(1)}$, we compute

$$\sin \theta_2 = \frac{1.247219}{\sqrt{2.333333^2 + 1.247219^2}} = 0.471405$$

$$\cos \theta_2 = \frac{2.333333}{\sqrt{2.333333^2 + 1.247219^2}} = 0.881917$$

so that, from Eqs. (5.188), we can write

$$\Theta_2 = \begin{bmatrix} 1 & 0 & 0 & 0 \\ 0 & 0.881917 & 0.471405 & 0 \\ 0 & -0.471405 & 0.881917 & 0 \\ 0 & 0 & 0 & 1 \end{bmatrix} \qquad \text{(d)}$$

Hence, premultiplying matrix (c) by matrix (d), we obtain

$$A_1^{(2)} = \Theta_2 A_1^{(1)} = \begin{bmatrix} 2.000000 & -7.505553 & -1.080123 & 0 \\ 0 & 2.645751 & 1.010153 & -0.058321 \\ 0 & 0 & 0.566947 & -0.109109 \\ 0 & 0 & -0.123718 & 0.357143 \end{bmatrix} \qquad \text{(e)}$$

Letting $k = 3$ in Eqs. (5.189) and using the elements $a_{33}^{(2)}$ and $a_{43}^{(2)}$ of the matrix $A_1^{(2)}$, we compute

$$\sin \theta_3 = \frac{-0.123718}{\sqrt{0.566947^2 + (-0.123718)^2}} = -0.213201$$

$$\cos \theta_3 = \frac{0.566947}{\sqrt{0.566947^2 + (-0.123718)^2}} = 0.977008$$

so that the third rotation matrix is

$$\Theta_3 = \begin{bmatrix} 1 & 0 & 0 & 0 \\ 0 & 1 & 0 & 0 \\ 0 & 0 & 0.977008 & -0.213201 \\ 0 & 0 & 0.213201 & 0.977008 \end{bmatrix} \tag{f}$$

Premultiplying matrix (e) by matrix (f), we have

$$R_1 = A_1^{(3)} = \Theta_3 A_1^{(2)} =$$

$$= \begin{bmatrix} 2.000000 & -7.505553 & -1.080123 & 0 \\ 0 & 2.645751 & 1.010153 & -0.058321 \\ 0 & 0 & 0.580288 & -0.182743 \\ 0 & 0 & 0 & 0.325669 \end{bmatrix} \tag{g}$$

and, using matrices (b), (d) and (f) in conjunction with Eq. (5.191) transposed, we obtain

$$Q_1 = \Theta_1^T \Theta_2^T \Theta_3^T = \begin{bmatrix} 0.500000 & 0.763763 & -0.398862 & -0.087039 \\ -0.866025 & 0.440959 & -0.230283 & -0.050252 \\ 0 & 0.471405 & 0.861640 & 0.188025 \\ 0 & 0 & -0.213201 & 0.977008 \end{bmatrix} \tag{h}$$

Finally, the first iteration step is concluded by introducing Eqs. (g) and (h) into the second of Eqs. (5.177) with $s = 1$ and computing

$$A_2 = R_1 Q_1 = \begin{bmatrix} 7.500000 & -2.291288 & 0 & 0 \\ -2.291288 & 1.642857 & 0.273551 & 0 \\ 0 & 0.273551 & 0.538961 & -0.069433 \\ 0 & 0 & -0.069433 & 0.318182 \end{bmatrix} \tag{i}$$

At this point, the question arises as to whether to proceed with the second iteration step using the matrix A_2 or to shift. To answer this question, we turn to criterion (5.196), which requires the solution of two

2×2 eigenvalue problems associated with the matrices in the bottom right corner of A_1 and A_2. The first eigenvalue problem has the form

$$\begin{vmatrix} 0.976190 - \mu & -0.123718 \\ -0.123718 & 0.357143 - \mu \end{vmatrix} = 0$$

and the eigenvalue closest to $a_{44}^{(1)} = 0.357143$ is $\mu_1 = 0.333333$. The second eigenvalue problem is

$$\begin{vmatrix} 0.538961 - \mu & -0.069433 \\ -0.069433 & 0.318182 - \mu \end{vmatrix} = 0$$

and the eigenvalue nearest $a_{44}^{(2)} = 0.318182$ is $\mu_2 = 0.298161$. Hence, criterion (5.196) yields

$$\left| \frac{\mu_2}{\mu_1} - 1 \right| = \left| \frac{0.298161}{0.333333} - 1 \right| = 0.105517 < \tfrac{1}{2}$$

so that a shift of origin at this time is well advised. Of course, the shift is $\mu_2 = 0.298161$, so that the new iteration cycle begins with the matrix

$$A_2 - \mu_2 I = \begin{bmatrix} 7.201839 & -2.291288 & 0 & 0 \\ -2.291288 & 1.344696 & 0.273551 & 0 \\ 0 & 0.273551 & 0.240800 & -0.069433 \\ 0 & 0 & -0.069433 & 0.020021 \end{bmatrix} \tag{j}$$

The iteration follows the pattern established in Eqs. (b)–(i). We shall dispense with the intermediate details and only list the results

$$R_2 = \begin{bmatrix} 7.557547 & -2.591130 & -0.082935 & 0 \\ 0 & 0.647370 & 0.338011 & -0.029339 \\ 0 & 0 & 0.128474 & -0.063768 \\ 0 & 0 & 0 & -0.077165 \end{bmatrix} \tag{k}$$

and

$$Q_2 = \begin{bmatrix} 0.952934 & 0.274782 & -0.107789 & -0.069236 \\ -0.303179 & 0.863677 & -0.338796 & -0.217618 \\ 0 & 0.422555 & 0.762572 & 0.489821 \\ 0 & 0 & -0.540441 & 0.841380 \end{bmatrix} \tag{l}$$

Hence, using the second of Eqs. (5.192) with $s = 2$, we obtain

$$A_3 = R_2 Q_2 + \mu_2 I = \begin{bmatrix} 8.285580 & -0.196269 & 0 & 0 \\ -0.196269 & 1.000108 & 0.054287 & 0 \\ 0 & 0.054287 & 0.430595 & 0.009277 \\ 0 & 0 & 0.009277 & 0.283719 \end{bmatrix}$$

(m)

and, comparing the values of the subdiagonal elements of A_3 with those of A_2, we conclude that the iteration converges rapidly.

Solving the eigenvalue problem associated with the 2×2 matrix in the bottom right corner of A_3, we obtain the new shift $\mu_3 = 0.283136$, so that the new iteration step will use the matrix

$$A_3 - \mu_3 I = \begin{bmatrix} 8.002444 & -0.196269 & 0 & 0 \\ -0.196269 & 0.716972 & 0.054287 & 0 \\ 0 & 0.054287 & 0.147459 & 0.009277 \\ 0 & 0 & 0.009277 & 0.000583 \end{bmatrix}$$

(n)

Omitting the details once again, we list

$$R_3 = \begin{bmatrix} 8.004850 & -0.213789 & -0.001331 & 0 \\ 0 & 0.714011 & 0.065325 & 0.000705 \\ 0 & 0 & 0.143207 & 0.009269 \\ 0 & 0 & 0 & -0.000017 \end{bmatrix}$$

(o)

and

$$Q_3 = \begin{bmatrix} 0.999699 & 0.024448 & -0.001860 & 0.000121 \\ -0.024519 & 0.996805 & -0.075848 & 0.004924 \\ 0 & 0.076031 & 0.995013 & -0.064593 \\ 0 & 0 & 0.064781 & 0.997901 \end{bmatrix}$$

(p)

so that

$$A_4 = R_3 Q_3 + \mu_3 I = \begin{bmatrix} 8.290819 & -0.017507 & 0 & 0 \\ -0.017507 & 0.999833 & 0.010888 & 0 \\ 0 & 0.010888 & 0.426229 & 0 \\ 0 & 0 & 0 & 0.283119 \end{bmatrix}$$

(r)

From the matrix A_4, it is clear that convergence has been achieved and that

$$\lambda_1 = 0.283119 \tag{s}$$

It is also clear that the iteration can continue now with the deflated matrix

171

A_4, i.e., the 3×3 matrix in the top left corner of Eq. (r).

Next, solving the eigenvalue problem associated with the 2×2 matrix in the bottom right corner of the deflated A_4 (the center 2×2 matrix of Eq. (r)), we obtain the shift $\mu_4 = 0.426022$. Keeping the same notation for the deflated matrices as for the original ones, we can write

$$A_4 - \mu_4 I = \begin{bmatrix} 7.864797 & -0.017507 & 0 \\ -0.017507 & 0.573811 & 0.010888 \\ 0 & 0.010888 & 0.000207 \end{bmatrix} \tag{t}$$

Triangularization of the above yields

$$R_4 = \begin{bmatrix} 7.864816 & -0.018784 & -0.000024 \\ 0 & 0.573875 & 0.010890 \\ 0 & 0 & 0 \end{bmatrix} \tag{u}$$

and

$$Q_4 = \begin{bmatrix} 0.999998 & 0.002226 & -0.000042 \\ -0.002226 & 0.999998 & -0.018973 \\ 0 & 0.018973 & 0.999820 \end{bmatrix} \tag{v}$$

so that

$$A_5 = R_4 Q_4 + \mu_4 I = \begin{bmatrix} 8.290860 & -0.001277 & 0 \\ -0.001277 & 1.000000 & 0 \\ 0 & 0 & 0.426022 \end{bmatrix} \tag{w}$$

Hence, the second eigenvalue is

$$\lambda_2 = 0.426022 \tag{x}$$

To obtain the two highest eigenvalues, we can deflate again. It turns out, however, that the only subdiagonal term remaining is too small to matter, so that the two highest eigenvalues are simply

$$\lambda_3 = 1.000000, \qquad \lambda_4 = 8.290860 \tag{y}$$

and note that the value of λ_4 is exactly the one obtained in Example 5.10.

5.14 The Cholesky algorithm

When the matrix $A = A_1$ is real and symmetric the QR method preserves symmetry. Indeed, in this case Eq. (5.179) with $s = 1$ becomes

$$A_2 = Q_1^T A_1 Q_1 \tag{5.197}$$

Moreover, as shown in Sec. 5.8, if A_1 is not only real and symmetric but also positive definite, then it is always possible to use the Cholesky decomposition and write

$$A_1 = L_1 L_1^T \tag{5.198}$$

Next, let us use the analogy with the QR method and define the matrix

$$\tilde{A}_2 = L_1^T L_1 = L_1^{-1} A_1 L_1 = L_1^T A_1 L_1^{-T} \tag{5.199}$$

where $L_1^{-T} = (L_1^T)^{-1}$. Clearly $\tilde{A}_2$ is symmetric. It is also positive definite because it is similar to A_1. Extending the analogy to $s > 1$, we can write

$$\begin{aligned}
\tilde{A}_s &= L_{s-1}^T L_{s-1} = L_{s-1}^{-1} \cdots L_2^{-1} L_1^{-1} A_1 L_1 L_2 \cdots L_{s-1} \\
&= L_{s-1}^T \cdots L_2^T L_1^T A_1 L_1^{-T} L_2^{-T} \cdots L_{s-1}^{-T}
\end{aligned} \tag{5.200}$$

and

$$L_1 L_2 \cdots L_s L_s^T \cdots L_2^T L_1^T = (L_1 L_2 \cdots L_s)(L_1 L_2 \cdots L_s)^T = A_1^s \tag{5.201}$$

where the latter represents the Cholesky decomposition of A_1^s. As s increases, $\tilde{A}_s$ tends to a diagonal matrix, so that the Cholesky decomposition can be used as the basis for an iteration scheme for the diagonalization of the matrix A_1. We shall refer to this scheme as the *Cholesky algorithm*. The Cholesky decomposition is known to possess great numerical stability, so that the same can be said about the Cholesky algorithm. As it turns out, the Cholesky algorithm is intimately related to the QR method.

Combining Eqs. (5.182) and (5.184), we obtain

$$(Q_1 Q_2 \cdots Q_s)(R_s \cdots R_2 R_1) = A_1^s \tag{5.202}$$

from which it follows that

$$(R_s \cdots R_2 R_1)^T (Q_1 Q_2 \cdots Q_s)^T (Q_1 Q_2 \cdots Q_s) (R_s \cdots R_2 R_1) = (A_1^s)^T A_1^s \tag{5.203}$$

and, recalling that the matrices $Q_1, Q_2, \ldots, Q_s$ are orthonormal and that A_1 is symmetric, we can reduce Eq. (5.203) to

$$(R_s \cdots R_2 R_1)^T (R_s \cdots R_2 R_1) = A_1^{2s} \tag{5.204}$$

which represents a Cholesky decomposition of A_1^{2s}. Moreover, if we replace s by $2s$ in Eq. (5.201), we obtain simply

$$(L_1 L_2 \cdots L_{2s})(L_1 L_2 \cdots L_{2s})^T = A_1^{2s} \tag{5.205}$$

which represents another Cholesky decomposition of A_1^{2s}. Because the two decompositions must be the same, we conclude that

$$L_1 L_2 \cdots L_{2s} = (R_s \cdots R_2 R_1)^T \qquad (5.206)$$

Next, let us replace s by $2s + 1$ in Eq. (5.200) and obtain

$$\tilde{A}_{2s+1} = (L_1 L_2 \cdots L_{2s})^T A_1 [(L_1 L_2 \cdots L_{2s})^T]^{-1} \qquad (5.207)$$

But, from Eqs. (5.177), we have

$$A_{s+1} = R_s A_s R_s^{-1} = \cdots = (R_s \cdots R_2 R_1) A_1 (R_s \cdots R_2 R_1)^{-1} \qquad (5.208)$$

so that, in view of Eqs. (5.206), we are led to the conclusion that

$$\tilde{A}_{2s+1} = A_{s+1} \qquad (5.209)$$

or, the $(2s + 1)$th matrix of the Cholesky algorithm is equal to the $(s + 1)$th matrix of the QR method. This implies that for every iteration step of the QR method one must perform two iteration steps for the Cholesky algorithm. But, because the amount of computation involved in one Cholesky algorithm step is considerably lower than in one step of the QR algorith, the Cholesky algorithm can produce a solution with much less effort.

The foregoing statements do not take into account shifts of origin. As might be expected, convergence of the Cholesky iteration can also be accelerated by means of shifts of origin. Because the Cholesky algorithm provides twice as many opportunities for shifting than the QR method, the convergence of the Cholesky algorithm can be accelerated to a larger extent than that of the QR method. In choosing the shifts, however, one must be careful not to destroy the positive definiteness of the matrix.

If the matrix A is tridiagonal, then the formula given in Sec. 5.8 can be simplified. Indeed, in this case all the elements of the matrices $L_1, L_2, \ldots, L_s$ are zero, with the exception of the main diagonal and the first subdiagonal elements. Hence, Eqs. (5.110) reduce to the simpler form

$$\begin{aligned}
l_{ii}^{(s)} &= [a_{ii}^{(s)} - (l_{i,i-1}^{(s)})^2]^{1/2}, \qquad i = 1, 2, \ldots, n \\
l_{ki}^{(s)} &= a_{ik}^{(s)} / l_{ii}^{(s)}, \qquad k = i+1; \qquad i = 1, 2, \ldots, n
\end{aligned} \qquad (5.210)$$

where $l_{10}^{(s)} = 0$.

Example 5.12

Use the Cholesky algorithm to calculate the matrix $\tilde{A}_3$ corresponding to the matrix A of Example 5.11 and verify that $\tilde{A}_3$ is equal to the matrix A_2 obtained by the QR method.

Using Eqs. (5.210), the nonzero elements of the matrix L_1 are

$l_{11}^{(1)} = \sqrt{a_{11}^{(1)}} = 1.000000$

$l_{21}^{(1)} = a_{12}^{(1)}/l_{11}^{(1)} = -1.732051$

$l_{22}^{(1)} = [a_{22}^{(1)} - (l_{21}^{(1)})^2]^{1/2} = [7.666667 - (-1.732051)^2]^{1/2} = 2.160247$

$l_{32}^{(1)} = a_{23}^{(1)}/l_{22}^{(1)} = 1.247219/2.160247 = 0.577350$ \hfill (a)

$l_{33}^{(1)} = [a_{33}^{(1)} - (l_{32}^{(1)})^2]^{1/2} = (0.976190 - 0.577350^2)^{1/2} = 0.801784$

$l_{43}^{(1)} = a_{34}^{(1)}/l_{33}^{(1)} = -0.123718/0.801784 = -0.154303$

$l_{44}^{(1)} = [a_{44}^{(1)} - (l_{43}^{(1)})^2]^{1/2} = [0.357143 - (-0.154303)^2]^{1/2} = 0.577350$

so that the matrix L_1 has the form

$$L_1 = \begin{bmatrix} 1.000000 & 0 & 0 & 0 \\ -1.732051 & 2.160247 & 0 & 0 \\ 0 & 0.577350 & 0.801784 & 0 \\ 0 & 0 & -0.154303 & 0.577350 \end{bmatrix} \qquad (b)$$

Hence, using Eq. (5.199), we obtain

$$\tilde{A}_2 = L_1^T L_1 = \begin{bmatrix} 4.000000 & -3.741657 & 0 & 0 \\ -3.741657 & 5.000000 & 0.462190 & 0 \\ 0 & 0.462190 & 0.666667 & -0.089087 \\ 0 & 0 & -0.089087 & 0.333333 \end{bmatrix}$$

$$(c)$$

Next, we compute

$l_{11}^{(2)} = \sqrt{a_{11}^{(2)}} = \sqrt{4.000000} = 2.000000$

$l_{21}^{(2)} = a_{12}^{(2)}/l_{11}^{(2)} = -3.741657/2.000000 = -1.870829$

$l_{22}^{(2)} = [a_{22}^{(2)} - (l_{21}^{(2)})^2]^{1/2} = [5.000000 - (-1.870829)^2]^{1/2} = 1.224745$

$l_{32}^{(2)} = a_{23}^{(2)}/l_{22}^{(2)} = 0.462190/1.224745 = 0.377965$ \hfill (d)

$l_{33}^{(2)} = [a_{33}^{(2)} - (l_{32}^{(2)})^2]^{1/2} = (0.666667 - 0.377965^2)^{1/2} = 0.723747$

$l_{43}^{(2)} = a_{34}^{(2)}/l_{33}^{(2)} = -0.089087/0.723747 = -0.123092$

$l_{44}^{(2)} = [a_{44}^{(2)} - (l_{43}^{(2)})^2]^{1/2} = [0.333333 - (-0.123092)^2]^{1/2} = 0.564076$

so that the matrix L_2 can be exhibited in the form

$$L_2 = \begin{bmatrix} 2.000000 & 0 & 0 & 0 \\ -1.870829 & 1.224745 & 0 & 0 \\ 0 & 0.377965 & 0.723747 & 0 \\ 0 & 0 & -0.123092 & 0.564076 \end{bmatrix} \qquad (e)$$

175

Letting $s = 3$ in Eq. (5.200) and using Eq. (e), we obtain

$$\tilde{A}_3 = L_2^T L_2 = \begin{bmatrix} 7.500000 & -2.291288 & 0 & 0 \\ -2.291288 & 1.642857 & 0.273551 & 0 \\ 0 & 0.273551 & 0.538961 & -0.069433 \\ 0 & 0 & -0.069433 & 0.318182 \end{bmatrix}$$

(f)

which, as expected, is identical to the matrix A_2 obtained in Example 5.11 by the QR method.

To complete the diagonalization of A, shifts of origin are a virtual necessity. The shifts are likely to be different than those of the QR method because the intermediate steps in the Cholesky algorithm provides an added opportunity for shifting. Indeed, convergence could have been accelerated in the present example by performing a shift of origin on $\tilde{A}_2$. To determine the shifts, some guidance can be obtained from Gerschgorin's theorem (Sec. 4.B), provided the off-diagonal elements of $\tilde{A}_s$ are relatively small.

5.15   Eigenvectors of a tridiagonal matrix

Let us consider a tridiagonal matrix A_k and write the equation

$$(A_k - \lambda I)\mathbf{x} = \mathbf{0} \tag{5.211}$$

Recalling that A_k has the form (5.144), Eq. (5.211) can be written explicitly as

$$\begin{aligned}
(\alpha_1 - \lambda)x_1 + \beta_2 x_2 &= 0 \\
\beta_2 x_1 + (\alpha_2 - \lambda)x_2 + \beta_3 x_3 &= 0 \\
&\cdots\cdots\cdots\cdots\cdots\cdots\cdots \\
\beta_n x_{n-1} + (\alpha_n - \lambda)x_n &= 0
\end{aligned} \tag{5.212}$$

If λ is a given eigenvalue of A_k, then one may expect the solution $x_1, x_2, \ldots, x_n$ of Eqs. (5.212) to represent the components of the eigenvector $\mathbf{x}$ belonging to λ. Letting $x_1 = 1$ arbitrarily and using the first $n-1$ of Eqs. (5.212), we obtain

$$x_r = (-1)^{r-1} p_{r-1}(\lambda)/\beta_2 \beta_3 \cdots \beta_r, \qquad r = 2, 3, \ldots, n \tag{5.213}$$

and we note that the last of Eqs. (5.213) is solved automatically. Indeed, introducing Eq. (5.213) into the last of Eqs. (5.212) and using Eqs. (5.173) with $i = n$, we can write

$$\frac{(-1)^{n-1}\beta_n}{\beta_2\beta_3\cdots\beta_{n-1}}\, p_{n-2}(\lambda) + \frac{(-1)^n(\alpha_n-\lambda)}{\beta_2\beta_3\cdots\beta_n}\, p_{n-1}(\lambda) = \frac{(-1)^n}{\beta_2\beta_3\cdots\beta_{n-1}}\, p_n(\lambda) = 0$$

(5.214)

because λ is a root of $p_n(\lambda) = 0$.

Contrary to expectations, however, although λ may be a very good approximation of an eigenvalue, the eigenvector x defined by Eqs. (5.213) may contain large errors (W1, Sec. 5.49). Hence, a different procedure for extracting the eigenvector of a tridiagonal matrix is highly desirable.

5.16   Inverse iteration

A method ideally suited for the computation of the eigenvectors of a tridiagonal matrix if the eigenvalues are known is the inverse iteration. Let us consider the system of nonhomogeneous equations

$$(A - \lambda I)x = c \tag{5.215}$$

where c is an arbitrary normalized vector. Next, let us assume that the vector c can be written as the linear combination

$$c = \sum_{i=1}^{n} \gamma_i x_i \tag{5.216}$$

where the vectors x_i $(i = 1, 2, \ldots, n)$ represent the set of orthonormal eigenvectors of A and γ_i are known constants. Then, the solution of Eq. (5.215) can be written in the form of the linear combination

$$x = \sum_{i=1}^{n} \delta_i x_i \tag{5.217}$$

where the coefficients δ_i need yet to be determined. Introducing Eqs. (5.216) and (5.217) into Eq. (5.215), we obtain

$$(A - \lambda I) \sum_{i=1}^{n} \delta_i x_i = \sum_{i=1}^{n} \delta_i A x_i - \lambda \sum_{i=1}^{n} \delta_i x_i = \sum_{i=1}^{n} (\lambda_i - \lambda)\delta_i x_i = \sum_{i=1}^{n} \gamma_i x_i$$

(5.218)

Multiplying Eq. (5.218) on the left by x_j^T and considering the orthonormality conditions, we obtain

$$\delta_i = \frac{\gamma_i}{\lambda_i - \lambda}, \qquad i = 1, 2, \ldots, n \tag{5.219}$$

so that the solution of Eq. (5.215) is

$$\mathbf{x} = \sum_{i=1}^{n} \frac{\gamma_i}{\lambda_i - \lambda} \mathbf{x}_i \qquad (5.220)$$

From Eqs. (5.216) and (5.220), we conclude that if λ is very close to a given eigenvalue, say λ_r, then the representation of the eigenvector $\mathbf{x}_r$ belonging to λ_r is much stronger in $\mathbf{x}$ than in $\mathbf{c}$. This suggests the following iteration scheme

$$(A - \lambda I)\mathbf{v}_p = \mathbf{v}_{p-1}, \qquad p = 1, 2, \ldots \qquad (5.221)$$

where $\mathbf{v}_0 = \mathbf{c}$ is an arbitrary initial trial vector. It follows that the pth iterated vector is

$$\mathbf{v}_p = \sum_{i=1}^{n} \frac{\gamma_i}{(\lambda_i - \lambda)^p} \mathbf{x}_i \qquad (5.222)$$

Now, if λ is close to the eigenvalue λ_r, then we can write

$$\mathbf{v}_p = \frac{1}{(\lambda_r - \lambda)^p} \left[\gamma_r \mathbf{x}_r + \sum_{\substack{i=1 \\ i \neq r}}^{n} \gamma_i \left(\frac{\lambda_r - \lambda}{\lambda_i - \lambda} \right)^p \mathbf{x}_i \right] \qquad (5.223)$$

and if, furthermore, we assume that $\gamma_r \neq 0$, for a sufficiently large p, we have

$$\mathbf{v}_p = \frac{1}{(\lambda_r - \lambda)^p} (\gamma_r \mathbf{x}_r + \boldsymbol{\varepsilon}_p) \qquad (5.224)$$

where $\boldsymbol{\varepsilon}_p$ is a vector with very small components compared to $\gamma_r \mathbf{x}_r$. Hence, in the limit, as $p \to \infty$, we obtain

$$\lim_{p \to \infty} \mathbf{v}_p = \frac{1}{(\lambda_r - \lambda)^p} \gamma_r \mathbf{x}_r \qquad (5.225)$$

so that, for λ close to λ_r, the process iterates to the eigenvector $\mathbf{x}_r$. This is true even in the case in which the initial trial vector $\mathbf{v}_0$ is highly deficient in $\mathbf{x}_r$. This process is known as *inverse iteration*.

   In obtaining the iterated vectors $\mathbf{v}_1, \mathbf{v}_2, \ldots$ by means of Eq. (5.221), one may be tempted to think that it is necessary to invert the matrix $A - \lambda I$, which would present a serious problem because $A - \lambda I$ is close to being singular. No inversion of $A - \lambda I$ is required, however, and in fact the vectors $\mathbf{v}_1, \mathbf{v}_2, \ldots$ are obtained by regarding Eq. (5.221) as representing a set of nonhomogeneous simultaneous equations, in which $\mathbf{v}_{p-1}$ plays the role of a known vector and $\mathbf{v}_p$ plays the role of a vector of unknowns. Solutions of nonhomogeneous equations can be obtained by the Gauss elimination method. Indeed, as shown in Secs. 5.2 and 5.3, Eq. (5.215) can be first reduced to the triangular form

$$U\mathbf{x} = \mathbf{b} \qquad (5.226)$$

where

$$U = P(A - \lambda I) \qquad (5.227)$$

is an upper triangular matrix and

$$\mathbf{b} = P\mathbf{c} \qquad (5.228)$$

The solution of Eq. (5.226) can be obtained by back substitution.

If λ is sufficiently close to a given eigenvalue λ_r, then convergence to $\mathbf{x}_r$ is extremely fast, sometimes in only one or two steps.

It should be pointed out that nowhere in the above development was the assumption made that A is tridiagonal. Hence, the method is valid for arbitrary matrices, not necessarily tridiagonal. However, the method is particularly efficient for tridiagonal matrices.

Example 5.13

Use the inverse iteration method to compute the eigenvector $\mathbf{x}_1$ of the tridiagonal matrix of Example 5.11 belonging to the eigenvalue of lowest modulus. Then, compute the actual eigenvector $\mathbf{u}_1$, i.e., the eigenvector corresponding to the original eigenvalue problem, as defined by Eqs. (a) and (b) of Example 5.8.

From Eq. (a) and (s) of Example 5.11, we obtain

$$A - \lambda_1 I = \begin{bmatrix} 0.716881 & -1.732051 & 0 & 0 \\ -1.732051 & 7.383548 & 1.247219 & 0 \\ 0 & 1.247219 & 0.693071 & -0.123718 \\ 0 & 0 & -0.123718 & 0.074024 \end{bmatrix} \qquad (a)$$

First, we wish to triangularize the matrix (a). To this end, we use the procedure developed in Sec. 5.3. To annihilate the element $a_{21} = -1.732051$ of the matrix $A - \lambda_1' I$, we use Eq. (5.20) and construct the transformation matrix

$$P_1 = \begin{bmatrix} 1 & 0 & 0 & 0 \\ \dfrac{1.732051}{0.716881} & 1 & 0 & 0 \\ 0 & 0 & 1 & 0 \\ 0 & 0 & 0 & 1 \end{bmatrix} \qquad (b)$$

Premultiplying the matrix $A - \lambda_1 I$ by P_1, we have

$$P_1(A - \lambda_1 I) = \begin{bmatrix} 0.716881 & -1.732051 & 0 & 0 \\ 0 & 3.198753 & 1.247219 & 0 \\ 0 & 1.247219 & 0.693071 & -0.123718 \\ 0 & 0 & -0.123718 & 0.074024 \end{bmatrix}$$

(c)

so that the next transformation matrix is simply

$$P_2 = \begin{bmatrix} 1 & 0 & 0 & 0 \\ 0 & 1 & 0 & 0 \\ 0 & -\dfrac{1.247219}{3.198753} & 1 & 0 \\ 0 & 0 & 0 & 1 \end{bmatrix}$$

(d)

Hence, the next step toward triangularization yields

$$P_2 P_1(A - \lambda_1 I) = \begin{bmatrix} 0.716881 & -1.732051 & 0 & 0 \\ 0 & 3.198753 & 1.247219 & 0 \\ 0 & 0 & 0.206743 & -0.123718 \\ 0 & 0 & -0.123718 & 0.074024 \end{bmatrix}$$

(e)

The third and final transformation matrix is

$$P_3 = \begin{bmatrix} 1 & 0 & 0 & 0 \\ 0 & 1 & 0 & 0 \\ 0 & 0 & 1 & 0 \\ 0 & 0 & \dfrac{0.123718}{0.206743} & 1 \end{bmatrix}$$

(f)

so that the desired upper triangular form is

$$U = P_3 P_2 P_1(A - \lambda_1 I) =$$

$$= \begin{bmatrix} 0.716881 & -1.732051 & 0 & 0 \\ 0 & 3.198753 & 1.247219 & 0 \\ 0 & 0 & 0.206743 & -0.123718 \\ 0 & 0 & 0 & -0.000009 \end{bmatrix}$$

(g)

To begin the iteration process, let us use

$$\mathbf{v}_0 = 10^{-5}[1 \quad 1 \quad 1 \quad 1]^T \tag{h}$$

which yields the set of simultaneous equations

$$
\begin{aligned}
0.716881x_1 - 1.732051x_2 &= 10^{-5} \\
3.198753x_2 + 1.247219x_3 &= 10^{-5} \\
0.206743x_3 - 0.123718x_4 &= 10^{-5} \\
-0.000009x_4 &= 10^{-5}
\end{aligned} \tag{i}
$$

Using back substitution, the solution of Eqs. (i) is

$$x_4 = -\frac{10^{-5}}{0.000009} = -10/9$$

$$x_3 = \frac{1}{0.206743}[10^{-5} + 0.123718(-10/9)] = -0.664857$$

$$x_2 = \frac{1}{3.198753}[10^{-5} - 1.247219(-0.664857)] = 0.259236$$

$$x_1 = \frac{1}{0.716881}(10^{-5} + 1.732051 \times 0.259236) = 0.626352 \tag{j}$$

Normalizing so as to render the last component equal to 10^{-5}, the new trial vector is

$$\mathbf{v}_1 = 10^{-5}[-0.563717 \quad -0.233312 \quad 0.598371 \quad 1]^T \tag{k}$$

so that the new set of simultaneous equations is

$$
\begin{aligned}
0.716881x_1 - 1.732051x_2 &= -0.563717 \times 10^{-5} \\
3.198753x_2 + 1.247219x_3 &= -0.233312 \times 10^{-5} \\
0.206743x_3 - 0.123718x_4 &= 0.598371 \times 10^{-5} \\
-0.000009x_4 &= 10^{-5}
\end{aligned} \tag{l}
$$

Using back substitution once again, we obtain the solution

$$
\begin{aligned}
x_4 &= -10/9, & x_3 &= -0.664876 \\
x_2 &= 0.259240, & x_1 &= 0.626339
\end{aligned} \tag{m}
$$

so that the new normalized vector is

$$\mathbf{v}_2 = 10^{-5}[-0.563705 \quad -0.233316 \quad 0.598388 \quad 1]^T \tag{n}$$

It turns out that one more iteration yields the same result, so that we shall accept the vector

$$\mathbf{x}_1 = [-0.563705 \quad -0.233316 \quad 0.598388 \quad 1]^T \tag{o}$$

as the eigenvector of the tridiagonal matrix A of Example 5.11 belonging to the eigenvalue $\lambda_1 = 0.283119$.

We observe that the coefficient of x_4 in the last of Eqs. (i) is very small, but that this fact did not hurt the solution. On the contrary, it made for fast convergence. Indeed, from the last three of Eqs. (j), we see that the smaller the coefficient of x_4 the faster the convergence. Clearly, in this example, it was sufficiently small to produce rapid convergence.

As shown in Sec. 5.11, the relation between the eigenvectors $\mathbf{u}_i$ of the original matrix and the eigenvectors $\mathbf{x}_i$ of the tridiagonal matrix is given by

$$\mathbf{u}_i = P\mathbf{x}_i, \qquad i = 1, 2, \ldots, n \tag{p}$$

where P is an orthonormal matrix (not to be confused with the transformation matrices P_1, P_2 and P_3 of the present example). This matrix was calculated in Example 5.9 for the Householder's method and has the form

$$P = \begin{bmatrix} 1 & 0 & 0 & 0 \\ 0 & -0.577350 & 0.771517 & 0.267261 \\ 0 & -0.577350 & -0.154303 & -0.801784 \\ 0 & -0.577350 & -0.617213 & 0.534522 \end{bmatrix} \tag{r}$$

Hence, using Eqs. (o) and (r), we compute

$$\mathbf{u}_1 = P\mathbf{x}_1 = [-0.563705 \quad 0.863633 \quad -0.759412 \quad 0.299894]^T \tag{s}$$

or, if the vector is normalized so that $\mathbf{u}_1^T \mathbf{u}_1 = 1$, we obtain

$$\mathbf{u}_1 = [-0.428544 \quad 0.656557 \quad -0.577326 \quad 0.227988]^T \tag{t}$$

We observe that the vector has three sign changes. This can be explained by the fact that the vector actually represents the fourth mode of vibration. Moreover, λ_1 is inversely proportional to ω_4^2, where $\omega_4 = 1.879384$ rad/s is the fourth natural frequency.

6

Response of discrete systems

6.1 Introduction

The problem of determining the response of discrete systems was introduced in Ch. 2 in the context of the free vibration of multi-degree-of-
freedom systems. In particular, it was shown in Sec. 2.5 that the response
to any initial excitation can be represented as a linear combination of the
system eigensolutions. This provided the motivation for the following
three chapters, in which the solution of the eigenvalue problem was
studied in detail. As shown in this chapter, the approach of Sec. 2.5 is
merely a special case of a more general approach.

The general response of a discrete system, i.e., the response to both
initial and external excitations, is governed by a set of simultaneous
ordinary differential equations. The set of simultaneous (coupled) equations can be rendered independent (uncoupled) by means of a linear
transformation that can be represented by the matrix of the system
eigenvectors. Because these eigenvectors are commonly referred to as
modal vectors, the transformation matrix is called the modal matrix and
the decoupling procedure itself is known as *modal analysis*. The independent ordinary differential equations resemble those of low-order systems
and can be solved with relative ease. The important feature of these
low-order systems is that they are described in terms of a single dependent variable.

In view of the above, this chapter begins with a study of efficient
techniques for solving low-order ordinary differential equations. To this
end, the impulse response and the convolution integral prove particularly
valuable, where the latter permits the derivation of the response to
external excitation in closed form. Quite often, however, it is desirable to
produce an explicit solution on a digital computer, in which case one must
discretize the system in time, giving rise to so-called *discrete-time systems*.
This subject receives special attention in this chapter. It should be pointed
out that the term 'discrete' used on different previous occasions referred
to space, whereas here it refers to time. Finally, various types of modal

analyses, especially tailored to vibrational systems ranging from un-damped nongyroscopic systems to general dynamical systems, are presented.

6.2 Linear systems. The superposition principle

Let us assume that the behavior of a given system can be described by the differential equation

$$G[x(t)] = F(t) \tag{6.1}$$

where $x(t)$ denotes the *system response* and $F(t)$ the *excitation*. Ordinarily, $x(t)$ represents a displacement and $F(t)$ an external force. The symbol G designates a differential operator, so that the left side of Eq. (6.1) represents a differential expression obtained as the result of G operating on $x(t)$. The operator G involves derivatives with respect to time, where the highest order derivative defines the order of the operator. It also involves certain system parameters describing the system physical properties, such as mass, damping and stiffness, where the parameters appear in the form of the coefficients of the differential expression. In effect, the operator G embodies all the system dynamic characteristics.

The relation between excitation and response defined by Eq. (6.1) can be represented schematically by the diagram depicted in Fig. 6.1. In the language of system analysis the excitation is referred to as *input* and the response as *output*. The system itself is often referred to as the 'black box', so that G gives the mathematical description of the black box. The object of this section is to give a precise definition of linearity and to focus attention on the far-reaching consequences of linearity in the determination of the system response.

Let us assume that two distinct excitations $F_1(t)$ and $F_2(t)$ act upon the system and denote by $x_1(t)$ and $x_2(t)$ the respective responses, so that

$$G[x_1(t)] = F_1(t), \qquad G[x_2(t)] = F_2(t) \tag{6.2}$$

Next, let us consider an excitation $F_3(t)$ consisting of a linear combination of $F_1(t)$ and $F_2(t)$ of the form

$$F_3(t) = c_1 F_1(t) + c_2 F_2(t) \tag{6.3}$$

Figure 6.1

where c_1 and c_2 are given constant scalars. If the response $x_3(t)$ to the excitation $F_3(t)$ is a linear combination of $x_1(t)$ and $x_2(t)$ of the same form, i.e., if

$$x_3(t) = c_1 x_1(t) + c_2 x_2(t) \tag{6.4}$$

the *system is linear*; otherwise the system is nonlinear. The fact that the system is linear implies that *the operator G is linear*, which can be expressed mathematically by

$$G[c_1 x_1(t) + c_2 x_2(t)] = c_1 G[x_1(t)] + c_2 G[x_2(t)] \tag{6.5}$$

A linear system can be easily recognized by the fact that G contains the dependent variable $x(t)$ and its time derivatives to the first and zero powers only.

Equations (6.3) and (6.4) indicate, that for a linear system, solutions to various excitations can be obtained separately and then added up linearly to obtain the combined response. This is the essence of the so-called *principle of superposition*, a very powerful principle that *applies to linear systems alone*. Indeed, the principle has no counterpart for nonlinear systems, which explains why the linear system theory is so much better developed than the nonlinear system theory. It is the superposition principle which permits us to obtain the response to external excitations and to initial excitations separately and then to combine them linearly.

As an example of a linear system, let us consider the mass-damper-spring system shown in Fig. 6.2a, where the damper and the spring are assumed to be linear, i.e., the damper exhibits a resisting force proportional to the velocity and the spring gives rise to a restoring force proportional to the displacement. The constants of proportionality are the damping coefficient c and the spring constant k, respectively. Figure 6.2b shows the free-body diagram for the system. Using Newton's second law and rearranging, we obtain the differential equation of motion

$$m \frac{d^2 x(t)}{dt^2} + c \frac{dx(t)}{dt} + kx(t) = F(t) \tag{6.6a}$$

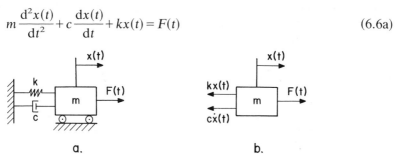

a.   b.

Figure 6.2

so that the operator G is a second-order operator having the form

$$G = m\frac{d^2}{dt^2} + c\frac{d}{dt} + k \tag{6.6b}$$

Clearly, G contains all the information necessary to describe the system characteristics.

6.3 Impulse response. The convolution integral

The differential operator G relates the response of a linear system to the excitation in an implicit manner. Although G contains all the necessary information concerning the system characteristics, the operator is not really a computational tool. In this section, we wish to introduce another relation between the excitation and response, one which is not only more explicit than the operator G but also more attractive computationally. This is the so-called *impulse response*, a very basic concept in system analysis.

Before we proceed with the derivation of the impulse response, it is necessary to introduce the *unit impulse*, or *Dirac's delta function*. The mathematical definition of the unit impulse is

$$\delta(t-a) = 0, \, t \neq a$$
$$\int_{-\infty}^{\infty} \delta(t-a)\,dt = 1 \tag{6.7}$$

and we note that the unit impulse has units time^{-1}. The unit impulse is depicted in Fig. 6.3 in the form of a thin rectangle of infinitestimal width ε and height $1/\varepsilon$. In the limit, as ε approaches zero, the width tends to zero while the height tends to infinity in a way that the area under the curve remains constant and equal to unity. Actually, the shape of the delta function is immaterial as long as the width is very thin and the area is equal to unity. In fact, the delta function should really resemble a 'spike'. Note that the unit impulse defined by Eqs. (6.7) is applied at $t = a$. A unit impulse applied at $t = 0$ is denoted by $\delta(t)$.

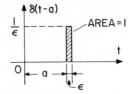

Figure 6.3

Figure 6.4

The *impulse response*, denoted by $h(t)$, is defined as the response of a system to a unit impulse applied at $t = 0$. Implied in the definition is the assumption that the initial excitation is zero. The relation between the unit impulse and the impulse response is shown schematically in the diagram of Fig. 6.4. Clearly, if the excitation is $\delta(t-a)$, i.e., if the unit impulse is delayed by the time interval a, then the response is $h(t-a)$, which is simply the impulse response delayed by the same time interval. The impulse response embodies all the system characteristics, so that it provides an alternative description of the system to that provided by the operator G. However, the impulse response is not merely a convenient way of describing the system characteristics but also a useful way, as it permits the synthesis of the response to any arbitrary excitation. Consistent with this idea, we can depict the relation between excitation and response by the diagram shown in Fig. 6.5, which can be regarded as an integral relation, as opposed to the differential relation of Fig. 6.1.

The unit impulse can be used for sampling purposes. For example, let us consider the arbitrary function $F(t)$, as shown in Fig. 6.6a. Figure 6.6b displays the product $F(t)\delta(t-a)$. Integration of the latter and consideration of definition (6.7) yields

$$\int_{-\infty}^{\infty} F(t)\delta(t-a)\,\mathrm{d}t = F(a) \int_{-\infty}^{\infty} \delta(t-a)\,\mathrm{d}t = F(a) \qquad (6.8)$$

Hence, the above operation provides a sample of the function $F(t)$ at $t = a$.

A function intimately related to the unit impulse is the *unit step function*. The mathematical definition of the unit step function is

$$u(t-a) = \begin{cases} 1 & \text{for } t > a \\ 0 & \text{for } t < a \end{cases} \qquad (6.9)$$

and it can be verified that the unit step function is the integral of the unit

Figure 6.5

187

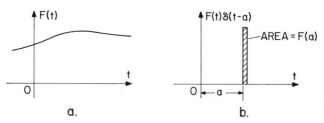

Figure 6.6

impulse. The unit step function is displayed in Fig. 6.7. The function is extremely useful in describing discontinuous functions. In this regard, we note that multiplication of an arbitrary function $f(t)$ by $u(t-a)$ has the effect of annihilating the portion of $f(t)$ corresponding to $t<a$ and leaving the portion corresponding to $t>a$ unchanged.

Next, let us assume that a linear system is subjected to the arbitrary excitation $F(t)$ and consider a thin rectangular area of width $\Delta\tau$ and height $F(\tau)$, as shown in Fig. 6.8. This particular increment of area can be regarded as an impulse of magnitude $F(\tau)\Delta\tau$ applied at $t=\tau$, so that the contribution to the system response attributable to this excitation is simply

$$\Delta x(t, \tau) = F(\tau)\Delta\tau h(t-\tau) \qquad (6.10)$$

where $h(t-\tau)$ is the impulse response delayed by $t=\tau$. Hence, approximating the arbitrary excitation $F(t)$ by a sequence of impulses $F(\tau)\Delta\tau$ and invoking the superposition principle, the response can be approximated by a collection of corresponding impulse responses, or

$$x(t) \approx \sum_{\tau} F(\tau)\Delta\tau h(t-\tau) \qquad (6.11)$$

The response can be rendered exact by letting $\Delta\tau \to 0$ and replacing the summation by integration, so as to obtain

$$x(t) = \int_{-\infty}^{t} F(\tau)h(t-\tau)\,\mathrm{d}\tau \qquad (6.12)$$

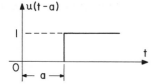

Figure 6.7

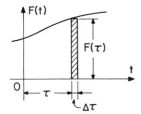

Figure 6.8

which is known as the *convolution integral.* It should be pointed out that Eq. (6.12) represents only the response to the excitation $F(t)$. The response to any possible initial excitation must be evaluated separately and added to it. The convolution integral is symmetric in $F(t)$ and $h(t)$. Indeed, it is not difficult to show that the integral can also be written in the form

$$x(t) = \int_{-\infty}^{t} F(t-\tau)h(\tau)\,d\tau \qquad (6.13)$$

The convolution integral can also be derived by means of the Laplace transformation method (see, for example, M6, Sec. B.5).

The evaluation of the convolution integral is no easy matter, except when the excitation function is relatively simple, as in the case of Example 6.2. In Sec. 6.4, we shall present a procedure for the evaluation of the integral on a digital computer.

Example 6.1

Consider the system of Fig. 6.2 and derive the impulse response for the case in which $m = 0$. Then, write a general expression for the system response in the form of a convolution integral.

Letting $m = 0$, $x(t) = h(t)$ and $F(t) = \delta(t)$ in Eq. (6.6a), the differential equation of motion for the system under consideration reduces to

$$c\dot{h}(t) + kh(t) = \delta(t) \qquad (a)$$

By assumption, the system is initially at rest, so that $h(0) = 0$. The solution of Eq. (a) can be obtained conveniently by the Laplace transform method (M6, Appendix B). The Laplace transform of $h(t)$ is defined as

$$\mathscr{L}h(t) = \bar{h}(s) = \int_{0}^{\infty} e^{-st}h(t)\,dt \qquad (b)$$

and, recalling that $h(0) = 0$, the Laplace transform of the first derivative

189

of $h(t)$ is

$$\mathscr{L}\dot{h}(t) = \int_0^\infty e^{-st}\dot{h}(t)\,dt = e^{-st}h(t)\big|_0^\infty + s\int_0^\infty e^{-st}h(t)\,dt$$

$$= s\bar{h}(s) \tag{c}$$

Moreover, in view of Eq. (6.8), we can write

$$\mathscr{L}\delta(t) = \int_0^\infty e^{-st}\delta(t)\,dt = 1 \tag{d}$$

Taking the Laplace transform of Eq. (a) and considering Eqs. (b)–(d), we obtain

$$c(s+\alpha)\bar{h}(s) = 1 \tag{e}$$

where

$$\alpha = k/c = T^{-1} \tag{f}$$

The constant T is sometimes referred to as the *time constant*. From Eq. (e), the Laplace transform of the impulse response is

$$\bar{h}(s) = \frac{1}{c(s+\alpha)} \tag{g}$$

The impulse response is obtained by taking the inverse Laplace transform of $\bar{h}(s)$, which has the general form

$$h(t) = \mathscr{L}^{-1}\bar{h}(s) = \frac{1}{2\pi i}\int_{\gamma-i\infty}^{\gamma+i\infty} e^{st}\bar{h}(s)\,ds \tag{h}$$

Equation (h) requires integration in the complex plane (defined by the subsidiary variable s) along a line parallel to the imaginary axis and at a distance γ away from it. It is more convenient, however, to obtain $h(t)$ by seeking a function whose transform $\bar{h}(s)$ is given by Eq. (g). From tables of Laplace transforms (M6, p. 468), we obtain the desired function

$$h(t) = \frac{1}{c}e^{-\alpha t} = \frac{1}{c}e^{-(k/c)t} \tag{i}$$

But the function $e^{-(k/c)t}$ is defined also for negative values of time, whereas there is clearly no excitation for $t < 0$. Hence, $h(t)$ must be set equal to zero for $t < 0$. This can be done conveniently by multiplying Eq. (i) by the unit step function applied at $t = 0$, so that the impulse response of the damper-spring system is

$$h(t) = \frac{1}{c} e^{-(k/c)t} u(t)$$ (j)

Next, let us insert Eq. (j) into the convolution integral, Eq. (6.12), and obtain

$$x(t) = \int_{-\infty}^{t} F(\tau) h(t-\tau) \, d\tau = \frac{1}{c} e^{-(k/c)t} \int_{-\infty}^{t} F(\tau) e^{(k/c)\tau} \, d\tau$$ (k)

which is the desired result.

Example 6.2

Obtain the impulse response for the system of Fig. 6.2, then use the convolution integral to derive the response to the excitation

$$F(t) = F_0 u(t)$$ (a)

where $F(t)$ is recognized as a step function of magnitude F_0 applied at $t = 0$.

The differential equation governing the behavior of the mass-damper-spring system of Fig. 6.2 is given by Eq. (6.6a). Letting $x(t) = h(t)$ and $F(t) = \delta(t)$ in that equation, we obtain the differential equation for the impulse response

$$m\ddot{h}(t) + c\dot{h}(t) + kh(t) = \delta(t)$$ (b)

which is subject to the initial conditions $h(0) = \dot{h}(0) = 0$. As in Example 6.1, we shall obtain the solution of Eq. (b) by the Laplace transform method. The Laplace transforms of $h(t)$, $\dot{h}(t)$ and $\delta(t)$ were calculated in Example 6.1, so that it remains for us to calculate the transform of $\ddot{h}(t)$. Recalling that the initial conditions are zero, we can write

$$\mathcal{L}\ddot{h}(t) = \int_{0}^{\infty} e^{-st} \ddot{h}(t) \, dt$$

$$= e^{-st} \dot{h}(t)\big|_{0}^{\infty} + s e^{-st} h(t)\big|_{0}^{\infty} + s^2 \int_{0}^{\infty} e^{-st} h(t) \, dt = s^2 \bar{h}(s)$$ (c)

Taking the Laplace transform of Eq. (b), we obtain

$$(ms^2 + cs + k)\bar{h}(s) = 1$$ (d)

so that the Laplace transform of the impulse response is

$$\bar{h}(s) = \frac{1}{ms^2 + cs + k} = \frac{1}{m(s^2 + 2\zeta\omega_n s + \omega_n^2)}$$ (e)

where we used the notation $2\zeta\omega_n = c/m$ and $\omega_n^2 = k/m$, in which ζ is the damping factor and ω_n is the natural frequency of the undamped oscillation.

From tables of Laplace transforms (M6, p. 468), we obtain

$$h(t) = \frac{1}{m\omega_d}\, e^{-\zeta\omega_n t} \sin \omega_d t \tag{f}$$

where $\omega_d = (1 - \zeta^2)^{1/2}\omega_n$ is the frequency of the damped oscillation. As in Example 6.1, we recall that $h(t)$ must be set equal to zero for $t < 0$, so that the impulse response of the mass-damper-spring system is actually

$$h(t) = \frac{1}{m\omega_d}\, e^{-\zeta\omega_n t} \sin \omega_d t\, \mathcal{u}(t) \tag{g}$$

Next, let us calculate the response of the system to the excitation $F(t)$, as given by Eq. (a). Introducing Eqs. (a) and (g) into the convolution integral, Eq. (6.13), we obtain

$$
\begin{aligned}
x(t) &= \int_{-\infty}^{t} F_0\, \mathcal{u}(t-\tau)\, \frac{1}{m\omega_d}\, e^{-\zeta\omega_n \tau} \sin \omega_d \tau\, \mathcal{u}(\tau)\, \mathrm{d}\tau \\
&= \frac{F_0}{m\omega_d} \int_0^t e^{-\zeta\omega_n \tau} \sin \omega_d \tau\, \mathrm{d}\tau
\end{aligned}
\tag{h}
$$

Using the identity

$$\sin \alpha = \frac{1}{2i}\,(e^{i\alpha} - e^{-i\alpha}) \tag{i}$$

Eq. (h) yields, after some algebraic operations,

$$
\begin{aligned}
x(t) &= \frac{F_0}{2im\omega_d} \int_0^t e^{-\zeta\omega_n \tau}(e^{i\omega_d \tau} - e^{-i\omega_d \tau})\, \mathrm{d}\tau \\
&= \frac{F_0}{2im\omega_d}\left[\frac{e^{(-\zeta\omega_n + i\omega_d)\tau}}{-\zeta\omega_n + i\omega_d} - \frac{e^{-(\zeta\omega_n + i\omega_d)\tau}}{-(\zeta\omega_n + i\omega_d)}\right]_0^t \\
&= \frac{F_0}{k}\left[1 - e^{-\zeta\omega_n t}\left(\cos \omega_d t + \frac{\zeta\omega_n}{\omega_d} \sin \omega_d t\right)\right] \mathcal{u}(t)
\end{aligned}
\tag{j}
$$

where the final expression for the response was multiplied by the unit step function $\mathcal{u}(t)$ to account for the fact that the response is zero for $t < 0$.

6.4   Discrete-time systems

In Secs. 6.2 and 6.3, we discussed the problem of deriving the response of a system to arbitrary excitation. Two alternative approaches were presented, the first involving the direct solution of a differential equation and the second the evaluation of a convolution integral, where the latter actually implies the first. For complicated excitation functions, both approaches must rely on numerical solutions. In this section, we shall discuss some concepts particularly suited for such numerical solutions.

Numerical solutions are carried out most efficiently on digital computers. This leads naturally to the concept of *discrete-time systems*, whereby the excitation and response are being treated as discrete functions of time, in contrast with *continuous-time systems*, in which they are continuous functions of time. The differential equations of continuous-time systems give way to difference equations in discrete-time systems and integrals to summations.

It is common in system analysis to refer to the excitation as the *input signal* and the response as the *output signal*. The type of signals discussed in Secs. 6.2 and 6.3 were described in terms of an independent variable, namely, the time t, which could take continuous values. For this reason, they are referred to as *continuous signals* and they are recognized as the most commonly encountered functions of time, such as that shown in Fig. 6.9a. On the other hand, if the independent variable t takes only discrete values t_k $(k = 0, \pm 1, \pm 2, \ldots)$, as shown in Fig. 6.9b, then t is said to be a *discrete-time variable* and the function is referred to as a *discrete signal*. Such signals do not arise naturally, but are the result of a certain discretization process, as discussed below. Digital computers work with discrete signals, which explains our interest in them.

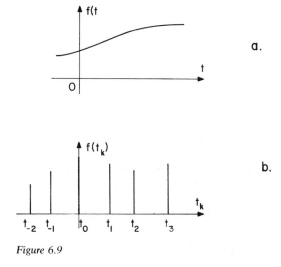

Figure 6.9

193

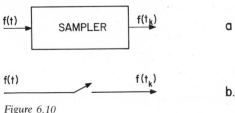

Figure 6.10

Let us consider the problem of converting a given continuous signal into a discrete one. This can be done by means of a *sampler*, as shown in Fig. 6.10a. The input to the sampler is the continuous signal $f(t)$ and the output is a sequence of numbers, spaced in time, which have the values equal to the continuous input signal at the sampling instances t_k. The sampler can be represented by the switch depicted in Fig. 6.10b, where the switch is open for all times except at the sampling instances t_k, when it closes instantaneously to permit the input signal to pass through. Ordinarily, the samplings are taken at equal time intervals, so that $t_k = kT$, where T is the *sampling period*.

To illustrate the idea, let us consider the numerical integration of the integral

$$g(t) = \int_0^t f(\tau)\, d\tau \tag{6.14}$$

The assumption is that the function $f(\tau)$ is such that no closed-form integral is known to exist. The sampler of Fig. 6.10 can be used to convert the continuous signal $f(\tau)$ into a discrete one. The result can be used to produce the piecewise constant function $f_k(\tau)$ given by

$$f_k(\tau) = f(kT), \qquad kT \le \tau \le kT + T \tag{6.15}$$

as shown in Fig. 6.11. Letting $t = nT$, the integral (6.14) can be approximated by

$$g(nT) = \int_0^{nT} f(\tau)\, d\tau \approx T \sum_{k=0}^{n-1} f(kT) \tag{6.16}$$

The summation on the right side of (6.16) requires the values $f(0), f(T), \ldots, f(nT - T)$. To save computer storage, however, it is desirable to work with a scheme which does not require retention of all these values. In this regard, let us consider the approximation

$$g(nT + T) = \int_0^{nT+T} f(\tau)\, d\tau = \int_0^{nT} f(\tau)\, d\tau + \int_{nT}^{nT+T} f(\tau)\, d\tau$$

$$\approx g(nT) + Tf(nT) \tag{6.17}$$

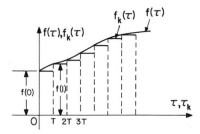

Figure 6.11

and we conclude that the scheme implicit in (6.17) does indeed save computer storage, because to evaluate $g(nT+T)$ it is only necessary to retain the previous value $g(nT)$ of the integral and the current sampled value $f(nT)$.

If the discrete signal is to be converted back into a continuous one, then the discrete signal must be passed through a *data hold circuit*. The simplest hold is the *zero-order hold*, defined mathematically by

$$f(\tau) = f(nT), \qquad nT \le \tau < nT + T \tag{6.18}$$

and it generates a *staircase* function of the type shown in Fig. 6.11.

It should be pointed out that for a signal to be accepted by a digital computer it is not sufficient that it be discrete, but it must also be a *digital signal*, i.e., a signal whose amplitude is restricted to a given set of values, generally binary. The original continuous signal and the signal generated by the sampler are both so-called *analog signals*. Hence, the discrete analog signal from the sampler must be digitized by means of an *analog-to-digital* (A/D) converter before it can be accepted by a digital computer, which implies quantization of the signal with respect to amplitude. Similarly, the discrete digital signal from the computer must be passed through a *digital-to-analog* (D/A) converter before it is fed into the hold circuit. The flow of signals is shown schematically in Fig. 6.12. In high-capacity computers the quantization error is insignificant. In such cases, the A/D and D/A conversions have relatively little mathematical significance and can be omitted from the diagram, with the understanding that the data conversion operations are implied.

Next, let us turn our attention to the simulation of the response of

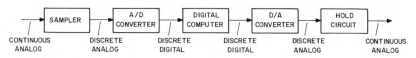

Figure 6.12`

the mass-damper-spring system on a digital computer. To this end, let us write Eq. (6.6) in the form

$$m\frac{d^2x(t)}{dt^2}\Big|_{t=nT} + c\frac{dx(t)}{dt}\Big|_{t=nT} + kx(nT) = F(nT) \tag{6.18}$$

and make the following approximations

$$\frac{dx(t)}{dt}\Big|_{t=nT} \approx \frac{1}{T}[x(nT) - x(nT-T)] \tag{6.19a}$$

$$\frac{d^2x(t)}{dt^2}\Big|_{t=nT} \approx \frac{1}{T^2}[x(nT) - 2x(nT-T) + x(nT-2T)] \tag{6.19b}$$

Introducing Eqs. (6.19) into Eq. (6.18), we obtain

$$\frac{m}{T^2}[x(nT) - 2x(nT-T) + x(nT-2T)]$$

$$+ \frac{c}{T}[x(nT) - x(nT-T)] + kx(nT) = F(nT) \tag{6.20}$$

This enables us to construct the following recursive relation

$$x(nT) = \frac{T^2}{m+cT+kT^2}\left[F(nT) + \frac{2m+cT}{T^2}x(nT-T) - \frac{m}{T^2}x(nT-2T)\right]$$
$$\tag{6.21}$$

which represents the response of the system to any arbitrary excitation. An important question is the selection of the sampling period T. Clearly, T must be sufficiently small that $x(nT)$ will yield a good approximation of $x(t)$ and yet not so small as to require an excessive amount of computer time. There are two factors that should be considered in selecting T. Both factors help determine an upper limit on T. The first consists of the external excitation $F(t)$. In particular, if $F(t)$ has significant high-frequency content, then T must be sufficiently small that the discretization will retain the essential character of the function. Hence, T should be equal to a fraction of the corresponding period. The second factor is internal and is related to the frequency of the damped oscillation $\omega_d = (1-\zeta^2)^{1/2}\omega_n$ of the continuous-time system, where $\omega_n = \sqrt{k/m}$ is the natural frequency of the undamped oscillation and $\zeta = c/2\sqrt{km}$ is the damping factor. For small ζ, the ratio T^2k/m should be much smaller than unity. The selection of a lower limit on T is a more subjective matter and the analyst has more freedom to use his own judgment.

The same problem can be solved in general form by means of the convolution integral. To this end, let us approximate the impulse response by a piecewise constant function, such as that shown in Fig. 6.11. Recalling the definition of the unit step function, Eq. (6.9), the impulse response can be approximated by

$$h(t) = \sum_{j=0}^{\infty} h_j[\mathscr{u}(t-jT) - \mathscr{u}(t-jT-T)] \qquad (6.22)$$

Letting $t = kT$ in Eq. (6.13) and using Eq. (6.22), the convolution integral becomes

$$x(kT) = \int_0^{kT} F(kT-\tau) \sum_{j=0}^{\infty} h_j[\mathscr{u}(\tau-jT) - \mathscr{u}(\tau-jT-T)]\,d\tau \qquad (6.23)$$

Introducing the change of variable $kT - \tau = \sigma$, Eq. (6.23) reduces to

$$x(kT) = \sum_{j=0}^{\infty} \int_{kT}^{0} F(\sigma)h_j[\mathscr{u}(kT-jT-\sigma) - \mathscr{u}[kT-jT-T-\sigma)]\,d(-\sigma)$$

$$= \sum_{j=0}^{\infty} h_j \int_{(k-j-1)T}^{(k-j)T} F(\sigma)\,d\sigma, \qquad k-j-1 \geq 0 \qquad (6.24)$$

Observing that

$$\int_{(k-j-1)T}^{(k-j)T} F(\sigma)\,d\sigma = TF(kT-jT-T) = TF_{k-j-1} \qquad (6.25)$$

Eq. (6.24) becomes finally

$$x_k = T \sum_{j=0}^{k-1} F_{k-j-1} h_j \qquad (6.26)$$

which represents a *convolution sum* and can be recognized as the discrete version of the convolution integral. Note that, in changing the upper limit of the summation over j in Eq. (6.26) from ∞ to $k-1$, use was made of the inequality in (6.24).

Formula (6.26) can be verified by means of a simple graphical procedure. To this end, let Fig. 6.13a represent the discretized version of the impulse response $h(\tau)$ and Fig. 6.13b the discretized version of the excitation $F(\tau)$. Figure 6.13c shows the function $F(kT-\tau)$, which is obtained by shifting $F(\tau)$ backward by the time interval kT to yield $F(kT+\tau)$ and then 'folding' with respect to the vertical axis so as to produce the mirror image $F(kT-\tau)$ of $F(kT+\tau)$. The convolution integral reduces to the integration with respect to τ of the function resulting from the multiplication of $h(\tau)$ of Fig. 6.13a by $F(kT-\tau)$ of Fig. 6.13c.

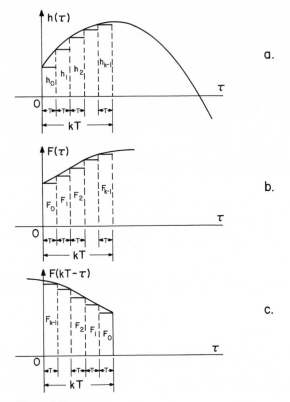

Figure 6.13

Because this resulting function is piecewise constant, we obtain

$$x_k = x(kT) = T(F_{k-1}h_0 + F_{k-2}h_1 + \cdots + F_1h_{k-2} + F_0h_{k-1})$$

$$= T \sum_{j=0}^{k-1} F_{k-j-1}h_j \tag{6.27}$$

which is the same as Eq. (6.26).

The convolution sum, Eq. (6.26) or (6.27), gives the response to the external excitation alone. The response to initial excitation must be obtained separately and added to it.

Example 6.3

Regard the system of Example 6.2 as a discrete-time system and use Eq. (6.26) to obtain the response to the excitation

$$F(t) = f_0 t\mathcal{u}(t) \tag{a}$$

where f_0 has units Ns^{-1}.

The excitation can be approximated by a staircase function given by

$$F_j = F(jT) = f_0 jT, \qquad j = 0, 1, 2, \ldots \tag{b}$$

The impulse response can also be approximated by a staircase function. From Eq. (g) of Example 6.2, we can write

$$h_j = h(jT) = \frac{1}{m\omega_d} e^{-\zeta\omega_n jT} \sin \omega_d jT, \qquad j = 0, 1, 2, \ldots \tag{c}$$

Hence, using Eq. (6.26), we derive the response in the form of the sequence

$$x_1 = x(T) = TF_0 h_0 = 0$$

$$x_2 = x(2T) = T(F_1 h_0 + F_0 h_1) = 0$$

$$x_3 = x(3T) = T(F_2 h_0 + F_1 h_1 + F_0 h_2) = \frac{f_0 T^2}{m\omega_d} e^{-\zeta\omega_n T} \sin \omega_d T$$

$$x_4 = x(4T) = T(F_3 h_0 + F_2 h_1 + F_1 h_2 + F_0 h_3) \tag{d}$$

$$= \frac{f_0 T^2}{m\omega_d} (2e^{-\zeta\omega_n T} \sin \omega_d T + e^{-2\zeta\omega_n T} \sin 2\omega_d T)$$

$$x_5 = x(5T) = T(F_4 h_0 + F_3 h_1 + F_2 h_2 + F_1 h_3 + F_0 h_4)$$

$$= \frac{f_0 T^2}{m\omega_d} (3e^{-\zeta\omega_n T} \sin \omega_d T + 2e^{-2\zeta\omega_n T} \sin 2\omega_d T + e^{-3\zeta\omega_n T} \sin 3\omega_d T)$$

The fact that both x_1 and x_2 are zero is due to the nature of the approximation, which takes both F_0 and h_0 equal to zero.

6.5   Response of undamped nongyroscopic systems

The differential equations governing the motion of an n-degree-of free-dom undamped nongyroscopic system are obtained by letting the gyroscopic matrix G, the damping matrix C and the circulatory matrix H in Eq. (2.22) be zero. The result is

$$M\ddot{\mathbf{q}}(t) + K\mathbf{q}(t) = \mathbf{Q}(t) \tag{6.28}$$

where M and K are $n \times n$ real symmetric matrices, the first being the inertia matrix and the second the stiffness matrix. Moreover, $\mathbf{q}(t)$ is the n-dimensional configuration vector and $\mathbf{Q}(t)$ is the associated generalized force vector. Whereas the inertia matrix M is positive definite, the stiffness matrix is assumed to be only positive semidefinite.

We shall obtain the solution of Eq. (6.28) by modal analysis, which amounts to reducing the response of an n-degree-of-freedom system to a linear combination of responses of n independent single-degree-of-freedom systems. To this end, we solve first the eigenvalue problem

$$\lambda M\mathbf{u} = K\mathbf{u}, \qquad \lambda = \omega^2 \tag{6.29}$$

It was shown in Sec. 3.3 that the solution of Eq. (6.29) consists of the modal matrix $U = [\mathbf{u}_1 \, \mathbf{u}_2 \cdots \mathbf{u}_n]$, where $\mathbf{u}_i$ $(i = 1, 2, \ldots, n)$ are the system eigenvectors, and the diagonal matrix of the eigenvalues $\Lambda = \text{diag}\,[\lambda_i] = \text{diag}\,[\omega_i^2]$, where ω_i $(i = 1, 2, \ldots, n)$ are the system natural frequencies. Because K is only positive semidefinite the system is semidefinite, so that some of the system eigenvectors represent rigid-body modes with associated natural frequencies equal to zero. The eigenvectors are orthogonal with respect to M (and K) and can be normalized so as to satisfy

$$U^T M U = I, \qquad U^T K U = \Lambda \tag{6.30}$$

In the sequel, we shall assume that there are r rigid-body modes and $n - r$ elastic modes.

By the expansion theorem, Eq. (3.54), the response of the system at any time t_1 can be expressed as a linear combination of the eigenvectors $\mathbf{u}_i$ $(i = 1, 2, \ldots, n)$. Denoting the coordinates of the vector $\mathbf{q}(t_1)$ with respect to the basis $\mathbf{u}_1, \mathbf{u}_2, \ldots, \mathbf{u}_n$ by $\eta_1(t_1), \eta_2(t_1), \ldots, \eta_n(t_1)$, respectively, we can write

$$\mathbf{q}(t_1) = \eta_1(t_1)\mathbf{u}_1 + \eta_2(t_1)\mathbf{u}_2 + \cdots + \eta_n(t_1)\mathbf{u}_n = \sum_{i=1}^{n} \eta_i(t_1)\mathbf{u}_i = U\boldsymbol{\eta}(t_1) \tag{6.31}$$

where $\boldsymbol{\eta}(t_1) = [\eta_1(t_1) \, \eta_2(t_1) \cdots \eta_n(t_1)]^T$ is the n-vector of coordinates. But t_1 is arbitrary, so that its value can be changed at will. Because Eq. (6.31) must hold for all other values of time, and because the solution must be continuous, we can replace t_1 by t and write in general

$$\mathbf{q}(t) = U\boldsymbol{\eta}(t) \tag{6.32}$$

Introducing Eq. (6.32) into Eq. (6.28), premultiplying the result by U^T and invoking Eqs. (6.30), we obtain

$$\ddot{\boldsymbol{\eta}}(t) + \Lambda\boldsymbol{\eta}(t) = \mathbf{N}(t) \tag{6.33}$$

where

$$\mathbf{N}(t) = U^T \mathbf{Q}(t) \tag{6.34}$$

is an n-vector of generalized forces associated with the vector $\boldsymbol{\eta}(t)$. Equation (6.33) represents a set of independent ode's having the form

$$\begin{aligned} \ddot{\eta}_i(t) &= N_i(t), & i &= 1, 2, \ldots, r \\ \ddot{\eta}_i(t) + \omega_i^2 \eta_i(t) &= N_i(t), & i &= r+1, r+2, \ldots, n \end{aligned} \tag{6.35}$$

The coordinates $\eta_i(t)$ $(i = 1, 2, \ldots, n)$ are referred to as *principal coordinates*, or *natural coordinates*. Because they are coordinates with respect to an orthonormal basis, they are also called *normal coordinates*.

By the superposition principle, the solutions of Eqs. (6.35) can be written in the form of linear combinations of the homogeneous solutions and the particular solutions, where the first represent physically the response to initial excitations and the second the response to external excitations. Letting the initial displacements and velocities be

$$\eta_i(0) = \eta_{i0}, \qquad \dot{\eta}_i(0) = \dot{\eta}_{i0}, \qquad i = 1, 2, \ldots, n \qquad (6.36)$$

and using results of Secs. 2.5 and 6.3, the solutions of Eqs. (6.35) can be shown to be

$$\eta_i(t) = \int_0^t \left[\int_0^\tau N_i(\sigma)\, d\sigma \right] d\tau + \eta_{i0} + \dot{\eta}_{i0} t, \qquad i = 1, 2, \ldots, r \qquad (6.37a)$$

$$\eta_i(t) = \frac{1}{\omega_i} \int_0^t N_i(t-\tau) \sin \omega_i \tau \, d\tau + \eta_{i0} \cos \omega_i t + \frac{\dot{\eta}_{i0}}{\omega_i} \sin \omega_i t,$$

$$i = r+1, r+2, \ldots, n \qquad (6.37b)$$

The initial displacement vector $\boldsymbol{\eta}_0$ and velocity vector $\dot{\boldsymbol{\eta}}_0$ can be expressed in terms of the original initial generalized displacement vector $\mathbf{q}_0 = \mathbf{q}(0)$ and velocity vector $\dot{\mathbf{q}}_0 = \dot{\mathbf{q}}(0)$. Indeed, using Eq. (6.32) in conjunction with the first of Eqs. (6.30), it can be verified easily that

$$\boldsymbol{\eta}_0 = U^T M \mathbf{q}_0, \qquad \dot{\boldsymbol{\eta}}_0 = U^T M \dot{\mathbf{q}}_0 \qquad (6.38)$$

The formal solution is completed by inserting Eqs. (6.37) into Eq. (6.32). Of course, for numerical evaluation on a digital computer, Eqs. (6.37) can be discretized in time by the approach presented in Sec. 6.4. The procedure for the determination of the solution of Eq. (6.32) in the form of a linear combination of the modal vectors multiplied by the principal coordinates is commonly referred to as the *classical modal analysis*.

On certain occasions, particularly for high-order systems, it is not necessary, nor is it feasible, to retain all the modal information in deriving the response. In this case, one does not solve the eigenvalue problem completely but only partially, in the sense that only a certain number of lower modes are sought and computed. If only p modes are retained, $p < n$, then the above modal analysis can still be used, except that U is now an $n \times p$ rectangular matrix and $\boldsymbol{\eta}$ is a p-vector. The matrices (6.30) retain their form, but I and Λ are now square matrices of order p instead of being of order n. The process of eliminating a given number of higher modes from the analysis is known as *modal truncation*, a subject discussed in more detail in Ch. 10.

Sometimes, in the course of design, it is necessary to change the mass and stiffness properties of the structure, so that the question arises whether it is necessary to repeat the above computations with the new mass and stiffness matrices. This question is of particular importance for high-order systems, for which an eigensolution is time-consuming and expensive to produce. The answer to this question lies in the magnitude of the changes in the system properties. Indeed, if the changes are relatively small, then they can be regarded as perturbations on the original system, so that the response of the newly designed system can be obtained by a perturbation technique basing the computations on the original eigensolution (R6).

Example 6.4

Consider the system of Fig. 4.3 and use modal analysis to derive the response to the excitation

$$F_1(t) = F_2(t) = 0, \qquad F_3(t) = \hat{F}_0 \delta(t) \tag{a}$$

where $\hat{F}_0$ is an impulsive force having units $N-s$ and $\delta(t)$ is the unit impulse.

The system equations of motion can be written in the matrix-vector form

$$M\ddot{u}(t) + Ku(t) = \mathbf{F}(t) \tag{b}$$

where

$$u(t) = [u_1(t) \ u_2(t) \ u_3(t)]^T \qquad \mathbf{F}(t) = [F_1(t) \ F_2(t) \ F_3(t)]^T \tag{c}$$

are the configuration vector and excitation vector, respectively, and

$$M = m \begin{bmatrix} 1 & 0 & 0 \\ 0 & 1 & 0 \\ 0 & 0 & 2 \end{bmatrix}, \qquad K = k \begin{bmatrix} 2 & -1 & 0 \\ -1 & 3 & -2 \\ 0 & -2 & 2 \end{bmatrix} \tag{d}$$

are the inertia and stiffness matrices, respectively.

To obtain the response by modal analysis, we must first solve the eigenvalue problem

$$KU = MU\Lambda \tag{e}$$

where U is the modal matrix and Λ is the diagonal matrix of the eigenvalues satisfying Eqs. (6.30). The eigenvalue problem (e) was solved in Example 5.7 by the Jacobi method with the result

$$U = \frac{1}{\sqrt{m}} \begin{bmatrix} 0.269108 & 0.878182 & -0.395445 \\ 0.500758 & 0.223148 & 0.836329 \\ 0.581731 & -0.299166 & -0.268493 \end{bmatrix} \tag{f}$$

and

$$\Lambda = \frac{k}{m} \begin{bmatrix} 0.139194 & 0 & 0 \\ 0 & 1.745898 & 0 \\ 0 & 0 & 4.114908 \end{bmatrix} \tag{g}$$

where the eigenvectors and eigenvalues have been rearranged according to the mode order. From Eq. (g), we obtain the natural frequencies

$$\omega_1 = \sqrt{0.139194\ k/m} = 0.373087\ \sqrt{k/m}$$

$$\omega_2 = \sqrt{1.745898\ k/m} = 1.321324\ \sqrt{k/m} \tag{h}$$

$$\omega_3 = \sqrt{4.114908 k/m} = 2.028524\ \sqrt{k/m}$$

Introducing the linear transformation

$$\mathbf{u}(t) = U\boldsymbol{\eta}(t) \tag{i}$$

into Eq. (b) and premultiplying the result by U^T, we obtain Eq. (6.33), in which

$$\mathbf{N}(t) = U^T \mathbf{F}(t) = \frac{\hat{F}_0\ \delta(t)}{\sqrt{m}} \begin{bmatrix} 0.581731 \\ -0.299166 \\ -0.268493 \end{bmatrix} \tag{j}$$

so that Eqs. (6.37b) yield

$$\eta_1(t) = \frac{1}{0.373087\ \sqrt{k/m}} \frac{0.581731\hat{F}_0}{\sqrt{m}} \int_0^t \delta(t-\tau) \sin 0.373087 \sqrt{\frac{k}{m}}\ \tau\ d\tau$$

$$= \frac{1.559237\hat{F}_0}{\sqrt{k}} \sin 0.373087 \sqrt{\frac{k}{m}}\ t$$

$$\eta_2(t) = \frac{1}{1.321324\ \sqrt{k/m}} \frac{-0.299166\hat{F}_0}{\sqrt{m}} \int_0^t \delta(t-\tau) \sin 1.321324 \sqrt{\frac{k}{m}}\ \tau\ d\tau$$

$$= -\frac{0.226414\hat{F}_0}{\sqrt{k}} \sin 1.321324 \sqrt{\frac{k}{m}}\ t \tag{k}$$

$$\eta_3(t) = \frac{1}{2.028524\ \sqrt{k/m}} \frac{-0.268493\hat{F}_0}{\sqrt{m}} \int_0^t \delta(t-\tau) \sin 2.028524 \sqrt{\frac{k}{m}}\ \tau\ d\tau$$

$$= -\frac{0.132359\hat{F}_0}{\sqrt{k}} \sin 2.028524 \sqrt{\frac{k}{m}}\ t$$

Finally, introducing Eqs. (f) and (k) into Eq. (i), we can write the displacement components

$$u_1(t) = \frac{\hat{F}_0}{\sqrt{mk}} \left(0.419603 \sin 0.373087 \sqrt{\frac{k}{m}}\, t \right.$$

$$-\, 0.198833 \sin 1.321324 \sqrt{\frac{k}{m}}\, t$$

$$\left. +\, 0.052341 \sin 2.028524 \sqrt{\frac{k}{m}}\, t \right)$$

$$u_2(t) = \frac{\hat{F}_0}{\sqrt{mk}} \left(0.780800 \sin 0.373087 \sqrt{\frac{k}{m}}\, t \right.$$

$$-\, 0.050524 \sin 1.321324 \sqrt{\frac{k}{m}}\, t \qquad (1)$$

$$\left. -\, 0.110696 \sin 2.028524 \sqrt{\frac{k}{m}}\, t \right)$$

$$u_3(t) = \frac{\hat{F}_0}{\sqrt{mk}} \left(0.907057 \sin 0.373087 \sqrt{\frac{k}{m}}\, t \right.$$

$$+\, 0.067735 \sin 1.321324 \sqrt{\frac{k}{m}}\, t$$

$$\left. +\, 0.035537 \sin 2.028524 \sqrt{\frac{k}{m}}\, t \right)$$

It is clear from Eqs. (1) that the motion is dominated by the first mode, as indicated by the fact that the harmonic terms with the frequency $\omega_1 = 0.373087 \sqrt{k/m}$ possess the largest coefficients. Because a force applied to the third mass tends to deform the system in a pattern resembling the first mode, this result is to be expected.

6.6   Response of undamped gyroscopic systems

In the absence of damping and circulatory forces, $C = H = 0$, Eq. (2.22) reduces to

$$M\ddot{\mathbf{q}}(t) + G\dot{\mathbf{q}}(t) + K\mathbf{q}(t) = \mathbf{Q}(t) \qquad (6.39)$$

where M, K, $\mathbf{q}(t)$ and $\mathbf{Q}(t)$ are as defined in Sec. 6.5. We shall assume here that M and K are not only real and symmetric but also that both are positive definite. In addition, we recall from Sec. 2.3 that G is a real skew symmetric gyroscopic matrix. Our object is to solve Eq. (6.39) by modal analysis. The classical modal analysis of Sec. 6.5 cannot be used in this case, however, because a transformation in the configuration space cannot uncouple the system of equations; a transformation in the state space can. In this section, we shall present a modal analysis for gyroscopic systems based on a transformation in the state space (M5).

Consistent with the approach first discussed in Sec. 2.5, let us introduce the $2n$-dimensional state vector

$$\mathbf{x}(t) = [\dot{\mathbf{q}}^T(t) \,\vdots\, \mathbf{q}^T(t)]^T \tag{6.40}$$

the associated $2n$-dimensional generalized force vector

$$\mathbf{X}(t) = [\mathbf{Q}^T(t) \,\vdots\, \mathbf{0}^T]^T \tag{6.41}$$

as well as the $2n \times 2n$ matrices

$$M^* = \begin{bmatrix} M & \vdots & 0 \\ \cdots & \cdots & \cdots \\ 0 & \vdots & K \end{bmatrix}, \qquad G^* = \begin{bmatrix} G & \vdots & K \\ \cdots & \cdots & \cdots \\ -K & \vdots & 0 \end{bmatrix} \tag{6.42}$$

where M^* is a real symmetric positive definite matrix and G^* is a real skew symmetric matrix. This enables us to replace the n second-order equations, Eq. (6.39), by the $2n$ first-order state equations

$$M^* \dot{\mathbf{x}}(t) + G^* \mathbf{x}(t) = \mathbf{X}(t) \tag{6.43}$$

To solve Eq. (6.43) by modal analysis, we consider first the eigen-value problem

$$sM^*\mathbf{x} + G^*\mathbf{x} = \mathbf{0} \tag{6.44}$$

The solution of Eq. (6.44) consists of n pairs of pure imaginary complex conjugate eigenvalues $s_r = i\omega_r$, $\bar{s}_r = -i\omega_r$ and n pairs of complex conjugate eigenvectors $\mathbf{x}_r = \mathbf{y}_r + i\mathbf{z}_r$, $\bar{\mathbf{x}}_r = \mathbf{y}_r - i\mathbf{z}_r$. As indicated in Sec. 2.5, the eigen-value problem (6.44) can be cast into the real symmetric form

$$K^*\mathbf{y}_r = \lambda_r M^*\mathbf{y}_r, \qquad K^*\mathbf{z}_r = \lambda_r M^*\mathbf{z}_r, \qquad \lambda_r = \omega_r^2,$$
$$r = 1, 2, \ldots, n \tag{6.45}$$

where

$$K^* = G^{*T}M^{*-1}G^* \tag{6.46}$$

is a real symmetric positive definite matrix. Chapter 5 provides ample justification as to why the real symmetric positive definite form (6.45) is to be preferred over the complex form (6.44) for computational purposes. From Eq. (6.45), we conclude that both the real part $\mathbf{y}_r$ and the imaginary part $\mathbf{z}_r$ of the eigenvector $\mathbf{x}_r$ satisfy the same eigenvalue problem, and in fact belong to the same eigenvalue $\lambda_r = \omega_r^2$. In spite of the fact that every

205

ω_r^2 has multiplicity two, because the system is positive definite, the eigenvalue problem can be described by a single real symmetric matrix, so that the eigenvectors $\mathbf{y}_r$ and $\mathbf{z}_r$ are linearly independent (see Sec. 3.3). In fact, all the system eigenvectors are orthogonal with respect to M^* and K^*. It will prove convenient to arrange the eigenvectors in the $2n \times 2n$ modal matrix

$$P = [\mathbf{y}_1 \, \mathbf{z}_1 \, \mathbf{y}_2 \, \mathbf{z}_2 \cdots \mathbf{y}_n \, \mathbf{z}_n] \tag{6.47}$$

and to normalize the matrix so as to satisfy

$$P^T M^* P = I \tag{6.48}$$

where I is a unit matrix of order $2n$. Then, it is easy to verify

$$P^T G^* P = A \tag{6.49}$$

where A is a block-diagonal matrix of the form

$$A = \text{block-diag} \begin{bmatrix} 0 & -\omega_r \\ \omega_r & 0 \end{bmatrix} \tag{6.50}$$

Next, let us consider the linear transformation

$$\mathbf{x}(t) = \sum_{r=1}^{n} [\xi_r(t)\mathbf{y}_r + \eta_r(t)\mathbf{z}_r] = P\mathbf{w}(t) \tag{6.51}$$

where

$$\mathbf{w}(t) = [\xi_1(t) \, \eta_1(t) \, \xi_2(t) \, \eta_2(t) \cdots \xi_n(t) \, \eta_n(t)]^T \tag{6.52}$$

Introducing Eq. (6.51) into Eq. (6.43), premultiplying the result by P^T and considering Eqs. (6.48)–(6.49), we obtain

$$\dot{\mathbf{w}}(t) + A\mathbf{w}(t) = \mathbf{W}(t) \tag{6.53}$$

where

$$\mathbf{W}(t) = P^T \mathbf{X}(t) \tag{6.54}$$

is a $2n$-dimensional generalized force vector. The vector can be written in the form

$$\mathbf{W}(t) = [Y_1(t) \, Z_1(t) \, Y_2(t) \, Z_2(t) \cdots Y_n(t) \, Z_n(t)] \tag{6.55}$$

where

$$Y_r(t) = \mathbf{y}_r^T \mathbf{X}(t), \qquad Z_r(t) = \mathbf{z}_r \mathbf{X}(t), \qquad r = 1, 2, \ldots, n \tag{6.56}$$

Although the matrix A does not have a classical Jordan form (see Sec. 3.2), the matrix can still be regarded as being in a Jordan form, because the general effect is the same. Indeed, the block-diagonal form (6.50) leads to Eq. (6.53), representing a set of n uncoupled second-order systems of the form

$$\begin{aligned}\dot{\xi}_r(t) - \omega_r \eta_r(t) &= Y_r(t) \\ \dot{\eta}_r(t) + \omega_r \xi_r(t) &= Z_r(t)\end{aligned}, \qquad r = 1, 2, \ldots, n \qquad (6.57)$$

which can be solved for every pair of conjugate generalized coordinates $\xi_r(t)$, $\eta_r(t)$ independently of any other pair.

The solution of Eqs. (6.57) can be obtained by the Laplace transform method. Letting $\bar{\xi}_r(s)$, $\bar{\eta}_r(s)$, $\bar{Y}_r(s)$ and $\bar{Z}_r(s)$ be the Laplace transforms of $\xi_r(t)$, $\eta_r(t)$, $Y_r(t)$ and $Z_r(t)$, respectively, and transforming both sides of Eqs. (6.57), we obtain

$$\begin{aligned}s\bar{\xi}_r(s) - \xi_r(0) - \omega_r \bar{\eta}_r(s) &= \bar{Y}_r(s) \\ s\bar{\eta}_r(s) - \eta_r(0) + \omega_r \bar{\eta}_r(s) &= \bar{Z}_r(s)\end{aligned}, \qquad r = 1, 2, \ldots, n \qquad (6.58)$$

where $\xi_r(0)$ and $\eta_r(0)$ are the initial values of $\xi_r(t)$ and $\eta_r(t)$ and they depend on the initial state vector $\mathbf{x}(0)$. Indeed, letting $t = 0$ in Eq. (6.51) premultiplying by $\mathbf{y}_r^T M^*$ and $\mathbf{z}_r^T M^*$, in turn, and taking advantage of the eigenvectors orthogonality, we obtain

$$\xi_r(0) = \mathbf{y}_r^T M^* \mathbf{x}(0), \qquad \eta_r(0) = \mathbf{z}_r^T M^* \mathbf{x}(0), \qquad r = 1, 2, \ldots, n \qquad (6.59)$$

Equations (6.58) represent a set of two simultaneous algebraic equations that can be solved readily for the transformed generalized coordinates $\bar{\xi}_r$ and $\bar{\eta}_r(s)$ with the result

$$\begin{aligned}\bar{\xi}_r(s) &= \frac{1}{s^2 + \omega_r^2}[s\bar{Y}_r(s) + \omega_r \bar{Z}_r(s) + s\xi_r(0) + \omega_r \eta_r(0)], \\ \bar{\eta}_r(s) &= \frac{1}{s^2 + \omega_r^2}[s\bar{Z}_r(s) - \omega_r \bar{Y}_r(s) + s\eta_r(0) - \omega_r \xi_r(0)],\end{aligned}$$
$$r = 1, 2, \ldots, n \qquad (6.60)$$

Using Borel's theorem (M2) and considering Eqs. (6.59), we can write the solution of Eqs. (6.60) in the form of the convolution integrals

$$\begin{aligned}\xi_r(t) &= \int_0^t [\mathbf{y}_r^T \mathbf{X}(\tau) \cos \omega_r(t - \tau) + \mathbf{z}_r^T \mathbf{X}(\tau) \sin \omega_r(t - \tau)] \, d\tau \\ &\quad + \mathbf{y}_r^T M^* \mathbf{x}(0) \cos \omega_r t + \mathbf{z}_r^T M^* \mathbf{x}(0) \sin \omega_r t\end{aligned}$$
$$r = 1, 2, \ldots, n \qquad (6.61)$$
$$\begin{aligned}\eta_r(t) &= \int_0^t [\mathbf{z}_r^T \mathbf{X}(\tau) \cos \omega_r(t - \tau) - \mathbf{y}_r^T \mathbf{X}(\tau) \sin \omega_r(t - \tau) \, d\tau \\ &\quad + \mathbf{z}_r^T M^* \mathbf{x}(0) \sin \omega_r t - \mathbf{y}_r^T M^* \mathbf{x}(0) \cos \omega_r t\end{aligned}$$

where it was assumed that the functions $Y_r(t)$ and $Z_r(t)$ are defined only for $t > 0$ and that they are identically zero for $t < 0$. This, of course, implies that the vector $\mathbf{X}(t)$ is defined only for $t > 0$ and is zero for $t < 0$. Finally, inserting Eqs. (6.61) into series (6.51), we obtain the complete solution for the state vector

$$
\begin{aligned}
\mathbf{x}(t) = \sum_{r=1}^{n} \Big\{ & \int_0^t [(\mathbf{y}_r \mathbf{y}_r^T + \mathbf{z}_r \mathbf{z}_r^T)\mathbf{X}(\tau) \cos \omega_r (t - \tau) \\
& + (\mathbf{y}_r \mathbf{z}_r^T - \mathbf{z}_r \mathbf{y}_r^T)\mathbf{X}(\tau) \sin \omega_r (t - \tau)]\, d\tau \\
& + (\mathbf{y}_r \mathbf{y}_r^T + \mathbf{z}_r \mathbf{z}_r^T) M^* \mathbf{x}(0) \cos \omega_r t + (\mathbf{y}_r \mathbf{z}_r^T - \mathbf{z}_r \mathbf{y}_r^T) M^* \mathbf{x}(0) \sin \omega_r t \Big\}
\end{aligned}
$$

(6.62)

which represents the response of an undamped linear gyroscopic system to any external forces and initial disturbances.

The above exposition follows closely that of Ref. M5. This reference contains also a numerical example consisting of a spinning rigid body with elastically connected point-masses.

Example 6.5

Consider the gyroscopic system of Example 4.2, let $\Omega = 1$ rad s^{-1} and obtain the response to the excitation

$$
Q_1(t) = Q_0\, u\,(t), \qquad Q_2(t) = 0
$$

(a)

The initial conditions are zero.

From Example 4.2, we obtain the matrices

$$
M^* = m \begin{bmatrix} 1 & 0 & 0 & 0 \\ 0 & 1 & 0 & 0 \\ 0 & 0 & 1 & 0 \\ 0 & 0 & 0 & 2 \end{bmatrix}, \qquad K^* = m \begin{bmatrix} 5 & 0 & 0 & 4 \\ 0 & 6 & -2 & 0 \\ 0 & -2 & 1 & 0 \\ 4 & 0 & 0 & 4 \end{bmatrix}
$$

(b)

so that, solving the eigenvalue problem (6.45), we obtain the natural frequencies

$$
\omega_1 = 0.546295 \text{ rad s}^{-1}, \qquad \omega_2 = 2.588738 \text{ rad s}^{-1}
$$

(c)

and the normalized modal matrix

$$
P = [\mathbf{y}_1\ \mathbf{z}_1\ \mathbf{y}_2\ \mathbf{z}_2]
$$

$$
= m^{-1/2} \begin{bmatrix} 0 & 0.515499 & 0 & 0.856890 \\ 0.331007 & 0 & 0.943628 & 0 \\ 0.943628 & 0 & -0.331007 & 0 \\ 0 & -0.605913 & 0 & 0.364513 \end{bmatrix}
$$

(d)

Moreover, from Eqs. (a), the excitation vector is

$$\mathbf{X}(t) = Q_0[u(t) \quad 0 \quad 0 \quad 0]^T \tag{e}$$

Introducing Eqs. (c)–(e) into Eq. (6.62), we obtain the response

$$\mathbf{x}(t) = \sum_{r=1}^{2} \int_0^t [(\mathbf{y}_r \mathbf{y}_r^T + \mathbf{z}_r \mathbf{z}_r^T)\mathbf{X}(\tau) \cos \omega_r(t-\tau)$$

$$+ (\mathbf{y}_r \mathbf{z}_r^T - \mathbf{z}_r \mathbf{y}_r^T)\mathbf{X}(\tau) \sin \omega_r(t-\tau)] \, d\tau$$

$$= \frac{Q_0}{m} \int_0^t \left\{ \left(\begin{bmatrix} 0 \\ 0.331007 \\ 0.943628 \\ 0 \end{bmatrix} \begin{bmatrix} 0 \\ 0.331007 \\ 0.943628 \\ 0 \end{bmatrix}^T \right. \right.$$

$$+ \begin{bmatrix} 0.515499 \\ 0 \\ 0 \\ -0.605913 \end{bmatrix} \begin{bmatrix} 0.515499 \\ 0 \\ 0 \\ -0.605913 \end{bmatrix}^T \right)$$

$$\times \begin{bmatrix} u(\tau) \\ 0 \\ 0 \\ 0 \end{bmatrix} \cos 0.546295(t-\tau) + \left(\begin{bmatrix} 0 \\ 0.331007 \\ 0.943628 \\ 0 \end{bmatrix} \begin{bmatrix} 0.515499 \\ 0 \\ 0 \\ -0.605913 \end{bmatrix}^T \right.$$

$$- \begin{bmatrix} 0.515499 \\ 0 \\ 0 \\ -0.605913 \end{bmatrix} \begin{bmatrix} 0 \\ 0.331007 \\ 0.943628 \\ 0 \end{bmatrix}^T \right) \begin{bmatrix} u(\tau) \\ 0 \\ 0 \\ 0 \end{bmatrix} \sin 0.546295(t-\tau)$$

$$+ \left(\begin{bmatrix} 0 \\ 0.943628 \\ -0.331007 \\ 0 \end{bmatrix} \begin{bmatrix} 0 \\ 0.943628 \\ -0.331007 \\ 0 \end{bmatrix}^T + \begin{bmatrix} 0.856890 \\ 0 \\ 0 \\ 0.364513 \end{bmatrix} \begin{bmatrix} 0.856890 \\ 0 \\ 0 \\ 0.364513 \end{bmatrix}^T \right)$$

$$\times \begin{bmatrix} u(\tau) \\ 0 \\ 0 \\ 0 \end{bmatrix} \cos 2.588738(t-\tau) + \left(\begin{bmatrix} 0 \\ 0.943628 \\ -0.331007 \\ 0 \end{bmatrix} \begin{bmatrix} 0.856890 \\ 0 \\ 0 \\ 0.364513 \end{bmatrix}^T \right.$$

$$- \begin{bmatrix} 0.856890 \\ 0 \\ 0 \\ 0.364513 \end{bmatrix} \begin{bmatrix} 0 \\ 0.943628 \\ -0.331007 \\ 0 \end{bmatrix}^T \right) \begin{bmatrix} u(\tau) \\ 0 \\ 0 \\ 0 \end{bmatrix} \sin 2.588738(t-\tau) \right\} \, d\tau$$

$$= \frac{Q_0}{m} \left\{ \begin{bmatrix} 0.486440 \\ 0 \\ 0 \\ -0.571756 \end{bmatrix} \sin 0.546295t + \begin{bmatrix} 0 \\ 0.312348 \\ 0.890434 \\ 0 \end{bmatrix} (1 - \cos 0.546295t) \right.$$

$$+ \begin{bmatrix} 0.283637 \\ 0 \\ 0 \\ 0.120656 \end{bmatrix} \sin 2.588738t + \begin{bmatrix} 0 \\ 0.312348 \\ -0.109566 \\ 0 \end{bmatrix} (1 - \cos 2.588738t) \left. \right\}$$

(f)

6.7   Response of damped systems

The equation governing the motion of a viscously damped nongyroscopic linear system is obtained from Eq. (2.22) by letting the gyroscopic matrix G and the circulatory matrix H be equal to zero, with the result

$$M\ddot{\mathbf{q}}(t) + C\dot{\mathbf{q}}(t) + K\mathbf{q}(t) = \mathbf{Q}(t) \tag{6.63}$$

The general solution of Eq. (6.63) requires a transformation from the configuration space to the state space, and we shall study this approach later in this section. At this point, however, we wish to examine the possibility of obtaining a solution of Eq. (6.63) by a transformation in the configuration space, where the transformation matrix is the classical modal matrix. It turns out that this possibility exists, albeit for very special cases.

Let us consider the linear transformation (6.32), in which U is the classical modal matrix as defined in Sec. 6.5. Introducing Eq. (6.32) into Eq. (6.63), premultiplying the result by U^T and considering the orthogonality relations (6.30), we obtain

$$\ddot{\boldsymbol{\eta}}(t) + C'\dot{\boldsymbol{\eta}}(t) + \Lambda\boldsymbol{\eta}(t) = \mathbf{N}(t) \tag{6.64}$$

where the vector $\mathbf{N}(t)$ is as given by Eq. (6.34) and

$$C' = U^T C U \tag{6.65}$$

is a real symmetric matrix, generally nondiagonal. Hence, in general, the classical modal matrix does not uncouple the equations of motion of a damped system. In the sequel, we shall examine some cases in which it does.

A special case of particular interest is that in which the damping matrix is a linear combination of the inertia matrix and the stiffness matrix, or

$$C = \alpha M + \beta K \tag{6.66}$$

where α and β are constant scalars, the first having units s^{-1} and the second having units s. For obvious reasons, this case is often referred to as *proportional damping*. Clearly, in this case the classical modal matrix does uncouple the equations of motion. Indeed, recalling Eqs. (6.30), we obtain

$$C' = \alpha I + \beta \Lambda \tag{6.67}$$

which is a diagonal matrix. For convenience, we shall introduce the notation

$$C' = \text{diag} \, [\alpha + \beta \omega_i^2] = \text{diag} \, [2\xi_i \omega_i] \tag{6.68}$$

so that, in this case, Eq. (6.64) represents the set of n independent equations

$$\ddot{\eta}_i(t) + 2\zeta_i \omega_i \dot{\eta}_i(t) + \omega_i^2 \eta_i(t) = N_i(t), \qquad i = 1, 2, \ldots, n \tag{6.69}$$

The particular solution of equations of the type (6.69) was discussed in Sec. 6.2, and can be given in the form of the convolution integral

$$\eta_i(t) = \int_0^t N_i(t-\tau) h_i(\tau) \, d\tau, \qquad i = 1, 2, \ldots, n \tag{6.70}$$

where $h_i(t)$ is the impulse response given by (see Example 6.2)

$$h_i(t) = \frac{1}{\omega_{di}} e^{-\zeta_i \omega_i t} \sin \omega_{di} t \, u(t), \qquad i = 1, 2, \ldots, n \tag{6.71}$$

in which $\omega_{di} = (1 - \zeta_i^2)^{1/2} \omega_i$ is the frequency of the damped oscillation in the ith mode and $u(t)$ is the unit step function applied at $t = 0$. Note that Eq. (6.71) implies that $\zeta_i < 1$. The complete solution of Eq. (6.69) is obtained by adding to Eq. (6.70) the homogeneous solution. Of course, the configuration vector $\mathbf{q}(t)$ is obtained by inserting $\eta_i(t)$ $(i = 1, 2, \ldots, n)$ into Eq. (6.32).

The special case of proportional damping can be generalized somewhat. Indeed, it can be shown (C1) that when $M^{-1/2} C M^{-1/2}$ can be written in the form of a series involving $M^{-1/2} K M^{-1/2}$ to fractional powers the modal matrix is once again capable of uncoupling the equations of motion. Such a case is seldom realized in practice, and we shall not pursue the subject any further.

When damping is not of the proportional type but is very small, the coupling terms in Eq. (6.64) are often ignored, which is tantamount to regarding the matrix C' as diagonal (although in reality it may not be diagonal). This approximation is frequently used even when damping is not small, so that it is advisable to verify in each case whether the approximation can be justified or not.

Before considering the general case of damping, we wish to consider one more case in which it is possible to use the classical modal analysis. In all our previous discussions, it was implied tacitly that damping was viscous. We note that viscous damping is responsible for resisting forces experienced by a body during motion in a viscous medium, where the forces are proportional to the relative velocity between the moving body and the resisting medium. Another type of damping results from internal friction in deformable bodies and is associated with the so-called hysteresis loop during cyclic stress. Such damping is commonly referred to as *structural damping*. It turns out that in the case of harmonic external excitation one can devise an analogy whereby structural damping can be treated as if it were viscous (M2, Secs. 9-4 and 9-5). Indeed, if the excitation in Eq. (6.63) is of the form

$$\mathbf{Q}(t) = \mathbf{Q}_0 e^{i\Omega t} \tag{6.72}$$

where $\mathbf{Q}_0$ is a constant vector and Ω is the frequency of excitation and if the system is known to possess structural damping, then Eq. (6.63) retains its form but the damping matrix has the form

$$C = \frac{1}{\pi \Omega} D \tag{6.73}$$

so that structural damping is inversely proportional to the excitation frequency. The matrix $(1/\pi\Omega)D$ is known as the *hysteretic damping matrix*.

It is customary to assume that the hysteretic damping matrix is proportional to the stiffness matrix, or

$$D = \pi \gamma K \tag{6.74}$$

where γ is a *structural damping factor*. Introducing Eqs. (6.72–74) into Eq. (6.63) and recognizing that in the case of harmonic excitation the steady-state response is such that

$$\dot{\mathbf{q}}(t) = i\Omega \mathbf{q}(t) \tag{6.75}$$

we obtain

$$M\ddot{\mathbf{q}}(t) + (1 + i\gamma)K\mathbf{q}(t) = \mathbf{Q}_0 e^{i\Omega t} \tag{6.76}$$

which can be solved by the classical modal analysis. The solution of Eq.

(6.76) is left as an exercise to the reader. The matrix $(1 + i\gamma)K$ is sometimes referred to as the *complex stiffness matrix*.

Next, we shall consider the general case of damping, so that we must return to the state-vector approach. From Sec. 6.6, we recall the definition of the $2n$-dimensional state vector $\mathbf{x}(t)$, Eq. (6.40), as well as the definition of the $2n$-dimensional generalized force vector $\mathbf{X}(t)$, Eq. (6.41). Then, considering the $2n \times 2n$ real symmetric matrices

$$M^* = \begin{bmatrix} M & 0 \\ \hline 0 & -K \end{bmatrix}, \qquad K^* = \begin{bmatrix} C & K \\ \hline K & 0 \end{bmatrix} \tag{6.77}$$

introduced for the first time in Sec. 2.5, we can reduce Eq. (6.63) to

$$M^* \dot{\mathbf{x}}(t) + K^* \mathbf{x}(t) = \mathbf{X}(t) \tag{6.78}$$

which represents a set of $2n$ simultaneous first-order ode's. The solution of Eq. (6.78) can be obtained by a modal analysis, not of the classic kind but one especially tailored for systems defined in terms of two real symmetric matrices M^* and K^* neither of which is positive definite, as shown in the sequel.

The eigenvalue problem corresponding to Eq. (6.78) has the form

$$\lambda M^* \mathbf{x} + K^* \mathbf{x} = \mathbf{0} \tag{6.79}$$

and was discussed in Sec. 3.6. The solution consists of $2n$ eigenvalues λ_i and $2n$ eigenvectors $\mathbf{x}_i$ ($i = 1, 2, \ldots, 2n$). We shall assume that all the eigenvalues and eigenvectors are complex. Because the matrices M^* and K^* are real, if λ_i is an eigenvalue, then $\bar{\lambda}_i$ is also an eigenvalue, and a similar statement can be made concerning the eigenvectors. Moreover, because M^* and K^* are symmetric, the eigenvectors are orthogonal with respect to both M^* and K^*. In view of this, we shall find it convenient to write the square matrix of the eigenvectors in the form

$$X = [\mathbf{x}_1 \ \bar{\mathbf{x}}_1 \ \mathbf{x}_2 \cdots \mathbf{x}_n \ \bar{\mathbf{x}}_n] \tag{6.80}$$

We shall assume that the eigenvectors are normalized so as to satisfy

$$X^T M^* X = I, \qquad X^T K^* X = -\Lambda \tag{6.81}$$

where Λ is the diagonal matrix of the eigenvalues.

Next, let us consider the expansion

$$\mathbf{x}(t) = \sum_{r=1}^{n} [\mathbf{x}_r z_r(t) + \bar{\mathbf{x}}_r \bar{z}_r(t)] = X\mathbf{z}(t) \tag{6.82}$$

where

$$\mathbf{z}(t) = [z_1(t) \ \bar{z}_1(t) \ z_2(t) \ \bar{z}_2(t) \cdots z_n(t) \ \bar{z}_n(t)]^T \tag{6.83}$$

213

6 Response of discrete systems

Introducing Eq. (6.82) into Eq. (6.78), premultiplying both sides by X^T and considering Eqs. (6.81), we obtain

$$\dot{\mathbf{z}}(t) = \Lambda \mathbf{z}(t) + \mathbf{Z}(t) \tag{6.84}$$

where

$$\mathbf{Z}(t) = X^T \mathbf{X}(t) \tag{6.85}$$

is a $2n$-dimensional generalized force vector. Equation (6.84) represents a set of $2n$ independent equations of the type

$$\dot{z}_r(t) = \lambda_r z_r(t) + Z_r(t), \qquad r = 1, 2, \ldots, n \tag{6.86a}$$

$$\dot{\bar{z}}_r(t) = \bar{\lambda}_r \bar{z}_r(t) + \bar{Z}_r(t), \qquad r = 1, 2, \ldots, n \tag{6.86b}$$

Of course, only Eqs. (6.86a) need be solved, as Eqs. (6.86b) do not contain any additional information. Note from Eq. (6.85) that

$$Z_r(t) = \mathbf{x}_r^T \mathbf{X}(t), \qquad r = 1, 2, \ldots, n \tag{6.87a}$$

$$\bar{Z}_r(t) = \bar{\mathbf{x}}_r^T \mathbf{X}(t), \qquad r = 1, 2, \ldots, n \tag{6.87b}$$

The particular solution of Eqs. (6.86a) can be written in terms of the convolution integral presented in Sec. 6.3. Hence, the complete solution is

$$z_r(t) = z_r(0)e^{\lambda_r t} + \int_0^t Z_r(\tau) h_r(t - \tau)\, d\tau$$

$$= z_r(0)e^{\lambda_r t} + \int_0^t Z_r(\tau) e^{\lambda_r (t-\tau)}\, d\tau,$$

$$r = 1, 2, \ldots, n \tag{6.88}$$

where

$$z_r(0) = \mathbf{x}_r^T M^* \mathbf{x}(0), \qquad r = 1, 2, \ldots, n \tag{6.89}$$

in which $\mathbf{x}(0)$ is the initial state vector. Inserting Eqs. (6.88–89) into Eq. (6.82), we obtain the general solution

$$\mathbf{x}(t) = \sum_{r=1}^{n} \left\{ \left[e^{\lambda_r t} \mathbf{x}_r^T M^* \mathbf{x}(0) + \int_0^t e^{\lambda_r (t-\tau)} \mathbf{x}_r^T \mathbf{X}(\tau)\, d\tau \right] \mathbf{x}_r \right.$$

$$\left. + \left[e^{\bar{\lambda}_r t} \bar{\mathbf{x}}_r^T M^* \mathbf{x}(0) + \int_0^t e^{\bar{\lambda}_r (t-\tau)} \bar{\mathbf{x}}_r^T \mathbf{X}(\tau)\, d\tau \right] \bar{\mathbf{x}}_r \right\}$$

$$= 2\, \mathrm{Re} \sum_{r=1}^{n} \left[e^{\lambda_r t} \mathbf{x}_r^T M^* \mathbf{x}(0) + \int_0^t e^{\lambda_r (t-\tau)} \mathbf{x}_r^T \mathbf{X}(\tau)\, d\tau \right] \mathbf{x}_r \tag{6.90}$$

Damped gyroscopic systems are considerably more complicated than either undamped gyroscopic systems or damped nongyroscopic systems and must be treated as general dynamical systems (see Sec. 6.8). When damping is sufficiently small that damping terms can be treated as perturbations on the undamped system, significant reduction in the computations can be realized. Indeed, basing the computations on the eigensolution of the undamped system, which is known to be real, the response can be produced by a perturbation approach working with real quantities alone (M9).

Example 6.6

The system shown in Fig. 6.14 is initially at rest. Obtain the response of the system to the excitation

$$Q_1(t) = 0, \qquad Q_2(t) = Q_0 \delta(t) \tag{a}$$

The system shown in Fig. 6.14 is initially at rest. Obtain the response of $\omega = 1$.

First, we solve the eigenvalue problem (6.79), where the matrices M^* and K^* are

$$M^* = m \begin{bmatrix} 1 & 0 & 0 & 0 \\ 0 & 2 & 0 & 0 \\ 0 & 0 & -5 & 4 \\ 0 & 0 & 4 & -4 \end{bmatrix}, \qquad K^* = m \begin{bmatrix} 0.4 & -0.2 & 5 & -4 \\ -0.2 & 0.2 & -4 & 4 \\ 5 & -4 & 0 & 0 \\ -4 & 4 & 0 & 0 \end{bmatrix} \tag{b}$$

The solution of the eigenvalue problem consists of the eigenvalues

$$\frac{\lambda_1}{\bar{\lambda}_1} = -0.222593 \pm 2.578255i$$

$$\frac{\lambda_2}{\bar{\lambda}_2} = -0.027406 \pm 0.545796i \tag{c}$$

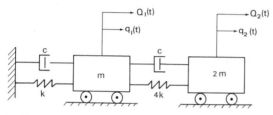

Figure 6.14

and the eigenvectors

$$
\frac{\mathbf{x}_1}{\bar{\mathbf{x}}_1} = m^{-1/2} \begin{bmatrix} 0.606770 \\ -0.257661 \\ -0.006335 \\ 0.008716 \end{bmatrix} \pm im^{-1/2} \begin{bmatrix} 0.035928 \\ 0.000395 \\ -0.234792 \\ 0.099184 \end{bmatrix}
$$

$$
\frac{\mathbf{x}_2}{\bar{\mathbf{x}}_2} = m^{-1/2} \begin{bmatrix} -0.364906 \\ -0.428675 \\ 0.023037 \\ 0.017020 \end{bmatrix} \pm im^{-1/2} \begin{bmatrix} -0.005717 \\ -0.012211 \\ 0.667411 \\ 0.784552 \end{bmatrix}
$$

(d)

where the eigenvectors have been normalized so as to satisfy Eqs. (6.81). In this regard, it should be pointed out that in general computer subroutines solving the eigenvalue problem (6.79) do not yield normalized eigenvectors satisfying Eqs. (6.81), so that it is necessary to normalize them. This can be done by multiplying a typical eigenvector $\mathbf{x}_r$ by a complex constant c_r and determining the constant from the formula

$$
c_r^2 = \frac{1}{\mathbf{x}_r^T M^* \mathbf{x}_r}, \qquad r = 1, 2, \ldots, n
$$

(e)

From Eqs. (a), we can write the excitation vector

$$
\mathbf{X}(t) = Q_0 [0 \quad \delta(t) \quad 0 \quad 0]^T
$$

(f)

so that Eq. (6.90) yields the response

$$
\mathbf{x}(t) = 2 \operatorname{Re} \sum_{r=1}^{2} \left[\int_0^t e^{\lambda_r(t-\tau)} \mathbf{x}_r^T \mathbf{X}(\tau) \, d\tau \right] \mathbf{x}_r
$$

$$
= 2Q_0 \operatorname{Re} \sum_{r=1}^{2} \left[x_{r2} \int_0^t e^{\lambda_r(t-\tau)} \delta(\tau) \, d\tau \right] \mathbf{x}_r
$$

$$
= \frac{2Q_0}{m} \{ e^{-0.222593t}
$$

$$
< (-0.257661 \cos 2.578255t - 0.000395 \sin 2.578255t)
$$

$$
\times \begin{bmatrix} 0.606770 \\ -0.257661 \\ -0.006335 \\ 0.008716 \end{bmatrix}
$$

$$
- (0.000395 \cos 2.578255t - 0.257661 \sin 2.578255t)
$$

$$\times \begin{bmatrix} 0.035928 \\ 0.000395 \\ -0.234792 \\ 0.099184 \end{bmatrix} > + e^{-0.027406t} < (-0.428675 \cos 0.545796t$$

$$+0.012211 \sin 0.545796t) \begin{bmatrix} -0.364906 \\ -0.428675 \\ 0.023037 \\ 0.017020 \end{bmatrix}$$

$$-(-0.012211 \cos 0.545796t$$

$$-0.428675 \sin 0.545796t) \begin{bmatrix} -0.005717 \\ -0.012211 \\ 0.667411 \\ 0.784552 \end{bmatrix} >\} \qquad (g)$$

6.8   Response of general dynamical systems

In Sec. 2.3, we derived the equation of motion of a general dynamical system in the form

$$M\ddot{\mathbf{q}}(t) + (G + C)\dot{\mathbf{q}}(t) + (K + H)\mathbf{q}(t) = \mathbf{Q}(t) \qquad (6.91)$$

where M, C and K are symmetric matrices and G and H are skew symmetric matrices. Although G is skew symmetric and C is symmetric, the matrix $G + C$ is arbitrary, and a similar statement can be made concerning $K + H$. Hence, no particular advantage accrues from the special form of the coefficient matrices and the system must be treated as a general one. The only assumption made here is that M and K are positive definite.

Introducing the $2n$-dimensional state vector $\mathbf{x}(t) = [\dot{\mathbf{q}}^T(t) \mid \mathbf{q}^T(t)]^T$ and the $2n$-dimensional excitation vector $\mathbf{X}(t) = [\mathbf{Q}^T(t) \mid \mathbf{0}^T]^T$, Eq. (6.91) can be transformed into

$$M^*\dot{\mathbf{x}}(t) + K^*\mathbf{x}(t) = \mathbf{X}(t) \qquad (6.92)$$

where

$$M^* = \begin{bmatrix} M & 0 \\ \hline 0 & K \end{bmatrix}, \qquad K^* = \begin{bmatrix} G + C & K + H \\ \hline -K & 0 \end{bmatrix} \qquad (6.93)$$

are real $2n \times 2n$ matrices. But, whereas M^* is a positive definite symmetric matrix, K^* is neither positive definite nor symmetric. Equation (6.92) can be rendered into a more convenient form by considering the Cholesky decomposition

$$M^* = LL^T \qquad (6.94)$$

Then, introducing the linear transformation

$$L^T\mathbf{x}(t) = \mathbf{u}(t), \qquad \mathbf{x}(t) = L^{-T}\mathbf{u}(t) \qquad (6.95)$$

where $L^{-T} = (L^T)^{-1} = (L^{-1})^T$, Eq. (6.92) can be reduced to

$$\dot{\mathbf{u}}(t) = A\mathbf{u}(t) + \mathbf{U}(t) \qquad (6.96)$$

in which

$$A = -L^{-1}K^*L^{-T} \qquad (6.97)$$

is a real nonsymmetric matrix and

$$\mathbf{U}(t) = L^{-1}\mathbf{X}(t) \qquad (6.98)$$

is a real vector.

The solution of Eq. (6.96) can be obtained by modal analysis, which amounts to the determination of the Jordan form for A. The eigenvalue problem associated with system (6.96) has the form

$$A\mathbf{u} = \lambda\mathbf{u} \qquad (6.99)$$

and was discussed in Sec. 3.7. The solution consists of $2n$ eigenvalues λ_i and $2n$ eigenvectors $\mathbf{u}_i$ $(i = 1, 2, \ldots, 2n)$. We shall assume that all the eigenvalues are distinct, so that the Jordan matrix is diagonal

$$\Lambda = \text{diag}\,[\lambda_i] \qquad (6.100)$$

The eigenvectors $\mathbf{u}_i$, known as right eigenvectors of A, can be arranged in the square matrix

$$U = [\mathbf{u}_1\,\mathbf{u}_2 \cdots \mathbf{u}_{2n}] \qquad (6.101)$$

The adjoint eigenvalue problem is

$$\mathbf{v}^T A = \lambda\mathbf{v}^T \qquad (6.102)$$

and its solution consists of the same eigenvalues λ_i as well as the left eigenvectors $\mathbf{v}_i$ $(i = 1, 2, \ldots, 2n)$. They also can be arranged in a square matrix, as follows:

$$V = [\mathbf{v}_1\,\mathbf{v}_2 \cdots \mathbf{v}_{2n}] \qquad (6.103)$$

It was shown in Sec. 3.7 that the set of eigenvectors $\mathbf{u}_i$ is orthogonal to the set of eigenvectors $\mathbf{v}_j$, a property known as biorthogonality. The eigenvectors can be normalized so as to satisfy

$$V^T U = U^T V = I \tag{6.104}$$

in which case the Jordan matrix is simply

$$V^T A U = \Lambda \tag{6.105}$$

Because the vectors $\mathbf{u}_i$ on the one hand and the vectors $\mathbf{v}_i$ on the other hand are linearly independent, either set of vectors can be taken as a basis for L^{2n}. Hence, we can assume that the solution of Eq. (6.96) has the form

$$\mathbf{u}(t) = \sum_{i=1}^{2n} \mathbf{u}_i z_i(t) = U\mathbf{z}(t) \tag{6.106}$$

where $\mathbf{z}(t)$ is a $2n$-vector with components $z_i(t)$. Introducing Eq. (6.106) into Eq. (6.96), premultiplying the result by V^T and considering Eqs. (6.104–105), we obtain

$$\dot{\mathbf{z}}(t) = \Lambda \mathbf{z}(t) + \mathbf{Z}(t) \tag{6.107}$$

in which

$$\mathbf{Z}(t) = V^T \mathbf{U}(t) \tag{6.108}$$

Equation (6.107) represents a set of $2n$ independent equations of the type (6.84). Assuming that all the eigenvalues and eigenvectors are complex conjugates, it follows that the independent equations consist of n pairs of complex conjugates, so that the solutions of these equations are as given by Eqs. (6.86). Recalling Eqs. (6.95) and (6.106), the actual response is

$$\mathbf{x}(t) = L^{-T} U\mathbf{z}(t) \tag{6.109}$$

The above decoupling procedure may not necessarily be the most efficient approach to the derivation of the response. It calls for the solution to the eigenvalue problem of a nonsymmetric matrix, and this is not nearly as desirable as that for a symmetric matrix. In general, it involves complex algebra, which implies more laborious computations. Moreover, the stability of the various computational algorithms cannot be taken for granted and care must be exercised in using them. As an alternative, one may wish to consider the direct solution of Eq. (6.96), i.e., a solution without decoupling. It is easy to verify that the homogeneous solution of Eq. (6.96) is

$$\mathbf{u}(t) = e^{At}\mathbf{u}(0) \tag{6.110}$$

The question remains as to how to obtain the particular solution. To this end, let us premultiply Eq. (6.96) by the nonsingular $2n \times 2n$ matrix $P(t)$ and write

$$P(t)\dot{\mathbf{u}}(t) = P(t)A\mathbf{u}(t) + P(t)\mathbf{U}(t) \tag{6.111}$$

Then, consider

$$\frac{\mathrm{d}}{\mathrm{d}t}[P(t)\mathbf{u}(t)] = \dot{P}(t)\mathbf{u}(t) + P(t)\dot{\mathbf{u}}(t)$$

$$= \dot{P}(t)\mathbf{u}(t) + P(t)A\mathbf{u}(t) + P(t)\mathbf{U}(t) \tag{6.112}$$

so that, letting

$$\dot{P}(t) = -P(t)A \tag{6.113}$$

which implies that

$$P(t) = e^{-At} \tag{6.114}$$

Eq. (6.112) reduces to

$$\frac{\mathrm{d}}{\mathrm{d}t}[P(t)\mathbf{u}(t)] = P(t)\mathbf{U}(t) \tag{6.115}$$

Equation (6.115) can be integrated to yield

$$P(t)\mathbf{u}(t) = P(0)\mathbf{u}(0) + \int_0^t P(\tau)\mathbf{U}(\tau)\,\mathrm{d}\tau \tag{6.116}$$

where, from Eq. (6.114), we conclude that

$$P(0) = I \tag{6.117}$$

and we note that Eq. (6.116) actually includes both the homogeneous and the particular solution. Hence, using Eq. (6.115), the complete solution of Eq. (6.96) is

$$\mathbf{u}(t) = P^{-1}(t)\mathbf{u}(0) + P^{-1}(t)\int_0^t P(\tau)U(\tau)\,\mathrm{d}\tau$$

$$= e^{At}\mathbf{u}(0) + \int_0^t e^{A(t-\tau)}\mathbf{U}(\tau)\,\mathrm{d}\tau \tag{6.118}$$

The matrix

$$\Phi(t, \tau) = e^{A(t-\tau)} \tag{6.119}$$

is often referred to as the *transition matrix*, and the question remains as to how to evaluate it. One approach is to expand the series

$$\Phi(t, \tau) = e^{A(t-\tau)} = I + (t-\tau)A + \frac{(t-\tau)^2}{2!} A^2 + \frac{(t-\tau)^3}{3!} A^3 + \cdots \quad (6.120)$$

Clearly, this is an infinite series and it must be truncated. The number of terms to be retained in the series depends on the product of the eigenvalue of A of highest modulus and the time interval $\Delta t = t - \tau$. To verify this statement, we recognize that Eq. (6.105) implies that

$$A = U\Lambda V^T \quad (6.121)$$

(note that $V^T = U^{-1}$) from which it follows that

$$\Phi(t, \tau) = I + (t-\tau)U\Lambda V^T + \frac{(t-\tau)^2}{2!} U\Lambda V^T U\Lambda V^T$$

$$+ \frac{(t-\tau)^3}{3!} U\Lambda V^T U\Lambda V^T U\Lambda V^T + \cdots \quad (6.122)$$

which, in view of Eq. (6.104), reduces to

$$\Phi(t, \tau) = U\left[I + (t-\tau)\Lambda + \frac{(t-\tau)^2}{2!} \Lambda^2 + \frac{(t-\tau)^3}{3!} \Lambda^3 + \cdots \right] V^T$$

$$= U e^{\Lambda(t-\tau)} V^T \quad (6.123)$$

where the expression inside the square brackets was recognized as $e^{\Lambda(t-\tau)}$.

If Δt is very large, then a large number of terms must be retained in series (6.120). To circumvent this problem, it is recommended that the interval Δt be divided into smaller subintervals and then use be made of the so-called 'group property', expressed by

$$\Phi(t, \tau) = \Phi(t, t_k)\Phi(t_k, t_{k-1}) \cdots \Phi(t_2, t_1)\Phi(t_1, \tau) \quad (6.124)$$

which permits the evaluation of the transition matrix by performing a continuous product of transition matrices corresponding to smaller time intervals.

6.9 Discrete-time model for general dynamical systems

In Sec. 6.8, we have shown how to evaluate the response of a general continuous-time system of order $2n$. In the process, we have introduced the concept of transition matrix relating the state of the system at one instant to the state at a different instant. The same response will be discussed now in the context of discrete-time systems.

Equation (6.118) can be reduced to a form more convenient for digital computation by the approach presented in Sec. 6.4. To this end, we assume that the components of the excitation vector $\mathbf{U}(t)$ can be approximated by piecewise constant functions of the type shown in Fig. 6.11. Letting $t = kT$, where T is the sampling time, Eq. (6.118) becomes

$$\mathbf{u}(kT) = e^{kTA}\mathbf{u}(0) + \int_0^{kT} e^{(kT-\tau)A}\mathbf{U}(\tau)\,d\tau \tag{6.125}$$

At the next sampling, we have

$$\mathbf{u}(kT+T) = e^{(kT+T)A}\mathbf{u}(0) + \int_0^{kT+T} e^{(kT+T-\tau)A}\mathbf{U}(\tau)\,d\tau \tag{6.126}$$

which can be rewritten as follows:

$$\mathbf{u}(kT+T) = e^{TA}\left[e^{kTA}\mathbf{u}(0) + \int_0^{kT} e^{(kT-\tau)A}\mathbf{U}(\tau)\,d\tau \right]$$
$$+ \int_{kT}^{kT+T} e^{(kT+T-\tau)A}\mathbf{U}(\tau)\,d\tau \tag{6.127}$$

Recalling that $\mathbf{U}(t)$ was assumed to be piecewise constant, we can write

$$\int_{kT}^{kT+T} e^{(kT+T-\tau)A}\mathbf{U}(\tau)\,d\tau \cong \left[\int_{kT}^{kT+T} e^{(kT+T-\tau)A}\,d\tau \right]\mathbf{U}(kT) \tag{6.128}$$

Next, let us introduce the change of variables $kT + T - \tau = t$, so that

$$\int_{kT}^{kT+T} e^{(kT+T-\tau)A}\,d\tau = \int_T^0 e^{tA}(-dt) = \int_0^T e^{tA}\,dt \tag{6.129}$$

Introducing Eqs. (6.128–129) into Eq. (6.127) and recalling Eq. (6.125), we obtain the sequence

$$\mathbf{u}_{k+1} = \Phi\mathbf{u}_k + \Gamma\mathbf{U}_k, \qquad k = 0, 1, 2, \ldots \tag{6.130}$$

where

$$\mathbf{u}_k = \mathbf{u}(kT), \qquad \mathbf{U}_k = \mathbf{U}(kT) \tag{6.131}$$

and

$$\Phi = e^{TA}, \qquad \Gamma = \int_0^T e^{tA}\,dt \tag{6.132}$$

Note that the first of Eqs. (6.132) really represents the transition matrix for the discrete-time system.

From the above, we conclude that the problem has been reduced to the evaluation of the matrix e^{TA}, a problem discussed in Sec. 6.8. Indeed, the matrix can be written in the form of the series

$$e^{TA} = I + TA + \frac{T^2}{2!} A^2 + \frac{T^3}{3!} A^3 + \cdots \qquad (6.133)$$

Recalling from Eq. (6.121) that $A = U \Lambda V^T$ and recognizing that $A^p = U \Lambda^p V^T$, where p is any integer, Eq. (6.133) can be rewritten as

$$e^{TA} = U \left(I + T\Lambda + \frac{1}{2!} T^2 \Lambda^2 + \frac{1}{3!} T^3 \Lambda^3 + \cdots \right) V^T \qquad (6.134)$$

Hence, by choosing T suficiently small that the product of the eigenvalue of A of largest modulus and T is a relatively small number, the series can be made to converge with only a limited number of terms.

6.10 Stability of motion in the neighborhood of equilibrium

In Sec. 2.3, we introduced the concepts of equilibrium points and of motion in the neighborhood of equilibrium points. In this section, we propose to bring these concepts into sharper focus.

Let us write the state equations of motion in the general form

$$M^* \dot{\mathbf{x}}(t) = \mathbf{F}[\mathbf{x}(t)] \qquad (6.135)$$

where M^* is the $2n \times 2n$ matrix given by the first of Eqs. (6.93), $\mathbf{x}(t) = [\dot{\mathbf{q}}^T(t) \vdots \mathbf{q}^T(t)]^T$ is the $2n$-dimensional state vector and $\mathbf{F}$ is a $2n$-dimensional force vector. The vector $\mathbf{F}$ is assumed to depend explicitly on the state vector but not on time. Then, an equilibrium point is defined as a constant vector $\mathbf{x}_e$ satisfying the vector equation

$$\mathbf{F}(\mathbf{x}_e) = \mathbf{0} \qquad (6.136)$$

In the general case in which $\mathbf{F}$ is a nonlinear function of $\mathbf{x}$, Eq. (6.136) admits a number of solutions. If $\mathbf{F}$ is a linear function of $\mathbf{x}$, then Eq. (6.136) admits only one solution.

Next, let us consider motion in the neighborhood of the equilibrium point $\mathbf{x}_e$. To this end, we introduce the transformation

$$\mathbf{x}(t) = \mathbf{x}_e + \mathbf{x}_p(t) \qquad (6.137)$$

where $\mathbf{x}_p(t)$ is a vector representing the perturbed motion from the equilibrium point $\mathbf{x}_e$. Expanding the vector $\mathbf{F}$ about $\mathbf{x}_e$, we obtain

$$\mathbf{F}(\mathbf{x}) = \mathbf{F}(\mathbf{x}_e) + S\mathbf{x}_p + \mathbf{0}(\mathbf{x}_p) \qquad (6.138)$$

where S is a $2n \times 2n$ matrix with the entries

$$s_{ij} = \frac{\partial F_i}{\partial x_j} \bigg|_{\mathbf{x}=\mathbf{x}_e}, \qquad i, j = 1, 2, \ldots, 2n \qquad (6.139)$$

and $\mathbf{0}(\mathbf{x}_p)$ is a vector consisting of nonlinear terms in $\mathbf{x}_p$, i.e., terms of degree two and higher in $\mathbf{x}_p$. Recalling Eq. (6.136), ignoring the nonlinear terms and introducing the notation

$$S = -K^* \tag{6.140}$$

Eq. (6.135) can be reduced to

$$M^*\dot{\mathbf{x}}_p(t) = -K^*\mathbf{x}_p(t) \tag{6.141}$$

Without loss of generality, the origin of the state space can be translated so as to make it coincident with the equilibrium point $\mathbf{x}_e$, in which case Eq. (6.141) becomes simply

$$M^*\dot{\mathbf{x}}(t) = -K^*\mathbf{x}(t) \tag{6.142}$$

Equation (6.142) can be recognized as Eq. (6.92) but with $\mathbf{X}(t) = \mathbf{0}$. It represents the linearized state equations of motion in the neighborhood of trivial equilibrium. Our object is to examine the character of the motion governed by Eq. (6.142) and under what conditions it is the same as the motion governed by Eq. (6.135).

Introducing the linear transformation (6.95), Eq. (6.142) can be reduced to

$$\dot{\mathbf{u}}(t) = A\mathbf{u}(t) \tag{6.143}$$

where A is given by Eq. (6.97). From Sec. 6.9, we conclude that the solution of Eq. (6.143) can be written in the form

$$\mathbf{u}(t) = e^{At}\mathbf{u}(0) \tag{6.144}$$

where

$$e^{At} = Ue^{\Lambda t}V^T \tag{6.145}$$

is the transition matrix, in which Λ is the matrix of the eigenvalues of A, U is the matrix of the right eigenvectors of A and V is the matrix of the left eigenvectors of A. Because U and V are constant, the nature of the motion $\mathbf{u}(t)$ depends on $e^{\Lambda t}$, and hence on the eigenvalues of A. If all the eigenvalues are distinct, then Λ is a diagonal matrix, so that the vector $\mathbf{u}(t)$ consists of a linear combination of vectors multiplied by exponential terms of the form $e^{\lambda_r t}(r = 1, 2, \ldots, 2n)$. Generally, the eigenvalues λ_r are complex,

$$\lambda_r = \alpha_r + i\beta_r, \qquad r = 1, 2, \ldots, 2n \tag{6.146}$$

so that

$$e^{\lambda_r t} = e^{\alpha_r t} e^{i\beta_r t}, \qquad r = 1, 2, \ldots, 2n \tag{6.147}$$

But $e^{i\beta_r t}$ is a complex vector of unit magnitude rotating counterclockwise in the complex plane with the angular velocity β_r. On the other hand, $e^{\alpha_r t}$ can be regarded as a time-dependent amplitude. Hence, the nature of the motion depends on the real part α_r of λ_r $(r = 1, 2, \ldots, 2n)$. In particular, if $\alpha_r < 0$, the term tends to zero as $t \to \infty$. But, for the motion to diverge, it is only necessary that a single exponential term tend to infinity. Motion that tends away from the equilibrium point as $t \to \infty$ is said to be *unstable*. Hence, *the motion in the neighborhood of the equilibrium is unstable if at least one of the system eigenvalues possess positive real part.* Motion that tends to the equilibrium point as $t \to \infty$ is said to be *asymptotically stable*. Hence, *the motion in the neighborhood of the equilibrium is asymptotically stable if all the system eigenvalues possess negative real parts.* There remains the question as to the case in which all the eigenvalues are pure imaginary, i.e., when they all possess zero real parts. In this case, the motion is pure oscillatory and it remains in the neighborhood of the equilibrium point, without exhibiting any secular trend. This implies that the motion neither tends to the equilibrium point nor does it tend away from it as $t \to \infty$. In this case, the motion is said to be merely *stable*.

When the system is unstable or asymptotically stable, the linearized system is said to exhibit *significant behavior*. When the system is merely stable, the linearized system is said to exhibit *critical behavior*. If the linearized system, Eq. (6.142), exhibits critical behavior, then its motion characteristics in the neighborhood of equilibrium are not necessarily the same as those of the nonlinear system and the nonlinear terms may have to be taken into account. A more extensive discussion of system stability can be found in M3 (Chs. 5 and 6).

It is often convenient to map the eigenvalues in the complex plane. Then, the imaginary axis represents the region of mere stability, the left half plane the region of asymptotic stability and the right half plane the region of instability, both these two latter regions excluding the imaginary axis (Fig. 6.15). Hence, the system exhibits critical behavior only in the region Re $\lambda = 0$ and it exhibits significant behavior in the regions Re $\lambda > 0$ and Re $\lambda < 0$, which include almost the entire complex plane. Clearly, critical behavior characterized by mere stability must be regarded as a borderline case, separating asymptotic stability from instability. Nevertheless, the case of critical behavior is a very important one in vibrations, as it represents the large class of conservative systems.

The counterpart of Eq. (6.143) for discrete systems is

$$\mathbf{u}_{k+1} = \Phi \mathbf{u}_k \tag{6.148}$$

225

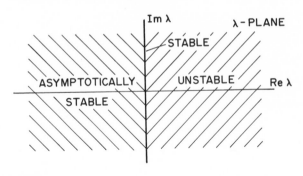

Figure 6.15

By analogy with Eq. (6.121), we can write

$$\Phi = U^* \Lambda^* V^{*T} \tag{6.149}$$

where Λ^* is the diagonal matrix of eigenvalues of Φ and U^* and V^* are the matrices of right and left eigenvectors of Φ, respectively. Then, introducing the linear transformation

$$\mathbf{u}_k = U^* \mathbf{z}_k, \qquad \mathbf{u}_{k+1} = U^* \mathbf{z}_{k+1} \tag{6.150}$$

Eq. (6.148) can be rewritten in the decoupled form

$$\mathbf{z}_{k+1} = \Lambda^* \mathbf{z}_k \tag{6.151}$$

Hence, the system behavior depends on the magnitudes $|\lambda_j^*|$ ($j = 1, 2, \ldots, 2n$) of the eigenvalues of Φ. Indeed, the system is *asymptotically stable* if $|\lambda_j^*| < 1$ *for all* j. The system is *unstable if at least one eigenvalue* λ_j *is such that* $|\lambda_j^*| > 1$. Moreover, the system is merely *stable if some eigenvalues are such that* $|\lambda_j^*| = 1$ *and the remaining ones are such that* $|\lambda_i^*| < 1, i \neq j$. If the eigenvalues are plotted in the complex plane, then asymptotic stability requires that all the eigenvalues lie inside the unit circle with the center at the origin of the complex plane, instability occurs if at least one eigenvalue lies outside the unit circle and mere stability occurs if some eigenvalues lie on the unit circle and the remaining ones lie inside the unit circle (Fig. 6.16).

Example 6.7

Derive the differential equation describing the motion of a uniform bar rotating about a vertical axis with the constant angular velocity Ω, as shown in Fig. 6.17. Then, formulate the state equations, identify the equilibrium positions and test the stability of motion in the neighborhood of the equilibrium positions.

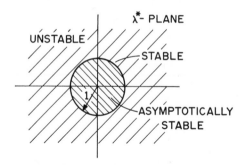

Figure 6.16

The bar is acted upon by gravitational and centrifugal forces. Taking moments about point 0, we obtain

$$I_0\ddot{\theta} = \int_0^L \Omega^2 \xi \sin\theta \, \xi \cos\theta \, \rho \, d\xi - \int_0^L \xi \sin\theta \, \rho g \, d\xi$$

$$= \tfrac{1}{3}\rho\Omega^2 L^3 \sin\theta \cos\theta - \tfrac{1}{2}\rho g L^2 \sin\theta \qquad\qquad (a)$$

where $I_0 = \rho L^3/3$ is the mass moment of inertia about 0, in which ρ is the mass per unit length and L is the length of the bar. Rearranging, Eq. (a) can be reduced to

$$\ddot{\theta} + \left(\frac{3}{2}\frac{g}{L} - \Omega^2 \cos\theta\right)\sin\theta = 0 \qquad\qquad (b)$$

Equation (b) can be transformed into two first-order state equations. Letting

$$\theta = x_1, \qquad \dot{\theta} = x_2 \qquad\qquad (c)$$

the state equations are

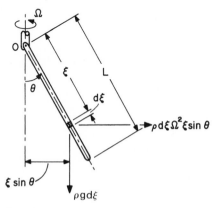

Figure 6.17

227

$$\dot{x}_1 = x_2$$

$$\dot{x}_2 = [\Omega^2 \cos x_1 - 3g/2L) \sin x_1 \tag{d}$$

The equilibrium points are defined as constant values of x_1 and x_2 that render the right side of Eqs. (d) equal to zero. it is easy to verify that there are three equilibrium points, namely,

$$E_1 : x_{e1} = x_{e2} = 0, \qquad E_2 : x_{e1} = \pi, \qquad x_{e2} = 0,$$

$$E_3 : x_{e1} = \cos^{-1} 3g/2L\Omega^2, \qquad x_{e2} = 0 \tag{e}$$

Of course, the first is the trivial equilibrium, but the other two are nontrivial equilibrium points. The linearized equations of motion in the neighborhood of equilibrium can be written in the matrix form

$$\dot{\mathbf{x}} = A\mathbf{x} \tag{f}$$

In the case of the equilibrium point E_1, the coefficient matrix is

$$A = \begin{bmatrix} 0 & 1 \\ \Omega^2 - 3g/2L & 0 \end{bmatrix} \tag{g}$$

in the case of the equilibrium point E_2, the coefficient matrix is

$$A = \begin{bmatrix} 0 & 1 \\ \Omega^2 + 3g/2L & 0 \end{bmatrix} \tag{h}$$

and in the case of the equilibrium point E_3, the coefficient matrix is

$$A = \begin{bmatrix} 0 & 1 \\ -\Omega^2 \sin^2 x_{e1} & 0 \end{bmatrix} \tag{i}$$

where, by the third of Eqs. (e), $x_{e1} = \cos^{-1} 3g/2L\Omega^2$.

In the case E_1, the eigenvalues of the matrix A are

$$\begin{matrix} \lambda_1 \\ \lambda_2 \end{matrix} = \pm(\Omega^2 - 3g/2L)^{1/2} \tag{j}$$

Hence, the equilibrium point E_1 is *stable if* $\Omega^2 < 3g/2L$ and *unstable if* $\Omega^2 > 3g/2L$. On the other hand, in the case E_2 the eigenvalues of A are

$$\begin{matrix} \lambda_1 \\ \lambda_2 \end{matrix} = \pm(\Omega^2 + 3g/2L)^{1/2} \tag{k}$$

so that the equilibrium point E_2 is *unstable*. Finally, in the case of the equilibrium point E_3, the eigenvalues of A are

$$\begin{matrix} \lambda_1 \\ \lambda_2 \end{matrix} = \pm i\Omega \sin x_{e1} \tag{l}$$

so that the equilibrium point E_3 is *stable*.

7

Vibration of continuous systems

7.1 Introduction

In Chs. 2–6, we have studied the vibration of systems with discrete properties. The spatial position of the variables describing the motion of discrete systems appears only implicitly, in the form of identifying indices, so that mathematically these variables depend only on time. As a result, the mathematical formulation is given by a set of simultaneous ordinary differential equations.

More often than not, discrete models are mere idealizations, and in fact the actual system properties are distributed in space. Systems with distributed properties are known as *distributed-parameter systems*, or *continuous systems*. The variables describing the motion of distributed systems depend explicitly on the spatial position, as well as on time. The vibration problem is defined over a given domain D and the mathematical formulation is in terms of one (or more) partial differential equation(s) to be satisfied over the domain D, as well as certain conditions to be satisfied at every point of the boundary S of the domain D.

Because of the difference in the mathematical formulation, different mathematical tools than those needed for discrete systems are needed for the treatment of the vibration of continuous systems. One significant difference between discrete and distributed systems is that, if the time dependence is eliminated from the system formulation, the behavior of discrete system is governed by a set of simultaneous algebraic equations, whereas that of distributed-parameter systems is described by a so-called *boundary-value problem*, consisting of a differential equation to be satisfied over D and boundary conditions to be satisfied at every point of S. Except for ideal cases, almost invariably defined by uniform mass and stiffness distributions, boundary-value problems defy closed-form solutions. This would seem to imply that the study of distributed-parameter systems is a wasted effort. This is not the case, however, as such a study is likely to enhance the understanding of the system behavior. Moreover, whereas closed-form solutions are desirable, they are not really necessary, and quite often approximate solutions are equally desirable. In producing approximate solutions, the knowledge gained from the study of continuous systems proves invaluable. It should be pointed out that approximate solutions for the vibration of continuous systems generally involve system discretization, and that many of the approximate methods differ only in the manner in which discretization is achieved.

The above discussion points to the fact that for the most part discrete systems and continuous systems represent merely different mathematical models of the same physical system. Hence, it should come as no surprise that, although the mathematical treatment is different, many of the concepts encountered in the study of discrete systems are encountered again in the study of distributed systems. These concepts include the natural modes of vibration, system symmetry, system positive definiteness and orthogonality of the natural modes, where the latter forms the basis for the expansion theorem and the closely related modal analysis.

In this chapter, we begin by deriving Lagrange's equation of motion for the vibration of continuous systems and the associated eigenvalue problem. Then, a discussion of the important class of self-adjoint systems follows. In this context, various classes of functions are defined, thus preparing the groundwork for a discussion in the next several chapters of approximate methods using finite series expansions. After presenting some typical examples of continuous systems, variational principles paralleling those for discrete systems are introduced. The chapter closes with an integral formulation of the eigenvalue problem and the derivation of the system response by modal analysis.

7.2   Lagrange's equation for continuous systems. Boundary-value problem

Our interest lies in the vibration of systems with distributed properties. The motion of such systems is described by variables depending not only on time but also on the spatial position. Because there are at least two independent variables, the motion of distributed systems is governed by partial differential equations (pde's) as opposed to discrete systems governed by ordinary differential equations (ode's). Although in general domains of extension of distributed systems are three-dimensional, many of the features characterizing distributed systems can be most conveniently discussed in terms of a one-dimensional spatial domain.

Let us consider a distributed system defined over the closed domain $0 \leq x \leq L$, where x is the spatial position of any material point of the system, and denote by $w(x, t)$ the dependent variable, which more often

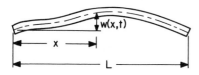

Figure 7.1

than not represents the displacement of the point from a given reference position (see Fig. 7.1). We shall assume that the kinetic energy of the system has the general expression

$$T(t) = \int_0^L \hat{T}(x, t) \, dx \tag{7.1}$$

where $\hat{T}(x, t)$ is the kinetic energy density having the functional form

$$\hat{T}(x, t) = \hat{T}(\dot{w}, \dot{w}') \tag{7.2}$$

where dots denote partial derivatives with respect to t and primes denote partial derivatives with respect to x. Note that the functional dependence of $\hat{T}$ on x and t is implicit through $w(x, t)$. Similarly, the virtual work is assumed to have the expression

$$\delta W(t) = \int_0^L \delta \hat{W}(x, t) \, dx \tag{7.3}$$

where the virtual work density $\delta \hat{W}(x, t)$ can be separated into

$$\delta \hat{W}(x, t) = \delta \hat{W}_c(x, t) + \delta \hat{W}_{nc}(x, t) \tag{7.4}$$

in which the subscripts c and nc refer to conservative and nonconservative work, respectively. The conservative virtual work is the negative of the variation in the potential energy (M3, Sec. 2.5), or

$$\delta \hat{W}_c(x, t) = -\delta \hat{V}(x, t) \tag{7.5}$$

where $\hat{V}(x, t)$ is the potential energy density, which for the purpose of this discussion is assumed to have the following functional dependence

$$\hat{V}(x, t) = \hat{V}(w, w', w'') \tag{7.6}$$

On the other hand, denoting the distributed nonconservative force by $p(x, t)$, the nonconservative virtual work density can be written as

$$\delta \hat{W}_{nc}(x, t) = p(x, t) \, \delta w(x, t) \tag{7.7}$$

where $\delta w(x, t)$ is the virtual displacement of any point in the domain.

231

The equations governing the motion of the system can be conveniently obtained by the extended Hamilton principle (M3, Sec. 2.7), which can be stated in the form

$$\int_{t_1}^{t_2} (\delta T + \delta W)\, dt = 0, \qquad \delta w = 0, \qquad 0 \le x \le L, \quad \text{at} \quad t = t_1, t_2 \qquad (7.8)$$

where t_1 and t_2 are arbitrary times. Introducing Eqs. (7.1–7.5) into Eq. (7.8), we obtain

$$\int_{t_1}^{t_2} \int_0^L (\delta \hat{L} + \delta \hat{W}_{nc})\, dx\, dt = 0, \qquad \delta w = 0, \qquad 0 \le x \le L, \quad \text{at} \quad t = t_1, t_2 \qquad (7.9)$$

where

$$\hat{L} = \hat{T} - \hat{V} = \hat{L}(w, w', w'', \dot{w}, \dot{w}') \qquad (7.10)$$

is the Lagrangian density. From Eq. (7.10), we can write the Lagrangian density variation

$$\delta \hat{L} = \frac{\partial \hat{L}}{\partial w} \delta w + \frac{\partial \hat{L}}{\partial w'} \delta w' + \frac{\partial \hat{L}}{\partial w''} \delta w'' + \frac{\partial \hat{L}}{\partial \dot{w}} \delta \dot{w} + \frac{\partial \hat{L}}{\partial \dot{w}'} \delta \dot{w}' \qquad (7.11)$$

so that, introducing Eqs. (7.7) and (7.11) into Eq. (7.9), we have

$$\int_{t_1}^{t_2} \int_0^L \left(\frac{\partial \hat{L}}{\partial w} \delta w + \frac{\partial \hat{L}}{\partial w'} \delta w' + \frac{\partial \hat{L}}{\partial w''} \delta w'' + \frac{\partial \hat{L}}{\partial \dot{w}} \delta \dot{w} + \frac{\partial \hat{L}}{\partial \dot{w}'} \delta \dot{w}' + p\, \delta w \right) dx\, dt = 0 \qquad (7.12)$$

The next step is to transform the integrand in Eq. (7.12) into one containing only δw, i.e., one that is free of $\delta w'$, $\delta w''$, $\delta \dot{w}$ and $\delta \dot{w}'$. This can be accomplished by integration by parts, both with respect to space and time. To this end, let us consider

$$\int_0^L \frac{\partial \hat{L}}{\partial w'} \delta w'\, dx = \frac{\partial \hat{L}}{\partial w'} \delta w \Big|_0^L - \int_0^L \frac{\partial}{\partial x} \left(\frac{\partial \hat{L}}{\partial w'} \right) \delta w\, dx \qquad (7.13a)$$

$$\int_0^L \frac{\partial \hat{L}}{\partial w''} \delta w''\, dx = \frac{\partial \hat{L}}{\partial w''} \delta w' \Big|_0^L - \frac{\partial}{\partial x} \left(\frac{\partial \hat{L}}{\partial w''} \right) \delta w \Big|_0^L + \int_0^L \frac{\partial^2}{\partial x^2} \left(\frac{\partial \hat{L}}{\partial w''} \right) \delta w\, dx \qquad (7.13b)$$

$$\int_{t_1}^{t_2} \frac{\partial \hat{L}}{\partial \dot{w}} \delta \dot{w}\, dt = \frac{\partial \hat{L}}{\partial \dot{w}} \delta w \Big|_{t_1}^{t_2} - \int_{t_1}^{t_2} \frac{\partial}{\partial t} \left(\frac{\partial \hat{L}}{\partial \dot{w}} \right) \delta w\, dt \qquad (7.13c)$$

$$\int_{t_1}^{t_2}\int_0^L \frac{\partial \hat{L}}{\partial \dot{w}'}\,\delta \dot{w}'\,dx\,dt = \int_{t_1}^{t_2}\left[\frac{\partial \hat{L}}{\partial \dot{w}'}\,\delta \dot{w}\Big|_0^L - \int_0^L \frac{\partial}{\partial x}\left(\frac{\partial \hat{L}}{\partial \dot{w}'}\right)\delta \dot{w}\,dx\right]dt$$

$$= \left[\frac{\partial \hat{L}}{\partial \dot{w}'}\,\delta w\Big|_0^L\,\Big|_{t_2}^{t_1} - \int_{t_1}^{t_2}\frac{\partial}{\partial t}\left(\frac{\partial \hat{L}}{\partial \dot{w}'}\right)\delta w\Big|_0^L\,dt\right.$$

$$-\left[\int_0^L \frac{\partial}{\partial x}\left(\frac{\partial \hat{L}}{\partial \dot{w}'}\right)\delta w\,dx\right]_{t_1}^{t_2}$$

$$+\int_{t_1}^{t_2}\int_0^L \frac{\partial^2}{\partial t\,\partial x}\left(\frac{\partial \hat{L}}{\partial \dot{w}'}\right)\delta w\,dx\,dt \qquad (7.13\text{d})$$

Introducing Eqs. (7.13) into Eq. (7.12), recalling that $\delta w = 0$ at $t = t_1,\ t_2$ and collecting terms, we can write

$$\int_{t_1}^{t_2}\left\{\int_0^L\left[\frac{\partial \hat{L}}{\partial w} - \frac{\partial}{\partial x}\left(\frac{\partial \hat{L}}{\partial w'}\right) + \frac{\partial^2}{\partial x^2}\left(\frac{\partial \hat{L}}{\partial w''}\right) - \frac{\partial}{\partial t}\left(\frac{\partial \hat{L}}{\partial \dot{w}}\right) + \frac{\partial^2}{\partial x\,\partial t}\left(\frac{\partial \hat{L}}{\partial \dot{w}'}\right) + p\right]\delta w\,dx\right.$$

$$\left.+\left[\frac{\partial \hat{L}}{\partial w'} - \frac{\partial}{\partial x}\left(\frac{\partial \hat{L}}{\partial w''}\right) - \frac{\partial}{\partial t}\left(\frac{\partial \hat{L}}{\partial \dot{w}'}\right)\right]\delta w\Big|_0^L + \frac{\partial \hat{L}}{\partial w''}\,\delta w'\Big|_0^L\right\}dt = 0 \qquad (7.14)$$

But the virtual displacement δw is arbitrary by definition, which implies that it can be assigned values at will, provided these values are compatible with the system constraints, such as geometric conditions at the end points. Letting $\delta w = \delta w' = 0$ at $x = 0$ and $x = L$, we conclude that Eq. (7.14) can be satisfied for all values of δw in the open domain $0 < x < L$ if and only if

$$\frac{\partial \hat{L}}{\partial w} - \frac{\partial}{\partial x}\left(\frac{\partial \hat{L}}{\partial w'}\right) + \frac{\partial^2}{\partial x^2}\left(\frac{\partial \hat{L}}{\partial w''}\right) - \frac{\partial}{\partial t}\left(\frac{\partial \hat{L}}{\partial \dot{w}}\right) + \frac{\partial^2}{\partial x\,\partial t}\left(\frac{\partial \hat{L}}{\partial \dot{w}'}\right) + p = 0,$$

$$0 < x < L \qquad (7.15)$$

Moreover, by writing

$$\left[\frac{\partial \hat{L}}{\partial w'} - \frac{\partial}{\partial x}\left(\frac{\partial \hat{L}}{\partial w''}\right) - \frac{\partial}{\partial t}\left(\frac{\partial \hat{L}}{\partial \dot{w}'}\right)\right]\delta w\Big|_0^L = 0 \qquad (7.16\text{a})$$

$$\frac{\partial \hat{L}}{\partial w''}\,\delta w'\Big|_0^L = 0 \qquad (7.16\text{b})$$

we take into account the possibilities that either $(\partial \hat{L}/\partial w') - [\partial(\partial \hat{L}/\partial w'')/\partial x] - [\partial(\partial \hat{L}/\partial \dot{w}')/\partial t]$ or δw on the one hand and either $\partial \hat{L}/\partial w''$ or $\delta w'$ on the other hand is zero at either of the ends $x = 0$ and $x = L$.

Equation (7.15) represents the *Lagrange differential equation of motion* for this continuous system and must be satisfied at every point of the open domain $0 < x < L$. On the other hand, Eqs. (7.16) can be satisfied in a variety of ways. Indeed, Eq. (7.16a) is satisfied if either

$$\frac{\partial \hat{L}}{\partial w'} - \frac{\partial}{\partial x}\left(\frac{\partial \hat{L}}{\partial w''}\right) - \frac{\partial}{\partial t}\left(\frac{\partial \hat{L}}{\partial \dot{w}'}\right) = 0 \quad \text{at} \quad x = 0, L \tag{7.17a}$$

or

$$w = 0 \quad \text{at} \quad x = 0, L \tag{7.17b}$$

and Eq. (7.16b) is satisfied if either

$$\frac{\partial L}{\partial w''} = 0 \quad \text{at} \quad x = 0, L \tag{7.18a}$$

or

$$w' = 0 \quad \text{at} \quad x = 0, L \tag{7.18b}$$

Equations (7.17) and (7.18) are known as *boundary conditions* and only two boundary conditions must be satisfied at either end, one from conditions (7.17) and the other from conditions (7.18). Clearly, the choice is not arbitrary but must reflect physical conditions at the two ends.

Problems such as those associated with the ordinary differential equations discussed in Ch. 6, in which the solutions propagate in time from given initial conditions, are known as *initial-value problems*. Problems in which the solutions must satisfy a differential equation in a given open domain and certain conditions on the boundaries of the domain are called *boundary-value problems*. The problem defined by Eqs. (7.15) and (7.17–18) is actually both an initial-value problem and a boundary-value problem at the same time, since its solution must satisfy both initial and boundary conditions. Nevertheless, because of its close identification with a boundary-value problem we shall refer to it as such.

The nature of the boundary conditions requires further elaboration. Indeed, the boundary conditions can be divided into two distinct classes, each reflecting different types of physical conditions, the first reflecting geometric constraints and the second force and moment balance at the boundaries. Perhaps this question can be better clarified by resorting to a more specific example.

Let us consider a bar in bending with one end free and with the other end attached rigidly to a shaft rotating uniformly with the angular velocity Ω (Fig. 7.2). For simplicity, we shall assume that the radius of the shaft is negligible. The interest lies in the vibration of the bar in the vertical plane containing the shaft. Note that the system can be regarded as simulating a helicopter blade. The kinetic energy has the expression

$$T(t) = \frac{1}{2} \int_0^L m(x)[\dot{w}(x, t)]^2 \, dx + \frac{1}{2} \int_0^L J(x)[\dot{w}'(x, t)]^2 \, dx \tag{7.19}$$

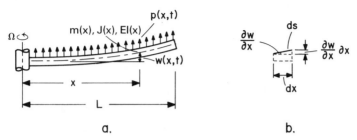

Figure 7.2

where $m(x)$ represents the mass of the bar per unit length and $J(x)$ is the mass moment of inertia of an element of bar of unit length about an axis normal to the plane of the figure and passing through the center line of the bar. Moreover, $\partial w / \partial t$ is recognized as the translational velocity and $\partial^2 w / \partial t \, \partial x$ as the angular velocity of the element in question. Accordingly, the first term on the right side of Eq. (7.19) represents the translational kinetic energy and the second represents the rotational kinetic energy.

The potential energy can be regarded as consisting of two parts, one due to bending and one due to the axial force acting through the shortening of the projection. Denoting an increment of length along the displaced axis by ds and the projection by dx, the potential energy can be written in the form

$$V(t) = \frac{1}{2} \int_0^L EI(x)[w''(x, t)]^2 \, dx + \int_0^L P(x, t)(ds - dx) \qquad (7.20)$$

where $EI(x)$ represents the flexural rigidity, in which E is the modulus of elasticity and $I(x)$ is the moment of inertia of the cross-sectional area about its centroidal axis, and $P(x, t)$ denotes the axial force, where in this case the latter arises from centrifugal effects. Anticipating that the shortening of the projection $ds - dx$ is a small quantity of second order, and wishing to retain only quadratic terms in the displacement and its derivatives, we shall assume that the centrifugal force on a differential element of mass is not affected by a change in its position. Hence, because the shaft rotates with the constant angular velocity Ω, the axial force is assumed to be

$$P(x, t) = P(x) = \int_x^L m(\xi) \Omega^2 \xi \, d\xi \qquad (7.21)$$

Moreover, from Fig. 7.2b, if we use the binomial expansion, the shortening of the projection can be approximated as follows:

$$ds - dx = \{(dx)^2 + [w'(x, t) \, dx]^2\}^{1/2} - dx \cong \tfrac{1}{2}[w'(x, t)]^2 \, dx \qquad (7.22)$$

235

where the customary small-motions assumption was made. Introducing Eqs. (7.21) and (7.22) into Eq. (7.20), the potential energy becomes

$$V(t) = \frac{1}{2} \int_0^L EI(x)[w''(x, t)]^2 \, dx + \frac{1}{2} \int_0^L \left[\int_x^L m(\xi)\Omega^2 \xi \, d\xi \right] [w'(x, t)]^2 \, dx$$

$$(7.23)$$

The virtual work is due to the blade's own weight. Although this work is conservative, we can treat it as nonconservative. Because the weight acts in the opposite direction to that of the virtual displacement, we have

$$\delta W_{nc} = -\int_0^L m(x)g \, \delta w(x, t) \, dx \qquad (7.24)$$

where g is the acceleration due to gravity.

At this point, we have all the ingredients necessary for the derivation of the boundary-value problem. From Eqs. (7.19) and (7.23), we obtain the Lagrangian density

$$\hat{L} = \tfrac{1}{2}m\dot{w}^2 + \tfrac{1}{2}J(\dot{w}')^2 - \tfrac{1}{2}EI(w'')^2 - \frac{1}{2}\left(\int_x^L m\Omega^2 \xi \, d\xi \right)(w')^2 \qquad (7.25)$$

whereas from eq. (7.24) we can write the distributed force

$$p = -mg \qquad (7.26)$$

and we observe that the force does not really depend on time. Next, let us evaluate the terms in Eq. (7.15) one by one, with the result

$$\frac{\partial \hat{L}}{\partial w} = 0$$

$$\frac{\partial}{\partial x}\left(\frac{\partial \hat{L}}{\partial w'} \right) = m\Omega^2 xw' - \left(\int_x^L m\Omega^2 \xi \, d\xi \right)w''$$

$$\frac{\partial^2}{\partial x^2}\left(\frac{\partial \hat{L}}{\partial w''} \right) = -\frac{\partial^2}{\partial x^2}(EIw'') \qquad (7.27)$$

$$\frac{\partial}{\partial t}\left(\frac{\partial \hat{L}}{\partial \dot{w}} \right) = m\ddot{w}$$

$$\frac{\partial^2}{\partial x \, \partial t}\left(\frac{\partial \hat{L}}{\partial \dot{w}'} \right) = \frac{\partial^2}{\partial x \, \partial t}(J\dot{w}') = \frac{\partial}{\partial x}(J\ddot{w}')$$

so that the differential equation of motion is

$$-m\Omega^2 xw' + \left(\int_x^L m\Omega^2 \xi \, d\xi \right)w'' - \frac{\partial^2}{\partial x^2}(EIw'') - m\ddot{w} + \frac{\partial}{\partial x}(J\ddot{w}') - mg = 0,$$

$$0 < x < L \quad (7.28)$$

Now, let us turn our attention to the boundary conditions. The necessary quantities are

$$\frac{\partial \hat{L}}{\partial w'} = -\left(\int_x^L m\Omega^2 \xi \, d\xi\right) w'$$

$$\frac{\partial}{\partial x}\left(\frac{\partial \hat{L}}{\partial w''}\right) = -\frac{\partial}{\partial x}(EIw'')$$

$$\frac{\partial}{\partial t}\left(\frac{\partial \hat{L}}{\partial \dot{w}'}\right) = J\ddot{w}'$$ (7.29)

$$\frac{\partial \hat{L}}{\partial w''} = -EIw''$$

so that Eq. (7.17a) takes the form

$$-\left(\int_x^L m\Omega^2 \xi \, d\xi\right) w' + \frac{\partial}{\partial x}(EIw'') - J\ddot{w}' = 0 \quad \text{at} \quad x = 0, L$$ (7.30)

whereas Eq. (7.18a) becomes simply

$$-EIw'' = 0 \quad \text{at} \quad x = 0, L$$ (7.31)

At the end $x = 0$, either Eq. (7.17b) or Eq. (7.30) and either Eq. (7.18b) or Eq. (7.31) must be satisfied. Because this end is fixed, it is easy to see that the boundary conditions are

$$w(0, t) = 0$$ (7.32a)

and

$$w'(x, t)\big|_{x=0} = 0$$ (7.32b)

which are clearly the result of geometric considerations. A choice similar to that for the end $x = 0$ must be made for the end $x = L$. Because $x = L$ is a free end, geometry makes no demands there. But, w and w' are generally not zero at a free end, which leads us to conclude that Eqs. (7.30) and (7.31) must be satisfied at $x = L$. Hence, we have

$$\frac{\partial}{\partial x}[EIw''(x, t)]\big|_{x=L} = J\ddot{w}'(x, t)\big|_{x=L}$$ (7.33a)

where we note that the first term in Eq. (7.30) is zero because the integral is zero, and

$$EIw''(x, t)\big|_{x=L} = 0$$ (7.33b)

Boundary conditions demanded by geometry are called *geometric* or

237

essential boundary conditions. Clearly, both boundary conditions (7.32) fall in this category. On the other hand, Eq. (7.33a) can be identified as the condition for shearing force balance and Eq. (7.33b) as the condition for bending moment balance. Such boundary conditions are known as *dynamic* or *natural boundary conditions.* We note that *geometric boundary conditions consist of terms involving derivatives with respect to x through first order only,* whereas the *dynamic boundary conditions consist of terms involving derivatives with respect to x through third order.* The above classification of boundary conditions has profound implications in the solution of boundary-value problems by approximate methods, as shown in Ch. 8.

The boundary conditions for a given end point can be of any type and of any mix. As shown earlier, for a fixed end there are only geometric boundary conditions and for a free end there are only dynamic boundary conditions. On the other hand, for a hinged end there is one geometric boundary condition and one dynamic boundary condition, the first stating that the displacement is zero and the second that the bending moment is zero. Equations (7.32) and (7.33) do not cover all possible forms of boundary conditions. Indeed, other forms of boundary conditions occur when ends contain discrete elements, such as lumped masses and springs. Hamilton's principle can easily produce the appropriate boundary conditions also in such cases, provided one does not fail to include in the kinetic energy terms accounting for the presence of lumped masses and in the potential energy terms due to springs.

The formulation of the boundary-value problem is not quite complete, as a complete solution of the problem requires the knowledge of initial conditions, to supplement the differential equation, Eq. (7.28), and the boundary conditions, Eqs. (7.32) and (7.33). The initial conditions have the general form

$$w(x, 0) = w_0(x) \tag{7.34a}$$

and

$$\dot{w}(x, t)\big|_{t=0} = \dot{w}_0(x) \tag{7.34b}$$

where $w_0(x)$ and $\dot{w}_0(x)$ are given functions of x.

An examination of Eqs. (7.28) and (7.32–34) reveals that the term $m(x)g$ will produce a time-invariant displacement $w_{st}(x)$, corresponding to a quasi-static equilibrium deformation pattern. With the understanding that the displacement $w(x, t)$ is measured relative to this static equilibrium, the term $m(x)g$ can be omitted from Eq. (7.28).

7.3   The eigenvalue problem

Let us consider the problem of helicopter blade vibration, as given by Eqs. (7.28) and (7.32–34). For simplicity, we shall assume that the displacement $w(x, t)$ is measured from the static equilibrium, so that the distributed force mg can be ignored from Eq. (7.28). Hence, in this case $w(x, t)$ represents the displacement from the equilibrium position, so that the latter is regarded as coinciding with the trivial solution. We shall assume that the equilibrium is stable. Because there are no external forces acting upon the system, the problem is one of *free vibration*.

Next, let us assume that the solution is separable in space and time. The physical implication of this assumption is that the system executes *synchronous motion*, i.e., one in which every point of the system executes the same motion in time. It follows that in synchronous motion the ratio of the displacements at two different points of the blade does not change with time. Mathematically, the assumption implies that the motion is separable in space and time. Hence, let us assume a solution of Eq. (7.28) in the form

$$w(x, t) = W(x)F(t) \tag{7.35}$$

where $W(x)$ depends on the spatial position alone and $F(t)$ depends on time alone. In the sequel, we shall omit the arguments x and t when no particular emphasis is necessary. Introducing Eq. (7.35) into Eq. (7.28), we obtain

$$\left[-m\Omega^2 x W' + \left(\int_x^L m\Omega^2 \xi \, d\xi \right) W'' - \frac{d^2}{dx^2} (EIW'') \right] F$$
$$= \left[mW - \frac{d}{dx} (JW') \right] \ddot{F},$$
$$0 < x < L \tag{7.36}$$

where now primes and dots denote total derivatives with respect to x and t, respectively. Similarly, Eqs. (7.32) yield

$$W(0)F = 0 \tag{7.37a}$$

and

$$W'|_{x=0} F = 0 \tag{7.37b}$$

whereas Eqs. (7.33) become

$$\left[\frac{d}{dx} (EIW'') \right]_{x=L} F = (JW')|_{x=L} F \tag{7.38a}$$

and

$$(EIW'')|_{x=L} F = 0 \tag{7.38b}$$

Dividing Eq. (7.36) through by $-F[mW - \mathrm{d}(JW')/\mathrm{d}x]$, we obtain

$$\frac{m\Omega^2 xW' - \left(\int_x^L m\Omega^2 \xi \, \mathrm{d}\xi\right) W'' + \mathrm{d}^2(EIW'')/\mathrm{d}x^2}{mW - \mathrm{d}(JW')/\mathrm{d}x} = -\frac{\ddot{F}}{F} \tag{7.39}$$

Because the left side of Eq. (7.39) depends on x alone and the right side on t alone, and x and t are independent, Eq. (7.39) can be satisfied if and only if the two sides of the equation are equal to a constant, a real constant. Equating the right side of Eq. (7.39) to λ, we can write

$$\ddot{F} + \lambda F = 0 \tag{7.40}$$

A solution of Eq. (7.40) can be assumed in the form

$$F(t) = A e^{st} \tag{7.41}$$

where A and s are constants. Introducing Eq. (7.41) into Eq. (7.40) and dividing through by $A e^{st}$, we obtain the *characteristic equation*

$$s^2 + \lambda = 0 \tag{7.42}$$

which has the two roots

$$\begin{matrix} s_1 \\ s_2 \end{matrix} = \pm\sqrt{-\lambda} \tag{7.43}$$

It follows that Eq. (7.40) has two solutions, one corresponding to the root s_1 and the other to the root s_2. Hence, the general solution of Eq. (7.40) is

$$F(t) = A_1 e^{s_1 t} + A_2 e^{s_2 t} = A_1 e^{\sqrt{-\lambda}t} + A_2 e^{-\sqrt{-\lambda}t} \tag{7.44}$$

The question remains as to the sign of λ. If λ is negative, then the exponents s_1 and s_2 are real, the first positive and the second negative. Because in this case the motion is unbounded, which is inconsistent with motion in the neighborhood of stable equilibrium, this possibility must be ruled out. If λ is positive, then we can introduce the notation $\lambda = \omega^2$ and obtain

$$\begin{matrix} s_1 \\ s_2 \end{matrix} = \pm i\omega \tag{7.45}$$

so that Eq. (7.45) becomes

$$F(t) = A_1 e^{i\omega t} + A_2 e^{-i\omega t} \tag{7.46}$$

and, because $F(t)$ must be real, $A_2 = \bar{A}_1$. Equation (7.46) represents harmonic oscillation with the frequency ω. Hence, we conclude that if synchronous motion is possible, then the time dependence is harmonic.
The question must be asked as to whether the frequency ω is arbitrary or not. To answer this question, we turn our attention to the left side of Eq. (7.39). Because the left side of Eq. (7.39) must also be equal to λ, we obtain

$$m\Omega^2 x W' - \left(\int_x^L m\Omega^2 \xi \, d\xi \right) W'' + \frac{d^2}{dx^2}(EIW'') = \lambda \left[mW - \frac{d}{dx}(JW') \right],$$

$$0 < x < L \quad (7.47)$$

Moreover, Eqs. (7.37) yield

$$W(0) = 0 \tag{7.48a}$$

and

$$W'|_{x=0} = 0 \tag{7.48b}$$

whereas, considering Eq. (7.40), Eqs. (7.38) reduce to

$$-\left[\frac{d}{dx}(EIW'') \right]_{x=L} = \lambda(JW')|_{x=L} \tag{7.49a}$$

and

$$(EIW'')|_{x=L} = 0 \tag{7.49b}$$

The problem of determining the parameter λ for which the differential equation (7.47) admits a nontrivial solution W, where the solution is subject to the boundary conditions (7.48–49), is known as the *characteristic-value problem* or *eigenvalue problem*. The eigenvalue problem (7.47–49) plays the same role for distributed systems as the eigenvalue problem (2.34) plays for discrete systems.
The solution of the eigenvalue problem (7.47–49) proceeds by first solving the differential equation (7.47). Because the equation is of fourth order, the solution must in general contain four constants of integration, in addition to the parameter λ, and there are only four boundary conditions available for the determination of these five constants. Hence, the solution cannot be unique. Use of the boundary conditions (7.48–49) yields four homogeneous equations with the constants of integration as unknowns. Three of these equations can be used to determine three of the constants in terms of the fourth, but the fourth cannot be determined uniquely. Instead, one obtains an equation for λ, generally transcendental, which is known as the *characteristic equation*, or *frequency equation*.

7 Vibration of continuous systems

There is a denumerably infinite set of values of λ satisfying the characteristic equation, namely, $\lambda_1, \lambda_2, \ldots$. They are called *characteristic values* or *eigenvalues*. Associated with these values there is a denumerably infinite set of functions $W_1, W_2, \ldots$ known as *characteristic functions* or *eigenfunctions*. Because one of the four constants of integration cannot be determined uniquely, the functions $W_1, W_2, \ldots$ can only be determined within a multiplying constant. The implication is that the shape of a given eigenfunction is unique but the amplitude is not. The square roots of the eigenvalues, $\omega_r = \sqrt{\lambda_r}$ $(r = 1, 2, \ldots)$, are the *natural frequencies* of the system, and, consistent with this, the functions W_r $(r = 1, 2, \ldots)$ are also referred to as *natural modes*. If they are normalized, thus rendering their amplitudes unique, then they are called *normal modes*. Hence, the answer to the question posed earlier is that synchronous motion is possible in an infinite number of ways. The motions consist simply of harmonic oscillations in the natural modes with frequencies equal to the natural frequencies. As we shall see later, every one of these natural motions can take place independently of the other.

Equations (7.47–49) represent but one example of an eigenvalue problem. Because of the particular form of the differential equation and the associated boundary conditions, its solution is unique, in the sense that the natural frequencies and mode shapes are unique. It should be perfectly obvious that if at least one boundary condition is different the solution will be different, even though the differential equation remains the same. The eigenvalue problem (7.47–49) belongs to a large class of eigenvalue problems, known as self-adjoint, and all the problems in this class exhibit similar characteristics. Hence, instead of attempting to produce a solution of Eqs. (7.47–49) at this point, we shall generalize the formulation, thus permitting us to establish properties of solution shared by the entire class of problems. The interesting part of this approach is that these properties can be established regardless of whether a solution of a given eigenvalue problem can be produced or not. In fact, as it turns out, no closed-form solution of the eigenvalue problem (7.47–49) is possible, so that the prospect of establishing properties of the solution, without actually having the solution, is very attractive. This is particularly true in view of the fact that knowledge of these properties can be used to produce approximate solutions.

7.4 Self-adjoint systems

In many cases, the eigenvalue problem does not lend itself to closed-form solution, so that one may wish to consider an approximate solution. This is certainly true in the case of the eigenvalue problem for the helicopter blade, Eqs. (7.47–49). To construct approximate solutions, some knowledge of the properties of the solution can be of great help. In this section we propose to examine ways of determining these properties, without actually having the solution. It turns out that the properties possessed by the eigenvalue problem (7.47–49) are shared by a large class of dynamical systems, namely the class of self-adjoint systems. In the sequel, we wish to introduce the concept of self-adjointness and to establish various properties of self-adjoint systems. To this end, it will be to our advantage to generalize the eigenvalue problem formulation. This in turn, necessitates the introduction of certain mathematical definitions.

Let w be a function of the spatial variable x and consider the linear homogeneous differential expression

$$Lw = A_0(x)w + A_1(x)\frac{dw}{dx} + \cdots + A_{2p}(x)\frac{d^{2p}w}{dx^{2p}} \tag{7.50}$$

where $A_0, A_1, \ldots, A_{2p}$ are known functions. The expression Lw is a linear homogeneous combination of the function w and its derivatives through order $2p$, where p is an integer, so that the differential expression is said to be of order $2p$. The symbol L represents a linear homogeneous differential operator of the type encountered in Sec. 6.2 and can be written in the form

$$L = A_0(x) + A_1(x)\frac{d}{dx} + \cdots + A_{2p}(x)\frac{d^{2p}}{dx^{2p}} \tag{7.51}$$

Hence, the differential expression (7.50) can be produced by letting L operate on w. The linearity of the operator L can be established in a manner analogous to that in Sec. 6.2. Note that the fact that the coefficient functions $A_0, A_1, \ldots, A_{2p}$ depend on x and are not constant does not affect the linearity of L.

Next, let M be an operator similar to L, but only of order $2q$, $q < p$, and write the differential equation

$$Lw = \lambda Mw \tag{7.52}$$

where λ is a parameter. Equation (7.52) is defined over the open interval $0 < x < L$. In addition to the differential equation (7.52), the function w must satisfy the boundary conditions

$$B_i w = \lambda C_i w, \qquad i = 1, 2, \ldots, p, \qquad x = 0, L \tag{7.53}$$

where B_i and C_i are linear homogeneous differential operators. The maximum order of B_i and C_i is $2p - 1$ and $2q - 1$, respectively. The

243

operators corresponding to the end point $x = 0$ are in general different from those at the end point $x = L$, although at times they can have the same form. The eigenvalue problem consists of determining the values of the parameter λ for which there are nontrivial functions w satisfying the differential equation (7.52) and the boundary conditions (7.53). Such values and functions are the system eigenvalues and eigenfunctions respectively.

To show the relevance of the general formulation given above, we shall identify the various differential operators in the eigenvalue problems (7.47–49), as follows:

$$L = m\Omega^2 x \frac{d}{dx} - \left[\left(\int_x^L m\Omega^2 \xi \, d\xi \right) - E \frac{d^2 I}{dx^2} \right] \frac{d^2}{dx^2} + 2E \frac{dI}{dx} \frac{d^3}{dx^3} + EI \frac{d^4}{dx^4}$$

$$(7.54a)$$

$$M = m - \frac{dJ}{dx} \frac{d}{dx} - J \frac{d^2}{dx^2}$$ (7.54b)

$$B_1 = 1, \qquad C_1 = 0, \qquad B_2 = \frac{d}{dx}, \qquad C_2 = 0 \quad \text{at} \quad x = 0 \qquad (7.54c)$$

$$B_1 = -E \frac{dI}{dx} \frac{d^2}{dx^2} - EI \frac{d^3}{dx^3}, \qquad C_1 = J \frac{d}{dx}, \qquad B_2 = EI \frac{d^2}{dx^2}, \qquad C_2 = 0$$

$$\text{at} \quad x = L \quad (7.54d)$$

It is obvious from the above that $p = 2$ and $q = 1$.

Although we formulated the eigenvalue problem (7.52–53) in terms of a one-dimensional domain, the same formulation is valid also for two- and three-dimensional domains (M2, Sec. 5-4). Of course, in such cases the operators L and M will involve in general partial derivatives with respect to the various spatial variables. Denoting by D the open domain over which the operators L and M are defined, the solution w must satisfy Eq. (7.52) at every point of D and Eqs. (7.53) at every point of S, where S is the boundary of D. Note that the operators B_i and C_i involve derivatives tangent and normal to S.

Before proceeding with the investigation of the properties of the solution of the eigenvalue problem (7.52–53), it may prove useful to introduce certain definitions concerning functions in general. In the first place, we shall assume that the functions of interest are real and piecewise smooth, i.e., that they are piecewise continuous and possess piecewise continuous first derivatives in the domain D: $0 \leq x \leq L$. Let us consider two such functions $f(x)$ and $g(x)$ and define the *inner product* of the two functions as

$$(f, g) = \int_0^L fg \, dx \qquad (7.55)$$

The functions f and g are said to be *orthogonal* if their inner product vanishes. The square root of the inner product of a function $f(x)$ with itself is called the *norm* of f and is denoted by $\|f\|$, where

$$\|f\|^2 = (f, f) = \int_0^L f^2 \, dx \qquad (7.56)$$

The existence of the norm simply implies that $\|f\|^2 < \infty$. Orthogonal functions with unit norm are said to be *orthonormal*. A function whose norm exists is said to be *square summable* in D, which implies that f^2 is integrable in the Lebesgue sense (C4, p. 108). Functions f such that $\|f\| < \infty$ are said to have *finite energy* and the space of such functions is denoted by $\mathcal{H}^0$, where the superscript indicates the order of the derivative of f required for finite energy, in this case zero.

A property intimately related to orthogonality is linear independence. Let us consider a set of n functions $\phi_1, \phi_2, \ldots, \phi_n$. Then, the set of functions $\phi_1, \phi_2, \ldots, \phi_n$ is said to be *linearly dependent* if a homogeneous relation with constant coefficients of the type

$$c_1\phi_1 + c_2\phi_2 + \cdots + c_n\phi_n = 0 \qquad (7.57)$$

exists without all the coefficients $c_1, c_2, \ldots, c_n$ being identically zero. If Eq. (7.57) can be satisfied only if all the coefficients $c_1, c_2, \ldots, c_n$ are identically zero, then the set of functions $\phi_1, \phi_2, \ldots, \phi_n$ is *linearly independent*. It is easy to show that orthogonality implies linear independence. Indeed, let us assume that the set of functions $\phi_1, \phi_2, \ldots, \phi_n$ in Eq. (7.57) is orthogonal. Multiplying Eq. (7.57) by ϕ_s and integrating over the interval $0 \le x \le L$, we obtain

$$\sum_{r=1}^n c_r \int_0^L \phi_s \phi_r \, dx = \sum_{r=1}^n c_r \|\phi_s\|^2 \delta_{rs} = c_s \|\phi_s\|^2 = 0, \qquad s = 1, 2, \ldots, n \qquad (7.58)$$

Because the norms cannot be zero, it follows that all the coefficient c_s ($s = 1, 2, \ldots, n$) must be zero, which is the condition for the set of functions $\phi_1, \phi_2, \ldots, \phi_n$ to be linearly independent. Hence, *orthogonal sets of functions are always linearly independent.* The converse is not necessarily true, however, as sets of independent functions need not be orthogonal, although they can be rendered so.

Next, let us consider an orthonormal system $\phi_1, \phi_2, \ldots$ and let f be any function. Moreover, let $c_1, c_2, \ldots$ be a set of constant coefficients given by

$$c_r = (f, \phi_r) = \int_0^L f\phi_r \, dx, \qquad r = 1, 2, \ldots \tag{7.59}$$

The coefficients so defined are known as the components of f with respect to the orthonormal system $\phi_1, \phi_2, \ldots$. Then, let us write the obvious relation

$$\left\| f - \sum_{r=1}^n c_n\phi_r \right\|^2 = \int_0^L \left(f - \sum_{r=1}^n c_r\phi_r \right)^2 dx \geq 0 \tag{7.60}$$

Expanding the integral in (7.60) and considering Eqs. (7.59), we obtain

$$\int_0^L f^2 \, dx - 2 \sum_{r=1}^n c_r \int_0^L f\phi_r \, dx + \sum_{r=1}^n c_r^2 = \|f\|^2 - \sum_{r=1}^n c_r^2 \geq 0 \tag{7.61}$$

from which it follows that

$$\|f\|^2 \geq \sum_{r=1}^n c_r^2 \tag{7.62}$$

But $\|f\|^2$ is independent of the number of terms in the series, so that

$$\|f\|^2 \geq \sum_{r=1}^\infty c_r^2 \tag{7.63}$$

which is known as *Bessel's inequality*. The inequality holds true for every orthonormal system and it proves that the sum of the squares of the coefficients c_r is always bounded.

The significance of the above results becomes apparent in the problem of approximating a given function $f(x)$ by a linear combination $\sum_{r=1}^n d_r\phi_r$, where ϕ_r are orthonormal functions, d_r are constant coefficients and n is fixed. To obtain the "best" approximation to $f(x)$, for a given set of functions ϕ_r and a given number n of terms, let us introduce the *mean square error*

$$M = \int_0^L \left(f - \sum_{r=1}^n d_r\phi_r \right)^2 dx \tag{7.64}$$

Then, we set our object to minimize the mean square error. Expanding the integral in (7.64) and considering Eqs. (7.59), we obtain

$$M = \int_0^L f^2 \, dx - 2 \sum_{r=1}^n d_r \int_0^L f\phi_r \, dx + \sum_{r=1}^n d_r^2$$

$$= \|f\|^2 - 2 \sum_{r=1}^n d_r c_r + \sum_{r=1}^n d_r^2 = \|f\|^2 + \sum_{r=1}^n (d_r - c_r)^2 - \sum_{r=1}^n c_r^2$$

(7.65)

It is obvious from Eq. (7.65) that M takes its minimum value when $d_r = c_r$ $(r = 1, 2, \ldots, n)$.

An approximation of any piecewise continuous function f by a series of orthonormal functions $\phi_1, \phi_2, \ldots$ is called an *approximation in the mean*. If by choosing n large enough the mean square error

$$\int_0^L \left(f - \sum_{r=1}^n c_r\phi_r \right)^2 dx$$

can be made less than any arbitrarily small positive number ε, then the set of functions $\phi_1, \phi_2, \ldots$ is said to be *complete*. For a complete set of functions, Bessel's inequality becomes an equality. Introducing the notation

$$f_n = \sum_{r=1}^n c_r\phi_r, \qquad n = 1, 2, \ldots$$

(7.66)

we shall state that if

$$\lim_{n \to \infty} \|f - f_n\| = 0$$

(7.67)

the sequence $f_1, f_2, \ldots$ *converges in the mean* to f.

Actually *the completeness of a set of functions does not require that the functions be orthonormal*, so that a system of functions is complete if every piecewise continuous function can be approximated in the mean to any desired degree of accuracy by a linear combination of the functions of the system.

We have special interest in the space of solutions of the eigenvalue problem (7.52–53). Because the differential equation (7.52) is of order $2p$, we shall denote this space by $\mathcal{H}_B^{2p}$ to indicate that the derivative $2p$ of w is required to have finite energy and that the function w satisfies the boundary conditions (7.53). We shall refer to functions belonging to $\mathcal{H}_B^{2p}$ as *comparison functions*. In short, comparison functions need only be $2p$-times differentiable and satisfy all the boundary conditions of the problem, Eqs. (7.53), but they need not satisfy the differential equation,

247

Eq. (7.52). The space of comparison functions is considerably larger than the space of eigenfunctions, and in fact it contains the latter as a subspace.

For the most part, eigenvalue problems have a simpler form than that given by Eqs. (7.52–53). In particular, M is more often than not a mere function and not a differential operator, and the parameter λ does not appear in the boundary conditions. As an example, note that the eigenvalue problem for the helicopter blade, Eqs. (7.47–49), reduces to this simpler case if the rotary inertia is ignored, $J = 0$. It should be pointed out that J is indeed very small for helicopter blades, so that it would be difficult to justify retention of the rotary inertia in the dynamic analysis. Its earlier inclusion was mainly for illustrative purposes. Hence, for simplicity of presentation, we shall confine ourself to the eigenvalue problem[*]

$$Lw = \lambda m w, \qquad 0 < x < L \tag{7.68a}$$

$$B_i w = 0, \qquad i = 1, 2, \ldots, p, \qquad x = 0, L \tag{7.68b}$$

where m is identified simply as the mass per unit length.

Next, let us consider two comparison functions u and v and introduce the inner product

$$(u, Lv) = \int_0^L u Lv \, dx \tag{7.69}$$

Then, the differential operator L, as well as the system (7.68), is said to be *self-adjoint* if the following statement holds true:

$$(u, Lv) = (v, Lu) \tag{7.70}$$

Whether or not Eq. (7.70) is satisfied can be ascertained by integration by parts. Such an integration by parts must take into account the boundary conditions (7.68b). The self-adjointness of a system implies certain symmetry of the eigenvalue problem. This symmetry can be demonstrated through integration of Eq. (7.69) by parts. The results can be written in the general form

$$\int_0^L u Lv \, dx = \int_0^L \sum_{k=0}^{p} a_k \frac{d^k u}{dx^k} \frac{d^k v}{dx^k} \, dx + \sum_{l=0}^{p-1} b_l \frac{d^l u}{dx^l} \frac{d^l v}{dx^l} \Bigg|_0^L \tag{7.71}$$

---

[*] For a discussion of the more general formulation (7.52–53), see M2, Secs. 5–8 and 5–9.

where a_k $(k = 1, 2, \ldots, p)$ and b_l $(l = 1, 2, \ldots, p-1)$ are in general functions of x. It will prove convenient to introduce the notation

$$[u, v] = \int_0^L \sum_{k=0}^p a_k \frac{d^k u}{dx^k} \frac{d^k v}{dx^k} \, dx + \sum_{l=0}^{p-1} b_l \frac{d^l u}{dx^l} \frac{d^l v}{dx^l} \bigg|_0^L \qquad (7.72)$$

where $[u, v]$ is often referred to as the *energy inner product*. It is clear from Eq. (7.72) that the energy inner product is symmetric in the functions u and v and their derivatives, which is a property shared by all self-adjoint systems. An examination of Eq. (7.72) reveals that the integrand contains derivatives of u and v through order p only, whereas the boundary term contains derivatives through order $p-1$ only, where the latter are characteristic of geometric boundary conditions. Hence, the question arises whether in defining the energy inner product (7.72) we cannot permit functions u and v that lie outside the space $\mathcal{H}_B^{2p}$. The answer is obvious. Indeed, Eq. (7.72) indicates that only the derivative p of u and v is required to have finite energy and that the functions need satisfy only the geometric boundary conditions, so that the space of functions can be enlarged accordingly. We shall denote this enlarged space by $\mathcal{H}_G^p$, where the superscript indicates that its members have derivatives of order p with finite energy and the subscript G indicates that its members satisfy the geometric boundary conditions. The space $\mathcal{H}_G^p$ is known as the *energy space* and functions belonging to the energy space are called *energy functions*, or *admissible functions*. Clearly, the space $\mathcal{H}_B^{2p}$ of comparison functions is a subspace of the space $\mathcal{H}_G^p$ of admissible functions. These two classes of functions play an important role in the process of obtaining solutions of the eigenvalue problem by approximate methods.

It may prove of interest to explain the reason for the term energy inner product for $[u, v]$. Letting $v = u$ in Eq. (7.72), we obtain

$$[u, u] = \int_0^L \sum_{k=0}^p a_k \left(\frac{d^k u}{dx^k}\right)^2 dx + \sum_{l=0}^{p-1} b_l \left(\frac{d^l u}{dx^l}\right)^2 \bigg|_0^L \qquad (7.73)$$

which can be interpreted as twice the maximum potential energy of the system whose eigenvalue problem is described by Eqs. (7.68). The inner product $[u, u]$ can be used to define another type of norm, known as the *energy norm* and denoted by

$$|u| = [u, u]^{1/2} \qquad (7.74)$$

Next, let us introduce the sequence

$$u_n = \sum_{r=1}^n c_r \phi_r, \qquad r = 1, 2, \ldots \qquad (7.75)$$

249

Then, if by choosing a large enough n the quantity $|u - u_r|^2$ can be made less than any arbitrarily small positive number ε, the set of functions $\phi_1, \phi_2, \ldots$ is said to be *complete in energy*. Moreover, the sequence of functions $u_1, u_2, \ldots$ is said to *converge in energy* to u if

$$\lim_{n \to \infty} |u - u_n| = 0 \tag{7.76}$$

In earlier discussions, we have made no assumption concerning the sign of λ in Eq. (7.68a). Clearly, the sign must depend on the nature of the system, as defined by the operator L. Indeed, if for any comparison function u

$$(u, Lu) = \int_0^L uLu \, \mathrm{d}x \geq 0 \tag{7.77}$$

and the equality sign holds if and only if u is identically zero, then the operator L, and hence the system, is said to be *positive definite*. In this case, *all the eigenvalues λ_r of the system (7.68) are positive*. This can be easily verified by multiplying Eq. (7.68a) by w and integrating over the interval $0 \leq x \leq L$. We recall that in the case of the helicopter blade of Sec. 7.3, λ_r was shown to be equal to ω_r^2, where ω_r is the rth natural frequency of the system. On physical grounds, we concluded there that λ was positive for stable motion, so that stability can be regarded as being tantamount to positive definiteness of a system, and vice-versa. On the other hand, if the equality sign in (7.77) can be realized even for $u \neq 0$, then the operator L, and hence the system, is only *positive semidefinite* and *there exists at least one eigenvalue equal to zero*. Positive semidefinite systems are characterized by the fact that they can undergo rigid body translation and/or rigid body rotation.

Next, we wish to prove one of the most important properties of self-adjoint systems, namely the orthogonality of the eigenfunctions. Let us consider two distinct solutions of the eigenvalue problem (7.68), denote them by λ_r, w_r and λ_s, w_s and write

$$Lw_r = \lambda_r m w_r \tag{7.78a}$$

$$Lw_s = \lambda_s m w_s \tag{7.78b}$$

Multiplying Eq. (7.78a) by w_s, Eq. (7.78b) by w_r and integrating over the interval $0 \leq x \leq L$, we obtain

$$\int_0^L w_s Lw_r \, \mathrm{d}x = \lambda_r \int_0^L m w_r w_s \, \mathrm{d}x \tag{7.79a}$$

$$\int_0^L w_r Lw_s \, \mathrm{d}x = \lambda_s \int_0^L m w_r w_s \, \mathrm{d}x \tag{7.79b}$$

Subtracting Eq. (7.79b) from (7.79a), we have

$$\int_0^L w_s L w_r \, dx - \int_0^L w_r L w_s \, dx = (\lambda_r - \lambda_s) \int_0^L m w_r w_s \, dx \qquad (7.80)$$

But for self-adjoint systems the left side of Eq. (7.80) is zero, because the two integrals are equal to one another. Assuming that the eigenvalues λ_r and λ_s are distinct, Eq. (7.80) yields

$$\int_0^L m w_r w_s \, dx = 0, \qquad \lambda_r \neq \lambda_s \qquad (7.81)$$

Hence, *for self-adjoint systems two eigenfunctions corresponding to distinct eigenvalues are orthogonal with respect to the function m*. If the eigenfunctions are weighted by multiplying them by $\sqrt{m}$, then the weighted eigenfunctions are orthogonal in the ordinary way

$$(\sqrt{m} w_r, \sqrt{m} w_s) = 0, \qquad \lambda_r \neq \lambda_s \qquad (7.82)$$

It follows immediately from Eqs. (7.79) that

$$\int_0^L w_s L w_r \, dx = \int_0^L w_r L w_s \, dx = 0, \qquad \lambda_r \neq \lambda_s \qquad (7.83)$$

or, *for self-adjoint systems the eigenfunctions are orthogonal also in the energy inner product*.

If a given eigenvalue λ_i has multiplicity m_i, where m_i must clearly be an integer, then there are exactly m_i independent eigenfunctions corresponding to λ_i. Moreover, any linear combination of these eigenfunctions is also an eigenfunction. These linear combinations of eigenfunctions can be taken so that the m_i resulting eigenfunctions are mutually orthogonal (with respect to the function m). Of course, they are orthogonal with respect to the remaining eigenfunctions. This permits us to regard all the eigenfunctions as orthogonal, independently of whether there is any multiplicity of eigenvalues or not. It will prove convenient to normalize the eigenfunctions so that they satisfy $\int_0^L m w_r^2 \, dx = 1$, in which case the eigenfunctions constitute a denumerably infinite orthonormal set satisfying

$$(\sqrt{m} w_r, \sqrt{m} w_s) = \int_0^L m w_r w_s \, dx = \delta_{rs}, \qquad r, s = 1, 2, \ldots \qquad (7.84a)$$

$$[w_r, w_s] = \int_0^L \sum_{k=0}^p a_k \frac{d^k w_r}{dx^k} \frac{d^k w_s}{dx^k} \, dx + \sum_{l=0}^{p-1} b_l \frac{d^l w_r}{dx^l} \frac{d^l x_s}{dx^l} \Big|_0^L = \lambda_r \delta_{rs},$$

$$r, s = 1, 2, \ldots \qquad (7.84b)$$

From our earlier discussion, it is obvious that *the eigenfunctions* w_r *($r = 1, 2, \ldots$) comprise a complete set.* This fact can be used to formulate the following *expansion theorem: Every function w with continuous Lw and satisfying the boundary conditions of the system can be expanded in an absolutely and uniformly convergent series in the eigenfunctions in the form*

$$w = \sum_{r=1}^{\infty} c_r w_r \qquad (7.85)$$

where the coefficients c_r are given by

$$c_r = \int_0^L m w w_r \, dx, \qquad r = 1, 2, \ldots \qquad (7.86)$$

Equations (7.85–86) constitute the expansion theorem for distributed self-adjoint systems and they are the counterpart of Eqs. (3.54–55) representing the expansion theorem for discrete systems with symmetric mass and stiffness matrices. The expansion theorem is of considerable importance in the vibration of continuous systems, as it forms the basis for the modal analysis for the determination of the response. Of course, in practice the response is approximated by a finite series, which raises the question as to how to truncate Eq. (7.85). We shall examine this question later in this text.

The expansion (7.85) is in terms of the system eigenfunctions and therein lies a serious drawback, as vibrating systems with an eigenvalue problem that lends itself to closed-form solution are very rare. For the most part, they represent very simple systems, typically (but not exclusively) with constant stiffness and mass distributions. When closed-form solutions are not feasible or possible, the expansion must be in terms of approximate eigenfunctions. Such approximate eigenfunctions are linear combinations of comparison functions or admissible functions, depending on how the problem is formulated. This question will receive ample attention in the next chapter. At this point it suffices to say that admissible functions are to be preferred because they are more plentiful and generally easier to work with than comparison functions. In this chapter we shall concern ourselves with closed-form solutions of the eigenvalue problem and in the next two chapters with approximate solutions.

7.5   Non-self-adjoint systems

Most systems encountered in structural dynamics belong to the class of self-adjoint systems, i.e. they exhibit the symmetry described by Eq. (7.72). However, some aeroelastic systems, such as those involving flutter, do not exhibit this symmetry, so that they must be considered non-self-adjoint (F5, Sec. 11.6).

Let us consider a linear homogeneous differential operator L. Then, it can be shown (F4, p. 43) that the operator L has always an *adjoint* L^* defined by

$$(v, Lu) = (L^*v, u) \qquad (7.87)$$

where u is any function in the domain of L and v any function in the domain of L^*. In particular, let u be an eigenfunction u_i of L and v an eigenfunction v_j of L^*, so that

$$Lu_i = \lambda_i u_i \qquad (7.88a)$$

$$L^*v_j = \lambda_j^* v_j \qquad (7.88b)$$

where λ_i and λ_j^* are real or complex eigenvalues of L and L^*, respectively. The set of eigenfunctions v_j $(j = 1, 2, \ldots)$ is said to be *adjoint* to the set of eigenfunctions u_i $(i = 1, 2, \ldots)$.

Next, let us multiply Eq. (7.88a) by v_j, Eq. (7.88b) by u_i and integrate over the interval $0 \le x \le L$ to obtain

$$(v_j, Lu_i) = \int_0^L v_j Lu_i \, dx = \lambda_i \int_0^L v_j u_i \, dx \qquad (7.89a)$$

$$(u_i, L^*v_j) = \int_0^L u_i L^*v_j \, dx = \lambda_j^* \int_0^L u_i v_j \, dx \qquad (7.89b)$$

Subtracting Eq. (7.89b) from (7.89a) and recalling Eq. (7.87), we have

$$(\lambda_i - \lambda_j^*) \int_0^L u_i v_j \, dx = 0 \qquad (7.90)$$

Hence, if $\lambda_i \ne \lambda_j^*$, we conclude that

$$(u_i, v_j) = \int_0^L u_i v_j \, dx = 0, \qquad \lambda_i \ne \lambda_j^*, \qquad i, j = 1, 2, \ldots \qquad (7.91)$$

or, *the eigenfunction of L corresponding to the eigenvalue λ_i is orthogonal to the eigenfunction of L^* corresponding to the eigenvalue λ_j^*, where λ_j^* is distinct from λ_i.* Equation (7.91) states that between the eigenfunctions u_i $(i = 1, 2, \ldots)$ and the adjoint set of eigenfunctions v_j $(j = 1, 2, \ldots)$ there exists a *biorthogonality* relation.

Let us now write Eq. (7.88a) in the form

$$(L - \lambda_i)u_i = 0 \qquad (7.92)$$

253

and form the obviously null inner product

$$(v_i, (L - \lambda_i)u_i) = 0 \qquad (7.93)$$

Expanding the left side of Eq. (7.93) and considering Eq. (7.87), we obtain

$$(v_i, Lu_i) - \lambda_i(v_i, u_i) = (L^*v_i, u_i) - \lambda(v_i, u_i) = ((L^* - \lambda_i)v_i, u_i) = 0 \qquad (7.94)$$

from which we conclude that u_i is orthogonal to every function of the form $(L^* - \lambda_i)v_i$. Assuming that $(L^* - \lambda_i)v_i$ represents any continuous function f, Eq. (7.94) yields $(f, u_i) = 0$. But this would imply that u_i is zero, which is not possible. Hence, the other alternative must be true, namely

$$(L^* - \lambda_i)v_i = 0 \qquad (7.95)$$

which proves that there exists an eigenfunction of L^* corresponding to the eigenvalue λ_i. This permits us to enunciate the following theorem: *The eigenvalues of the operator L coincide with the eigenvalues of the operator L^*.* Of course, the eigenfunctions of L differ from the eigenfunctions of L^*.

It will prove convenient to normalize the two sets of eigenfunctions so as to satisfy $\int_0^L u_i v_i \, dx = 1$ $(i = 1, 2, \ldots)$, in which case the two sets become *biorthonormal*, a property expressed by

$$(u_i, v_j) = \int_0^L u_i v_j \, dx = \delta_{ij}, \qquad i, j = 1, 2, \ldots \qquad (7.96)$$

The two sets of eigenfunctions are assumed to be complete, so that every square summable function f can be expanded in a series in either set of eigenfunctions. First, let us expand the series in the eigenfunctions u_i, or

$$f = \sum_{i=1}^{\infty} \alpha_i u_i \qquad (7.97)$$

Then, forming the inner product (v_j, f) and considering Eqs. (7.96), we obtain the coefficients

$$\alpha_i = (v_i, f), \qquad i = 1, 2, \ldots \qquad (7.98)$$

Equations (7.97) and (7.98) constitute one version of the *expansion theorem for non-self-adjoint systems*. A second version can be obtained by expanding the series in terms of the eigenfunctions v_j. The result is

$$f = \sum_{j=1}^{\infty} \beta_j v_j \qquad (7.99)$$

where the coefficients β_j have the expression

$$\beta_j = (u_j, f), \qquad j = 1, 2, \ldots \tag{7.100}$$

As a matter of interest, let us write

$$Lf = \sum_{i=1}^{\infty} \alpha_i L u_i = \sum_{i=1}^{\infty} \alpha_i \lambda_i u_i \tag{7.101}$$

so that Lf can also be expanded in a series in the eigenfunctions u_i. Similarly,

$$L^*f = \sum_{j=1}^{\infty} \beta_j L^* v_j = \sum_{j=1}^{\infty} \beta_j \lambda_j v_j \tag{7.102}$$

with an analogous conclusion.

If the two operators coincide, $L^* = L$, then the operator L is said to be self-adjoint and the two sets of eigenfunctions are the same.

Non-self-adjoint problems do not generally lend themselves to closed-form solution, and further discussion of this subject will be postponed until the next chapter.

7.6 Vibration of rods, shafts and strings

Rods in axial vibration, shafts in torsional vibration and strings in transverse vibration represent entirely analogous dynamical systems from a mathematical point of view. Indeed, the vibration of all three systems can be described by essentially the same partial differential equation, namely a second-order equation in space. Moreover, the boundary conditions for all three have the same structure (M2, Secs. 5–6, 5–7 and 5–9). Hence, we shall present a single unified mathematical formulation and simply identify the system parameters for each case.

Let us assume that the dynamical system under consideration has the kinetic energy

$$T(t) = \frac{1}{2} \int_0^L m(x) \dot{w}^2(x, t) \, dx \tag{7.103}$$

the potential energy

$$V(t) = \frac{1}{2} \int_0^L \{s(x)[w'(x, t)]^2 + k(x) w^2(x, t)\} \, dx + \tfrac{1}{2} K w^2(L, t) \tag{7.104}$$

and the nonconservative virtual work

$$\delta W_{nc}(t) = \int_0^L p(x, t) \, \delta w(x, t) \, dx \tag{7.105}$$

In the case of a rod in axial vibration, $w(x, t)$ is the axial displacement, $m(x)$ is the mass per unit length, $s(x)$ is the axial stiffness $EA(x)$, in which E is the modulus of elasticity and $A(x)$ is the cross-sectional area, $k(x)$ is a distributed axial spring, K is a discrete axial spring acting at $x = L$ and $p(x, t)$ is a distributed axial external force. For a shaft in torsional vibration, $w(x, t)$ is the angular displacement $\theta(x, t)$, $m(x)$ is the polar mass moment of inertia $I(x)$ per unit length, $s(x)$ is the torsional stiffness $GJ(x)$, in which G is the shear modulus and $J(x)$ is the cross-sectional area moment of inertia, $k(x)$ is a distributed torsional spring, K is a discrete torsional spring acting at $x = L$ and $p(x, t)$ is a distributed external torque. Finally, in the case of the string, $w(x, t)$ is the transverse displacement, $m(x)$ is the mass per unit length, $s(x)$ is the string tension $T(x)$, $k(x)$ is a distributed transverse spring, K is a discrete transverse spring acting at $x = L$ and $p(x, t)$ is a distributed transverse external force. All the springs are assumed to exhibit linear behavior. To help visualizing the problem, Fig. 7.3 shows a string in transverse vibration. Note that the distributed spring $k(x)$ can be regarded as simulating an elastic foundation.

The equations of motion will be obtained by means of the formulation of Sec. 7.2. However, the formulation must be altered somewhat, to account for the presence of the discrete spring at the end $x = L$. It is not difficult to verify that the extended Hamilton principle can be written in the form

$$\int_{t_1}^{t_2} \left[\int_0^L (\delta \hat{L} + \delta \hat{W}_{nc}) \, dx + \delta L_0 \right] dt = 0 \qquad \begin{array}{l} \delta w = 0, \qquad 0 \le x \le L, \\ \text{at} \quad t = t_1, t_2 \end{array} \qquad (7.106)$$

where $\hat{L}$ is the Lagrangian density

$$\hat{L} = \tfrac{1}{2} m \dot{w}^2 - \tfrac{1}{2} s (w')^2 - \tfrac{1}{2} k w^2 \qquad (7.107)$$

L_0 is a discrete component of the Lagrangian

$$L_0 = -\tfrac{1}{2} K w^2 (L, t) \qquad (7.108)$$

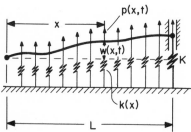

Figure 7.3

and $\delta \hat{W}_{nc}$ is the nonconservative virtual work density

$$\delta \hat{W}_{nc} = p \, \delta w \tag{7.109}$$

Following the pattern used in Sec. 7.2, we obtain

$$\int_{t_1}^{t_2} \left\{ \int_0^L \left[\frac{\partial \hat{L}}{\partial w} - \frac{\partial}{\partial x} \left(\frac{\partial \hat{L}}{\partial w'} \right) - \frac{\partial}{\partial t} \left(\frac{\partial \hat{L}}{\partial \dot{w}} \right) + p \right] \delta w \, dx \right.$$

$$\left. + \left(\frac{\partial \hat{L}}{\partial w'} + \frac{\partial L_0}{\partial w} \right) \delta w \Big|_{x=L} - \frac{\partial \hat{L}}{\partial w'} \, \delta w \Big|_{x=0} \right\} dt = 0 \tag{7.110}$$

so that the usual argument concerning the arbitrariness of the virtual displacement δw yields the general Lagrange equation

$$\frac{\partial \hat{L}}{\partial w} - \frac{\partial}{\partial x} \left(\frac{\partial \hat{L}}{\partial w'} \right) - \frac{\partial}{\partial t} \left(\frac{\partial \hat{L}}{\partial \dot{w}} \right) + p = 0, \qquad 0 < x < L \tag{7.111}$$

and the boundary conditions

$$\frac{\partial \hat{L}}{\partial w'} \, \delta w = 0 \quad \text{at} \quad x = 0 \tag{7.112a}$$

$$\left(\frac{\partial \hat{L}}{\partial w'} + \frac{\partial L_0}{\partial w} \right) \delta w = 0 \quad \text{at} \quad x = L \tag{7.112b}$$

Introducing Eqs. (7.107–109) into Eq. (7.111), we obtain the explicit differential equation

$$-kw + \frac{\partial}{\partial x} (sw') - m\ddot{w} + p = 0, \qquad 0 < x < L \tag{7.113}$$

and, assuming that the end $x = 0$ is fixed, we conclude that the boundary conditions are

$$w(0, t) = 0 \tag{7.114a}$$

$$sw' + Kw = 0 \quad \text{at} \quad x = L \tag{7.114b}$$

Note that boundary condition (7.114a) is geometric and boundary condition (7.114b) is dynamic. In the case of axial vibration, the discrete spring effect is to produce normal stresses on the cross-sectional area whose resultant is equal to the force in the spring but acting in the opposite direction. Similarly, in the case of torsional vibration, there are shearing stresses resulting in a torque, and in the case of transverse string vibration, the string must have a nonzero slope resulting in a transverse component of force.

Next, let us consider the free vibration case, $p = 0$, and derive the eigenvalue problem. To this end, we use the method of separation of variables presented in Sec. 7.3 and let $w(x, t) = W(x)F(t)$, where $F(t)$ is harmonic with frequency ω. Standard application of the method yields the differential equation

$$-\frac{d}{dx}(sW') + kW = \omega^2 mW, \qquad 0 < x < L \tag{7.115}$$

and the boundary conditions

$$W(0) = 0 \tag{7.116a}$$

$$sW' + KW = 0 \quad \text{at} \quad x = L \tag{7.116b}$$

To establish general properties of the eigensolution, we must first identify the various differential operators. Comparing the eigenvalue problem (7.115–116) with the one in operator form introduced in Sec. 7.4, Eqs. (7.68), we conclude that

$$L = -\frac{d}{dx}\left(s\frac{d}{dx}\right) + k, \qquad \lambda = \omega^2$$

$$B_1 = 1 \quad \text{at} \quad x = 0 \tag{7.117}$$

$$B_1 = s\frac{d}{dx} + K \quad \text{at} \quad x = L$$

Of course, the order of L is two, so that $p = 1$.

The first task is to check whether the system is self-adjoint. Letting u and v be two comparison functions and integrating by parts, we obtain

$$\int_0^L uLv \, dx = \int_0^L u\left[-\frac{d}{dx}\left(s\frac{dv}{dx}\right) + kv\right] dx$$

$$= -us\frac{dv}{dx}\bigg|_0^L + \int_0^L \left(s\frac{du}{dx}\frac{dv}{dx} + kuv\right) dx \tag{7.118}$$

From Eq. (7.116b), we can write

$$-sv' = Kv \quad \text{at} \quad x = L \tag{7.119}$$

so that, recalling Eq. (7.116a), Eq. (7.118) can be rewritten in the form

$$\int_0^L uLv \, dx = Kuv|_{x=L} + \int_0^L (su'v' + kuv) \, dx \tag{7.120}$$

The right side of Eq. (7.120) is clearly symmetric in u and v and their first derivatives and is recognized as the energy inner product for the system.

It follows that the operator L, and hence the system, is self-adjoint. We conclude immediately that the system eigenfunctions are orthogonal. Moreover, letting $v = u$ in Eq. (7.120), we obtain the obvious inequality

$$\int_0^L uLu \, dx = Ku^2|_{x=L} + \int_0^L [s(u')^2 + ku^2] \, dx > 0 \qquad (7.121)$$

so that the system is positive definite, and hence all the eigenvalues are positive. This result has already been anticipated.

The above conclusions have been reached with ease and without knowledge of the eigensolution. This is no ordinary feat in view of the fact that the eigenvalue problem does not lend itself to closed-form solution. The difficulty can be traced to the fact that the differential equation has coefficients that depend on the spatial variable x, and very few such problems admit closed-form solutions. In the next chapter, we shall explore methods for the solution of problems with space-dependent coefficients. In this section, however, we shall moderate our goals and seek a solution for the special case in which all the coefficients are constant. Letting s, k and m all be constant, we can rewrite Eq. (7.115) in the form

$$W'' + \beta^2 W = 0, \qquad 0 < x < L \qquad (7.122)$$

where

$$\beta^2 = (\omega^2 m - k)/s \qquad (7.123)$$

The solution of Eq. (7.122) is simply

$$W(x) = C_1 \sin \beta x + C_2 \cos \beta x \qquad (7.124)$$

Boundary condition (7.116a) shows that $C_2 = 0$. On the other hand, boundary condition (7.116b) yields

$$s\beta C_1 \cos \beta L + KC_1 \sin \beta L = 0 \qquad (7.125)$$

so that this boundary condition is incapable of producing the other constant of integration, as expected. Equation (7.125) does, however, represent the characteristic equation for the system, which can be written in the more convenient form

$$\tan \beta L = -\frac{s}{KL} \beta L \qquad (7.126)$$

Equation (7.126) is a transcendental equation whose solution must be obtained numerically. The solution consists of the denumerably infinite set of eigenvalues $\beta_r L$ ($r = 1, 2, \ldots$). Associated with these eigenvalues, we have the eigenfunctions

$$W_r(x) = C_r \sin \beta_r x, \qquad r = 1, 2, \dots \tag{7.127}$$

Using Eq. (7.123), we obtain the natural frequencies

$$\omega_r = \sqrt{(s\beta_r^2 + k)\,m}, \qquad r = 1, 2, \dots \tag{7.128}$$

The above eigenfunctions are orthogonal, but not orthonormal. If the eigenfunctions are normalized so as to satisfy $\int_0^L m W_r^2\,dx = 1$, then the coefficients C_r can be shown to have the values

$$C_r = 2\sqrt{\beta_r/m\,(2\beta_r L - \sin 2\beta_r L)}, \qquad r = 1, 2, \dots \tag{7.129}$$

The first three eigenfunctions are plotted in Fig. 7.4.

   We note that the distributed spring k affects neither the eigenvalues $\beta_r L$ nor the eigenfunctions $W_r(x)$. Indeed, they remain the same for $k = 0$. On the other hand, k does raise the natural frequencies, as can be concluded from Eq. (7.128), albeit this effect tends to vanish for higher modes.

   The solution of the characteristic equation (7.126) can also be obtained graphically, as shown in Fig. 7.5. We observe that as $r \to \infty$ the eigenvalues approach odd multiples of $\pi/2$ and the eigenfunctions tend to $W_r(x) = C_r \sin (2r-1)\pi x/2L$, which is more typical of a system with the end $x = L$ free. Hence, the effect of the discrete spring K diminishes as the mode number increases.

   Finally, we wish to examine the orthogonality of modes. This orthogonality cannot be established by inspection, as the eigenfunctions at

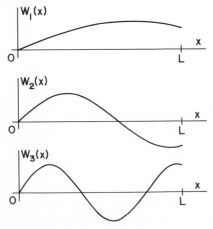

Figure 7.4

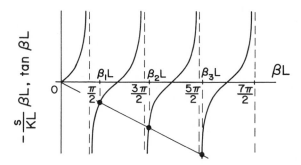

Figure 7.5

hand are not simple trigonometric functions, at least not the lower eigenfunctions. Because $m(x) = m = \text{const}$, we can verify orthogonality by examining the integral

$$\int_0^L \sin \beta_r x \sin \beta_s x \, dx = \frac{1}{2} \int_0^L \left[\cos (\beta_r - \beta_s)x - \cos (\beta_r + \beta_s)x \right] dx$$

$$= \frac{1}{2(\beta_r^2 - \beta_s^2)} \left[(\beta_r + \beta_s) \sin (\beta_r - \beta_s)x - (\beta_r - \beta_s) \sin (\beta_r + \beta_s)x \right]\Big|_0^L$$

$$= \frac{1}{\beta_r^2 - \beta_s^2} (\beta_s \sin \beta_r L \cos \beta_s L - \beta_r \sin \beta_s L \cos \beta_r L) \qquad (7.130)$$

To establish that the right side of Eq. (7.130) is indeed zero, we must invoke the characteristic equation, Eq. (7.126). Recognizing that $\beta_r L$ and $\beta_s L$ are solutions of Eq. (7.126), we can rewrite Eq. (7.130) in the form

$$\int_0^L \sin \beta_r x \sin \beta_s x \, dx = \frac{1}{\beta_r^2 - \beta_s^2} \left(-\beta_s \cos \beta_s L \frac{s}{KL} \beta_r L \cos \beta_r L \right.$$

$$\left. + \beta_r \cos \beta_r L \frac{s}{KL} \beta_s L \cos \beta_s L \right) = 0, \quad r \neq s \qquad (7.131)$$

which verifies the orthogonality of the eigenfunctions. In other cases, the eigenfunctions are not so simple and this verification cannot be made so readily. Of course, verification is not really necessary, as the fact that a system is self-adjoint guarantees orthogonality.

7.7   Bending vibration of bars

In Secs. 7.2 and 7.3, we formulated the boundary-value problem and eigenvalue problem for the bending vibration of a rotating helicopter blade. In this section, we wish to pursue this problem in more detail. As pointed our earlier, the rotary inertia J has only a minimal effect on the vibration of helicopter blades, so that it can be safely ignored. Letting $J = 0$, the differential equation, Eq. (7.47), reduces to

$$\frac{d^2}{dx^2}(EIW'') - \frac{d}{dx}\left[\left(\int_x^L m\Omega^2\xi \, d\xi\right)W'\right] = \lambda m W, \qquad \lambda = \omega^2, \quad 0 < x < L \tag{7.132}$$

where the first two terms in Eq. (7.47) have been combined into one. Moreover, from Eqs. (7.48–49), we obtain the boundary conditions

$$W(0) = 0 \tag{7.133a}$$

$$W'(0) = 0 \tag{7.133b}$$

$$EIW''|_{x=L} = 0 \tag{7.133c}$$

$$-\frac{d}{dx}(EIW'')|_{x=L} = 0 \tag{7.133d}$$

The eigenvalue problem (7.132–133) admits no closed-form solution. This is true even when both the flexural rigidity EI and the mass per unit length m are constant, as the differential equation continues to possess coefficients depending on x because of the axial force. We shall consider approximate solutions of the problem in the next two chapters, but in this section we wish to examine the self-adjointness and positive definiteness of the system.

The eigenvalue problem (7.132–133) is of the form (7.68), in which the operator L has the expression

$$L = \frac{d^2}{dx^2}\left(EI\frac{d^2}{dx^2}\right) - \frac{d}{dx}\left[\left(\int_x^L m\Omega^2\xi \, d\xi\right)\frac{d}{dx}\right] \tag{7.134}$$

to examine the self-adjointness of L, let us consider two admissible functions u and v. Because u and v satisfy all four boundary conditions (7.133), we can write

$$\begin{aligned}(u, Lv) &= \int_0^L uLv \, dx = \int_0^L u\left\{(EIv'')'' - \left[\left(\int_x^L m\Omega\xi \, d\xi\right)v'\right]'\right\} dx \\ &= u(EIv'')'|_0^L - u'(EIv'')|_0^L - u\left(\int_x^L m\Omega^2\xi \, d\xi\right)v'|_0^L \\ &\quad + \int_0^L \left[u''EIv'' + u'\left(\int_x^L m\Omega^2\xi \, d\xi\right)v'\right] dx \\ &= \int_0^L \left[EIu''v'' + \left(\int_x^L m\Omega^2\xi \, d\xi\right)u'v'\right] dx \end{aligned} \tag{7.135}$$

where all the boundary terms have vanished. Note that the boundary term $u(\int_x^L m\Omega^2\xi \, d\xi)v'|_{x=L}$ is zero by virtue of the fact that the integral, representing the horizontal tensile force at point x due to centrifugal effects, is zero at $x = L$. Because the extreme right side of Eq. (7.135) is symmetric in the first and second derivatives of u and v, the system is self-adjoint, so that the system eigenfunctions can be safely assumed to be orthogonal. We note, in passing, that the symmetric expression in (7.135) is simply the energy inner product for the system, as can be concluded from Eq. (7.72). Moreover, by letting $v = u$ in Eq. (7.135), we obtain the inequality

$$[u, u] = \int_0^L uLu \, dx = \int_0^L \left[EI(u'')^2 + \left(\int_x^L m\Omega^2\xi \, d\xi \right)(u')^2 \right] dx > 0$$

$$(7.136)$$

so that the system is positive definite, with the implication that all the system eigenvalues are positive. The inequality can be explained by the fact that the system does not admit a solution $W = \text{const}$, corresponding to rigid-body translation.

As a matter of interest, let us consider the case in which the end $x = 0$ of the blade is not fixed but *hinged*. In the helicopter field, such a blade is referred to as being 'articulated'. The differential equation (7.132) and boundary conditions (7.133a), (7.133c) and (7.133d) remain the same, but boundary condition (7.133b) must be replaced by

$$EIW''|_{x=0} = 0 \tag{7.137}$$

reflecting the fact that a hinge cannot carry bending moments. On the other hand, the elastic line can have a nonzero slope at $x = 0$. Because now a solution $W = x$ is possible, the question arises as to whether the presence of the hinge renders the system only positive semidefinite. Although this solution does have the appearance of a rigid-body mode, the frequency associated with this mode is not zero. This is confirmed by the fact that a substitution of $u = x$ in (7.136) does not cause the integral to vanish. Indeed, the integral remains positive, so that the frequency associated with this mode is not zero. Hence, this rigid-body rotation is oscillatory in nature. To obtain the natural frequency associated with this motion, let us introduce $W = x$ in Eq. (7.132). Then, recalling that $\lambda = \omega^2$, where ω is the frequency of the harmonic motion, we conclude that $\omega = \Omega$. The implication is that the articulated rotor blade admits a natural mode in the form of rigid-body rotation with associated frequency equal to the rotor angular velocity Ω. This is the first natural mode of the system and is known as the 'flapping mode'.

Another case of interest is that of the *simple nonrotating cantilever bar*, obtained by letting $\Omega = 0$ in Eq. (7.132). Even this simpler problem does not admit a closed-form solution, with the notable exception of that of the uniform bar, $EI = \text{const}$, $m = \text{const}$. In this case, the differential equation reduces to

$$W'''' - \beta^4 W = 0, \qquad \beta^4 = \omega^2 m/EI, \qquad 0 < x < L \qquad (7.138)$$

The boundary conditions (7.133) remain essentially the same, although a slight simplification can be obtained by recognizing that the boundary condition (7.133d) is equivalent to $W'''(L) = 0$. The general solution of Eq. (7.138) is

$$W(x) = C_1 \sin \beta x + C_2 \cos \beta x + C_3 \sinh \beta x + C_4 \cosh \beta x \qquad (7.139)$$

Using boundary conditions (7.133), we obtain the characteristic equation

$$\cos \beta L \cosh \beta L = -1 \qquad (7.140)$$

which yields a denumerably infinite set of solutions $\beta_r L$ $(r = 1, 2, \ldots)$, as well as the eigenfunctions

$$W_r(x) = A_r[(\sin \beta_r L - \sinh \beta_r L)(\sin \beta_r x - \sinh \beta_r x)$$
$$+ (\cos \beta_r L + \cosh \beta_r L)(\cos \beta_r x - \cosh \beta_r x)], \qquad r = 1, 2, \ldots \quad (7.141)$$

The system is self-adjoint and positive definite, so that the eigenfunctions are orthogonal and the eigenvalues are all positive. The eigenfunctions given by Eq. (7.141) have not been normalized. Indeed, normalization in this case is a very tedious task. The first three natural modes are shown in Fig. 7.6.

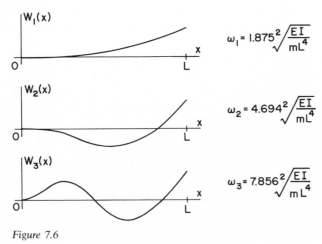

Figure 7.6

When the end $x = 0$ of a nonrotating bar $(\Omega = 0)$ is *hinged*, instead of being fixed, while the end $x = L$ remains *free*, the system ceases to be positive definite. In this case, the system is only positive semidefinite, so that a zero natural frequency exists. It is the frequency associated with the rigid-body rotational mode. This mode, which is oscillatory for a spinning rotor, $\Omega \neq 0$, is no longer so in this case.

Another case of special interest is that of a *free-free* bar. In this case, there are two zero natural frequencies. They are associated with two rigid-body modes, namely translation and rotation of the bar as a whole (M2, pp. 163–166). Note that a missile in free flight can at times be regarded as a free-free bar.

The simplest case of a bar in bending is the *simply-supported uniform bar*, i.e. a bar hinged at both ends. The differential equation for the eigenvalue problem remains in the form (7.138), but the boundary conditions are

$$W(0) = W''(0) = W(L) = W''(L) = 0 \qquad (7.142)$$

Introducing boundary conditions (7.142) into solution (7.139), we obtain the characteristic equation

$$\sin \beta L = 0 \qquad (7.143)$$

which yields the eigenvalues

$$\beta_r L = r\pi, \qquad r = 1, 2, \ldots \qquad (7.144)$$

and the natural modes

$$W_r(x) = \sqrt{\frac{2}{mL}} \sin \frac{r\pi x}{L}, \qquad r = 1, 2, \ldots \qquad (7.145)$$

which are clearly orthonormal. The mode shapes are easy to visualize and there is no need to plot them. They are the simplest modes encountered. The natural frequencies are

$$\omega_r = \beta_r^2 \sqrt{\frac{EI}{m}} = (r\pi)^2 \sqrt{\frac{EI}{mL^4}}, \qquad r = 1, 2, \ldots \qquad (7.146)$$

so that the higher frequencies are integer multiples of the fundamental frequency.

7.8   Two-dimensional problems

Two-dimensional problems introduce a new element into the boundary-value problems, namely, the shape of the boundary. For two-dimensional

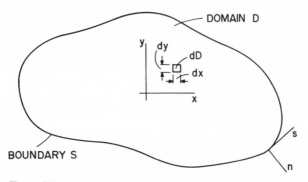

Figure 7.7

domains, there is the additional problem of choosing the type of coordinates to be used. The choice of coordinates consists of cartesian, polar, elliptical, etc., and is not entirely arbitrary, as the best choice is generally dictated by the shape of the boundary S. This is so because the boundary conditions involve in general derivatives along the tangent s and the normal n to the boundary (Fig. 7.7). For example, in the case of a rectangular domain D the natural choice is cartesian coordinates, for a circular domain one should use polar coordinates, etc. The question remains as to the choice of coordinates if the domain does not have a regular shape. In this case, there is virtually no hope for a closed-form solution and one should consider an approximate solution, in which case the choice of coordinates may depend on the method of solution. For the most part, one is advised to use cartesian coordinates, unless there is a good reason for making a different choice.

A potentially critical question in two-dimensional problems is the smoothness of the boundary S, particularly if the boundary has corners. In such cases, the solution can exhibit singularities and special care must be exercised in producing an approximate solution (S2, Ch. 8).

One of the simplest two-dimensional problems in vibrations is that of a membrane. The equations of motion of a membrane can be conveniently obtained by Hamilton's principle, Eq. (7.8). For simplicity, let us consider a rectangular membrane extending over the open domain D: $0 < x < a$, $0 < y < b$ bounded by S: $x = 0$, a; $y = 0$, b. Denoting the transverse displacement by $w(x, y, t)$, the kinetic energy is

$$T(t) = \frac{1}{2} \iint_D \rho \dot{w}^2 \, dx \, dy \tag{7.147}$$

where ρ is the mass per unit area of membrane. The variation in the kinetic energy is simply

$$\delta T = \iint\limits_{D} \rho \dot{w} \, \delta \dot{w} \, dx \, dy \qquad (7.148)$$

The virtual work can be written in the form

$$\delta W = -\delta V + \delta W_{nc} \qquad (7.149)$$

where V is the potential energy and δW_{nc} is the nonconservative virtual work. The potential energy density is proportional to the change in the area of a differential element of the membrane, the constant of proportionality being equal to the tension T. The change in area of the differential element $dx \, dy$ can be shown to be

$$\left[1 + \left(\frac{\partial w}{\partial x}\right)^2 + \left(\frac{\partial w}{\partial y}\right)^2\right]^{1/2} dx \, dy - dx \, dy \cong \frac{1}{2}\left[\left(\frac{\partial w}{\partial x}\right)^2 + \left(\frac{\partial w}{\partial y}\right)^2\right] dx \, dy$$

$$(7.150)$$

so that the potential energy has the form

$$V(t) = \frac{1}{2} \iint\limits_{D} T\left[\left(\frac{\partial w}{\partial x}\right)^2 + \left(\frac{\partial w}{\partial y}\right)^2\right] dx \, dy \qquad (7.151)$$

Hence,

$$\delta V = \iint\limits_{D} T\left(\frac{\partial w}{\partial x} \delta \frac{\partial w}{\partial x} + \frac{\partial w}{\partial y} \delta \frac{\partial w}{\partial y}\right) dx \, dy \qquad (7.152)$$

Moreover, assuming that the membrane is subjected to a distributed transverse force $p(x, y, t)$, the nonconservative virtual work is simply

$$\delta W_{nc}(t) = \iint\limits_{D} p \, \delta w \, dx \, dy \qquad (7.153)$$

Introducing Eqs. (7.148), (7.149), (7.152) and (7.153) into Eq. (7.8), we obtain

$$\int_{t_1}^{t_2} \left\{ \iint\limits_{D} \left[\rho \dot{w} \, \delta \dot{w} - T\left(\frac{\partial w}{\partial x} \delta \frac{\partial w}{\partial x} + \frac{\partial w}{\partial y} \delta \frac{\partial w}{\partial y}\right) + p \, \delta w \right] dx \, dy \right\} dt = 0,$$

$$\delta w = 0 \quad \text{at} \quad t = t_1, t_2 \quad (7.154)$$

A number of integrations by parts yield

$$\int_{t_1}^{t_2} \left\{ \iint\limits_{D} \left[-\rho \ddot{w} + \frac{\partial}{\partial x}\left(T\frac{\partial w}{\partial x}\right) + \frac{\partial}{\partial y}\left(T\frac{\partial w}{\partial y}\right) + p \right] \delta w \, dx \, dy \right.$$

$$\left. - \int_0^b T\frac{\partial w}{\partial x} \delta w \Big|_0^a dy - \int_0^a T\frac{\partial w}{\partial y} \delta w \Big|_0^b dx \right\} dt = 0 \quad (7.155)$$

7 Vibration of continuous systems

Using the standard argument concerning the arbitrariness of δw, we obtain the differential equation

$$\frac{\partial}{\partial x}\left(T\frac{\partial w}{\partial x}\right) + \frac{\partial}{\partial y}\left(T\frac{\partial w}{\partial y}\right) + p = \rho\ddot{w} \quad \text{in} \quad D \tag{7.156}$$

and the boundary conditions

$$T\frac{\partial w}{\partial x}\,\delta w = 0 \quad \text{on} \quad x = 0, a \tag{7.157a}$$

$$T\frac{\partial w}{\partial y}\,\delta w = 0 \quad \text{on} \quad y = 0, b \tag{7.157b}$$

Boundary conditions (7.157a) imply that either $T(\partial w/\partial x)$ or w is zero on $x = 0$, a and a similar statement can be made in relation with boundary conditions (7.157b). A different type of boundary condition arises when a segment of the boundary is supported elastically. For example, if the boundary $x = a$ is attached to a distributed spring $k_a(y)$, then the corresponding boundary condition is

$$\left(k_a w + T\frac{\partial w}{\partial x}\right)\delta w = 0 \quad \text{on} \quad x = a \tag{7.158}$$

Of course, in this case one has a boundary term in the potential energy of the form

$$\frac{1}{2}\int_0^b k_a(y)w^2(a, y, t)\,dy \tag{7.159}$$

Next, let us consider a *uniform membrane* subjected to a *constant tension* T and with *all its boundaries free.*[*] In this case, the differential equation reduces to

$$T\nabla^2 w + p = \rho\ddot{w} \quad \text{in } D \tag{7.160}$$

where

$$\nabla^2 = \frac{\partial^2}{\partial x^2} + \frac{\partial^2}{\partial y^2} \tag{7.161}$$

is the Laplace operator, and the boundary conditions become

$$T\frac{\partial w}{\partial x} = 0 \quad \text{on} \quad x = 0, a \tag{7.162a}$$

$$T\frac{\partial w}{\partial y} = 0 \quad \text{on} \quad y = 0, b \tag{7.162b}$$

---

[*] Note that these boundary conditions imply that the membrane can slide freely up and down a rectangular box, while the walls of the box exert constant tension on the membrane.

Letting $p = 0$ and assuming that w is separable in space and time, we have

$$w(x, y, t) = W(x, y)F(t) \qquad (7.163)$$

where $F(t)$ is harmonic with frequency ω, we obtain the eigenvalue problem defined by the differential equation

$$\nabla^2 W + \beta^2 W = 0, \qquad \beta^2 = \omega^2 \rho/T, \quad \text{in } D \qquad (7.164)$$

and the boundary conditions

$$\frac{\partial W}{\partial x} = 0 \quad \text{on} \quad x = 0, a \qquad (7.165a)$$

$$\frac{\partial W}{\partial y} = 0 \quad \text{on} \quad y = 0, b \qquad (7.165b)$$

Using the method of separation of variables to separate the x and y coordinates, the general solution of Eq. (7.164) can be shown to be (M2, p. 170)

$$W(x, y) = A_1 \sin \alpha x \sin \gamma y + A_2 \sin \alpha x \cos \gamma y + A_3 \cos \alpha x \sin \gamma y$$
$$+ A_4 \cos \alpha x \cos \gamma y \qquad (7.166)$$

where

$$\alpha^2 + \gamma^2 = \beta^2 \qquad (7.167)$$

Application of the boundary conditions (7.165) yields the characteristic equations

$$\sin \alpha a = 0, \qquad \sin \gamma b = 0 \qquad (7.168)$$

which have the solutions

$$\begin{aligned}\alpha_m a &= m\pi, \qquad m = 0, 1, 2, \ldots \\ \gamma_n b &= n\pi, \qquad n = 0, 1, 2, \ldots \end{aligned} \qquad (7.169)$$

so that there is a double infinity of eigenvalues. These are

$$\beta_{mn} = \sqrt{\alpha_m^2 + \gamma_n^2} = \pi\sqrt{\left(\frac{m}{a}\right)^2 + \left(\frac{n}{b}\right)^2}, \qquad m, n = 0, 1, 2, \ldots \qquad (7.170)$$

Similarly, the eigenfunctions can be shown to be

$$W_{00}(x, y) = \sqrt{\frac{1}{\rho ab}}$$

$$W_{mn}(x, y) = \sqrt{\frac{2}{\rho ab}} \cos \frac{m\pi x}{a} \cos \frac{n\pi y}{b}, \qquad m, n = 0, 1, 2, \ldots ; m + n > 0 \qquad (7.171)$$

269

which have been normalized so as to satisfy $\iint_D \rho W_{mn}^2 \, dx \, dy = 1$. Recalling that $\omega^2 = \beta^2 T/\rho$, we obtain the natural frequencies

$$\omega_{00} = 0$$

$$\omega_{mn} = \pi \sqrt{\left[\left(\frac{m}{a}\right)^2 + \left(\frac{n}{b}\right)^2\right]\frac{T}{\rho}}, \qquad m, n = 0, 1, 2, \ldots ; m+n > 0 \tag{7.172}$$

The first mode is, of course, the rigid-body mode, in which the membrane moves parallel to the x, y plane. Figure 7.8 shows the modes W_{01} and W_{11}.

When the ratio $R = a/b$ is a rational number, there are repeated natural frequencies $\omega_{mn} = \omega_{rs}$ every time the integers m, n and r, s satisfy the equation

$$m^2 + n^2 R^2 = r^2 + s^2 R^2 \tag{7.173}$$

in such cases, two distinct eigenfunctions, W_{mn} and W_{rs}, correspond to the same eigenvalue, so that any linear combination of W_{mn} and W_{rs} is also an eigenfunction. These cases are called *degenerate*. For example, if $R = 4/3$, then $\omega_{35} = \omega_{54}$, $\omega_{83} = \omega_{46}$, etc. For a square membrane, $R = 1$, we obtain repeated natural frequencies $\omega_{mn} = \omega_{nm}$ for every pair m, n

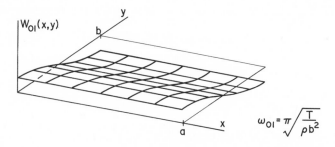

$$\omega_{01} = \pi \sqrt{\frac{T}{\rho b^2}}$$

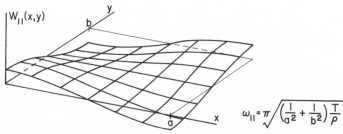

$$\omega_{11} = \pi \sqrt{\left(\frac{1}{a^2} + \frac{1}{b^2}\right)\frac{T}{\rho}}$$

Figure 7.8

$(m \neq n)$. In degenerate cases, a large variety of nodal patterns can occur (C4, p. 302). In fact, the nodal lines are not necessarily straight lines, as they are in nondegenerate cases.

As a matter of interest, let us consider the problem of a *circular uniform membrane free around its entire boundary*. Of course, in this case, it is advisable to use the polar coordinates r and θ. The differential equation remains in the form (7.164), but the Laplacian operator must be expressed in terms of polar coordinates, i.e.

$$\nabla^2 = \frac{\partial^2}{\partial r^2} + \frac{1}{r}\frac{\partial}{\partial r} + \frac{1}{r^2}\frac{\partial^2}{\partial \theta^2} \tag{7.174}$$

Moreover, boundary conditions (7.165) reduce to

$$\frac{\partial W}{\partial r} = 0 \quad \text{on} \quad r = a \tag{7.175}$$

where a is the radius of the membrane.

Using the method of separation of variables, we can write a solution in the form

$$W(r, \theta) = R(r)\Theta(\theta) \tag{7.176}$$

which separates Eq. (7.164) into the two equations

$$\frac{d^2\Theta}{d\theta^2} + m^2\Theta = 0, \qquad 0 \leq \theta < 2\pi \tag{7.177a}$$

$$\frac{d^2R}{dr^2} + \frac{1}{r}\frac{dR}{dr} + \left(\beta^2 - \frac{m^2}{r^2}\right)R = 0, \qquad 0 \leq r < a \tag{7.177b}$$

where m^2 must be a positive constant so as to permit the solution of Eq. (7.177a) to be harmonic in θ. Moreover, because the solutions at θ and $\theta + 2\pi j$ $(j = 1, 2, \ldots)$ must be identical for every θ, it follows that m must be an integer. Equation (7.177b) is a Bessel equation and its solution consists of Bessel functions. It can be verified (M2, p. 174) that the general solution (7.176) has the form

$$W_m(r, \theta) = J_m(\beta r)(A_{1m} \sin m\theta + A_{2m} \cos m\theta)$$

$$+ Y_m(\beta r)(A_{3m} \sin m\theta + A_{4m} \cos m\theta), \qquad m = 0, 1, 2, \ldots \tag{7.178}$$

where $J_m(\beta r)$ and $Y_m(\beta r)$ are Bessel functions of order m and of the first and second kind, respectively.

Because the solution must be finite everywhere, and the Bessel functions of the second kind are infinite at $r = 0$, we must set the coefficients A_{3m} and A_{4m} equal to zero, so that the solution can be reduced accordingly. It can be shown that (B5, p. 93)

$$\frac{d}{dr} J_m(\beta r) = \frac{\beta}{2} [J_{m-1}(\beta r) - J_{m+1}(\beta r)], \qquad m = 0, 1, 2, \ldots \qquad (7.179)$$

so that use of the boundary condition (7.175) yields the infinite set of characteristic equations

$$J_{m-1}(\beta a) - J_{m+1}(\beta a) = 0, \qquad m = 0, 1, 2, \ldots \qquad (7.180)$$

The solutions of Eq. (7.180) for every m represent the eigenvalues $\beta_{mn}a$, so that there is a double infinity of eigenvalues. For each eigenvalue $\beta_{mn}a$ there are two eigenfunctions, one corresponding to $\sin m\theta$ and the other to $\cos m\theta$, except for $m = 0$ when there is only one. Hence, the modes corresponding to $m \neq 0$ are degenerate, which should come as no surprise because of the symmetry of the membrane. The modes can be written in the form

$$W_{0n}(r, \theta) = W_{0n}(r) = A_{0n}J_0(\beta_{0n}r), \qquad n = 0, 1, 2, \ldots \qquad (7.181a)$$

$$\left.\begin{aligned} W_{mnc}(r, \theta) = A_{mnc}J_m(\beta_{mn}r) \cos m\theta \\ W_{mns}(r, \theta) = A_{mns}J_m(\beta_{mn}r) \sin m\theta \end{aligned}\right\} \quad m, n = 1, 2, \ldots \qquad (7.181b)$$

and we note that the modes corresponding to $m = 0$ possess axial symmetry. We also note that the modes W_{mns} are similar to the modes W_{mnc} but they are displaced by $\pi/2$ relative to them.

The system is self-adjoint but only positive semidefinite. Hence, the eigenfunctions are orthogonal and at least one zero natural frequency exists. In fact, there is only one zero frequency, namely $\omega_{00} = 0$, and it corresponds to the rigid-body mode in which the membrane translates parallel to itself. This can be easily verified by letting $m = 0$ in Eq. (7.180), which yields the characteristic equation

$$J_{-1}(\beta_{0n}a) - J_1(\beta_{0n}a) = -2J_1(\beta_{0n}a) = 0 \qquad (7.182)$$

The roots of $J_1(\beta_{0n}a)$ are (B5, p. ix) $\beta_{00}a = 0$, $\beta_{01}a = 3.832$, $\beta_{02}a = 7.016$, etc., so that the first eigenfunction $W_{00}(r, \theta)$ is simply a constant. The remaining eigenfunctions are

$$W_{01}(r) = A_{01}J_0\left(3.832\frac{r}{a}\right)$$

$$W_{02}(r) = A_{02}J_0\left(7.016\frac{r}{a}\right) \qquad (7.182)$$

. . . .

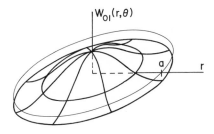

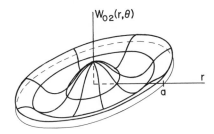

Figure 7.9

The first two modes are plotted in Fig. 7.9. Note that W_{01} has one circular nodal line, W_{02} has two circular nodal lines, etc.

The eigenvalues corresponding to $m = 1$ must satisfy the characteristic equation

$$J_0(\beta_{1n}a) - J_2(\beta_{1n}a) = 0, \qquad n = 1, 2, \ldots \qquad (7.184)$$

The roots of Eq. (7.184) are $\beta_{11}a = 1.841$, $\beta_{12}a = 5.331$, etc., so that the eigenfunctions have the form

$$\begin{aligned}
W_{11c}(r, \theta) &= A_{11c} \\
W_{11s}(r, \theta) &= A_{11s}
\end{aligned} J_1\left(1.841\frac{r}{a}\right)\begin{array}{l}\cos \theta \\ \sin \theta\end{array}$$

$$\begin{aligned}
W_{12c}(r, \theta) &= A_{12c} \\
W_{12s}(r, \theta) &= A_{12s}
\end{aligned} J_1\left(5.331\frac{r}{a}\right)\begin{array}{l}\cos \theta \\ \sin \theta\end{array} \qquad (7.185)$$

....

The modes $W_{11c}(r, \theta)$ and $W_{12c}(r, \theta)$ are shown in Fig. 7.10.

The modes corresponding to $m = 2, 3, \ldots$ can be obtained in the same fashion.

The vibration of plates is of one level of complexity higher than that of membranes, as the differential equation involves the biharmonic

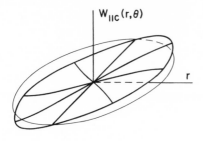

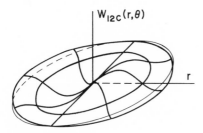

Figure 7.10

operator instead of the Laplace operator and the boundary conditions are generally more complicated. A discussion of the subject can be found in the text by Meirovitch (M2, Sec. 5–12).

7.9   Variational characterization of the eigenvalues

In Ch. 4, we demonstrated that the eigenvalues and eigenvectors of a real symmetric matrix can be characterized by a variational problem consisting of rendering the value of Rayleigh's quotient stationary. The same approach proves extremely attractive in the case of distributed-parameter systems, particularly in the solution of the eigenvalue problem by approximate methods, as the net effect is the substitution of a variational problem for a differential problem.

Let us consider a self-adjoint system described by the eigenvalue problem and define the Rayleigh quotient as

$$R(u) = \frac{[u, u]}{(\sqrt{m}u, \sqrt{m}u)} \tag{7.186}$$

where $[u, u]$ is given by Eq. (7.73), and is recognized as the square of the energy norm $|u|$, and $(\sqrt{m}u, \sqrt{m}u)$ is the square of the norm $\|\sqrt{m}u\|$ of the function $\sqrt{m}u$, in which u is a trial function. By the expansion theorem,

Eq. (7.85), we can express u in the series

$$u = \sum_{r=1}^{\infty} c_r w_r \qquad (7.187)$$

where w_r $(r = 1, 2, \ldots)$ are the system eigenfunctions. Introducing Eq. (7.187) into Eq. (7.186), we obtain

$$R(c_1, c_2, \ldots) = \frac{\displaystyle\sum_{r=1}^{\infty} \sum_{s=1}^{\infty} c_r c_s [w_r, w_s]}{\displaystyle\sum_{r=1}^{\infty} \sum_{s=1}^{\infty} c_r c_s (\sqrt{m} w_r, \sqrt{m} w_s)} \qquad (7.188)$$

But the eigenfunctions of a self-adjoint system are orthogonal and they can be normalized so as to satisfy $\int_0^L m w_r^2 \, \mathrm{d}x = 1$ $(r = 1, 2, \ldots)$, in which case they satisfy Eqs. (7.84). Introducing Eqs. (7.84) into Eq. (7.188), Rayleigh's quotient reduces to

$$R(c_1, c_2, \ldots) = \frac{\displaystyle\sum_{r=1}^{\infty} c_r^2 \lambda_r}{\displaystyle\sum_{r=1}^{\infty} c_r^2} \qquad (7.189)$$

The first variation of Rayleigh's quotient is

$$\delta R = \frac{\partial R}{\partial c_1} \delta c_1 + \frac{\partial R}{\partial c_2} \delta c_2 + \cdots = \sum_{i=1}^{\infty} \frac{\partial R}{\partial c_i} \delta c_i \qquad (7.190)$$

At a stationary point of R, the variation δR must vanish, which yields the stationarity conditions

$$\frac{\partial R}{\partial c_i} = 0, \qquad i = 1, 2, \ldots \qquad (7.191)$$

Introducing Eq. (7.189) into conditions (7.191), we obtain

$$\frac{\partial R}{\partial c_i} = \frac{\left(\displaystyle\sum_{r=1}^{\infty} 2 c_r \frac{\partial c_r}{\partial c_i} \lambda_r\right) \displaystyle\sum_{r=1}^{\infty} c_r^2 - \left(\displaystyle\sum_{r=1}^{\infty} 2 c_r \frac{\partial c_r}{\partial c_i}\right) \displaystyle\sum_{r=1}^{\infty} c_r^2 \lambda_r}{\left(\displaystyle\sum_{r=1}^{\infty} c_r^2\right)^2}$$

$$= \frac{2 c_i \lambda_i \displaystyle\sum_{r=1}^{\infty} c_r^2 - 2 c_i \displaystyle\sum_{r=1}^{\infty} c_r^2 \lambda_r}{\left(\displaystyle\sum_{r=1}^{\infty} c_r^2\right)^2} = \frac{2 c_i \displaystyle\sum_{r=1}^{\infty} (\lambda_i - \lambda_r) c_r^2}{\left(\displaystyle\sum_{r=1}^{\infty} c_r^2\right)^2} = 0,$$

$$i = 1, 2, \ldots \quad (7.192)$$

275

If u coincides with one of the eigenfunctions, say $u = w_i$, then $c_r = c_i \delta_{ir}$ $(r = 1, 2, \ldots)$, so that every term in the series on the numerator of (7.192) vanishes except for the term corresponding to $r = i$, and this latter term vanishes because $\lambda_i - \lambda_r = 0$ for $r = i$. Hence, conditions (7.192) are satisfied when the trial function coincides with an eigenfunction, so that *Rayleigh's quotient has stationary points at the system eigenfunctions*. These are the only stationary points of Rayleigh's quotient. Letting $u = c_i w_i$ in Eq. (7.189), we conclude that

$$R(w_i) = \lambda_i, \qquad i = 1, 2, \ldots \tag{7.193}$$

or *the stationary values of Rayleigh's quotient are precisely the system eigenvalues*. The above result is not entirely unexpected, as similar results were obtained in the case of a self-adjoint discrete system.

The above variational characterization can be shown to be equivalent to the eigenvalue problem, Eqs. (7.68). Indeed, let us assume that the trial function u has the form of the varied function $w_i + \varepsilon v$, where v is an a priori given function and ε is a small parameter. Introducing $u = w_i + \varepsilon v$ into Eq. (7.186) and recalling Eq. (7.193), we have

$$
\begin{aligned}
R(w_i + \varepsilon v) &= \frac{[w_i + \varepsilon v, \, w_i + \varepsilon v]}{(\sqrt{m}(w_i + \varepsilon v), \, \sqrt{m}(w_i + \varepsilon v))} \\[2mm]
&= \frac{[w_i, \, w_i] + 2\varepsilon[w_i, \, v] + \varepsilon^2[v, \, v]}{(\sqrt{m}w_i, \, \sqrt{m}w_i) + 2\varepsilon(\sqrt{m}w_i, \, \sqrt{m}v) + \varepsilon^2(\sqrt{m}v, \, \sqrt{m}v)} \\[2mm]
&= R(w_i) + 2\varepsilon \frac{[w_i, \, v](\sqrt{m}w_i, \, \sqrt{m}w_i) - [w_i, \, w_i](\sqrt{m}w_i, \, \sqrt{m}v)}{(\sqrt{m}w_i, \, \sqrt{m}w_i)^2} \\[2mm]
&\quad + 0(\varepsilon^2) \\[2mm]
&= \lambda_i + 2\varepsilon \frac{[w_i, \, v] - \lambda_i(\sqrt{m}w_i, \, \sqrt{m}v)}{(\sqrt{m}w_i, \, \sqrt{m}w_i)} + 0(\varepsilon^2) \tag{7.194}
\end{aligned}
$$

But for a fixed function v, $R(w_i + \varepsilon v)$ depends only on ε. If $R(w_i + \varepsilon v)$ is to have a stationary value at w_i, then $R(w_i + \varepsilon v)$ cannot vary linearly with ε, so that the coefficient of 2ε must vanish. In view of Eqs. (7.71–72), this condition implies that

$$\int_0^L v(Lw_i - \lambda_i m w_i) \, dx = 0 \tag{7.195}$$

or the function v must be orthogonal to every function $Lw_i - \lambda_i m w_i$. Choosing the function v to be equal to $Lw_i - \lambda_i m w_i$, we conclude that Eq.

(7.195) can be satisfied if and only if

$$Lw_i - \lambda_i m w_i = 0 \qquad (7.196)$$

which proves that the stationarity of Rayleigh's quotient is equivalent to the solution of the eigenvalue problem (7.68).

As in Sec. 4.2, it is easy to show that if the eigenvalues are ordered so as to satisfy $\lambda_1 \le \lambda_2 \le \ldots$, then Rayleigh's quotient is an upper bound for the lowest eigenvalue

$$R(u) \ge \lambda_1 \qquad (7.197)$$

Moreover, if the trial function u is orthogonal to the first s eigenfunctions, then Rayleigh's quotient is an upper bound for λ_{s+1}, or

$$R(u) \ge \lambda_{s+1}, \qquad (u, w_i) = 0, \qquad i = 1, 2, \ldots, s \qquad (7.198)$$

Of course, this characterization of the eigenvalues has the drawback that it requires the eigenfunctions w_i $(i = 1, 2, \ldots, s)$, which quite often are not available.

A characterization of λ_{s+1} that is independent of the eigenfunctions w_i $(i = 1, 2, \ldots, s)$ can be obtained by the same approach as that used in Sec. 4.4. Hence, by analogy, we can write

$$\lambda_{s+1} = \max \min R(u), \qquad (u, v_i) = 0, \qquad i = 1, 2, \ldots, s \qquad (7.199)$$

where v_i are arbitrary functions. This *maximum–minimum characterization of the eigenvalues* is due to Courant and Fischer and can be stated as follows: *The eigenvalue λ_{s+1} of the system described by Eqs. (7.68) is the maximum value which can be given to* min $R(u)$ *by the imposition of the s constraints* $(u, v_i) = 0$ $(i = 1, 2, \ldots, s)$.

The above properties of Rayleigh's quotient can also be demonstrated geometrically by means of a construction similar to that of Secs. 4.2 and 4.3. Indeed, the equation

$$[u, u] = 1 \qquad (7.200)$$

can be interpreted geometrically as defining an ellipsoid in infinitely many dimensions. On that ellipsoid, Rayleigh's quotient takes values according to

$$R(u) = \frac{1}{(\sqrt{m}u, \sqrt{m}u)} = \frac{1}{\|\sqrt{m}u\|^2} \qquad (7.201)$$

From Eqs. (7.189) and (7.200), we conclude that the semi-major axis of the ellipsoid is in the direction of the first eigenfunction w_r. But from Eq. (7.201), we observe that $\|\sqrt{m}w_1\| = \sqrt{1/\lambda_1}$, so that the intersection of the axis w_1 and the ellipsoid defines the lowest eigenvalue λ_1. The constraint $(u, w_1) = 0$ can be interpreted as defining a plane normal to w_1. The intersection of this plane and the ellipsoid consists of an ellipsoid of one dimension lower than the ellipsoid $[u, u] = 1$. The semi-major axis of the ellipsoid of constraint is along the direction of the eigenfunction w_2 and it defines the second eigenvalue λ_2. If the plane of constraint is not orthogonal to w_1 but to some other direction defined by v_1, then the intersection of the semi-major axis and the ellipsoid of constraint will define an eigenvalue $\tilde{\lambda}_1$, where $\lambda_1 \le \tilde{\lambda}_1 \le \lambda_2$, which represents the geometrical interpretation of the Courant and Fischer maximum–minimum principle for the case of one constraint. Similar interpretation can be advanced for the case of more than one constraint.

There remains the question of the nature of the trial functions u. Clearly, functions belonging to the space $\mathcal{H}_B^{2p}$, i.e., the space of comparison functions, are to be admitted, because they are in the domain of the operator L and satisfy all the boundary conditions. This space is not complete, however, and it can be completed by admitting functions in the space $\mathcal{H}_G^p$, which is the space of admissible functions, or energy functions. Indeed, Rayleigh's quotient, Eq. (7.186), is well defined for functions that are p-times differentiable and satisfy all the geometric boundary conditions. We recall that the energy inner product $[u, u]$ takes into account the dynamic boundary conditions automatically.

The variational approach discussed above forms the basis for the Rayleigh-Ritz method for determining approximate solutions of the eigenvalue problem.

7.10   Integral formulation of the eigenvalue problem

We recall from Sec. 3.2 that if λ and $\mathbf{x}$ are an eigenvalue and an eigenvector of a nonsingular matrix A, then λ^{-1} and $\mathbf{x}$ are an eigenvalue and eigenvector of the matrix A^{-1}. Hence, at least in principle, one has the choice of solving either the eigenvalue problem of A or that of the inverse A^{-1}. If A is regarded as an operator, then $K = A^{-1}$ can be regarded as the inverse operator, which indeed it is. In vibration of discrete systems, A is the flexibility matrix and K is the stiffness matrix, where A and K are known to be the inverse of one another. The question is whether such a reciprocal relation exists also for distributed systems.

Let us consider the differential equation

$$Lw(x) = f(x), \qquad 0 < x < L \tag{7.202}$$

where L is a positive definite self-adjoint differential operator of the

7.10   *Integral formulation of the eigenvalue problem*

type introduced in Sec. 7.4. The interest lies in the existence of the inverse relation

$$Af(x) = w(x) \qquad (7.203)$$

where A is an operator which is to be interpreted as the inverse of L. Because L is positive definite, such an operator A must exist and is likely to be an *integral operator*. We shall assume that the integral operator is defined by

$$w = Af = \int_0^L a(x, \xi) f(\xi) \, d\xi \qquad (7.204)$$

where the function $a(x, \xi)$ is known as the *kernel of the integral operator* A. It is also known as the *Green's function* or *influence function* for the system. The reciprocal relation between the operators L and A is demonstrated by Courant and Hilbert (C4, Sec. 5.14).

It will prove of interest to provide some physical interpretation for the above concepts. To this end, let us consider the problem of time-independent deformation of an elastic system under some distributed load. Without loss of generality, we shall consider the simple cantilever bar shown in Fig. 7.11. Denoting by $a(x, \xi)$ the displacement at point x due to a unit load at point ξ, the total displacement $w(x)$ at point x can be written in the form of the integral

$$w(x) = \int_0^L a(x, \xi) f(\xi) \, d\xi \qquad (7.205)$$

which has precisely the same form as Eq. (7.204). Recognizing that the analogous concept for discrete systems is called a flexibility influence coefficient, the term influence function for $a(x, \xi)$ is fully justified. By Maxwell's reciprocity theorem, the displacement at point x due to a unit load at point ξ is equal to the displacement at point ξ due to a unit load at point x, or

$$a(x, \xi) = a(\xi, x) \qquad (7.206)$$

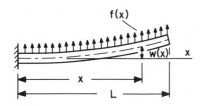

Figure 7.11

279

which proves that the influence function is symmetric in x and ξ. The self-adjointness of the system implies the symmetry of the Green's function (C4, Sec. 5.14), and vice-versa.

To relate the above concepts to our eigenvalue problem, let us consider a system in free vibration. Using d'Alembert's principle (M3, Sec. 2.6), the load is simply the inertia force, so that the differential equation of motion replacing Eq. (7.202) is

$$Lw = -m\ddot{w} \tag{7.207}$$

Because in free vibration the oscillation is harmonic, we can write

$$\ddot{w} = -\omega^2 w \tag{7.208}$$

where ω is the frequency of oscillation, so that Eq. (7.207) can be rewritten as

$$Lw = \lambda mw, \qquad \lambda = \omega^2 \tag{7.209}$$

where now w is only the time-independent part of the displacement. Comparing Eqs. (7.202) and (7.209), we conclude that

$$f = \lambda mw \tag{7.210}$$

Introducing Eq. (7.210) into Eq. (7.205), we obtain

$$w(x) = \lambda \int_0^L a(x, \xi) m(\xi) w(\xi) \, d\xi \tag{7.211}$$

which represents the *eigenvalue problem in integral form*.

To bring the analogy between continuous and discrete systems into sharper focus, we recall from Sec. 2.5 that the eigenvalue problem for a conservative multi-degree-of-freedom system can be written in the form

$$K\mathbf{u} = \lambda M\mathbf{u} \tag{7.212}$$

where K is the stiffness matrix mentioned earlier and M is the mass matrix. Premultiplying both sides of Eq. (7.212) by $K^{-1} = A$, where A is the flexibility matrix, we obtain

$$\mathbf{u} = \lambda A M \mathbf{u} \tag{7.213}$$

Assuming that M is diagonal, we can write Eq. (7.213) in the index notation

$$u_i = \lambda \sum_{j=1}^n a_{ij} m_j u_j, \qquad i = 1, 2, \ldots, n \tag{7.214}$$

which is the discrete counterpart of Eq. (7.211). In fact, we can obtain

Eq. (7.211) from Eq. (7.214) through a limiting process whereby summation is replaced by integration.

Equation (7.211) can be used as the basis for an iterative solution of the eigenvalue problem in the form

$$w_{k+1}(x) = \lambda \int_0^L a(x, \xi) m(\xi) w_k(\xi) \, d\xi, \qquad k = 0, 1, \ldots \tag{7.215}$$

The interesting part about this iteration process is that if the first trial function $w_0(x)$ is an admissible function, then the resulting function $w_1(x)$ is a comparison function. Note that, in integrating Eq. (7.215), one must split the interval $0 \le x \le L$ into two intervals, namely $0 \le \xi \le x$ and $x \le \xi \le L$.

As an illustration, let us consider a uniform cantilever bar, $EI =$ const, $m =$ const. The influence function can be shown to be

$$a(x, \xi) = \begin{cases} \dfrac{1}{2EI} \xi^2(x - \tfrac{1}{3}\xi), & \xi < x \\[2mm] \dfrac{1}{2EI} x^2(\xi - \tfrac{1}{3}x), & \xi > x \end{cases} \tag{7.216}$$

Letting the initial trial function be in the form of the admissible function

$$w_0(x) = x^2 \tag{7.217}$$

Eq. (7.215) yields the new trial function

$$
\begin{aligned}
w_1(x) &= \frac{\lambda m}{2EI} \left[\int_0^x \xi^2(x - \tfrac{1}{3}\xi)\xi^2 \, d\xi + \int_x^L x^2(\xi - \tfrac{1}{3}x)\xi^2 \, d\xi \right] \\
&= \frac{\lambda m}{2EI} \left[\left(x\frac{\xi^5}{5} - \frac{1}{3}\frac{\xi^6}{6} \right)\Big|_0^x + x^2\left(\frac{\xi^4}{4} - \frac{1}{3}x\frac{\xi^3}{3} \right)\Big|_x^L \right] \\
&= \frac{\lambda m}{2EI} [L^3(\tfrac{1}{4}x^2 L - \tfrac{1}{9}x^3) + \tfrac{1}{180}x^6] \tag{7.218}
\end{aligned}
$$

It can be easily verified that $w_1''(L) = w_1'''(L) = 0$, so that w_1 satisfies the dynamic boundary conditions, in addition to the geometric boundary conditions $w_1(0) = w_1'(0) = 0$. Hence, $w_1(x)$ is a comparison function.

Although convergence is relatively rapid, the process is quite tedious, so that iteration based on Eq. (7.215) cannot really be regarded as a computational tool.

7.11   The response problem

The boundary-value problem of Sec. 7.2 can be generalized by writing

$$Lw(x, t) + m(x)\ddot{w}(x, t) = p(x, t), \qquad 0 < x < L \tag{7.219}$$

where L is a self-adjoint spatial differential operator of order $2p$. The displacement $w(x, t)$ must satisfy the boundary conditions

$$B_i w(x, t) = 0, \qquad i = 1, 2, \ldots, p, \qquad x = 0, L \qquad (7.220)$$

as well as the initial conditions

$$w(x, 0) = w_0(x), \qquad \dot{w}(x, 0) = v_0(x) \qquad (7.221)$$

where $w_0(x)$ and $v_0(x)$ are given functions. There remains the question as to what type of spatial dependence can be admitted for the external force $p(x, t)$. The first reaction is to require that functions describing $p(x, t)$ have finite energy, i.e., that they belong to the space $\mathcal{H}^0$, but this requirement turns out to be unduly restrictive. Indeed, this would rule out concentrated forces, which can be represented by the spatial Dirac delta function $\delta(x - x_0)$, a singular function. Because the displacements must belong only to the energy space $\mathcal{H}_G^p$, and this condition is derived from the requirement that the integral $\int_0^L wLw \, dx$ exist, the consistent requirement on p would be that the integral $\int_0^L wp \, dx$ exist. But this latter integral is finite even when p is proportional to $\delta(x - x_0)$, so that such singular functions will be accepted. On the other hand, functions of higher singularities, such as the unit doublet $\delta'(x - x_0)$ will be rejected as the integral $\int_0^L w\delta'(x - x_0) \, dx$ cannot be guaranteed to be finite.

The solution of Eq. (7.219), in conjunction with the boundary conditions (7.220) and initial conditions (7.221), can be obtained by modal analysis. Such a modal analysis follows the pattern established in Sec. 6.5 for discrete systems. The eigenvalue problem for the system (7.219–220) has the form (7.68), and its solution consists of the eigenvalues $\lambda_r = \omega_r^2$ and the eigenfunctions w_r ($r = 1, 2, \ldots$). We shall assume that the eigenfunctions have been normalized so as to satisfy

$$\int_0^L mw_r w_s \, dx = \delta_{rs}, \qquad \int_0^L w_s Lw_r \, dx = \omega_r^2 \delta_{rs}, \qquad r, s = 1, 2, \ldots \quad (7.222)$$

In the sequel, we shall consider the eigenvalues and eigenfunctions as known.

Using the expansion theorem, Eq. (7.85), we can write the system response at a particular instant $t = t_1$ in the form of the series

$$w(x, t_1) = \eta_1(t_1)w_1(x) + \eta_2(t_1)w_2(x) + \cdots = \sum_{r=1}^{\infty} \eta_r(t_1)w_r(x) \qquad (7.223)$$

where $\eta_r(t_1)$ play the role of the expansion coefficients c_r. The solution at

$t = t_2$ will have the same form as (7.223) but with the coefficients $\eta_r(t_2)$ replacing the coefficients $\eta_r(t_1)$. To a small time increment $\Delta t = t_2 - t_1$, there correspond small increments $\Delta \eta_r = \eta_r(t_2) - \eta_r(t_1)$, which implies that the coefficients η_r change in a continuous manner as time unfolds. Hence, regarding $\eta_r(t)$ as continuous functions of time, generally called *normal coordinates*, we can write

$$w(x, t) = \sum_{r=1}^{\infty} \eta_r(t) w_r(x) \tag{7.224}$$

Introducing solution (7.224) into Eq. (7.219), we obtain

$$\sum_{r=1}^{\infty} \eta_r(t) L w_r(x) + \sum_{r=1}^{\infty} \ddot{\eta}_r(t) m w_r(x) = p(x, t), \qquad 0 < x < L \tag{7.225}$$

Multiplying Eq. (7.225) by $w_s(x)$ and integrating over the interval $0 \le x \le L$, we have

$$\sum_{r=1}^{\infty} \eta_r(t) \int_0^L w_s L w_r \, dx + \sum_{r=1}^{\infty} \ddot{\eta}_r(t) \int_0^L m w_r w_s \, dx = \int_0^L w_s p \, dx \tag{7.226}$$

Recalling Eqs. (7.222) and introducing the notation

$$\int_0^L w_s p \, dx = N_s(t), \qquad s = 1, 2, \ldots \tag{7.227}$$

Eq. (7.226) reduces to the infinite set of independent equations

$$\ddot{\eta}_s(r) + \omega_s^2 \eta_s(t) = N_s(t), \qquad s = 1, 2, \ldots \tag{7.228}$$

where the coordinates $\eta_s(t)$ are subject to the initial conditions $\eta_s(0)$ and $\dot{\eta}_s(0)$ $(s = 1, 2, \ldots)$. These initial normal coordinates and velocities are related to the spatial initial conditions. Indeed, introducing Eq. (7.224) into Eqs. (7.221), we obtain

$$\sum_{r=1}^{\infty} \eta_r(0) w_r(x) = w_0(x), \qquad \sum_{r=1}^{\infty} \dot{\eta}_r(0) w_r(x) = v_0(x) \tag{7.229}$$

Multiplying both sides of Eqs. (7.229) by $m w_s$, integrating over the interval $0 \le x \le L$ and considering the first of Eqs. (7.222), we have

$$\eta_s(0) = \int_0^L m w_s w_0 \, dx, \qquad \dot{\eta}_s(0) = \int_0^L m w_s v_0 \, dx, \qquad s = 1, 2, \ldots \tag{7.230}$$

Equations (7.228) are precisely of the form encountered in Sec. 6.5, so

283

that their solution has the general expression

$$\eta_s(t) = \frac{1}{\omega_s} \int_0^t N_s(t-\tau) \sin \omega_s \tau \, \mathrm{d}\tau + \eta_s(0) \cos \omega_s t + \frac{\dot\eta_s(0)}{\omega_s} \sin \omega_s t,$$

$$s = 1, 2, \ldots \quad (7.231)$$

The formal solution is completed by introducing Eqs. (7.231) into Eq. (7.224).

8

Discretization of continuous systems

8.1 Introduction

In Ch. 7, we have discussed various problems associated with the vibration of continuous systems, and in particular boundary-value problems. We have found the task of deriving the equations of motion appreciably simpler than the one of solving them. In fact, closed-form solutions to boundary-value problems are possible only in relatively few cases, almost invariably (but not exclusively) characterized by uniform stiffness and mass distributions and by simple boundary conditions. Closed-form solutions are often desirable because they afford valuable insight into the system behavior, through ready access to natural frequencies and easy visualization of modal patterns. In many cases, however, the effort devoted to the search of a closed-form solution may be so great and the chances of success so cloudy, that a closed-form solution must be regarded as unfeasible. In still more cases, closed-form solutions are simply not possible. This points to the desirability of studying ways of producing approximate solutions, which is the object of this chapter.

The difficulties in obtaining closed-form solutions to problems of distributed-parameter systems can be traced to the inherent difficulty of solving partial differential equations, with or without space-dependent coefficients, and in satisfying boundary conditions, particularly for two- and three-dimensional problems. These difficulties can be circumvented by eliminating the spatial dependence from the problems through discretization in space, but this implies certain approximations. There are two major classes of discretization procedures, one based on expansion of the solution in a finite series of given functions, and the second consisting of simply lumping the system properties. The first is more analytical in nature and the second is more intuitive in character. Some hybrids exist, but their use may not be very desirable (Sec. 9.7). Discretization essentially transforms vibration problems described by partial differential equations into problems described by sets of simultaneous ordinary differential equations. Equivalently, it reduces the associated eigenvalue problems from differential ones to algebraic ones.

There are two classes of discretization schemes based on series expansions, namely, Rayleigh-Ritz type methods and weighted residual methods. Rayleigh-Ritz type methods are based on a given variational principle and are applicable to self-adjoint problems. By contrast, weighted residual methods are more general in scope. They do not require a variational principle and are applicable to both self-adjoint and non-self-adjoint problems.

In this chapter, we discuss the Rayleigh-Ritz method and the closely related assumed-modes method, as well as a variety of weighted residual methods, including perhaps the most widely used one, namely, the Galerkin method. A brief discussion of the lumped-parameter method employing influence coefficients is also presented. The chapter ends with the derivation of the system response by approximate methods.

The finite element method is also a Rayleigh-Ritz method and it really belongs in this chapter. Because the procedural details differ from those of the classical Rayleigh-Ritz method, and the method has acquired an identity of its own, its treatment in a separate chapter (Ch. 9) seems justified.

8.2 The Rayleigh-Ritz method

We shall be concerned with the solution of the eigenvalue problem

$$Lw = \lambda mw, \qquad 0 < x < L \tag{8.1a}$$

$$B_i w = 0, \qquad i = 1, 2, \ldots, p \text{ at } x = 0, L \tag{8.1b}$$

where L is a self-adjoint differential operator of order $2p$. Particular interest lies in the case in which no closed-form solution of the eigenvalue problem (8.1) exists, or is feasible, so that the only possible alternative is an approximate solution. In this section, we shall seek an approximate solution by a variational method, namely, the Rayleigh-Ritz method.

We have shown in Sec. 7.9 that solving Eqs. (8.1) is equivalent to rendering Rayleigh's quotient for the system stationary. It turns out that this variational approach represents an attractive alternative to the solution of a differential problem. Hence, instead of seeking a solution of Eqs. (8.1), we shall seek the stationary values of Rayleigh's quotient

$$R(w) = \frac{[w, w]}{(\sqrt{m}w, \sqrt{m}w)} \tag{8.2}$$

where w is a trial function. As indicated in Sec. 7.9, we shall look for trial functions in the energy space κ_G^p. In fact, we shall not consider the entire space κ_G^p, but only a finite-dimensional subspace S^n of κ_G^p, where n is a finite integer. Denoting a function in S^n approximating w by w^n, we reduce the problem of rendering $R(w)$ stationary to that of determining pairs w^n and $\Lambda^n = r(w^n)$ such that

$$R(w^n) = \frac{[w^n, w^n]}{(\sqrt{m}w^n, \sqrt{m}w^n)} \tag{8.3}$$

is stationary.

Let us select a sequence of elements $\phi_1, \phi_2, \ldots, \phi_n, \ldots$ that satisfy the following two conditions: (i) Any n elements $\phi_1, \phi_2, \ldots, \phi_n$ are linearly independent, and (ii) The sequence $\phi_1, \phi_2, \ldots, \phi_n, \ldots$ is complete in energy, which implies that for any element w in κ_G^p and any small $\varepsilon > 0$, there is a large enough number N and constants $\alpha_1, \alpha_2, \ldots$ such that

$$\left\| w - \sum_{i=1}^{N} \alpha_i \phi_i \right\| < \varepsilon \tag{8.4}$$

Stated slightly differently, Condition (ii) implies that the sequence $\alpha_1 \phi_1, \alpha_1 \phi_1 + \alpha_2 \phi_2, \ldots$ converges in energy to w.

The Rayleigh-Ritz method consists of choosing n functions $\phi_1, \phi_2, \ldots, \phi_n$ from a complete set and regarding them as a basis for the n-dimensional subspace of S^n of κ_G^p. This implies that any element w^n in S^n can be expanded as

$$w^n = \sum_{i=1}^{n} a_i \phi_i \tag{8.5}$$

where the coefficients $a_1, a_2, \ldots, a_n$ are yet to be determined. Inserting Eq. (8.5) into (8.3), we obtain

$$R(a_1, a_2, \ldots, a_n) = \frac{\left[\sum\limits_{i=1}^{n} a_i \phi_i, \sum\limits_{j=1}^{n} a_j \phi_j \right]}{\left(\sqrt{m} \sum\limits_{i=1}^{n} a_i \phi_i, \sqrt{m} \sum\limits_{j=1}^{n} a_j \phi_j \right)}$$

$$= \frac{\sum\limits_{i=1}^{n} \sum\limits_{j=1}^{n} a_i a_j [\phi_i, \phi_j]}{\sum\limits_{i=1}^{n} \sum\limits_{j=1}^{n} a_i a_j (\sqrt{m}\phi_i, \sqrt{m}\phi_j)} \tag{8.6}$$

8 Discretization of continuous systems

Next, let us introduce the notation

$$[\phi_i, \phi_j] = \int_0^L \sum_{k=0}^p a_k \frac{d^k\phi_i}{dx^k} \frac{d^k\phi_j}{dx^k} dx + \sum_{l=0}^{p-1} b_l \frac{d^l\phi_i}{dx^l} \frac{d^l\phi_j}{dx^l}\bigg|_0^L = k_{ij},$$

$$i, j = 1, 2, \ldots, n \quad (8.7a)$$

$$(\sqrt{m}\phi_i, \sqrt{m}\phi_j) = \int_0^L m\phi_i\phi_j \, dx = m_{ij}, \qquad i, j = 1, 2, \ldots, n \quad (8.7b)$$

where k_{ij} are called *stiffness coefficients* and m_{ij} are known as *mass coefficients*. We note that the coefficients are symmetric

$$k_{ij} = k_{ji}, m_{ij} = m_{ji}, i, j = 1, 2, \ldots, n \quad (8.8)$$

which is a consequence of the fact that the system is self-adjoint. Substituting Eqs. (8.7) into Eq. (8.6), we can write Rayleigh's quotient in the form

$$R(a_1, a_2, \ldots, a_n) = \frac{N(a_1, a_2, \ldots, a_n)}{D(a_1, a_2, \ldots, a_n)} \quad (8.9)$$

where

$$N(a_1, a_2, \ldots, a_n) = \sum_{i=1}^n \sum_{j=1}^n k_{ij}a_ia_j \quad (8.10a)$$

$$D(a_1, a_2, \ldots, a_n) = \sum_{i=1}^n \sum_{j=1}^n m_{ij}a_ia_j \quad (8.10b)$$

are the numerator and denominator of Rayleigh's quotient, respectively, and they are related to the potential and kinetic energy of an n-degree-of-freedom system, as can also be concluded from earlier work by Lord Rayleigh (R3, p. 88). Hence, the net result of approximating the solution w by the finite series (8.5) is to transform the distributed system into a discrete system with n degrees of freedom.

The conditions for the stationarity of Rayleigh's quotient are

$$\frac{\partial R}{\partial a_r} = 0, \qquad r = 1, 2, \ldots, n \quad (8.11)$$

so that, using Eq. (8.9), we obtain

$$\frac{\partial R}{\partial a_r} = \frac{(\partial N/\partial a_r)D - (\partial D/\partial a_r)N}{D^2} = \frac{(\partial N/\partial a_r) - \Lambda^n(\partial D/\partial a_r)}{D} = 0,$$

$$r = 1, 2, \ldots, n \quad (8.12)$$

288

where Λ^n is the stationary value of Rayleigh's quotient. From Eqs. (8.10) we can write

$$\frac{\partial N}{\partial a_r} = \sum_{i=1}^{n} \sum_{j=1}^{n} k_{ij} \left(\frac{\partial a_i}{\partial a_r} a_j + a_i \frac{\partial a_j}{\partial a_r} \right) = \sum_{i=1}^{n} \sum_{j=1}^{n} k_{ij} (\delta_{ir} a_j + a_i \delta_{jr})$$

$$= \sum_{j=1}^{n} k_{rj} a_j + \sum_{i=1}^{n} k_{ir} a_i = 2 \sum_{j=1}^{n} k_{rj} a_j, \qquad r = 1, 2, \ldots, n \qquad (8.13a)$$

$$\frac{\partial D}{\partial a_r} = \sum_{i=1}^{n} \sum_{j=1}^{n} m_{ij} \left(\frac{\partial a_i}{\partial a_r} a_j + a_i \frac{\partial a_j}{\partial a_r} \right) = \sum_{i=1}^{n} \sum_{j=1}^{n} m_{ij} (\delta_{ir} a_j + a_i \delta_{jr})$$

$$= \sum_{j=1}^{n} m_{rj} a_j + \sum_{i=1}^{n} m_{ir} a_i = 2 \sum_{j=1}^{n} m_{rj} a_j, \qquad r = 1, 2, \ldots, n \qquad (8.13b)$$

where advantage was taken of the symmetry of the coefficients k_{ij} and m_{ij} $(i, j = 1, 2, \ldots, n)$. A combination of Eqs. (8.12) and (8.13) yields

$$\sum_{j=1}^{n} (k_{ij} - \Lambda^n m_{ij}) a_j = 0, \qquad i = 1, 2, \ldots, n \qquad (8.14)$$

which constitute a set of n simultaneous homogeneous algebraic equations called *Galerkin's equations*. They are also referred to as a *Ritz system*. Equations (8.14) represent the eigenvalue problem for the n-degree-of-freedom discrete system, so that the problem of rendering Rayleigh's quotient, Eq. (8.3), stationary is equivalent to solving the eigenvalue problem (8.14). Hence, the solution of the differential eigenvalue problem (8.1) has been reduced ultimately to the solution of an algebraic eigenvalue problem. The reduction process itself is referred to as *spatial discretization*.

Equations (8.14) can be written in the matrix form

$$K\mathbf{a} = \Lambda^n M\mathbf{a} \qquad (8.15)$$

where $K = K^T$ and $M = M^T$ are $n \times n$ real symmetric matrices with their elements equal to the stiffness and mass coefficients and they are called the *stiffness matrix* and *mass matrix*, respectively, and $\mathbf{a}$ is an n-vector of coefficients. The eigenvalue problem (8.15) is precisely of a form encountered earlier (see Eq. (2.38)). Properties of and computational methods for the solution of the algebraic eigenvalue problem (8.15) have been discussed extensively in Chs. 3–5. The mass matrix M is positive definite by definition, and if the operator L is positive definite, then so is the stiffness matrix K. Hence, all the eigenvalues Λ_r^n $(r = 1, 2, \ldots, n)$ are real and positive. Clearly, all the eigenvectors $\mathbf{a}_r$ $(r = 1, 2, \ldots, n)$ are real. The eigenvalues Λ_r^n represent approximations to the first n eigenvalues of the system (8.1). On the other hand, using Eq. (8.5), we can write the approximations to the first n eigenfunctions w_r of the system (8.1) in the form

$$w_r^n = \sum_{i=1}^n a_{ir}\phi_i, \qquad r = 1, 2, \ldots, n \tag{8.16}$$

where a_{ir} is the ith component of the rth vector $\mathbf{a}_r$. We shall refer to Λ_r^n and w_r^n ($r = 1, 2, \ldots, n$) as *Ritz eigenvalues* and *Ritz eigenfunctions*, respectively. The question remains as to how the Ritz eigenvalues and eigenfunctions relate to the actual eigenvalues and eigenfunctions. We propose to address this question in the sequel.

Let us assume that the eigenvalues of (8.1) and (8.15) are so ordered as to satisfy $\lambda_1 \le \lambda_2 \le \cdots$ and $\Lambda_1^n \le \Lambda_2^n \le \cdots \le \Lambda_n^n$, respectively. Because the system of admissible functions $\phi_1, \phi_2, \ldots$ is complete, one can obtain the solution of the eigenvalue problem (8.1) by letting $n \to \infty$ in series (8.5), at least in principle. But n is only a finite number, so that the Ritz system can be regarded as being obtained from the actual system by imposing the constraints

$$a_{n+1} = a_{n+2} = \cdots = 0 \tag{8.17}$$

Because λ_1 is the minimum value of Rayleigh's quotient over the whole energy space κ_G^p and Λ_1^n is the minimum value over the subspace S^n of κ_G^p, it follows that

$$\Lambda_1^n \ge \lambda_1 \tag{8.18}$$

To examine how the higher eigenvalues of the Ritz system and those of the actual system relate to each other, let us invoke the maximum–minimum principle of Sec. 7.9. By imposing the constraint $(w, \psi_1) = 0$ on the actual system, we can write

$$\lambda_2 = \max \min R(w), \qquad (w, \psi_1) = 0 \tag{8.19}$$

On the other hand, by imposing the same constraint on the Ritz system, we have

$$\Lambda_2^n = \max \min R(w), \qquad (w, \psi_1) = 0 \quad \text{and} \quad (w, \phi_j) = 0,$$
$$j = n+1, n+2, \ldots \tag{8.20}$$

Because the space of constraint used for calculating Λ_n^2 is of lower dimension than that used for calculating λ_2, we conclude that

$$\Lambda_2^n \ge \lambda_2 \tag{8.21}$$

generalizing the above results, we can write

$$\Lambda_r^n \geq \lambda_r, \qquad r = 1, 2, \ldots, n \tag{8.22}$$

so that *the Ritz eigenvalues bound the actual eigenvalues from above*. How well the Ritz eigenvalues approximate the actual eigenvalues depends on the number and on the nature of the admissible functions ϕ_r. In general, however, the lower eigenvalues are better approximations than the higher eigenvalues. In fact, the lowest eigenvalue provides the best approximation, with the higher approximations losing accuracy as the eigenvalue number increases, culminating in the higher eigenvalue providing the poorest approximation. In many cases, the higher Ritz eigenvalues are unreliable. As the number of terms in the series (8.5) is increased, the errors tend to decrease, or at least they do not increase. The most dramatic improvement takes place in the higher eigenvalues, as the lower ones have less room for improvement. The type of functions ϕ_r used can have a profound effect on how close the Ritz eigenvalues are to the actual eigenvalues, although this effect should diminish as n increases. Because the admissible functions ϕ_r must belong to a complete set, the effect should vanish as $n \to \infty$. Unfortunately, the Ritz eigenfunctions w_r^n ($r = 1, 2, \ldots, n$) do not lend themselves to such meaningful comparative analysis. However, it can be stated in general that the Ritz eigenvalues are better approximations to the actual eigenvalues than the Ritz eigenfunctions are to the actual eigenfunctions, a fact that can be attributed to the stationarity of Rayleigh's quotient.

The Rayleigh-Ritz method calls for the use of the sequence

$$w^1 = a_1 \phi_1$$
$$w^2 = a_1 \phi_1 + a_2 \phi_2 \tag{8.23}$$
$$w^3 = a_1 \phi_1 + a_2 \phi_2 + a_3 \phi_3$$
$$\cdots\cdots\cdots\cdots\cdots\cdots$$

obtained by increasing the number of terms in the series (8.5) continuously, solving the Ritz system of equations (8.15) and observing the improvement in the computed eigenvalues. The computation is stopped when a desired number of eigenvalues and eigenvectors reach sufficient accuracy, i.e., when an additional term does not produce meaningful improvement. Of course, the number of accurate Ritz eigenvalues and eigenfunctions is generally significantly smaller than the number of terms in the series (8.5), a fact to be kept in mind when trying to decide how large n should be.

The case in which w is approximated by only one term is not without interest. It arises when the object is merely to estimate the fundamental frequency of the distributed-parameter system (8.1). In this case, one must choose a function ϕ_1 resembling the first eigenfunction as closely as possible. This task is not as difficult as it may seem, because in many cases reasonably good estimates of the first eigenfunction can be obtained by letting ϕ_1 be the static deformation pattern of the structure under its own weight. Of course, in this case there are no coefficients to be concerned with and there is no discrete eigenvalue problem to be solved, as Rayleigh's quotient is simply the ratio of two numbers. The procedure for estimating the fundamental frequency by using a single trial function is known as *Rayleigh's energy method*.

One attractive feature of the Rayleigh-Ritz method is that in increasing the number of terms in the series (8.5) the previously calculated stiffness and mass coefficients do not change, so that one need calculate only one additional row for the matrices K and M. Obviously, there is also an additional column, but because of symmetry no calculation is necessary, as the additional column is simply the transpose of the additional row. Hence, if K^n and M^n denote the stiffness and mass matrices corrresponding to n terms in (8.15) and K^{n+1} and M^{n+1} are those corresponding to $n+1$ terms, then we have

$$K^{n+1} = \begin{bmatrix} & & & x \\ & [K^n] & & x \\ & & & x \\ x & x & x & x \end{bmatrix}, \qquad M^{n+1} = \begin{bmatrix} & & & x \\ & [M^n] & & x \\ & & & x \\ x & x & x & x \end{bmatrix} \qquad (8.24)$$

where the x's denote the additional elements. This feature appears even more remarkable if we recognize that it permits a quantitative substantiation of the qualitative statements made earlier concerning the nature of the approximate eigenvalues. Indeed, denoting by Λ^{n+1} the Ritz eigenvalues corresponding to the system defined by the matrices K^{n+1} and M^{n+1}, where $\Lambda_1^{n+1} \le \Lambda_2^{n+1} \le \cdots \le \Lambda_{n+1}^{n+1}$, we can invoke the inclusion principle of Sec. 4.5 and state that

$$\Lambda_1^{n+1} \le \Lambda_1^n \le \Lambda_2^{n+1} \le \Lambda_2^n \le \cdots \le \Lambda_n^n \le \Lambda_{n+1}^{n+1} \qquad (8.25)$$

But, because the admissible functions $\phi_1, \phi_2, \ldots$ are from a complete set, as n increases the mean square error $\|w - w_r^n\|^2$ approaches zero for every r. In view of the stationary of Rayleigh's quotient, inequalities (8.25) indicate that *as n increases, approximate eigenvalues approach the actual eigenvalues monotonically from above.* Moreover, with each additional term there is a new estimated eigenvalue at the high end of the spectrum. Hence, we can write

$$\lim_{n \to \infty} \Lambda_r^n = \lambda_r, \qquad r = 1, 2, \ldots, n \qquad (8.26)$$

The rate of convergence of the solution depends to a large extent on the nature of the set of admissible functions ϕ_1, ϕ_2, ... This fact together with the uncertainty as to what makes a good set of functions and how to go about selecting the functions have tended to inhibit the use of the Rayleigh-Ritz method. In the sequel, we propose to advance some guidelines that may prove useful in selecting functions.

As pointed out earlier, the admissible functions ϕ_1, ϕ_2, . . . should be linearly independent and they must be complete in the energy space κ_G^p. Moreover, it is highly desirable that the functions be such that the computational effort be kept to a minimum. As a first step, one can investigate existing functions. There are various sets of functions in existence that are known to be independent, such as power series, trigonometric functions, Bessel functions, Legendre polynomials, Tchebycheff polynomials, etc. Some of these functions are orthogonal, but the orthogonality may not be the same as that required by the problem. Of course, orthogonality is not required, but if the functions are such that $(\sqrt{m}\phi_r, \sqrt{m}\phi_s) = \delta_{rs}$, then substantial computational savings accrue, as in this case the mass matrix is simply the identity matrix. Any set of independent functions can be rendered orthogonal by the Gram-Schmidt orthogonalization process (C4, p. 50), but the effort may be difficult to justify, particularly when the object is to shift routine work from the analyst to the computer. Whether or not the functions are complete in energy may not be so easy to ascertain, but for the most part one can assume that they are.

Perhaps a more desirable alternative is to generate a set of admissible functions. This can be done by solving a simpler eigenvalue problem related to the problem (8.1). Then, the eigenfunctions of the simpler related problem can be used as admissible functions for the original problem. For example, if the operator L in Eq. (8.1) contains space-dependent coefficients and the function m is also space-dependent, then a simpler eigenvalue problem would be in terms of a space-independent operator L_C, obtained from L by letting all the coefficients be space-independent, and a constant m, where the latter can be taken as the average over the interval $0 \leq x \leq L$. In solving the simplified eigenvalue problem, only the geometric boundary conditions need be satisfied, and the dynamic boundary conditions can be ignored. Functions thus obtained represent a complete set and are unlikely to cause the energy inner product of the original problem to diverge, so that they can be assumed to belong to the energy space κ_G^p. In other cases, the problem can be simplified by omitting terms involving space-dependent coefficients entirely. As an illustration, let us consider the eigenvalue problem of the helicopter blade of Sec. 7.7, as expressed by Eqs. (7.132–133). As pointed out in Sec. 7.7, that problem does not lend itself to closed-form solution. But the simplified eigenvalue problem, obtained by letting $EI = \text{const}$, $m = \text{const}$ and $\Omega = 0$ in Eqs. (7.132–133) does admit a closed-form solution. In fact, the eigenfunctions of this latter problem are precisely the simple cantilever modes given by Eq. (7.141). Hence, if the original helicopter blade problem is to be solved by the Rayleigh-Ritz method, then the simple cantilever modes will prove to be a suitable set of admissible functions. The simple cantilever modes do have a drawback, however. They are not simple functions themselves, so that they are not very desirable from a computational point of view.

   A Rayleigh-Ritz method that circumvents many of the above problems is the finite element method. In fact, one of the advantages of the finite element method lies in working with very simple and computationally desirable admissible functions.

Example 8.1

Let us consider the rod in axial vibration shown in Fig. 8.1, derive the eigenvalue problem and obtain a Rayleigh-Ritz solution for $n = 1, 2, 3, 4$. The system parameters are as follows:

$$EA(x) = \frac{6EA}{5}\left[1 - \frac{1}{2}\left(\frac{x}{L}\right)^2\right], \qquad K = 0.2\frac{EA}{L}$$

$$m(x) = \frac{6m}{5}\left[1 - \frac{1}{2}\left(\frac{x}{L}\right)^2\right]$$

(a)

The kinetic energy for the system is

$$T(t) = \frac{1}{2}\int_0^L m(x)\dot{u}^2(x, t)\,dx$$

(b)

and the potential energy is

$$V(t) = \frac{1}{2}\int_0^L EA(x)[u'(x, t)]^2\,dx + \frac{1}{2}Ku^2(L, t)$$

(c)

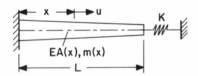

Figure 8.1

From Sec. 7.6, we can write the eigenvalue problem consisting of the differential equation

$$-\frac{d}{dx}\left[EA(x)\frac{dU(x)}{dx}\right] = \omega^2 m(x)U(x), \qquad 0 < x < L \tag{d}$$

and the boundary conditions

$$U(0) = 0 \tag{e}$$

$$EAU' + KU = 0 \quad \text{at} \quad x = L \tag{f}$$

where $EA(x)$ and $m(x)$ are given by Eqs. (a). Hence, from Secs. 7.4 and 7.9, we conclude that Rayleigh's quotient has the form

$$R(U) = \frac{[U, U]}{(\sqrt{m}U, \sqrt{m}U)} = \frac{\displaystyle\int_0^L EA(x)[U'(x)]^2\,dx + KU^2(L)}{\displaystyle\int_0^L m(x)U^2(x)\,dx} \tag{g}$$

A Rayleigh-Ritz solution of the eigenvalue problem described above has the form

$$U(x) = \sum_{i=1}^n a_i\phi_i \tag{h}$$

where $\phi_i(x)$ $(i = 1, 2, \ldots, n)$ are admissible functions. We shall generate these functions by simply solving the eigenvalue problem obtained by letting $EA(x) = EA = \text{const}$, $m(x) = m = \text{const}$ and by dropping the term KU from the boundary condition (f). The resulting eigenvalue problem is given by the differential equation

$$U'' + \beta^2 U = 0, \qquad \beta^2 = \omega^2 m/EA \tag{i}$$

and the boundary conditions $U(0) = U'(L) = 0$. The solution of Eq. (i) can be shown to consist of the eigenfunctions $U_r(x) = A_r \sin(2r-1)\pi x/2L$ $(r = 1, 2, \ldots)$. Hence, we shall use the admissible functions

$$\phi_i(x) = \sin\frac{(2i-1)\pi x}{2L}, \qquad i = 1, 2, 3, 4 \tag{j}$$

Using Eqs. (8.7), we obtain the stiffness and mass coefficients

$$k_{ij} = [\phi_i, \phi_j] = \int_0^L EA(x)\phi_i'(x)\phi_j'(x)\,dx + K\phi_i(L)\phi_j(L)$$

$$= \frac{6EA}{5}\frac{(2i-1)\pi}{2L}\frac{(2j-1)\pi}{2L}\int_0^L \left[1 - \frac{1}{2}\left(\frac{x}{L}\right)^2\right]$$

$$\times \cos\frac{(2i-1)\pi x}{2L}\cos\frac{(2j-1)\pi x}{2L}dx + K\sin\frac{(2i-1)\pi}{2}\sin\frac{(2j-1)\pi}{2}$$

$$\tag{k}$$

$$m_{ij} = (\sqrt{m}\phi_i, \sqrt{m}\phi_j) = \int_0^L m(x)\phi_i(x)\phi_j(x)\,dx$$

$$= \frac{6m}{5}\int_0^L \left[1 - \frac{1}{2}\left(\frac{x}{L}\right)^2\right]\sin\frac{(2i-1)\pi x}{2L}\sin\frac{(2j-1)\pi x}{2L}dx$$

Performing the above integrations, the coefficients can be shown to have the expressions

$$k_{ij} = \frac{EA}{20L}[7 + \tfrac{5}{2}(2i-1)^2\pi^2], \qquad i = j$$

$$\tag{l}$$

$$k_{ij} = \frac{EA}{5L}(-1)^{i+j}\left\{1 + \tfrac{3}{4}(2i-1)(2j-1)\left[\frac{1}{(i+j-1)^2} - \frac{1}{(i-j)^2}\right]\right\}, \qquad i \neq j$$

$$m_{ij} = \frac{mL}{2}\left[1 - \frac{6}{5}\frac{1}{(2i-1)^2\pi^2}\right], \qquad i = j$$

$$\tag{m}$$

$$m_{ij} = \frac{3mL}{5\pi^2}(-1)^{i+j-1}\left[\frac{1}{(i-j)^2} + \frac{1}{(i+j-1)^2}\right], \qquad i \neq j$$

Letting $n = 1$ in Eq. (8.15), we obtain

$$\frac{EA}{20L}(7 + \tfrac{5}{2}\pi^2)a_1^1 = \Lambda_1^1\frac{mL}{2}\left(1 - \frac{6}{5}\frac{1}{\pi^2}\right)a_1^1 \tag{n}$$

yielding the estimated first eigenvalue

$$\Lambda_1^1 = \frac{7 + 5\pi^2/2}{1 - 6/5\pi^2}\frac{EA}{10mL^2} = 3.605816\frac{EA}{mL^2} \tag{o}$$

There is no eigenvector in this case, and the scalar a_1^1 is inconsequential and can be taken as unity. Hence, the estimated first natural frequency and natural mode are

$$\omega_1^1 = 1.898899\sqrt{\frac{EA}{mL^2}}, \qquad w_1^1 = \sin\frac{\pi x}{2L} \tag{p}$$

For $n = 2$, Eq. (8.15) yields the eigenvalue problem

$$\frac{EA}{20L}\begin{bmatrix} 7+5\pi^2/2 & 11/4 \\ 11/4 & 7+45\pi^2/2 \end{bmatrix}\begin{bmatrix} a_1^2 \\ a_2^2 \end{bmatrix}$$

$$= \Lambda^2 \frac{mL}{2\pi^2}\begin{bmatrix} \pi^2-6/5 & 3/2 \\ 3/2 & \pi^2-2/15 \end{bmatrix}\begin{bmatrix} a_1^2 \\ a_2^2 \end{bmatrix} \qquad (q)$$

which has the solution

$$\Lambda_1^2 = 3.601450\,\frac{EA}{mL^2}, \qquad \Lambda_2^2 = 23.860220\,\frac{EA}{mL^2}$$

$$\mathbf{a}_1^2 = \begin{bmatrix} 0.999901 \\ 0.014070 \end{bmatrix}, \qquad \mathbf{a}_2^2 = \begin{bmatrix} -0.185108 \\ 0.982718 \end{bmatrix} \qquad (r)$$

so that the estimated first two natural frequencies and modes are

$$\omega_1^2 = 1.897749\,\sqrt{\frac{EA}{mL^2}}, \qquad \omega_2^2 = 4.884692\,\sqrt{\frac{EA}{mL^2}}$$

$$w_1^2 = 0.999901 \sin\frac{\pi x}{2L} + 0.014070 \sin\frac{3\pi x}{2L} \qquad (s)$$

$$w_2^2 = -0.185108 \sin\frac{\pi x}{2L} + 0.982718 \sin\frac{3\pi x}{2L}$$

where it should be reiterated that the superscript 2 merely indicates the number of terms in the series (n). Similarly, for $n = 3$ we obtain the estimated first three natural frequencies and modes

$$\omega_1^3 = 1.896942\,\sqrt{\frac{EA}{mL^2}}, \qquad \omega_2^3 = 4.883993\,\sqrt{\frac{EA}{mL^2}},$$

$$\omega_3^3 = 7.968519\,\sqrt{\frac{EA}{mL^2}}$$

$$w_1^3 = 0.999861 \sin\frac{\pi x}{2L} + 0.015225 \sin\frac{3\pi x}{2L} - 0.006814 \sin\frac{5\pi x}{2L} \qquad (t)$$

$$w_2^3 = -0.185985 \sin\frac{\pi x}{2L} + 0.982475 \sin\frac{5\pi x}{2L} - 0.012781 \sin\frac{5\pi x}{2L}$$

$$w_3^3 = 0.078747 \sin\frac{\pi x}{2L} - 0.127673 \sin\frac{5\pi x}{2L} + 0.988685 \sin\frac{5\pi x}{2L}$$

297

and for $n = 4$ we obtain

$$\omega_1^4 = 1.896424 \sqrt{\frac{EA}{mL^2}}, \qquad \omega_2^4 = 4.883993 \sqrt{\frac{EA}{mL^2}},$$

$$\omega_3^4 = 7.965769 \sqrt{\frac{EA}{mL^2}}, \qquad \omega_4^4 = 11.082449 \sqrt{\frac{EA}{mL^2}}$$

$$w_1^4 = 0.999845 \sin \frac{\pi x}{2L} + 0.015475 \sin \frac{3\pi x}{2L} - 0.007441 \sin \frac{5\pi x}{2L}$$

$$+ 0.003862 \sin \frac{7\pi x}{2L}$$

$$w_2^4 = -0.185968 \sin \frac{\pi x}{2L} + 0.982472 \sin \frac{3\pi x}{2L} - 0.012815 \sin \frac{5\pi x}{2L} \qquad \text{(u)}$$

$$- 0.000194 \sin \frac{7\pi x}{2L}$$

$$w_3^4 = 0.079737 \sin \frac{\pi x}{2L} - 0.128712 \sin \frac{3\pi x}{2L} + 0.988110 \sin \frac{5\pi x}{2L}$$

$$- 0.026703 \sin \frac{7\pi x}{2L}$$

$$w_4^4 = -0.042834 \sin \frac{\pi x}{2L} + 0.055534 \sin \frac{3\pi x}{2L} - 0.106995 \sin \frac{5\pi x}{2L}$$

$$+ 0.991783 \sin \frac{7\pi x}{2L}$$

From Eqs. (p), (s), (t) and (u), we conclude that the convergence is extremely rapid, which indicates that the admissible functions $\phi_i = \sin(2i - 1)\pi x/2L$ $(i = 1, 2, \ldots)$ resemble closely the actual eigenfunctions.

8.3   The assumed-modes method

The assumed-modes method is a procedure for the discretization of a distributed-parameter system prior to the derivation of the equations of motion. As in the Rayleigh-Ritz method, the solution is assumed in the form of a finite series of space-dependent admissible functions, but the coefficients are time-dependent generalized coordinates instead of being constant. This series is substituted in the kinetic and potential energy, thus reducing them to discrete form, and the equations of motion are derived by means of Lagrange's equations. The assumed-modes method is also a variational method, as Lagrange's equations are really the product of rendering the integral $\int_{t_1}^{t_2} L \, dt$ stationary (M3, Secs. 2.7 and 2.8), where L is the system Lagrangian.

Let us consider a distributed-parameter system and write the kinetic energy

$$T(t) = \frac{1}{2} \int_0^L m \dot{w}^2 \, dx \tag{8.27}$$

and the potential energy

$$V(t) = \frac{1}{2} \int_0^L \sum_{k=0}^{p} a_k \left(\frac{\partial^k w}{\partial x^k} \right)^2 dx + \frac{1}{2} \sum_{l=0}^{p-1} b_l \left(\frac{\partial^l w}{\partial x^l} \right)^2 \Big|_0^L \tag{8.28}$$

where the various quantities are as defined in Ch. 7. The object is to approximate the displacement w by w^n, where w^n is a function in the n-dimensional subspace S^n of κ_G^p. Choosing n admissible functions $\phi_1, \phi_2, \ldots, \phi_n$ from a complete set in κ_G^p, where the functions are of the same type as those used in Sec. 8.2, we can expand the series

$$w^n(x, t) = \sum_{i=1}^{n} \phi_i(x) q_i(t) \tag{8.29}$$

where $q_i(t)$ are generalized coordinates. Introducing Eq. (8.29) into Eq. (8.27), we obtain the discretized kinetic energy

$$T(t) = \frac{1}{2} \int_0^L m \left(\sum_{i=1}^{n} \phi_i \dot{q}_i \right) \left(\sum_{j=1}^{n} \phi_j \dot{q}_j \right) dx = \frac{1}{2} \sum_{i=1}^{n} \sum_{j=1}^{n} \dot{q}_i \dot{q}_j \int_0^L m \phi_i \phi_j \, dx$$

$$= \frac{1}{2} \sum_{i=1}^{n} \sum_{j=1}^{n} m_{ij} \dot{q}_i \dot{q}_j \tag{8.30}$$

where

$$m_{ij} = m_{ji} = \int_0^L m \phi_i \phi_j \, dx, \qquad i, j = 1, 2, \ldots, n \tag{8.31}$$

are symmetric mass coefficients. Moreover, inserting Eq. (8.29) into Eq.

299

8 Discretization of continuous systems

(8.28), we obtain the discretized potential energy

$$
\begin{aligned}
V(t) &= \frac{1}{2} \int_0^L \sum_{k=0}^n a_k \left(\sum_{i=1}^n \frac{d^k \phi_i}{dx^k} q_i \right) \left(\sum_{j=1}^n \frac{d^k \phi_j}{dx^k} q_j \right) \\
&\quad + \frac{1}{2} \sum_{l=0}^{p-1} b_l \left(\sum_{i=1}^n \frac{d^l \phi_i}{dx^l} q_i \right) \left(\sum_{j=1}^n \frac{d^l \phi_j}{dx^l} q_j \right) \Bigg|_0^L \\
&= \frac{1}{2} \sum_{i=1}^n \sum_{j=1}^n q_i q_j \left(\int_0^L \sum_{k=0}^p a_k \frac{d^k \phi_i}{dx^k} \frac{d^k \phi_j}{dx^k} \, dx \right. \\
&\quad \left. + \sum_{l=0}^{p-1} b_l \frac{d^l \phi_i}{dx^l} \frac{d^l \phi_j}{dx^l} \Bigg|_0^L \right) \\
&= \frac{1}{2} \sum_{i=1}^n \sum_{j=1}^n k_{ij} q_i q_j
\end{aligned}
\tag{8.32}
$$

where

$$
k_{ij} = k_{ji} = \int_0^L \sum_{k=0}^p a_k \frac{d^k \phi_i}{dx^k} \frac{d^k \phi_j}{dx^k} \, dx + \sum_{l=0}^{p-1} b_l \frac{d^l \phi_i}{dx^l} \frac{d^l \phi_j}{dx^l} \Bigg|_0^L ,
$$
$$
i, j = 2, \ldots, n \tag{8.33}
$$

are symmetric stiffness coefficients. Note that the mass and stiffness coefficients are identical to those obtained in the preceding section by the Rayleigh-Ritz method.

Lagrange's equations of motion for a discrete conservative system have the form

$$
\frac{d}{dt} \left(\frac{\partial L}{\partial \dot{q}_r} \right) - \frac{\partial L}{\partial q_r} = 0, \qquad r = 1, 2, \ldots, n \tag{8.34}
$$

where $L = T - V$ is the system Langrangian. Introducing Eqs. (8.31) and (8.33) into Eq. (8.33), we obtain the equations of motion

$$
\sum_{j=1}^n (m_{rj} \ddot{q}_j + k_{rj} q_j) = 0, \qquad r = 1, 2, \ldots, n \tag{8.35}
$$

Letting the solution be harmonic, so that $\ddot{q}_j = -\omega^2 q_j = -\Lambda^n q_j$, Eqs. (8.35) yield the eigenvalue problem

$$
\sum_{j=1}^n (k_{rj} - \Lambda^n m_{rj}) q_j = 0, \qquad r = 1, 2, \ldots, n \tag{8.36}
$$

which is identical to that obtained by the Rayleigh-Ritz method, Eq. (8.14).

300

8.4   The method of weighted residuals

The variational method for the solution of the eigenvalue problem (8.1) is based on the stationarity of Rayleigh's quotient, where the quotient represents a functional related to the differential equation and the boundary conditions. There are problems for which variational methods are not applicable, however, so that some method more general in scope is desirable. Such a method is the *method of weighted residuals,* which turns out to be a blanket name for a number of seemingly dissimilar approximate methods. In contrast with the Rayleigh-Ritz method, the method of weighted residuals works directly with the differential equation and boundary conditions. The method is applicable to differential equations in general, although our special interest here lies in eigenvalue problems.

Let us consider a differential eigenvalue problem, such as that described by Eqs. (8.1). For generality, however, no assumption is made concerning the self-adjointness of L, and the operator can in fact be non-self-adjoint. We assume that the eigenvalue problem does not lend itself to closed-form solution and that we are interested in an approximate solution. To this end, we consider a trial function $w(x)$ from the space κ_B^{2p}. In general, the trial function does not satisfy Eq. (8.1a), so that a measure of the error introduced by substituting the trial function for the actual solution can be written in the form

$$R(w, x) = Lw - \lambda m w \tag{8.37}$$

where $R(w, x)$ is known as the *residual.* It is clearly a function of the trial function w and it depends on the position x. If the trial function were the eigenfunction w_i and λ the eigenvalue λ_i, then the residual would be zero. Next, let us consider a *test function,* or *weighting function* from the space κ^0 and define the *weighted residual*

$$vR = v(Lw - \lambda m w) \tag{8.38}$$

If v is orthogonal to R, then the integral of the weighted residual is zero

$$(v, R) = \int_0^L v(Lw - \lambda m w) \, dx = 0 \tag{8.39}$$

Because the weighted residual is integrated once, we can actually accept weighting functions from the space κ^{-1}. This special case will be discussed later in this section.

The requirements on the trial function w can be lowered through some integration by parts, but this increases the requirements on the weighting function v, so that v can no longer be from the space κ^0. More

301

specifically, s integrations by parts would require that w be only in the space κ^{2p-s} and v in the space κ^s. In the process of integrating by parts, the boundary conditions must be taken into account, so that w need satisfy only boundary conditions of order less than $2p - s$. In particular, if $s = p$, then the function w need be only from the space κ_G^p.

The method of weighted residuals consists of assuming an approximate solution of Eqs. (8.1) in the form

$$w^n = \sum_{j=1}^{n} a_j \phi_j \tag{8.40}$$

where $\phi_1, \phi_2, \ldots, \phi_n$ are n functions chosen from a complete set of comparison functions. The functions are regarded as a basis for an n-dimensional subspace S^n of κ_B^{2p}. Moreover, it consists of choosing n functions $\psi_1, \psi_2, \ldots, \psi_n$ from another complete set and regarding them as a basis for an n-dimensional space V^n. The spaces S^n and V^n will be referred to as the *trial space* and *test space*, respectively. Then, the coefficients a_j $(j = 1, 2, \ldots, n)$ are determined by imposing the conditions that the functions ψ_i $(i = 1, 2, \ldots, n)$ be orthogonal to the weighted residual $R = R(w^n, x)$ or

$$(\psi_i, R) = \int_0^L \psi_i (Lw^n - \Lambda^n m w^n)\, dx = 0, \qquad i = 1, 2, \ldots, n \tag{8.41}$$

where Λ^n is the approximate eigenvalue associated with w^n. The question arises as to the reason why Eqs. (8.41) lead to a solution converging to that of the differential eigenvalue problem (8.1). The explanation lies in the facts that the functions ψ_i $(i = 1, 2, \ldots, n)$ are from a complete set and that the function $Lw^n - \Lambda^n m w^n$ is orthogonal to every ψ_i. Indeed, as the number of test functions ψ_i is increased without bounds, the only possible way for the function $Lw^n - \Lambda^n m w^n$ to be orthogonal to the complete set is for the function itself to be identically zero, or

$$\lim_{n \to \infty} (Lw^n - \Lambda^n m w^n) = Lw - \lambda m w = 0 \tag{8.42}$$

Introducing Eq. (8.40) into Eq. (8.41), we obtain the algebraic eigenvalue problem

$$(\psi_i, R) = \int_0^L \psi_i \left(\sum_{j=1}^{n} a_j L\phi_j - \Lambda^n \sum_{j=1}^{n} a_j m\phi_j \right) dx$$

$$= \sum_{j=1}^{n} (k_{ij} - \Lambda^n m_{ij})a_j = 0, \qquad i = 1, 2, \ldots, n \tag{8.43}$$

where

$$k_{ij} = (\psi_i, L\phi_j) = \int_0^L \psi_i L\phi_j \, dx, \qquad i, j = 1, 2, \ldots, n \qquad (8.44a)$$

$$m_{ij} = (\psi_i, m\phi_j) = \int_0^L m\psi_i\phi_j \, dx, \qquad i, j = 1, 2, \ldots, n \qquad (8.44b)$$

are constant coefficients. The coefficients k_{ij} are generally not symmetric, regardless of whether L is self-adjoint or not, and neither are the coefficients m_{ij}. The algebraic eigenvalue problem (8.43) can be cast in matrix form. To this end, let us introduce the vector $\mathbf{a} = [a_1 \, a_2 \cdots a_n]^T$ and the matrices $K = [k_{ij}]$ and $M = [m_{ij}]$, so that Eqs. (8.43) can be rewritten in the compact form

$$K\mathbf{a} = \Lambda^n M\mathbf{a} \qquad (8.45)$$

where K and M are generally nonsymmetric. Solution of the eigenvalue problem (8.45) was discussed in Chs. 3–5.

As mentioned earlier, there are a number of different methods of weighted residuals, the difference being in the nature of the weighting functions. In the sequel, we shall discuss some of the most important ones.

i. *Galerkin's method*

This is perhaps the most widely used of the weighted residuals methods. In Galerkin's method, the weighting functions coincide with the trial functions, or

$$\psi_i = \phi_i, \qquad i = 1, 2, \ldots, n \qquad (8.46)$$

so that the coefficients k_{ij} and m_{ij} become

$$k_{ij} = (\phi_i, L\phi_j) = \int_0^L \phi_i L\phi_j \, dx, \qquad i, j = 1, 2, \ldots, n \qquad (8.47a)$$

$$m_{ij} = (\phi_i, m\phi_j) = \int_0^L m\phi_i\phi_j \, dx, \qquad i, j = 1, 2, \ldots, n \qquad (8.47b)$$

But, whereas the coefficients m_{ij} are symmetric, $m_{ij} = m_{ji}$, the coefficients k_{ij} are in general not symmetric.

The functions $\phi_1, \phi_2, \ldots, \phi_n$ are comparison functions, so that they belong to the space κ_B^{2p}. Hence, one can integrate Eq. (8.47a) by parts p times to lower the order of derivatives from $2p$ to p. But, unless the

303

8 Discretization of continuous systems

operator L is self-adjoint, the resulting expression will not be symmetric in ϕ_i and ϕ_j and their derivatives, which is consistent with the fact that for a non-self-adjoint system the coefficients k_{ij} are not symmetric. As a simple illustration, let us consider the non-self-adjoint eigenvalue problem

$$-\frac{d}{dx}\left(s\frac{dw}{dx}\right)+r\frac{dw}{dx}=\lambda mw, \qquad 0<x<L \tag{8.48a}$$

$$w(0)=w'(L)=0 \tag{8.48b}$$

The coefficient k_{ij} is simply

$$k_{ij}=(\phi_i, L\phi_j)=\int_0^L \phi_i\left[-\frac{d}{dx}\left(s\frac{d\phi_j}{dx}\right)+r\frac{d\phi_j}{dx}\right]dx \tag{8.49}$$

which, upon one integration by parts and consideration of the boundary conditions (8.48b), reduces to

$$k_{ij}=\int_0^L (s\phi_i'\phi_j'+r\phi_i\phi_j')\,dx \tag{8.50}$$

which is clearly not symmetric in ϕ_i and ϕ_j and their first derivatives.

In the special case in which L is self-adjoint, the coefficients k_{ij} do become symmetric, $k_{ij}=k_{ji}$. In this case, p integration by parts of Eq. (8.47a) yield

$$k_{ij}=k_{ji}=[\phi_i, \phi_j] \tag{8.51}$$

where now the functions ϕ_i need be from the energy space κ_G^p only. Hence, *for self-adjoint systems, the Galerkin and the Rayleigh-Ritz methods yield the same results.* Returning to the example discussed above, if $r=0$, then the system becomes self-adjoint and the coefficients k_{ij} become symmetric in ϕ_i' and ϕ_j', as can be concluded from Eq. (8.50).

ii. *The collocation method*

The collocation method is another weighted residuals method that has seen a great deal of use. In this case, the weighting functions are spatial Dirac delta functions

$$\psi_i = \delta(x-x_i), \qquad i=1,2,\ldots,n \tag{8.52}$$

which are really from the space κ^{-1}. As pointed out earlier, this is permissible. The points $x=x_i$ are chosen in advance. Introducing Eq.

304

(8.52) into Eq. (8.41), we obtain

$$\int_0^L \delta(x - x_i)(Lw^n - \Lambda^n mw^n)\,dx$$

$$= (Lw^n - \Lambda^n mw^n)\big|_{x=x_i} = 0, \qquad i = 1, 2, \ldots, n \quad (8.53)$$

Equations (8.53) indicate that the differential equation is satisfied at n preselected locations $x = x_i$ $(i = 1, 2, \ldots, n)$ throughout the interval $0 < x < L$, which explains the name of the method. The convergence of the method can be explained by the fact that as n increases the differential equation is satisfied at an increasing number of points. Ultimately, if the number of point x_i is infinitely large, one can regard the differential equation as being satisfied at every point of the interval $0 < x < L$.

Introducing Eqs. (8.52) into Eqs. (8.44), we obtain the coefficients

$$k_{ij} = \int_0^L \delta(x - x_i)L\phi_j\,dx = L\phi_j(x_i), \qquad i, j = 1, 2, \ldots, n \quad (8.54a)$$

$$m_{ij} = \int_0^L \delta(x - x_i)m\phi_i\,dx = m(x_i)\phi_j(x_i), \qquad i, j = 1, 2, \ldots, n \quad (8.54b)$$

It is clear from Eqs. (8.54) that neither the coefficients k_{ij} nor the coefficients m_{ij} are symmetric. The curious thing about this method is that, unlike the Galerkin method, the coefficients k_{ij} are not symmetric even when L is self-adjoint. Hence, the eigenvalue problem is in the form (8.45), in which K and M are not symmetric. Nevertheless, if L is self-adjoint, the solution of the eigenvalue problem (8.45) exhibits one of the characteristics of a self-adjoint system. In particular, the eigenvalues Λ_r^n are real. The eigenvectors are also real, but there are two biorthogonal sets of eigenvectors (see Sec. 3.7).

In the case of two- and three-dimensional domains D, the boundary S also has an infinity of points. Hence, the possibility exists for the method to be extended to cover the entire closed domain $\bar{D} = D \cup S$. Indeed, in this case the trial functions need not satisfy the boundary conditions exactly. In fact, there are three options available, depending on the trial functions:

1. *Interior method*—The trial functions ϕ_i satisfy the boundary conditions but not the differential equation.
2. *Boundary method*—The trial functions ϕ_i satisfy the differential equation but not the boundary conditions.
3. *Mixed method*—The trial functions ϕ_i satisfy neither the differential equation nor the boundary conditions.

The interior method is the classical method presented above. The boundary method is not as frivolous as it may appear, particularly for three-dimensional problems in which the exact satisfaction of the boundary conditions may present serious difficulties. Clearly, the mixed method is designed for very difficult problems, in which neither the differential equation nor the boundary conditions can be satisfied exactly.

The obvious advantage of the collocation method lies in the ease with which the coefficients k_{ij} and m_{ij} can be evaluated. Weighing against the method is the fact that one must solve a nonsymmetric eigenvalue problem even for a self-adjoint system.

iii. *The method of least squares*

The method of least squares has a hybrid character, in the sense that it is both a variational method and a weighted residual method. Because the method is applicable also to cases for which a classical variational principle does not exist, its classification as a weighted residual method seems justified. The method consists of minimizing the square of the norm of the residual, or

$$\|R\|^2 = \int_0^L R^2 \, \mathrm{d}x = \text{minimum} \tag{8.55}$$

where R is the residual, as given by Eq. (8.38). No assumption is made concerning the differential operator L, so that the operator can be self-adjoint or non-self-adjoint. Assuming a solution in the form (8.40), Eq. (8.55) becomes

$$\|R(a_1, a_2, \ldots, a_n)\|^2 = \int_0^L R^2(a_1, a_2, \ldots, a_n) \, \mathrm{d}x = \text{minimum} \tag{8.56}$$

where the unknown coefficients $a_1, a_2, \ldots, a_n$ are determined in the process of rendering $\|R\|^2$ a minimum. The conditions for a minimum are

$$\frac{\partial}{\partial a_i} \|R\|^2 = 2 \int_0^L \frac{\partial R}{\partial a_i} R \, \mathrm{d}x = 0, \qquad i = 1, 2, \ldots, n \tag{8.57}$$

so that the method of least squares is a weighted residual method in which the weighting functions are

$$\psi_i = \frac{\partial R}{\partial a_i}, \qquad i = 1, 2, \ldots, n \tag{8.58}$$

Introducing solution (8.40) into Eq. (8.38) and using (8.57), we

obtain

$$\int_0^L \left[\frac{\partial}{\partial a_i} \left(\sum_{j=1}^n a_j L\phi_j - \Lambda^n \sum_{j=1}^n a_j m\phi_j \right) \right] \left(\sum_{j=1}^n a_j L\phi_j - \Lambda^n \sum_{j=1}^n a_j m\phi_j \right) dx$$

$$= \sum_{j=1}^n a_j \int_0^L [(L\phi_i)(L\phi_j) - \Lambda^n(L\phi_i)(m\phi_j) - \Lambda^n(m\phi_i)(L\phi_j)$$

$$+ (\Lambda^n)^2 (m\phi_i)(m\phi_j)]\, dx = \sum_{j=1}^n a_j [(L\phi_i, L\phi_j)$$

$$- \Lambda^n(L\phi_i, m\phi_j) - \Lambda^n(m\phi_i, L\phi_j) + (\Lambda^n)^2(m\phi_i, m\phi_j)] = 0,$$

$$i = 1, 2, \ldots, n \quad (8.59)$$

where the inner product notation is self-evident. Equations (8.59) represent the algebraic eigenvalue problem for the method of least squares.
Introducing the $n \times n$ matrices

$$K = [k_{ij}] = [(L\phi_i, L\phi_j)]$$

$$H = [h_{ij}] = [(L\phi_i, m\phi_j)] \qquad i, j = 1, 2, \ldots, \quad (8.60)$$

$$M = [m_{ij}] = [(m\phi_i, m\phi_j)]$$

and the n-vector

$$\mathbf{a} = [a_1\ a_2 \cdots a_n]^T \quad (8.61)$$

The eigenvalue problem can be written in the matrix form

$$[K - \Lambda^n(H + H^T) + (\Lambda^n)^2 M]\mathbf{a} = \mathbf{0} \quad (8.62)$$

The eigenvalue problem (8.62) involves both Λ^n and $(\Lambda^n)^2$, so that it
is not in the standard form (8.45). To reduce it to the form (8.45), let us
introduce the auxiliary vector

$$\mathbf{b} = \Lambda^n \mathbf{a} \quad (8.63)$$

and define the $2n$-vector of coefficients

$$\mathbf{a}^* = [\mathbf{b}^T \mid \mathbf{a}^T]^T \quad (8.64)$$

Moreover, let us define the $2n \times 2n$ matrices

$$K^* = \begin{bmatrix} H + H^T & -K \\ -K & 0 \end{bmatrix}, \qquad M^* = \begin{bmatrix} M & 0 \\ 0 & -K \end{bmatrix} \quad (8.65)$$

which permits the reduction of the eigenvalue problem (8.62) to the

8 Discretization of continuous systems

standard form

$$K^* \mathbf{a}^* = \Lambda^n M^* \mathbf{a}^* \tag{8.66}$$

Hence, the method of least squates requires the solution of an eigenvalue problem of twice the order of that required by the Galerkin method or the collocation method. This fact has tended to inhibit the use of the method in problems of structural dynamics.

As a matter of interest, we note that the formulation (8.65–66) is similar to that for the damped system of Sec. 2.5, with the exception that the damping matrix is replaced here by $-(H + H^T)$.

iv. *The method of subdomains*

In this method, the domain D is divided into n smaller subdomains D_i $(i = 1, 2, \ldots, n)$, $D = \sum_{i=1}^{n} D_i$, and the test functions are chosen in the form

$$\psi_i = \begin{cases} 1 \text{ for } x \text{ in } D_i \\ 0 \text{ for } x \text{ not in } D_i \end{cases} \tag{8.67}$$

Inserting Eq. (8.67) into Eq. (8.41) and recalling Eq. (8.40), we obtain

$$\int_{D_i} \left(\sum_{j=1}^{n} a_j L\phi_j - \Lambda^n \sum_{j=1}^{n} a_j m\phi_j \right) dx = 0, \qquad i = 1, 2, \ldots, n \tag{8.68}$$

so that the average residual is zero over each of the subdomains D_i. As the number of subdomains increases, the eigenvalue problem is satisfied on the average in smaller and smaller subdomains, the expectation being that in the limit the eigenvalue problem will be satisfied everywhere. The eigenvalue problem remains in the form (8.43), where

$$k_{ij} = \int_{D_i} L\phi_j \, dx, \qquad i, j = 1, 2, \ldots, n \tag{8.69a}$$

$$m_{ij} = \int_{D_i} m\phi_j \, dx, \qquad i, j = 1, 2, \ldots, n \tag{8.69b}$$

Hence, the coefficients are nonsymmetric even in the case of a self-adjoint problem. The application of the method of subdomains to vibration problems has been scant.

v. *The method of moments*

The method of moments employs the weighting functions

$$\psi_i = x^{i-1}, \qquad i = 1, 2, \ldots, n \tag{8.70}$$

308

so that Eq. (8.41), in conjunction with solution (8.40), becomes

$$\int_0^L x^{i-1} \left(\sum_{j=1}^n a_j L \phi_j - \Lambda^n \sum_{j=1}^n a_j m \phi_j \right) dx = 0, \qquad i = 1, 2, \ldots, n \qquad (8.71)$$

The equation corresponding to $i = 1$ implies that the average residual over the entire domain is zero. On the other hand, the equations for $i = 2, 3, \ldots, n$ imply that successively higher moments of the residuals, beginning with the first moment and ending with $(n-1)$st moment, are required to be zero, which explains the name of the met od. From Eq. (8.71), we conclude that the eigenvalue problem remains in the form (8.43), where the matrix coefficients are

$$k_{ij} = \int_0^L x^{i-1} L \phi_j \, dx, \qquad i, j = 1, 2, \ldots, n \qquad (8.72a)$$

$$m_{ij} = \int_0^L x^{i-1} m \phi_j \, dx, \qquad i, j = 1, 2, \ldots, n \qquad (8.72b)$$

Here also, the coefficients are generally nonsymmetric even when the problem is self-adjoint. The method of moments has been developed for application to boundary layer problems and there seem to be virtually no applications to vibration problems.

In view of the above discussion, our preference for non-self-adjoint eigenvalue problems is for the Galerkin method.

Example 8.2

Consider the system of Example 8.1, use the comparison functions

$$\phi_i(x) = \sin \beta_i x, \qquad i = 1, 2, \ldots \qquad (a)$$

where β_i are such that ϕ_i satisfy the dynamic boundary condition at $x = L$, and solve the eigenvalue problem by the collocation method. Estimate the number of points to be used in order for the method to yield estimated lower natural frequencies comparable in accuracy to those obtained in Example 8.1 by the Rayleigh-Ritz method.

For the functions ϕ_i to be comparison functions, they must satisfy the boundary condition at $x = L$, or

$$EA(x) \frac{d\phi_i(x)}{dx} \bigg|_{x=L} = -K\phi_i(L) \qquad (b)$$

Using Eqs. (a) of Example 8.1, Eq. (b) reduces to

$$\tan \beta_i L = -3\beta_i L, \qquad i = 1, 2, \ldots \qquad (c)$$

8   Discretization of continuous systems

yielding the roots

$$\beta_1 L = 1.758164, \qquad \beta_2 L = 4.781983, \qquad \beta_3 L = 7.896171$$

$$\beta_4 L = 11.025777, \qquad \beta_5 L = \cdots \tag{d}$$

The stiffness coefficients are obtained from Eq. (8.54a), where in this particular case the operator L has the expression

$$L = -\frac{d}{dx}\left[EA(x)\frac{d}{dx}\right] = \frac{6EA}{5}\left\{\frac{x}{L^2}\frac{d}{dx} - \left[1 - \frac{1}{2}\left(\frac{x}{L}\right)^2\right]\frac{d^2}{dx^2}\right\} \tag{e}$$

Hence, the stiffness coefficients are

$$k_{ij} = \frac{6EA}{5}\left\{\frac{x_i}{L^2}\beta_j \cos\beta_j x_i + \left[1 - \frac{1}{2}\left(\frac{x_i}{L}\right)^2\right]\beta_j^2 \sin\beta_j x_i\right\}$$

$$i, j = 1, 2, \ldots \quad \text{(f)}$$

Moreover, using Eq. (8.54b), the mass coefficients are

$$m_{ij} = \frac{6m}{5}\left[1 - \frac{1}{2}\left(\frac{x_i}{L}\right)^2\right]\sin\beta_j x_i, \qquad i, j = 1, 2, \ldots \tag{g}$$

Letting $n = 2$ and choosing $x_1 = L/3$ and $x_2 = 2L/3$, the stiffness and mass matrices can be shown to be

$$K = \frac{EA}{L^2}\begin{bmatrix} 2.523489 & 25.864930 \\ 3.204803 & -4.811262 \end{bmatrix}$$

$$M = m\begin{bmatrix} 0.626820 & 1.133027 \\ 0.860132 & -4.328438 \end{bmatrix} \tag{h}$$

Solving the associated eigenvalue problem, we obtain the estimated first two natural frequencies

$$\omega_1^2 = 1.940773\sqrt{\frac{EA}{mL^2}}, \qquad \omega_2^2 = 5.018765\sqrt{\frac{EA}{mL^2}} \tag{i}$$

and we note that they are somewhat higher than those obtained in Example 8.1 for $n = 2$.

As n increases, convergence is relatively slow. For $n = 14$, we obtain the estimated first two natural frequencies

$$\omega_1^{14} = 1.896734\sqrt{\frac{EA}{mL^2}}, \qquad \omega_2^{14} = 4.889894\sqrt{\frac{EA}{mL^2}} \tag{j}$$

which are comparable to those obtained by the Rayleigh-Ritz method with $n = 4$.

310

Example 8.3

Solve the eigenvalue problem of Example 8.1 by the least squares method for $n = 1, 2, 3, 4$ and tabulate the eigenvalues. Use the same comparison functions as in Example 8.2.

Using Eqs. (8.60), in conjunction with the comparison functions $\phi_i = \sin \beta_i x$ $(i = 1, 2, 3, 4)$ we obtain

$$k_{ij} = \int_0^L (L\phi_i)(L\phi_j)\, dx = \left(\frac{6EA}{5L^2}\right)^2 \int_0^L \left\{ \beta_i x \cos \beta_i x \right.$$

$$+ \left[1 - \frac{1}{2}\left(\frac{x}{L}\right)^2 \right] (\beta_i L)^2 \sin \beta_i x \right\} \left\{ \beta_j x \cos \beta_j x \right.$$

$$+ \left[1 - \frac{1}{2}\left(\frac{x}{L}\right)^2 \right] (\beta_j L)^2 \sin \beta_j x \right\} dx \qquad i, j = 1, 2, 3, 4 \qquad (a)$$

$$h_{ij} = \int_0^L (L\phi_i)(m\phi_j)\, dx = \frac{6EA}{5L^2}\frac{6m}{5} \int_0^L \left\{ \beta_i x \cos \beta_i x \right.$$

$$+ \left[1 - \frac{1}{2}\left(\frac{x}{L}\right)^2 \right] (\beta_i L)^2 \sin \beta_i x \right\} \left[1 - \frac{1}{2}\left(\frac{x}{L}\right)^2 \right] \sin \beta_j x \, dx$$

$$m_{ij} = \int_0^L (m\phi_i)(m\phi_j)\, dx = \left(\frac{6m}{5}\right)^2 \int_0^L \left[1 - \frac{1}{2}\left(\frac{x}{L}\right)^2 \right]^2 \sin \beta_i x \sin \beta_j x \, dx$$

Inserting the values of β_i $(i = 1, 2, 3, 4)$ from Example 8.2, we compute the coefficient matrices

$$K = \begin{bmatrix} 6.121020 & 11.297895 & -6.208816 & 5.377565 \\ & 280.487600 & 180.534456 & -66.282126 \\ \text{symm} & & 2{,}035.542997 & 919.705326 \\ & & & 7{,}683.912402 \end{bmatrix}$$

$$H + H^T = \begin{bmatrix} 3.305894 & 2.834824 & -1.157978 & 0.907666 \\ & 23.718532 & 9.656858 & -2.705852 \\ \text{symm} & & 64.464491 & 21.210356 \\ & & & 125.578282 \end{bmatrix} \qquad (b)$$

$$M = \begin{bmatrix} 0.451131 & 0.147588 & -0.029719 & 0.013258 \\ & 0.510198 & 0.124533 & -0.024149 \\ \text{symm} & & 0.514000 & 0.120471 \\ & & & 0.514993 \end{bmatrix}$$

Table I

n	$\|\omega_1^n\| \sqrt{\dfrac{mL^2}{EA}} \underline{/\psi_1^n}$	$\|\omega_2^n\| \sqrt{\dfrac{mL^2}{EA}} \underline{/\psi_2^n}$	$\|\omega_3^n\| \sqrt{\dfrac{mL^2}{EA}} \underline{/\psi_3^n}$	$\|\omega_4^n\| \sqrt{\dfrac{mL^2}{EA}} \underline{/\psi_4^n}$
1	$1.919244 \underline{/\pm2°\ 56.9'}$			
2	$1.902000 \underline{/\pm1°\ 45.8'}$	$4.913135 \underline{/\pm1°\ 35.3'}$		
3	$1.897603 \underline{/\pm1°\ 9.8'}$	$4.893105 \underline{/\pm1°\ 1.8'}$	$7.998692 \underline{/\pm1°\ 9.3'}$	
4	$1.896054 \underline{/\pm0°\ 51.5'}$	$4.887541 \underline{/\pm0°\ 42.5'}$	$7.976459 \underline{/\pm0°\ 46.6'}$	$11.112914 \underline{/\pm0°\ 54.7'}$

Inserting the above matrices into Eqs. (8.65–66) and solving the eigenvalue problem for $n = 1, 2, 3, 4$, we obtain complex eigenvalues, where the imaginary parts are small and diminishing with increasing n. To permit a comparison with the results of Example 8.1, we tabulate the natural frequencies by magnitude $|\omega_i^n|$ and phase angle ψ_i^n, as shown in Table I.

Except for the bothersome presence of phase angles, the magnitudes of the natural frequencies compare reasonably well with the natural frequencies obtained by the Rayleigh-Ritz method. On balance, however, there appears to be no reason for choosing to use the least squares method over the Rayleigh-Ritz method. Indeed, the computation of the coefficient matrices is more tedious for the least squares method and the order of the eigenvalue problem is twice as large as that in the Rayleigh-Ritz method. This is in addition to the fact that the matrices in the least squares method are not positive definite, so that many of the efficient algorithms for real symmetric positive definite matrices cannot be used.

8.5   Flutter of a cantilever aircraft wing

An example of a non-self-adjoint problem is the combined bending and torsional vibration of a cantilever wing in steady air flow (Fig. 8.2). We shall denote the bending deflection by $w(x, t)$ and the torsional rotation about the elastic axis* by $\theta(x, t)$, where w is assumed positive downward and θ is assumed positive if the leading edge is up. The angle θ is known as local angle of attack. For simplicity, the elastic axis is assumed to be straight and coincident with the x-axis. The distance between the leading edge and the elastic axis is denoted by y_0, the distance between the elastic

---

* The elastic axis is defined as the locus of the shear centers of the cross sections, where a shear center is a point such that a shearing force acting through it produces pure bending (no torsion) and a moment about it produces pure torsion (no bending).

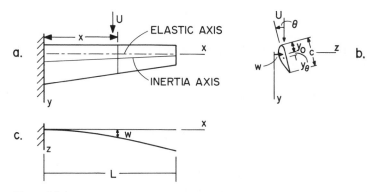

Figure 8.2

axis and the inertia axis by y_θ and the chord length by c, all three quantities being functions of x. The speed of the air flow relative to the wing is assumed to be constant and is denoted by U.

The free vibration of the wing, in which all forces except the aerodynamic forces are ignored, can be shown to be (F5, Sec. 6.3)

$$\frac{\partial^2}{\partial x^2}\left(EI\frac{\partial^2 w}{\partial x^2}\right) + m\frac{\partial^2 w}{\partial t^2} + my_\theta\frac{\partial^2 \theta}{\partial t^2} + \frac{\rho U^2}{2}c\frac{dC_L}{d\theta}$$

$$\times\left[\theta + \frac{1}{U}\frac{\partial w}{\partial t} + \frac{c}{U}\left(\frac{3}{4} - \frac{y_0}{c}\right)\frac{\partial \theta}{\partial t}\right] = 0, \qquad 0 < x < L \qquad (8.73a)$$

$$-\frac{\partial}{\partial x}\left(GJ\frac{\partial \theta}{\partial x}\right) + my_\theta\frac{\partial^2 w}{\partial t^2} + I_\theta\frac{\partial^2 \theta}{\partial t^2} - \frac{\rho U^2}{2}c^2$$

$$\times\left\{-\frac{c\pi}{8U}\frac{\partial \theta}{\partial t} + \left(\frac{y_0}{c} - \frac{1}{4}\right)\frac{dC_L}{d\theta}\left[\theta + \frac{1}{U}\frac{\partial w}{\partial t} + \frac{c}{U}\left(\frac{3}{4} - \frac{y_0}{c}\right)\frac{\partial \theta}{\partial t}\right]\right\} = 0,$$

$$0 < x < L \quad (8.73b)$$

where EI and GJ are the bending and torsional stiffness, respectively, m is the mass per unit length, I_θ is the mass moment of inertia per unit length, ρ is the air density, and C_L is the local lift coefficient. The aerodynamic forces and moments are obtained by using the so-called quasi-steady, 'strip theory', whereby the local lift coefficient C_L is proportional to the instantaneous angle of attack θ. The derivative $dC_L/d\theta$ is considered to be constant, with a theoretical value of 2π for incompressible flow and an experimental value of somewhat less than 2π. Moreover, the aerodynamic analysis is subject to the quasi-steady assumption (F5, p. 192), which implied that only the instantaneous deformation is important and the history of the motion may be neglected. Otherwise, certain time

313

integrals would appear in Eqs. (8.73). Equations (8.73) are linear and homogeneous, with coefficients depending on the spatial position x as well as on the parameter U. The displacements w and θ are subject to the boundary conditions.

$$w = \frac{\partial w}{\partial x} = \theta = 0 \quad \text{at} \quad x = 0 \tag{8.74a}$$

$$EI\frac{\partial^2 w}{\partial x^2} = \frac{\partial}{\partial x}\left(EI\frac{\partial^2 w}{\partial x^2}\right) = \frac{\partial}{\partial x}\left(GJ\frac{\partial \theta}{\partial x}\right) = 0 \quad \text{at} \quad x = L \tag{8.74b}$$

Our first objective is to derive the eigenvalue problem. Because the system is not self-adjoint, we cannot assume that the eigenvalues are real. Hence, let us assume a solution in the form

$$w(x, t) = W(x)e^{\lambda t}, \qquad \theta(x, t) = \Theta(x)e^{\lambda t} \tag{8.75}$$

where λ is generally complex. Introducing Eqs. (8.75) into Eqs. (8.73) and dividing through by $e^{\lambda t}$, we obtain the ordinary differential equations

$$(EIW'')'' + \frac{\rho U^2}{2}c\frac{dC_L}{d\theta}\Theta + \lambda\frac{\rho U}{2}c\frac{dC_L}{d\theta}\left[W + c\left(\frac{3}{4} - \frac{y_0}{c}\right)\Theta\right]$$
$$+ \lambda^2 m(W + y_\theta\Theta) = 0, \qquad 0 < x < L \tag{8.76a}$$

$$-(GJ\Theta')' - \frac{\rho U^2}{2}c^2\left(\frac{y_0}{c} - \frac{1}{4}\right)\frac{dC_L}{d\theta}\Theta - \lambda\frac{\rho U}{2}c^2\left\{\left(\frac{y_0}{c} - \frac{1}{4}\right)\frac{dC_L}{d\theta}W\right.$$
$$+ c\left[\left(\frac{y_0}{c} - \frac{1}{4}\right)\left(\frac{3}{4} - \frac{y_0}{c}\right)\frac{dC_L}{d\theta} - \frac{\pi}{8}\right]\Theta\right\} + \lambda^2(my_\theta W + I_\theta\Theta) = 0,$$
$$0 < x < L \tag{8.76b}$$

The boundary conditions (8.74) retain the same form, except that w and θ are replaced by W and Θ, respectively, and partial derivatives of w and θ with respect to x by total derivatives.

No closed-form solution of Eqs. (8.76) is possible, so that we shall seek an approximate solution. Before proceeding with the solution, however, it will prove instructive to examine the effect of the air flow speed U on the eigenvalue λ. For $U = 0$, Eqs. (8.76) reduce to

$$(EIW'')'' + \lambda^2 m(W + y_\theta\Theta) = 0, \qquad 0 < x < L \tag{8.77a}$$

$$-(GJ\Theta')' + \lambda^2(my_\theta W + I_\theta\Theta) = 0, \qquad 0 < x < L \tag{8.77b}$$

The system (8.77) can be shown to be self-adjoint and positive definite. Because the definition of self-adjointness introduced in Ch. 7 was in terms of a single dependent variable, the proof is somewhat indirect and it consists of reducing Eqs. (8.77) to discrete form. Hence, let us consider Galerkin's method and assume a solution of Eqs. (8.77) in the form

$$W = \sum_{j=1}^{n} a_j \phi_j, \qquad \Theta = \sum_{j=n+1}^{n+m} a_j \phi_j \qquad (8.78)$$

where ϕ_j are comparison functions. The first n need be four times differentiable and satisfy the boundary conditions $\phi_j(0) = \phi_j'(0) = EI\phi_j''|_{x=L} = (EI\phi_j'')'|_{x=L} = 0$ and the remaining m need be twice differentiable and satisfy the boundary conditions $\phi_j(0) = GJ\phi_j'|_{x=L} = 0$. The independent pure bending modes and pure torsional modes of the cantilever wing are suitable comparison functions. Introducing Eqs. (8.78) into Eqs. (8.77), we can write

$$\sum_{j=1}^{n} a_j (EI\phi_j'')'' + \lambda^2 \sum_{j=1}^{n} a_j m\phi_j + \lambda^2 \sum_{j=n+1}^{n+m} a_j m y_\theta \phi_j = 0 \qquad (8.79a)$$

$$-\sum_{j=n+1}^{n+m} a_j (GJ\phi_j')' + \lambda^2 \sum_{j=1}^{n} a_j m y_\theta \phi_j + \lambda^2 \sum_{j=n+1}^{n+m} a_j I_\theta \phi_j = 0 \qquad (8.79b)$$

Next, multiply Eq. (8.79a) by ϕ_i $(i = 1, 2, \ldots, n)$ and Eq. (8.79b) by ϕ_i $(i = n+1, n+2, \ldots, n+m)$, integrate both results over the interval $0 \leq x \leq L$, and obtain the algebraic eigenvalue problem

$$\sum_{j=1}^{n+m} k_{ij} a_j + \lambda^2 \sum_{j=1}^{n+m} m_{ij} a_j = 0, \qquad i = 1, 2, \ldots, n+m \qquad (8.80)$$

$$k_{ij} = \int_0^L \phi_i (EI\phi_j'')'' \, dx = \int_0^L EI\phi_i''\phi_j'' \, dx = k_{ij}, \qquad i, j = 1, 2, \ldots, n$$

$$k_{ij} = k_{ji} = 0, \qquad i = 1, 2, \ldots, n; j = n+1, n+2, \ldots, n+m \qquad (8.81a)$$

$$k_{ij} = -\int_0^L \phi_i (GJ\phi_j')' \, dx = \int_0^L GJ\phi_i'\phi_j' \, dx = k_{ji},$$
$$i, j = n+1, n+2, \ldots, n+m$$

and

$$m_{ij} = m_{ji} = \int_0^L m\phi_i\phi_j \, dx, \qquad i, j = 1, \ldots, n$$

$$m_{ij} = m_{ji} = \int_0^L m y_\theta \phi_i \phi_j \, dx, \qquad i = 1, 2, \ldots, n;$$
$$j = n+1, n+2, \ldots, n+m \qquad (8.81b)$$

$$m_{ij} = m_{ji} = \int_0^L I_\theta \phi_i \phi_j \, dx, \qquad i, j = n+1, n+2, \ldots, n+m$$

are symmetric stiffness and mass coefficients. Equations (8.80) can be written in the matrix form

$$Ka = -\lambda^2 Ma \qquad (8.82)$$

where K and M are positive definite symmetric matrices, so that *for $U = 0$ the system is self-adjoint and positive definite.* Hence, the eigenvalue $-\lambda^2$ must be real and positive. This leads us to the conclusion that $\lambda = \pm i\omega$, indicating that the free vibration is oscillatory with the frequency ω. Because our immediate interest is mainly in a qualitative investigation of the system behaviour as U varies, we shall not pursue the eigenvalue problem (8.82) any further.

The eigenvalue λ is a continuous function of the air speed U. When U is not zero, but infinitesimally small, the exponent λ is no longer pure imaginary but complex, $\lambda = \alpha + i\omega$. Of course, to investigate this case we must return to the original non-self-adjoint system, Eqs. (8.76). It can be shown (F5, p. 196) that for sufficiently small U and for $(\mathrm{d}C_L/\mathrm{d}\theta) < 2\pi$ the wing is losing energy to the surrounding air, so that the motion is damped oscillatory, and hence asymptotically stable. The clear implication is that α is negative. As U increases, α can become positive, so that at the point at which α changes sign (see Fig. 8.3) the motion ceases to be damped oscillatory and becomes unstable. The air speed corresponding to $\alpha = 0$ is known as critical speed and denoted by U_{cr}. There are many critical values of U but, because in actual flight U increases from an initially zero value, the lowest critical value is the most important. One can distinguish between two critical cases, depending on the value of ω. When $\alpha = 0$ and $\omega = 0$ the wing is said to be in critical *divergent* condition. When $\alpha = 0$ and $\omega \neq 0$ the wing is said to be in critical *flutter* condition.

The above qualitative discussion can be substantiated by a more quantitative analysis. To this end, we must derive and solve the complete non-self-adjoint eigenvalue problem. Introducing solution (8.78) into Eqs. (8.76), multiplying Eq. (8.76a) by ϕ_i $(i = 1, 2, \ldots, n)$ and Eq. (8.76b) by ϕ_i $(i = n+1, n+2, \ldots, n+m)$ and integrating both results over the interval $0 < x < L$, we obtain the eigenvalue problem

$$[K + U^2 H + \lambda UL + \lambda^2 M]\mathbf{a} = \mathbf{0} \qquad (8.83)$$

Figure 8.3

where the matrices K and M are the same symmetric matrices derived earlier and whose elements are given by Eqs. (8.81). On the other hand, the matrices H and L are not symmetric. Their elements can be shown to have the expressions

$$h_{ij} = 0, \qquad i, j = 1, 2, \ldots, n$$

$$h_{ij} = \frac{\rho}{2} \frac{dC_L}{d\theta} \int_0^L c\phi_i\phi_j \, dx, \qquad i = 1, 2, \ldots, n; \; j = n+1, n+2, \ldots, n+m$$

$$h_{ij} = 0, \qquad i = n+1, n+2, \ldots, n+m; \; j+1, 2, \ldots, n$$
$$\text{(8.84a)}$$

$$h_{ij} = -\frac{\rho}{2} \frac{dC_L}{d\theta} \int_0^L c^2\left(\frac{y_0}{c} - \frac{1}{4}\right)\phi_i\phi_j \, dx, \qquad i, j = n+1, n+2, \ldots, n+m$$

$$l_{ij} = \frac{\rho}{2} \frac{dC_L}{d\theta} \int_0^L c\phi_i\phi_j \, dx, \qquad i, j = 1, 2, \ldots, n$$

$$l_{ij} = \frac{\rho}{2} \frac{dC_L}{d\theta} \int_0^L c^2\left(\frac{3}{4} - \frac{y_0}{c}\right)\phi_i\phi_j \, dx, \qquad i = 1, 2, \ldots, n;$$

$$j = n+1, n+2, \ldots, n+m$$
$$\text{(8.84b)}$$

$$l_{ij} = -\frac{\rho}{2} \frac{dC_L}{d\theta} \int_0^L c^2\left(\frac{y_0}{c} - \frac{1}{4}\right)\phi_i\phi_j \, dx, \qquad i = n+1, n+2, \ldots, n+m;$$

$$j = 1, 2, \ldots, n$$

$$l_{ij} = \frac{\rho}{2} \int_0^L c^3\left[\frac{\pi}{8} - \left(\frac{y_0}{c} - \frac{1}{4}\right)\left(\frac{3}{4} - \frac{y_0}{c}\right)\frac{dC_L}{d\theta}\right]\phi_i\phi_j \, dx,$$

$$i, j = n+1, n+2, \ldots, n+m$$

Using the same procedure as in Sec. 8.4, we can reduce the eigenvalue problem (8.83) to the standard form

$$K^*\mathbf{a}^* = \lambda M^*\mathbf{a}^* \tag{8.85}$$

where

$$\mathbf{a}^* = [\mathbf{a}^T \,\vdots\, \mathbf{b}^T]^T = [\mathbf{a}^T \,\vdots\, \lambda\mathbf{a}^T]^T \tag{8.86}$$

is a $2(n+m)$-vector and

$$K^* = \left[\begin{array}{c:c} 0 & 1 \\ \hline -(K+U^2H) & -UL \end{array}\right], \qquad M^* = \left[\begin{array}{c:c} 1 & 0 \\ \hline 0 & M \end{array}\right] \tag{8.87}$$

are $2(n+m) \times 2(n+m)$ matrices. The critical value U_{cr} of interest here is the lowest value of U for which $\alpha = \operatorname{Re}\lambda = 0$. Flutter occurs if at U_{cr}, $\omega = \operatorname{Im}(\lambda) \neq 0$. For $\omega = 0$, divergence occurs. To compute U_{cr}, one must solve the eigenvalue problem (8.85) repeatedly for increasing values of U. For small values of U, all the eigenvalues λ_r $(r = 1, 2, \ldots, 2n + 2m)$ have negative real parts. The first value of U at which the real part of an eigenvalue reduces to zero is U_{cr}.

A first estimate of U_{cr} can be obtained by approximating W and Θ by means of a single term, $n = m = 1$. Then, letting $\lambda = i\omega$ in Eq. (8.85), U_{cr} is the value of U which permits the solution of the determinantal equation

$$\det \begin{bmatrix} i\omega & 0 & -1 & 0 \\ 0 & i\omega & 0 & -1 \\ k_{11} & U^2 h_{12} & i\omega m_{11} + U l_{11} & i\omega m_{12} + U l_{12} \\ 0 & k_{22} + U^2 h_{22} & i\omega m_{12} + U l_{21} & i\omega m_{22} + U l_{22} \end{bmatrix} = 0 \qquad (8.88)$$

Equation (8.88) yields

$$\begin{aligned}
&\omega^4 (m_{11} m_{22} - m_{12}^2) - i\omega^3 U[m_{11} l_{22} + m_{22} l_{11} - m_{12}(l_{12} + l_{21})] \\
&\quad - \omega^2 [U^2 (l_{11} l_{22} - l_{12} l_{21} - h_{12} m_{12} + h_{22} m_{11}) + k_{22} m_{11} + k_{11} m_{22}] \\
&\quad - i\omega U[U^2 (h_{12} l_{21} + h_{22} l_{11}) + k_{22} l_{11} - k_{11} l_{22}] + k_{11}(U^2 h_{22} + k_{22}) = 0
\end{aligned}$$
$$(8.89)$$

Equating the imaginary part of Eq. (8.89) to zero, we obtain

$$\omega^2 = \frac{U^2 (h_{12} l_{21} + h_{22} l_{11}) + k_{22} l_{11} - k_{11} l_{22}}{m_{12}(l_{12} + l_{21}) - (m_{11} l_{22} + m_{22} l_{11})} \qquad (8.90)$$

so that, substituting Eq. (8.90) into the real part of Eq. (8.89), we can write the quadratic equation in U^2

$$AU^4 + BU^2 + C = 0 \qquad (8.91)$$

where

$$\begin{aligned}
A = {}&(h_{12} l_{21} + h_{22} l_{11})\{(h_{12} l_{21} + h_{22} l_{11})(m_{11} m_{22} - m_{12}^2) \\
&- (l_{11} l_{22} - l_{12} l_{21} - h_{12} m_{12} + h_{22} m_{11})[m_{12}(l_{12} + l_{21}) \\
&- (m_{11} l_{22} + m_{22} l_{11})]\} \\[4pt]
B = {}&2(h_{12} l_{21} + h_{22} l_{11})(k_{22} l_{11} - k_{11} l_{22})(m_{11} m_{22} - m_{12}^2) \\
&- [(h_{12} l_{21} + h_{22} l_{11})(k_{22} m_{11} + k_{11} m_{22}) \\
&+ (k_{22} l_{11} - k_{11} l_{22})(l_{11} l_{22} - l_{12} l_{21} - h_{12} m_{12} \\
&+ h_{22} m_{11})][m_{12}(l_{12} + l_{21}) - (m_{11} l_{22} + m_{22} l_{11})] \\
&+ k_{11} h_{22}[m_{12}(l_{12} + l_{21}) - (m_{11} l_{22} + m_{22} l_{11})]^2
\end{aligned}$$

$$C = (k_{22}l_{11} - k_{11}l_{22})^2(m_{11}m_{22} - m_{12}^2) - (k_{22}l_{11} - k_{11}l_{22})(k_{22}m_{11}$$
$$+ k_{11}m_{22})[m_{12}(l_{12} + l_{21}) - (m_{11}l_{22} + m_{22}l_{11})]$$
$$+ k_{11}k_{22}[m_{12}(l_{12} + l_{21}) - (m_{11}l_{22} + m_{22}l_{11})]^2 \tag{8.92}$$

The solution of Eq. (8.91) is

$$U^2 = -\frac{B}{2A} \pm \frac{1}{2A}\sqrt{B^2 - 4AC} \tag{8.93}$$

The critical value U_{cr} is the smallest positive value of U that can be obtained from Eq. (8.93).

Additional physical insights into the problem of dynamic aeroelastic instability of a cantilever aircraft wing can be found in the text by Dowell (D2, pp. 76–81).

8.6   Integral formulation of the method of weighted residuals

In Sec. 7.10, we have shown that the eigenvalue problem for a self-adjoint system can be written in the integral form

$$w(x) = \lambda \int_0^L a(x, \xi)m(\xi)w(\xi)\,d\xi \tag{8.94}$$

where $a(x, \xi)$ is the symmetric Green's function or influence function. Under consideration is the case in which no closed-form solution of Eq. (8.94) exists and the interest lies in an approximate solution, particularly by the method of weighted residuals.

Let us consider an approximate solution Λ^n, w^n of the eigenvalue problem (8.94) and define the residual

$$R(w^n, x) = w^n(x) - \Lambda^n \int_0^L a(x, \xi)m(\xi)w^n(\xi)\,d\xi \tag{8.95}$$

We shall assume an approximate solution in the form of the series

$$w^n(x) = \sum_{j=1}^n a_j\phi_j(x) \tag{8.96}$$

where ϕ_j $(j = 1, 2, \ldots, n)$ are n members from a complete set of comparison functions. The functions ϕ_j form a basis for an n-dimensional subspace S^n of the space κ_B^{2p}. Introducing Eq. (8.96) into Eq. (8.95), we obtain

$$R(a_1, a_2, \ldots, a_n, x) = \sum_{j=1}^n a_j\phi_j - \Lambda^n \sum_{j=1}^n a_j \int_0^L a(x, \xi)m(\xi)\phi_j(\xi)\,d\xi \tag{8.97}$$

8 Discretization of continuous systems

Next, let us consider n test functions ψ_i $(i = 1, 2, \ldots, n)$ from another complete set and regard them as a basis for an n-dimensional space V^n. Then, the coefficients a_j $(j = 1, 2, \ldots, n)$ are determined by insisting that the integrals of the weighted residuals $\psi_i R$ over the interval $0 < x < L$ be zero, or

$$(\psi_i, R) = \int_0^L \psi_i \left[\sum_{j=1}^n a_j \phi_j - \Lambda^n \sum_{j=1}^n a_j \int_0^L a(x, \xi) m(\xi)\, \phi_j(\xi)\, d\xi \right] dx = 0,$$

$$i = 1, 2, \ldots, n \quad (8.98)$$

Introducing the notation

$$m_{ij} = \int_0^L \psi_i \phi_j\, dx, \qquad i, j = 1, 2, \ldots, n \qquad (8.99a)$$

$$h_{ij} = \int_0^L \psi_i \left[\int_0^L a(x, \xi) m(\xi) \phi_j(\xi)\, d\xi \right] dx, \qquad i, j = 1, 2, \ldots, n \qquad (8.99b)$$

Eqs. (8.98) take the form

$$\sum_{j=1}^n (m_{ij} - \Lambda^n h_{ij}) a_j = 0, \qquad i = 1, 2, \ldots, n \qquad (8.100)$$

which represent the algebraic eigenvalue problem for the system, replacing the integral eigenvalue problem (8.94). The coefficients m_{ij} and h_{ij} are generally not symmetric, in spite of the fact that the system is self-adjoint.

The coefficients m_{ij} and h_{ij} depend on the trial functions ϕ_j $(j = 1, 2, \ldots, n)$ and the test functions ψ_i $(i = 1, 2, \ldots, n)$. As pointed out in Sec. 8.4, the various methods of weighted residuals differ in their choice of the test function ψ_i. In particular, in Galerkin's method, the test functions coincide with the trial functions, so that in the differential formulation the problem can be symmetrized. In the integral formulation also, Galerkin's method yields nonsymmetric coefficients, but the problem can once again be symmetrized. Indeed, choosing the test functions

$$\psi_i(x) = m(x)\phi_i(x), \qquad i = 1, 2, \ldots, n \qquad (8.101)$$

the coefficients m_{ij} and h_{ij} become symmetric

$$m_{ij} = m_{ji} = \int_0^L m\phi_i\phi_j\, dx, \qquad i, j = 1, 2, \ldots, n \qquad (8.102a)$$

$$h_{ij} = h_{ji} = \int_0^L m\phi_i \left[\int_0^L a(x, \xi) m(\xi)\phi_j(\xi)\, d\xi \right] dx,$$

$$i, j = 1, 2, \ldots, n \quad (8.102b)$$

Introducing the n-dimensional vector $\mathbf{a} = [a_1, a_2 \cdots a_n]^T$ and the $n \times n$ matrices $M = [m_{ij}]$, $H = [h_{ij}]$, the eigenvalue problem (8.100) can be written in the matrix form

$$M\mathbf{a} = \Lambda^n H\mathbf{a} \qquad (8.103)$$

which can be solved by one of the methods described in Ch. 5. If the matrix coefficients are given by Eqs. (8.102), then the matrices M and H are symmetric and significant computational advantage accrues. Whereas M can be identified as the mass matrix, H is not a stiffness matrix. Because the evaluation of the coefficients h_{ij} involves double integration, the process can become tedious, so that the value of the method as a computational tool is limited.

8.7   Lumped-parameter method employing influence coefficients

Let us consider the system of Fig. 8.4 and divide the interval $0 < x < L$ into n segments of length Δx_j $(j = 1, 2, \ldots, n)$. The increments Δx_j need not have the same length. Then, letting $x = x_i$, $\xi = x_j$ and $d\xi = \Delta x_j$, and replacing integration by summation, Eq. (8.94) can be approximated by

$$w(x_i) \cong \lambda \sum_{j=1}^{n} a(x_i, x_j) m(x_j) w(x_j) \, \Delta x_j, \qquad i = 1, 2, \ldots, n \qquad (8.104)$$

Moreover, letting

$$w(x_i) = w_i, \qquad a(x_i, x_j) = a_{ij}, \qquad m(x_j) \, \Delta x_j = m_j \qquad (8.105)$$

where a_{ij} are recognized as flexibility influence coefficients, Eqs. (8.104) yield the algebraic eigenvalue problem

$$w_i = \lambda \sum_{j=1}^{n} a_{ij} m_j w_j, \qquad i = 1, 2, \ldots, n \qquad (8.106)$$

which represents a discretized version of Eq. (8.94).

The above process is known as *lumping* and is one of the oldest and simplest approaches to the derivation of the discrete eigenvalue problem.

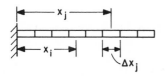

Figure 8.4

321

It is somewhat similar in concept to collocation. The matrix form of Eq. (8.106) is

$$w = \lambda A M w \qquad (8.107)$$

where w is the displacement vector, A is the symmetric flexibility matrix and M is the diagonal mass matrix. The problem can be symmetrized by introducing the vector

$$u = M^{1/2} w \qquad (8.108)$$

yielding the standard eigenvalue problem

$$u = \lambda A' u \qquad (8.109)$$

where

$$A' = M^{1/2} A M^{1/2} \qquad (8.110)$$

is a real symmetric matrix. The eigenvalue problem (8.109) can be solved by any of the algorithms discussed in Ch. 5.

8.8   System response by approximate methods

We shall be concerned with the response of distributed-parameter systems for which the eigenvalue problem does not lend itself to closed-form solution. Invariably, the approach to such problems consists of discretization in one form or another, so that ultimately partial differential equations are reduced to sets of ordinary differential equations. The nature of the ordinary differential equations is characterized by the coefficient matrices and depends on the type of distributed-parameter system. In this section, we shall discuss several types, paralleling the discussion of discrete systems of Ch. 6.

i. *Undamped nongyroscopic systems*

Let us consider a distributed-parameter system described by the differential equation

$$L w(x, t) + m(x) \frac{\partial^2 w(x, t)}{\partial t^2} = f(x, t), \qquad 0 < x < L \qquad (8.111)$$

where L is a self-adjoint operator of order $2p$. The displacement $w(x, t)$ is subject to the boundary conditions

$$B_i w = 0, \qquad i = 1, 2, \ldots, p, \qquad x = 0, L \qquad (8.112)$$

where B_i are differential operators of maximum order $2p - 1$.

8.8 System response by approximate methods

We shall seek an approximate solution of Eq. (8.111) by the method of weighted residuals. To this end, let

$$w^n = \sum_{j=1}^{n} \phi_j(x) q_j(t) \tag{8.113}$$

where $\phi_j(x)$ are n known comparison functions from a complete set and $q_j(t)$ are n generalized coordinates. Introducing Eq. (8.113) into Eq. (8.111), we can define the residual

$$R(x, t) = Lw^n(x, t) + m(x) \frac{\partial^2 w^n(x, t)}{\partial t^2} - f(x, t)$$

$$= \sum_{j=1}^{n} q_j(t) L\phi_j(x) + \sum_{j=1}^{n} \ddot{q}_j(t) m(x) \phi_j(x) - f(x, t) \tag{8.114}$$

Next, let us choose a set of n weighting functions $\psi_i(x)$ $(i = 1, 2, \dots, n)$ and insist that the residual R be orthogonal to all n functions ψ_i, so that

$$\int_0^L \psi_i R \, dx = \sum_{j=1}^{n} q_j \int_0^L \psi_i L\phi_j \, dx + \sum_{j=1}^{n} \ddot{q}_j \int_0^L m\psi_i\phi_j \, dx$$

$$- \int_0^L \psi_i f \, dx, \qquad i = 1, 2, \dots, n \tag{8.115}$$

Introducing the notation

$$k_{ij} = \int_0^L \psi_i L\phi_j \, dx, \qquad i, j = 1, 2, \dots, n \tag{8.116a}$$

$$m_{ij} = \int_0^L m\psi_i\phi_j \, dx, \qquad i, j = 1, 2, \dots, \tag{8.116b}$$

$$Q_i(t) = \int_0^L \psi_i f \, dx, \qquad i = 1, 2, \dots, n \tag{8.116c}$$

Eqs. (8.115) reduce to

$$\sum_{j=1}^{n} m_{ij} \ddot{q}_j(t) + \sum_{j=1}^{n} k_{ij} q_j(t) = Q_i(t), \qquad i = 1, 2, \dots, n \tag{8.117}$$

which represent n simultaneous ordinary differential equations with constant coefficients. They can be written in the matrix form

$$M\ddot{\mathbf{q}}(t) + K\mathbf{q}(t) = \mathbf{Q}(t) \tag{8.118}$$

where the notation is obvious.

The task of selecting the test functions ψ_i remains. Of course, the

323

choice differs for different methods of weighted residuals. Because the operator L is self-adjoint, it will prove advantageous to choose $\psi_i = \phi_i$ $(i = 1, 2, \ldots, n)$, which implies opting for the Galerkin method. As a result, both matrices M and K are symmetric. In fact, n integrations by parts permit us to write the coefficients k_{ij} in the form of the energy inner products

$$k_{ij} = [\phi_i, \phi_j], \qquad i, j = 1, 2, \ldots, n \tag{8.119}$$

so that the ϕ_i need be only admissible functions instead of comparison functions. The solution of Eq. (8.118) was discussed in Sec. 6.5.

ii. *Undamped gyroscopic systems*

Quite often, when the equilibrium of a system consists of uniform rotation, the perturbed motion exhibits so-called gyroscopic effects, characterized by conservative terms that are linear in the velocities. Gyroscopic effects are encountered in the spinning of a spacecraft, spinning of a helicopter rotor, whirling of a shaft, etc. They are also encountered in nonrotating systems, such as vibrating pipes containing flowing liquid. In the case of a spinning spacecraft, or a helicopter rotor, the motion involves rotational angular displacements in addition to elastic displacements. In the case of a whirling elastic shaft, the motion is described by two bending displacements about orthogonal axes.

Let us consider the case of a spinning spacecraft, in which the system rotates with uniform angular velocity when disturbed slightly. Denoting by $\theta_i(t)$ $(i = 1, 2, 3)$ the angular perturbations from the state of steady spin and by $w(x, t)$ the elastic displacement from a given rotating frame, we can write the kinetic energy in the general functional form

$$T = T(\theta_i, \dot{\theta}_i, w, \dot{w}) \tag{8.120}$$

Moreover, let us assume that the potential energy can be written in the form of the energy inner product

$$V = \tfrac{1}{2}[w, w] \tag{8.121}$$

and that the nonconservative virtual work has the form

$$\delta W_{nc} = \boldsymbol{\Theta}^T \, \delta\boldsymbol{\theta} + \int_0^L f \, \delta w \, dx \tag{8.122}$$

where $\boldsymbol{\Theta}$ is a torque vector and f a distributed force, and we note that $\boldsymbol{\Theta}$ includes contributions from f.

We shall find it convenient to discretize the system before proceeding with the derivation of the equations of motion. To this end, we propose to use the assumed-modes method and approximate the elastic displacement w by the series

$$w^n = \sum_{j=4}^{3+n} \phi_j(x)\zeta_j(t) \tag{8.123}$$

where ϕ_j are known admissible functions and ζ_j are generalized coordinates. Regarding the angular motions as small perturbations, we can introduce the $(3+n)$-dimensional configuration vector

$$\mathbf{q}(t) = [\boldsymbol{\Theta}^T(t) \mid \boldsymbol{\zeta}^T(t)]^T \tag{8.124}$$

which permits us to write the kinetic energy in the form

$$T = \tfrac{1}{2}\dot{\mathbf{q}}^T M \dot{\mathbf{q}} + \mathbf{q}^T F \dot{\mathbf{q}} + \tfrac{1}{2}\mathbf{q}^T K_T \mathbf{q} \tag{8.125}$$

in which the matrices, M, F and K_T have the elements

$$m_{jk} = \frac{\partial^2 T}{\partial \dot{q}_j\, \partial \dot{q}_k}, \qquad f_{jk} = \frac{\partial^2 T}{\partial q_j\, \partial \dot{q}_k}, \qquad k_{Tjk} = \frac{\partial^2 T}{\partial q_j\, \partial q_k},$$

$$j, k = 1, 2, \ldots, 3+n \quad (8.126)$$

where the various partial derivatives are evaluated at the equilibrium state. Note that the coefficients m_{jk} and k_{Tjk} are symmetric. Similarly, the potential energy becomes

$$V = \tfrac{1}{2}\mathbf{q}^T K_V \mathbf{q} \tag{8.127}$$

where K_V is a matrix with the elements in the first three rows and columns equal to zero and with the remaining elements given by

$$k_{Vjk} = [\phi_j, \phi_k], \qquad j, k = 4, 5, \ldots, 3+n \tag{8.128}$$

The coefficients k_{Vjk} are also symmetric. Moreover, introducing Eq. (8.123) into Eq. (8.122), we obtain the nonconservative virtual work

$$\delta W_{nc} = \mathbf{Q}^T \, \delta \mathbf{q} \tag{8.129}$$

where $\mathbf{Q}$ is the generalized force vector

$$\mathbf{Q} = [\boldsymbol{\Theta}^T \mid \mathbf{Z}^T]^T \tag{8.130}$$

in which $\mathbf{Z}$ is an n-vector with the components

$$Z_j = \int_0^L f\phi_j \, \mathrm{d}x, \qquad j = 4, 5, \ldots, 3+n \tag{8.131}$$

8 Discretization of continuous systems

Lagrange's equations can be written in the symbolic vector form

$$\frac{d}{dt}\left(\frac{\partial L}{\partial \dot{\mathbf{q}}}\right) - \frac{\partial L}{\partial \mathbf{q}} = \mathbf{Q} \tag{8.132}$$

where the derivative of L with respect to a vector represents a vector with its components equal to the derivatives of L with respect to the vector components, in which $L = T - V$ is the Lagrangian. Introducing Eqs. (8.125) and (8.127) into Eq. (8.132), we obtain a set of simultaneous ordinary differential equations of motion having the vector form

$$M\ddot{\mathbf{q}}(t) + G\dot{\mathbf{q}}(t) + K\mathbf{q}(t) = \mathbf{Q}(t) \tag{8.133}$$

where

$$G = F^T - F = -G^T \tag{8.134}$$

is a skew symmetric gyroscopic matrix and

$$K = K_V - K_T \tag{8.135}$$

is a symmetric stiffness matrix, including elastic and centrifugal effects. The solution of Eq. (8.133) was discussed in Sec. 6.6.

iii. Damped nongyroscopic systems

The vibration of a viscously damped distributed system can be written in the general form

$$Lw(x, t) + \frac{\partial}{\partial t}[Cw(x, t)] + m(x)\frac{\partial^2 w(x, t)}{\partial t^2} = f(x, t), \qquad 0 < x < L \tag{8.136}$$

where L is a self-adjoint stiffness operator, C is a self-adjoint damping operator and $f(x, t)$ is a nonconservative distributed force excluding the damping force. Once again, the displacement w is subject to boundary conditions of the type (8.112).

Equation (8.136) resembles Eq. (8.111), with the exception that Eq. (8.136) contains the additional damping term $\partial[Cw(x, t)]/\partial t$. Letting the solution of Eq. (8.136) have the form (8.113), in which ϕ_i are admissible functions and using Galerkin's method, we obtain the set of ordinary differential equations

$$M\ddot{\mathbf{q}}(t) + C\dot{\mathbf{q}}(t) + K\mathbf{q}(t) = \mathbf{Q}(t) \tag{8.137}$$

The matrices M and K are given by Eqs. (8.116b) (with ψ_i replaced by ϕ_i) and Eqs. (8.119), respectively. On the other hand, C is a damping matrix with the elements

$$c_{ij} = \int_0^L \phi_i C \phi_j \, dx, \qquad i, j = 1, 2, \ldots, n \tag{8.138}$$

No confusion should arise from the fact that we used the same notation for the damping operator C and the damping matrix C. Note that, in calculating the matrices K and C, appropriate integrations by parts are implied so as to permit the use of admissible functions instead of comparison functions. Of course, all three matrices M, C and K are symmetric.

The solution of Eq. (8.137) was discussed in Sec. 6.7. The assumption of proportional damping of Sec. 6.7 implies that the damping operator is given by the linear combination

$$C = \alpha m + \beta L \tag{8.139}$$

where α and β are constant scalars.

If damping is of structural type and if $f(x, t)$ is harmonic, then the differential equation of motion has the form

$$(1 + i\gamma)Lw(x, t) + m(x)\frac{\partial^2 w(x, t)}{\partial t^2} = f_0(x)e^{i\Omega t} \tag{8.140}$$

where γ is a structural damping factor. This is a case in which the analogy with the viscous damping is valid, so that Eq. (8.137) can be used, provided the damping matrix is taken as

$$C = \frac{\lambda}{\Omega} K \tag{8.141}$$

This is simply a special case of proportional damping in which $\alpha = 0$ and $\beta = \gamma/\Omega$.

If damping is not of the proportional type, the matrix C will bear no resemblance to either the matrix M or the matrix K, so that the general approach of Sec. 6.7 must be used.

9

The finite element method

9.1 Introduction

The finite element method was first suggested by Courant in what has come to be regarded as a classic paper (C3). However, the development of the method as practiced today can be traced to an independent effort by Turner et al (T1) and Melosh (M12), who were painfully aware of the inadequacy of the classical methods in analyzing complex structures. Although the method was originally developed for the analysis of structures, its applicability to other fields of engineering has been widely recognized.

There is a certain degree of irony in the fact that the finite element method, designed to work where such classical methods as the Rayleigh–Ritz method have proved wanting, was soon identified as a Rayleigh–Ritz method itself. This must be considered as a happy turn of events, however, as the entire methematical theory developed for the Rayleigh–Ritz method could be called upon to provide a solid mathematical foundation to the finite element method.

Because the finite element method can be regarded as a Rayleigh–Ritz method, the question arises as to what sets it apart from the classical Rayleigh–Ritz method and makes it such a success. The answer can perhaps be summarized in one word, namely, versatility. In the classical Rayleigh–Ritz method, the solution is approximated by a finite series of admissible functions that are defined over the entire domain. For systems with complex geometry, the task of producing admissible functions is often hopeless. In other cases, even when admissible functions can be generated, these functions can be so complicated and difficult to work with as to inhibit many ambitious analysts. It is the beauty of the finite element method that it can accommodate complicated geometries and make the task easier for the analyst at the same time.

The idea behind the finite element method is disarmingly simple. Instead of defining the admissible functions over the entire domain, one needs to define them only over relatively small subdomains, called finite elements. This not only permits the application of the method to systems with complicated geometries, but also permits the use of very simple admissible functions. Indeed, for the most part, the admissible functions are low-degree polynomials, known as interpolation functions. Low-degree polynomials are computationally attractive, as integrals involving such polynomials can be evaluated in closed form, thus eliminating errors that may result from numerical integration.

Unfortunately, the phenomenal success of the finite element method has had some undesirable side effects also. The method has come to be regarded as the almost standard tool of structural analysis, pushing aside many other methods of analysis, including the classical Rayleigh–Ritz method. This apparent paradox can be explained by the fact that one can learn to apply the finite element method without adopting the Rayleigh–Ritz point of view, or even without being aware that it is a Rayleigh–Ritz method. Yet, it is important to recognize that the finite element method is not always the most indicated one to use. In fact, when the geometry is relatively simple and there are no sudden changes in the system properties, such as the stiffness and mass distributions, the finite element method is plainly inferior to the classical Rayleigh–Ritz method. This is so because the finite element method tends to require a large number of degrees of freedom for a certain desired accuracy, when compared to the classical Rayleigh–Ritz method. Whereas a large number of degrees of freedom may not necessarily imply more work for the analyst, it does require large computer capacity and it can lead to numerical difficulties. As a matter of fact, the development of a number of computational algorithms for large-order matrices (see Ch. 10) can be attributed to needs arising from the use of the finite element method. Hence, whereas the finite element methods is considerably more versatile than the classical Rayleigh–Ritz method, and it can yield results where the latter fails, one is advised to keep an open mind as to the choice of the method of analysis.

This chapter concentrates on the theoretical and computational aspects of the finite element method, particularly as pertaining to the eigenvalue problem. Its purpose is to provide the reader with a fundamental understanding as well as an elementary working knowledge of the method.

9.2 Second-order problems. Linear elements

In Sec. 7.6, we have shown that the vibration problems of strings, rods and shafts all lead to eigenvalue problems similar in structure, so that they can all be covered by a unified formulation. In particular, the eigenvalue problem was defined by the second-order differential equation

$$-(sw')' + kw = \omega^2 mw, \qquad 0 < x < L \tag{9.1}$$

and some typical boundary conditions

$$w(0) = 0 \tag{9.2a}$$

$$sw' + Kw = 0, \qquad x = L \tag{9.2b}$$

where boundary condition (9.2a) is geometric and it states that the end $x = 0$ is fixed and boundary condition (9.2b) is natural and it implies that the end $x = L$ is restrained by means of a spring of stiffness K. The eigenvalue problem (9.1–2) can be cast in the general form

$$Lw = \lambda mw, \qquad 0 < x < L \tag{9.3a}$$

$$B_1 w = 0, \qquad x = 0, L \tag{9.3b}$$

where

$$L = -\frac{d}{dx}\left(s\frac{d}{dx}\right) + k, \qquad \lambda = \omega^2$$

$$B_1 = 1 \quad \text{at} \quad x = 0 \tag{9.4}$$

$$B_1 = s\frac{d}{dx} + K \quad \text{at} \quad x = L$$

Clearly, L is of order two. It can be shown that L is self-adjoint and positive definite.

If we were to seek an approximate solution of the eigenvalue problem (9.1–2), we would look for a solution of the form

$$w^n = \sum_{j=1}^{n} a_j \phi_j \tag{9.5}$$

where ϕ_j are n trial functions from a complete set forming a basis for a subspace S^n of the space κ_B^2. In short, ϕ_j are n independent comparison functions, i.e., they are twice differentiable and satisfy all the boundary conditions. We have shown in Sec. 7.9, however, that the differential problem, Eqs. (9.3), can be replaced by the variational problem consisting of rendering stationary the Rayleigh quotient

$$R = \frac{[w, w]}{(\sqrt{m}w, \sqrt{m}w)} \tag{9.6}$$

where $[w, w]$ is the energy inner product and $(\sqrt{m}w, \sqrt{m}w)$ is a weighted inner product. In the problem at hand, the energy inner product is

$$[w, w] = \int_0^L [s(w')^2 + kw^2] \, dx + Kw^2(L) \qquad (9.7)$$

and the weighted inner product is

$$(\sqrt{m}w, \sqrt{m}w) = \int_0^L mw^2 \, dx \qquad (9.8)$$

If now we consider a solution of the variational problem in the form (9.5), we conclude that the trial functions ϕ_i need be only from the energy space κ_G^1, i.e., *they need be only admissible functions*, and hence be differentiable only once and satisfy the geometric boundary conditions alone.

In the Rayleigh–Ritz method, we would choose a set of admissible functions ϕ_i from a known set of complete functions, such as trigonometric functions, Bessel functions, Legendre polynomials, etc., or we would generate a set by solving a simpler eigenvalue problem related to the eigenvalue problem (9.1–2). The latter may be obtained by assuming that s, k and m are constant and that K is zero. Considering the admissible functions as known, introducing Eq. (9.5) into Eq. (9.6) and invoking the conditions for the stationarity of Rayleigh's quotient, we would obtain the algebraic eigenvalue problem

$$\sum_{j=1}^n k_{ij}a_j = \Lambda^n \sum_{j=1}^n m_{ij}a_j, \qquad i = 1, 2, \ldots, n \qquad (9.9)$$

where

$$k_{ij} = [\phi_i, \phi_j] = \int_0^L (s\phi_i'\phi_j' + k\phi_i\phi_j) \, dx + K\phi_i(L)\phi_j(L),$$

$$i, j = 1, 2, \ldots, n \qquad (9.10a)$$

$$m_{ij} = (\sqrt{m}\phi_i, \sqrt{m}\phi_j) = \int_0^L m\phi_i\phi_j \, dx, \qquad i, j = 1, 2, \ldots, n \qquad (9.10b)$$

Examining Eqs. (9.10), we observe that if s, k and m are relatively complicated functions of x, the generation of the stiffness coefficients k_{ij} and mass coefficients m_{ij} may be unduly tedious and time consuming, even when the admissible functions are simple trigonometric functions. Hence, the question arises whether it is not possible to construct admissible functions capable of reducing the work involved in the evaluation of k_{ij} and m_{ij} in a substantial way.

For the problem at hand, the admissible functions ϕ_i must be from the energy space κ_G^1, which implies that the functions ϕ_i must be merely continuous. This rules out piecewise constant functions, but piecewise linear functions are admissible, provided they contain no discontinuities. Moreover, these piecewise linear functions need not be defined over the entire domain $D : 0 \leq x \leq L$ but only over certain subdomains of D and can be zero identically everywhere else. Such functions are referred to as a *local basis*. *The basic idea of the finite element method is to work with admissible functions ϕ_i that constitute a local basis and have as simple a form as possible*, consistent with the requirement that they belong to the energy space. Hence, let us divide the domain D into n subdomains, referred to as 'finite elements', and consider the piecewise linear functions shown in Fig. 9.1. For simplicity, we have chosen finite elements of equal length, $h = L/n$, but this is not necessary, and in some cases involving rapid variation in the system parameters it may not even be desirable. The boundaries between elements, $x_j = jh$, are called *nodes*, which is unfortunate terminology in view of the fact that in vibrations nodes refer to points of zero displacement. Nevertheless, this term is so entrenched in the finite element terminology that we shall adopt it with the proper caveat. Although the finite element method can be regarded as a Rayleigh–Ritz method, with the difference being more in form than in substance, when comparing the two methods we shall refer to the traditional approach discussed in Ch. 8 as the *classical Rayleigh–Ritz method* to distinguish it from the finite element approach.

The functions $\phi_j(x)$ depicted in Fig. 9.1 are referred to as *linear elements*, or more pictorially as *roof functions*. The function $\phi_j(x)$ extends over two elements, $(j-1)h \leq x \leq (j+1)h$, except for the function $\phi_n(x)$, which extends over the element $(n-1)h \leq x \leq nh = L$. The latter function permits us to take into account the fact that the end $x = L$ is not fixed but elastically restrained by the spring K. A significant feature of functions ϕ_j extending over two elements, such as the roof functions just described, is that they are *nearly orthogonal*. Indeed, ϕ_j is orthogonal to every other function ϕ_i except for $i = j \pm 1$. This near orthogonality holds regardless

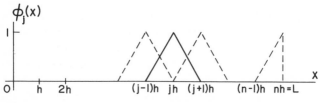

Figure 9.1

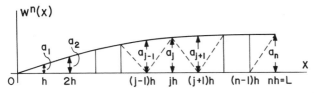

Figure 9.2

of any weighting functions. All the functions ϕ_i can be regarded as being normalized in the sense that they have unit amplitudes, which has significant physical implications. Indeed, introducing the normalized roof functions into Eq. (9.5), we can plot the function w^n as shown in Fig. 9.2. The implication is clear: *the coefficients a_i are simply the displacements of the nodes $x_i = jh$*. This is in direct contrast with the classical Rayleigh–Ritz method, in which the coefficients a_j have more mathematical than physical meaning. Perhaps a more significant difference between the classical Rayleigh–Ritz method and the finite element method is that in the classical Rayleigh–Ritz method the space S^{n+1} is obtained by simply adding the function ϕ_{n+1} to the basis of S^n, without affecting this basis. The implication is that the space S^n is a subspace of the space S^{n+1}. As a result, the coefficients k_{ij} and m_{ij} ($i, j = 1, 2, \ldots, n$) are not affected by the addition of the function ϕ_{n+1} to the space. By contrast, in the finite element method the space S^{n+1} is obtained in general by redividing the domain D into $n+1$ elements instead of n, so that *the entire set of admissible functions changes*, in the sense that the functions ϕ_i are defined over smaller subdomains. Hence, even though the shape of the functions remains the same, *the addition of one dimension to the space S^n causes all the coefficients k_{ij} and m_{ij} to change*. On the other hand, the finite element method has the advantages that the admissible functions are nonzero only over small subdomains of D and that the functions themselves represent low-degree polynomials. As a result, the calculation of the coefficients k_{ij} and m_{ij} lends itself to systematization, so that an increase in the number of elements can be accommodated without difficulties. In fact, the success of the finite element method can be attributed to the parallel development of the digital computer, as the two together have vastly increased the problem solving capability of the structural analyst.

The calculation of the coefficients k_{ij} and m_{ij} can perhaps be simplified by considering the finite elements separately. Figure 9.3 shows the jth element and the associated nodal displacements. Our object is to evaluate the Rayleigh quotient by calculating the contribution of every finite element separately and combining the results, where the latter

333

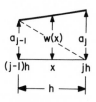

Figure 9.3

process is known in the finite element terminology as 'assembling' the finite elements. Hence, let us write Rayleigh's quotient in the form

$$R = \sum_{j=1}^{n} N_j \Big/ \sum_{j=1}^{n} D_j \tag{9.11}$$

where

$$N_j = \int_{(j-1)h}^{jh} [s(w')^2 + kw^2]\,dx + \delta_{jn} K w^2(L), \qquad j = 1, 2, \ldots, n \tag{9.12a}$$

$$D_j = \int_{(j-1)h}^{jh} m w^2\,dx, \qquad j = 1, 2, \ldots, n \tag{9.12b}$$

Considering Fig. 9.3 and using the similarity of triangles, the displacement $w(x)$ in the interval $(j-1)h \le x \le jh$ can be expressed in terms of the nodal coordinates a_{j-1} and a_j as follows:

$$w(x) = \frac{jh - x}{h}\,a_{j-1} + \frac{x - (j-1)h}{h}\,a_j, \qquad j = 1, 2, \ldots, n \tag{9.13}$$

where $a_0 = 0$.

Because of the localized nature of the calculations associated with the element j, it will prove convenient to work with *local coordinates*. To this end, let us introduce the notation

$$\frac{jh - x}{h} = j - \frac{x}{h} = \xi \tag{9.14a}$$

$$\frac{x - (j-1)h}{h} = \frac{x}{h} - (j-1) = 1 - \xi \tag{9.14b}$$

which can be regarded as nondimensional coordinates known in the finite element terminology as *natural coordinates,* another unfortunate term in view of the different connotation the term natural coordinates has in vibrations. We note that ξ represents the right side of ϕ_{j-1} and $1 - \xi$ represents the left side of ϕ_j. Introducing the notation

$$L_1 = \xi, \qquad L_2 = 1 - \xi \tag{9.15}$$

and considering Eqs. (9.14), we can write Eq. (9.13) in the form

$$w(x) = L_1 a_{j-1} + L_2 a_j \tag{9.16}$$

so that L_1 and L_2 play the role of admissible functions for the element j. Because a node is shared by two elements, there is really only one admissible function per element. The functions L_1 and L_2 can be identified as *linear interpolation functions*. They are also known as *shape functions*. For linear elements they coincide with the natural coordinates. Introducing Eqs. (9.15–16) into Eq. (9.12), we can write

$$N_j = \begin{bmatrix} a_{j-1} \\ a_j \end{bmatrix}^T K_j \begin{bmatrix} a_{j-1} \\ a_j \end{bmatrix}, \qquad j = 1, 2, \ldots, n \tag{9.17a}$$

$$D_j = \begin{bmatrix} a_{j-1} \\ a_j \end{bmatrix}^T M_j \begin{bmatrix} a_{j-1} \\ a_j \end{bmatrix}, \qquad j = 1, 2, \ldots, n \tag{9.17b}$$

where K_j and M_j $(j = 1, 2, \ldots, n)$ are 2×2 matrices with the entries

$$k_{j11} = \frac{1}{h} \int_0^1 [s(\xi) + h^2 \xi^2 k(\xi)] \, d\xi$$

$$k_{j12} = k_{j21} = \frac{1}{h} \int_0^1 [-s(\xi) + h^2 \xi(1-\xi) k(\xi)] \, d\xi, \qquad j = 1, 2, \ldots, n$$

$$\tag{9.18a}$$

$$k_{j22} = \frac{1}{h} \int_0^1 [s(\xi) + h^2 (1-\xi)^2 k(\xi)] \, d\xi + \delta_{jn} K$$

$$m_{j11} = h \int_0^1 m(\xi) \xi^2 \, d\xi$$

$$m_{j12} = m_{j21} = h \int_0^1 m(\xi) \xi(1-\xi) \, d\xi, \qquad j = 1, 2, \ldots, n \tag{9.18b}$$

$$m_{j22} = h \int_0^1 m(\xi)(1-\xi)^2 \, d\xi$$

Actually, because $a_0 = 0$, we must strike out the first row and column from the matrices K_1 and M_1, so that we are really left with scalars. We have treated K_1 and M_1 as matrices for the purpose of standardization, with a view to the task of computer programming. Note that, for sufficiently small h, the parameters $s(\xi)$, $k(\xi)$ and $m(\xi)$ can be regarded as constant over the element.

The matrices K_j and M_j $(j = 1, 2, \ldots, n)$ are known as *element stiffness* and *mass matrices*, respectively. Introducing Eqs. (9.17) into Eq.

9 The finite element method

(9.11), the assembly process results in the Rayleigh quotient

$$R = \frac{\mathbf{a}^T K \mathbf{a}}{\mathbf{a}^T M \mathbf{a}} \tag{9.19}$$

where $\mathbf{a} = [a_1\, a_2 \cdots a_n]^T$ is the *nodal vector* and K and M are *global stiffness* and *mass matrices*, respectively, and are obtained from the element matrices K_j and M_j $(j = 1, 2, \ldots, n)$ in the schematic form

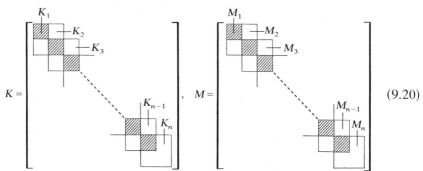

$$\tag{9.20}$$

The shaded areas denote entries that are the sum of the overlapping entries. Note that the use of a local basis, with the admissible functions extending over two adjacent finite elements, results in banded matrices K and M, where the bandwidth is equal to three.

The requirement that Rayleigh's quotient, Eq. (9.19), be stationary yields the real symmetric eigenvalue problem

$$K\mathbf{a} = \Lambda^n M \mathbf{a} \tag{9.21}$$

which was discussed in ample detail in Chs. 3–5.

The finite element solution w^n in κ_G^1 converges in the energy norm as $n \to \infty$, although there is some question as to the rate of convergence. A proof of convergence of such a solution is provided in the text by Strang and Fix (S2, p. 47). Error bounds for the solution will be discussed later in this chapter.

As an illustration, let us assume that the system (9.1) has constant parameters, $s = \text{const}$, $k = \text{const}$, $m = \text{const}$, and derive the stiffness and mass matrices. Under these circumstances, Eqs. (9.18) yield the local stiffness and mass matrices

$$K_j = \frac{s}{h} \begin{bmatrix} 1 & -1 \\ -1 & 1 \end{bmatrix} + \frac{1}{6} hk \begin{bmatrix} 2 & 1 \\ 1 & 2 \end{bmatrix} + \delta_{jn} K \begin{bmatrix} 0 & 0 \\ 0 & 1 \end{bmatrix}, \qquad j = 1, 2, \ldots, n \tag{9.22a}$$

$$M_j = \frac{1}{6} hm \begin{bmatrix} 2 & 1 \\ 1 & 2 \end{bmatrix}, \qquad j = 1, 2, \ldots, n \tag{9.22b}$$

Performing the assembly indicated by (9.20), we obtain

$$
K = \frac{s}{h}
\begin{bmatrix}
2 & -1 & 0 & \cdots & 0 & 0 \\
 & 2 & -1 & \cdots & 0 & 0 \\
 & & 2 & \cdots & 0 & 0 \\
 & \text{Symm} & & \cdots\cdots & & \\
 & & & \cdots\cdots & & \\
 & & & & 2 & -1 \\
 & & & & & 1
\end{bmatrix}
+ \frac{hk}{6}
\begin{bmatrix}
4 & 1 & 0 & \cdots & 0 & 0 \\
 & 4 & 1 & \cdots & 0 & 0 \\
 & & 4 & & 0 & 0 \\
 & \text{Symm} & & \cdots\cdots & & \\
 & & & \cdots\cdots & & \\
 & & & & 4 & 1 \\
 & & & & & 2
\end{bmatrix}
$$

$$
+ K
\begin{bmatrix}
0 & 0 & 0 & \cdots & 0 & 0 \\
 & 0 & 0 & \cdots & 0 & 0 \\
 & & 0 & \cdots & 0 & 0 \\
 & \text{Symm} & & \cdots\cdots & & \\
 & & & \cdots\cdots & & \\
 & & & & 0 & 0 \\
 & & & & & 1
\end{bmatrix}
\tag{9.23a}
$$

and

$$
M = \frac{hm}{6}
\begin{bmatrix}
4 & 1 & 0 & \cdots & 0 & 0 \\
 & 4 & 1 & \cdots & 0 & 0 \\
 & & 4 & \cdots & 0 & 0 \\
 & \text{Symm} & & \cdots\cdots & & \\
 & & & \cdots\cdots & & \\
 & & & & 4 & 1 \\
 & & & & & 2
\end{bmatrix}
\tag{9.23b}
$$

and we observe that the matrices K and M are banded with bandwidth equal to three, as pointed out earlier. Note that the symbol K on the right side of Eq. (9.23a) is merely a scalar.

Example 9.1

Solve the eigenvalue problem of Example 8.1 by the finite element method using linear elements. Determine the approximate number of elements to be taken so as to obtain first two estimated natural frequencies having the same accuracy as those obtained in Example 8.1 by the classical Rayleigh–Ritz method.

The element stiffness and mass matrices are derived by using Eqs. (9.18). Introducing

$$
s(\xi) = \frac{6EA}{5}\left[1 - \frac{1}{2}\left(\frac{h}{L}\right)^2(j - \xi)^2\right], \qquad K = \frac{EA}{5L}
$$

$$
m(\xi) = \frac{6m}{5}\left[1 - \frac{1}{2}\left(\frac{h}{L}\right)^2(j - \xi)^2\right]
\tag{a}
$$

337

into Eqs. (9.18) and integrating, we obtain the element matrices

$$K_j = \frac{6EA}{5L}\left[n - \frac{1}{6n}(1-3j+3j^2)\right]\begin{bmatrix} 1 & -1 \\ -1 & 1 \end{bmatrix} + \delta_{jn}\frac{EA}{5L}\begin{bmatrix} 0 & 0 \\ 0 & 1 \end{bmatrix},$$

$$j = 1, 2, \ldots, n \qquad (b)$$

$$M_j = \frac{mL}{5n}\begin{bmatrix} 2 & 1 \\ 1 & 2 \end{bmatrix} - \frac{mL}{100n}\begin{bmatrix} 2(6-15j+10j^2) & 3-10j+10j^2 \\ 3-10j+10j^2 & 2(1-5j+10j^2) \end{bmatrix},$$

$$j = 1, 2, \ldots, n \qquad (c)$$

in which we set $L/h = n$. Using the scheme (9.20), where we recall that the first row and column must be removed from K_1 and M_1, the global stiffness and mass matrices can be shown to be

$$K = \frac{6EAn}{5L}\begin{bmatrix} 2 & -1 & 0 \cdots 0 & 0 \\ & 2 & -1 \cdots 0 & 0 \\ & & 2 \cdots 0 & 0 \\ & \text{Symm} & \cdots\cdots\cdots\cdots \\ & & & 2 & 1 \\ & & & & 1+1/6n \end{bmatrix}$$

$$-\frac{EA}{5Ln}\begin{bmatrix} 8 & -7 & 0 & \cdots & 0 & 0 \\ & 26 & -19 & \cdots & 0 & 0 \\ & & 56 & \cdots & 0 & 0 \\ & \text{Symm} & & \cdots\cdots\cdots\cdots\cdots\cdots\cdots \\ & & & & 2(4-6n+3n^2) & -(1-3n+3n^2) \\ & & & & & 1-3n+3n^2 \end{bmatrix} \qquad (d)$$

and

$$M = \frac{mL}{5n}\begin{bmatrix} 4 & 1 & 0\cdots 0 & 0 \\ & 4 & 1\cdots 0 & 0 \\ & & 4\cdots 0 & 0 \\ & \text{Symm} & \cdots\cdots\cdots \\ & & & 4 & 1 \\ & & & & 2 \end{bmatrix}$$

$$-\frac{mL}{100n^3}\begin{bmatrix} 44 & 23 & 0 & \cdots & 0 & 0 \\ & 164 & 63 & \cdots & 0 & 0 \\ & & 364 & \cdots & 0 & 0 \\ & \text{Symm} & & \cdots\cdots\cdots\cdots\cdots\cdots\cdots \\ & & & & 2(37-50n+20n^2) & 3-10n+10n^2 \\ & & & & & 2(1-5n+10n^2) \end{bmatrix}$$

$$(e)$$

The eigenvalue problem associated with the matrices K and M, Eqs. (d) and (e), has been solved for various values of n. First two estimated natural frequencies having the same accuracy as those obtained by the Rayleigh–Ritz method with $n = 4$ are obtained in the present case for $n = 30$. These value are

$$\omega_1^{30} = 1.894771 \sqrt{\frac{EA}{mL^2}}, \qquad \omega_2^{30} = 4.888352 \sqrt{\frac{EA}{mL^2}} \tag{f}$$

and we observe that ω_1^{30} is slightly lower than ω_1^4 obtained in Example 8.1, whereas ω_2^{30} is slightly higher than ω_2^4.

9.3 Higher-degree elements. Interpolation functions

The piecewise linear functions used in Sec. 9.2 are the simplest possible polynomials, so that the question arises whether one cannot obtain better approximations by using polynomials of higher degree as trial functions. Higher-degree polynomials can be generated by the Lagrangian interpolation formula (H4, p. 150). Because our interest will be confined to quadratic and cubic elements alone, we shall generate the elements directly, without the use of the formula.

To introduce the ideas, let us return to the *linear elements* of Sec. 9.2 and write

$$L_i = c_{i1} + c_{i2}\xi, \qquad i = 1, 2 \tag{9.24}$$

where L_1 represent the right side of ϕ_{j-1} and L_2 the left side of ϕ_j. Hence, L_1 must be equal to one at the node $\xi = 1$ and to zero at the node $\xi = 0$, where we recall from Eq. (9.14a) that

$$\xi = j - x/h \tag{9.25}$$

is a nondimensional natural coordinate. Imposing the conditions stated above, we obtain

$$L_1(1) = c_{11} + c_{12} = 1, \qquad L_1(0) = c_{11} = 0$$
$$L_2(0) = c_{21} = 1, \qquad L_2(1) = c_{21} + c_{22} = 0 \tag{9.26}$$

which yield

$$L_1 = \xi, \qquad L_2 = 1 - \xi \tag{9.27}$$

and we observe that the functions L_1 and L_2 are identical to those obtained in Sec. 9.2.

Quadratic elements can be generated from the expression

$$L_i = c_{i1} + c_{i2}\xi + c_{i3}\xi^2, \qquad i = 1, 2, 3 \tag{9.28}$$

339

9 The finite element method

Because the element had only two nodes and there are three coefficients, we must introduce an *internal node*. In view of this, we shall refer to the nodes at the element nodes as *external nodes*. For simplicity, we shall take the internal node at $\xi = 1/2$. Every quadratic element will be set equal to one at one node and equal to zero at the other two nodes, so that

$$L_1(1) = c_{11} + c_{12} + c_{13} = 1, \qquad L_1(1/2) = c_{11} + \tfrac{1}{2}c_{12} + \tfrac{1}{4}c_{13} = 0,$$
$$L_1(0) = c_{11} = 0$$
$$L_2(1) = c_{21} + c_{22} + c_{23} = 0, \qquad L_2(1/2) = c_{21} + \tfrac{1}{2}c_{22} + \tfrac{1}{4}c_{23} = 1,$$
$$L_2(0) = c_{21} = 0 \tag{9.29}$$
$$L_3(1) = c_{31} + c_{32} + c_{33} = 0, \qquad L_3(1/2) = c_{31} + \tfrac{1}{2}c_{32} + \tfrac{1}{4}c_{33} = 0,$$
$$L_3(0) = c_{31} = 1$$

Solving Eqs. (9.29) for the coefficients c_{ik} $(i, k = 1, 2, 3)$ and introducing the coefficients into Eqs. (9.28), we obtain

$$L_1 = \xi(2\xi - 1), \qquad L_2 = 4\xi(1 - \xi), \qquad L_3 = 1 - 3\xi + 2\xi^2 \tag{9.30}$$

The functions L_i $(i = 1, 2, 3)$ are plotted in Fig. 9.4.

Next, we wish to calculate the stiffness and mass matrices for the quadratic elements. To this end, we express the displacement w in terms of the quadratic elements in the general form

$$w = L_1 a_{j-1} + L_2 a_{j-1/2} + L_3 a_j = \mathbf{L}^T \mathbf{a}_j \tag{9.31}$$

where $\mathbf{L}$ and $\mathbf{a}_j$ are three-dimensional vectors given by

$$\mathbf{L} = [L_1 \quad L_2 \quad L_3]^T \tag{9.32a}$$
$$\mathbf{a}_j = [a_{j-1} \quad a_{j-1/2} \quad a_j]^T \tag{9.32b}$$

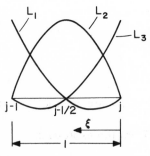

Figure 9.4

340

Then, recognizing that L_i ($i = 1, 2, 3$) are functions of ξ, we can write

$$\frac{\mathrm{d}w}{\mathrm{d}x} = \frac{\mathrm{d}\mathbf{L}^T}{\mathrm{d}x}\,\mathbf{a}_j = -\frac{1}{h}\frac{\mathrm{d}\mathbf{L}^T}{\mathrm{d}\xi}\,\mathbf{a}_j = -\frac{1}{h}\mathbf{L}'^T\mathbf{a}_j \tag{9.33}$$

where $\mathbf{L}' = \mathrm{d}\mathbf{L}/\mathrm{d}\xi$. Introducing Eqs. (9.25), (9.31) and (9.33) into Eqs. (9.12), we can write N_j and D_j in the form of the matrix products

$$N_j = \mathbf{a}_j^T K_j \mathbf{a}_j, \qquad j = 1, 2, \ldots, n \tag{9.34a}$$

$$D_j = \mathbf{a}_j^T M_j \mathbf{a}_j, \qquad j = 1, 2, \ldots, n \tag{9.34b}$$

where K_j and M_j are the element stiffness and mass matrices, respectively, namely, 3×3 matrices of the form

$$K_j = h \int_0^1 \left(\frac{s}{h^2}\mathbf{L}'\mathbf{L}'^T + k\mathbf{L}\mathbf{L}^T\right)\mathrm{d}\xi + \delta_{jn}\mathbf{K}\mathbf{L}(0)\mathbf{L}^T(0),$$

$$j = 1, 2, \ldots, n \tag{9.35a}$$

$$M_j = h \int_0^1 m\mathbf{L}\mathbf{L}^T\,\mathrm{d}\xi, \qquad j = 1, 2, \ldots, n \tag{9.35b}$$

The global stiffness and mass matrices can be assembled in the same way as in Sec. 9.2. The matrices have the schematic form

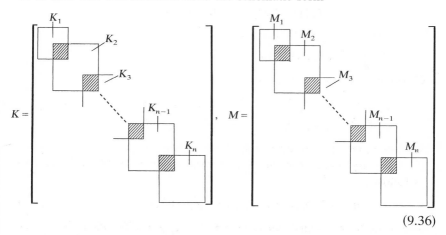

$$\tag{9.36}$$

Note that nonoverlapping elements on the main diagonal correspond to the interior nodes, as interior nodes are not shared by other elements. The bandwidth of K and M is four.

Considering once again the case in which the parameters s, k and m are constant and introducing Eqs. (9.30), in conjunction with Eq. (9.32a), into Eqs. (9.35), we obtain the element matrices

$$K_j = \frac{s}{3h} \begin{bmatrix} 7 & -8 & 1 \\ -8 & 16 & -8 \\ 1 & -8 & 7 \end{bmatrix} + \frac{hk}{30} \begin{bmatrix} 4 & 2 & -1 \\ 2 & 16 & 2 \\ -1 & 2 & 4 \end{bmatrix} + \delta_{jn} K \begin{bmatrix} 0 & 0 & 0 \\ 0 & 0 & 0 \\ 0 & 0 & 1 \end{bmatrix},$$

$$j = 1, 2, \ldots, n \quad (9.37a)$$

$$M_j = \frac{hm}{30} \begin{bmatrix} 4 & 2 & -1 \\ 2 & 16 & 2 \\ -1 & 2 & 4 \end{bmatrix}, \quad j = 1, 2, \ldots, n \quad (9.37b)$$

The assembling of the matrices K and M follows the pattern given in (9.36). Note that once again, because $a_0 = 0$, the first row and column must be removed from K_1 and M_1.

The same procedure can be used to derive *cubic elements*. Such elements require *two internal nodes*, so that there are four nodes per element. Letting the cubic be described by

$$L_i = c_{i1} + c_{i2}\xi + c_{i3}\xi^2 + c_{i4}\xi^3, \quad i = 1, 2, 3, 4 \quad (9.38)$$

and imposing the conditions

$$\begin{aligned} L_1(1) &= 1, & L_1(\tfrac{2}{3}) &= L_1(\tfrac{1}{3}) = L_1(0) = 0 \\ L_2(\tfrac{2}{3}) &= 1, & L_2(1) &= L_2(\tfrac{1}{3}) = L_2(0) = 0 \\ L_3(\tfrac{1}{3}) &= 1, & L_3(1) &= L_3(\tfrac{2}{3}) = L_3(0) = 0 \\ L_4(0) &= 1, & L_4(1) &= L_4(\tfrac{2}{3}) = L_4(\tfrac{1}{3}) = 0 \end{aligned} \quad (9.39)$$

we obtain the cubic elements

$$\begin{aligned} L_1 &= \tfrac{1}{2}\xi(2 - 9\xi + 9\xi^2), & L_2 &= -\tfrac{9}{2}\xi(1 - 4\xi + 3\xi^2) \\ L_3 &= \tfrac{9}{2}\xi(2 - 5\xi + 3\xi^2), & L_4 &= 1 - \tfrac{11}{2}\xi + 9\xi^2 - \tfrac{9}{2}\xi^3 \end{aligned} \quad (9.40)$$

The functions are plotted in Fig. 9.5.

The displacement w can be expressed in terms of the cubic elements as follows:

$$w = L_1 a_{j-1} + L_2 a_{j-2/3} + L_3 a_{j-1/3} + L_4 a_j = \mathbf{L}^T \mathbf{a}_j \quad (9.41)$$

where the four-dimensional vectors $\mathbf{L}$ and $\mathbf{a}_j$ are self-evident. The calculation of the element stiffness and matrices for the cubic elements is left as an exercise to the reader. Of course, they will be 4×4 matrices.

In general, quadratic elements will yield better approximations than linear elements and cubic elements better than quadratic elements, with the exception of cases in which the solution w does not possess a

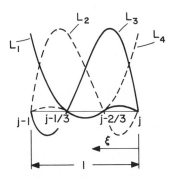

Figure 9.5

continuous derivative. This is the case when there are point loads or discontinuous distributed loads present. These are points of internal singularity, at which points the solution must exhibit discontinuity of derivative. For convergence, it is advised to choose points of singularity as external nodes. Elements with continuous derivatives at nodes are discussed in the next section, in connection with the bending vibration of bars.

Example 9.2

Solve the problem of Example 9.1 by using first quadratic elements and then cubic elements.

From Eqs. (9.30), we write the vector of nondimensional quadratic elements

$$\mathbf{L} = [\xi(2\xi - 1) \quad 4\xi(1 - \xi) \quad 1 - 3\xi + 2\xi^2]^T \tag{a}$$

Inserting Eq. (a) into Eqs. (9.35) and using the parameters of Example 9.1, we obtain the element stiffness and mass matrices

$$K_j = \frac{2EAn}{5L}\begin{bmatrix} 7 & -8 & 1 \\ -8 & 16 & -8 \\ 1 & -8 & 7 + \delta_{jn}/2n \end{bmatrix}$$

$$-\frac{EA}{25Ln}\begin{bmatrix} 23 - 55j + 35j^2 & -26 + 60j - 40j^2 & 3 - 5j + 5j^2 \\ -26 + 60j - 40j^2 & 32 - 80j + 80j^2 & -6 + 20j - 40j^2 \\ 3 - 5j + 5j^2 & -6 + 20j - 40j^2 & 3 - 15j + 35j^2 \end{bmatrix},$$

$$j = 1, 2, \ldots, n \tag{b}$$

and

$$M_j = \frac{mL}{25n}\begin{bmatrix} 4 & 2 & -1 \\ 2 & 16 & 2 \\ -1 & 2 & 4 \end{bmatrix}$$

$$-\frac{mL}{700n^3}\begin{bmatrix} 44-98j+56j^2 & 24-56j+28j^2 & -5+14j-14j^2 \\ 24-56j+28j^2 & 64-224j+224j^2 & -4+28j^2 \\ -5+14j-14j^2 & -4+28j^2 & 2-14j+56j^2 \end{bmatrix}$$

$$j = 1, 2, \ldots, n \quad \text{(c)}$$

Using the scheme (9.36), the global stiffness and mass matrices can be shown

$$K = \frac{2EAn}{5L}\begin{bmatrix} 16 & -8 & 0 & \cdots & 0 & 0 & 0 \\ & 14 & -8 & \cdots & 0 & 0 & 0 \\ & & 16 & \cdots & 0 & 0 & 0 \\ & \text{Symm} & & \cdots\cdots\cdots\cdots\cdots \\ & & & & 14 & -8 & 1 \\ & & & & & 16 & -8 \\ & & & & & & 7+\tfrac{1}{2}n \end{bmatrix}$$

$$-\frac{EA}{25\,Ln}\begin{bmatrix} 32 & -26 & 0 & \cdots & 0 \\ & 26 & -6 & \cdots & 0 \\ & & 32 & \cdots & 0 \\ & \text{Symm} & & \cdots\cdots\cdots\cdots \\ & & & & 76-140n+70n^2 \end{bmatrix}$$

$$\begin{bmatrix} & 0 & & 0 \\ & 0 & & 0 \\ & 0 & & 0 \\ \cdots\cdots\cdots\cdots\cdots\cdots\cdots\cdots \\ -26+60n-40n^2 & & 3-5n+5n^2 \\ 32-80n+80n^2 & & -6+20n-40n^2 \\ & & 3-15n+35n^2 \end{bmatrix} \quad \text{(d)}$$

and

$$M = \frac{mL}{25n}\begin{bmatrix} 16 & 2 & 0 & \cdots & 0 & 0 & 0 \\ & 8 & 2 & \cdots & 0 & 0 & 0 \\ & & 16 & \cdots & 0 & 0 & 0 \\ & \text{Symm} & & \cdots\cdots\cdots\cdots \\ & & & & 8 & 2 & -1 \\ & & & & & 16 & 2 \\ & & & & & & 4 \end{bmatrix}$$

$$-\frac{mL}{700n^3}\begin{bmatrix} 64 & 28 & 0 & \cdots & 0 \\ & 116 & 24 & \cdots & 0 \\ & & 512 & \cdots & 0 \\ & \text{Symm} & & \cdots\cdots\cdots\cdots\cdots\cdots \\ & & & & 116-224n+112n^2 \end{bmatrix}$$

$$\begin{bmatrix} 0 & 0 \\ 0 & 0 \\ 0 & 0 \\ \cdots\cdots\cdots\cdots\cdots\cdots\cdots\cdots \\ 24-56n+28n^2 & -5+14n-14n^2 \\ 64-224n+224n^2 & -4+28n^2 \\ & 2-14n+56n^2 \end{bmatrix} \qquad \text{(e)}$$

We observe that matrices K and M are of order $2n$.

Solving the eigenvalue problem associated with K and M, Eqs. (d) and (e), for various values of n, we conclude that the first two estimated natural frequencies comparable in value to those computed in Example 8.1 by the classical Rayleigh–Ritz method with $n=4$ are obtained for $n=12$. These values are

$$\omega_1^{24} = 1.894544 \sqrt{\frac{EA}{mL^2}}, \qquad \omega_2^{24} = 4.883427 \sqrt{\frac{EA}{mL^2}} \qquad \text{(f)}$$

both values being slightly lower than ω_1^4 and ω_2^4, respectively, computed in Example 8.1.

For cubic elements, we use the vector

$$\mathbf{L} = [\tfrac{1}{2}\xi(2-9\xi+9\xi^2) \qquad -\tfrac{9}{2}\xi(1-4\xi+3\xi^2)$$
$$\tfrac{9}{2}\xi(2-5\xi+3\xi^2) \qquad 1-\tfrac{11}{2}\xi+9\xi^2-\tfrac{9}{2}\xi^3]^T \qquad \text{(g)}$$

Following the established procedure, we derive first the element and then the global stiffness and mass matrices, where the latter are of order $3n$. Omitting the details, we merely state that four cubic elements yield

$$\omega_1^{12} = 1.894544 \sqrt{\frac{EA}{mL^2}}, \qquad \omega_2^{12} = 4.883386 \sqrt{\frac{EA}{mL^2}} \qquad \text{(h)}$$

which compare favorably with Eqs. (f), obtained by using twelve quadratic elements.

In conclusion, to compute the first two natural frequencies with the same accuracy, the classical Rayleigh–Ritz method requires an eigensolution of order four, the finite element method using cubic elements requires an eigensolution of order twelve and that using quadratic elements one of order twenty-four. Hence, for the example at hand, the classical Rayleigh–Ritz method proves superior to the finite element method, and the finite element method using cubic elements proves superior to that using quadratic elements.

9.4 Fourth-order problems

Let us consider the bending vibration of the rotating helicopter blade introduced in Sec. 7.2. Ignoring rotary inertia. Rayleigh's quotient remains in the form (9.6), where the numerator represents the energy inner product

$$[w, w] = \int_0^L [EI(w'')^2 + P(w')^2] \, \mathrm{d}x \qquad (9.42)$$

in which EI is the flexural rigidity and P is the axial force. The denominator is given by Eq. (9.8). The presence of the second derivative in the energy inner product implies that an approximate solution of the eigenvalue problem must be from the space κ_G^2. Hence, linear elements must be ruled out. In fact, quadratic elements must also be ruled out, because both the displacement and its first derivative must be continuous at both nodes, so that four constants are necessary to define the polynomial. Quadratic elements are defined by only three components. Hence, the lowest degree polynomials admissible are cubics.

In view of the above, we shall use as nodal coordinates the displacements w_{j-1} and w_j as well as the rotations θ_{j-1} and θ_j, as shown in Fig. 9.6. Hence, we shall write the displacement w in the form

$$w = L_1 w_{j-1} + L_2 h \theta_{j-1} + L_3 w_j + L_4 h \theta_j, \qquad (j-1)h \le x \le jh \qquad (9.43)$$

For convenience, we shall continue to work with the nondimensional natural coordinate $\xi = j - x/h$, so that we consider the cubic polynomials

$$L_i = c_{i1} + c_{i2}\xi + c_{i3}\xi^2 + c_{i4}\xi^3, \qquad i = 1, 2, 3, 4 \qquad (9.44)$$

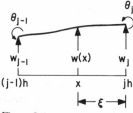

Figure 9.6

where the coefficients c_{ik} ($i, k = 1, 2, 3, 4$) are determined by insisting that w and w' take the values w_{j-1} and θ_{j-1} at $x = (j-1)h$, or $\xi = 1$, and the values w_j and θ_j at $x = jh$, or $\xi = 0$.

Imposing the conditions

$$
\begin{aligned}
L_1(1) &= 1, & L_1'(1) &= L_1(0) = L_1'(0) = 0 \\
L_2'(1) &= -1, & L_2(1) &= L_2(0) = L_2'(0) = 0 \\
L_3(0) &= 1, & L_3(1) &= L_3'(1) = L_3'(0) = 0 \\
L_4'(0) &= -1, & L_4(1) &= L_4'(1) = L_4(0) = 0
\end{aligned}
\tag{9.45}
$$

We obtain the interpolation functions

$$
\begin{aligned}
L_1 &= 3\xi^2 - 2\xi^3, & L_2 &= \xi^2 - \xi^3 \\
L_3 &= 1 - 3\xi^2 + 2\xi^2, & L_4 &= -\xi + 2\xi^2 - \xi^3
\end{aligned}
\tag{9.46}
$$

which are known as *Hermite cubics*. They are plotted in Fig. 9.7.

It will prove convenient to introduce the notation

$$
w = \mathbf{L}^T \mathbf{a}_j, \qquad (j-1)h \le x \le jh
\tag{9.47}
$$

where

$$
\mathbf{L} = [L_1 \quad L_2 \quad L_3 \quad L_4]^T
\tag{9.48a}
$$

$$
\mathbf{a}_j = [w_{j-1} \quad h\theta_{j-1} \quad w_j \quad h\theta_j]^T
\tag{9.48b}
$$

so that, following the procedure of Sec. 9.3, we can write

$$
\frac{dw}{dx} = -\frac{1}{h}\mathbf{L}'^T \mathbf{a}_j, \qquad \frac{d^2w}{dx^2} = \frac{1}{h^2}\mathbf{L}''^T \mathbf{a}_j, \qquad (j-1)h \le x \le jh
\tag{9.49}
$$

where primes indicate derivatives with respect to ξ. Hence, the numerator

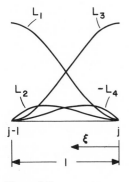

Figure 9.7

347

and denominator of Rayleigh's quotient can be written in the form

$$[w, w] = \int_0^L \left[EI \left(\frac{d^2 w}{dx^2} \right)^2 + P \left(\frac{dw}{dx} \right)^2 \right] dx$$

$$= \sum_{j=1}^{n} \int_{(j-1)h}^{jh} \left[EI \left(\frac{d^2 w}{dx^2} \right)^2 + P \left(\frac{dw}{dx} \right)^2 \right] dx = \sum_{j=1}^{n} \mathbf{a}_j^T K_j \mathbf{a}_j \qquad (9.50a)$$

$$(\sqrt{m} w, \sqrt{m} w) = \int_0^L m w^2 \, dx = \sum_{j=1}^{n} \int_{(j-1)h}^{jh} m w^2 \, dx = \sum_{j=1}^{n} \mathbf{a}_j^T M_j \mathbf{a}_j \qquad (9.50b)$$

where

$$K_j = h \int_0^1 \left(\frac{EI_j}{h^4} \mathbf{L}'' \mathbf{L}''^T + \frac{P_j}{h^2} \mathbf{L}' \mathbf{L}'^T \right) d\xi, \qquad j = 1, 2, \ldots, n \qquad (9.51a)$$

$$M_j = h \int_0^1 m_j \mathbf{L} \mathbf{L}^T \, d\xi, \qquad j = 1, 2, \ldots, n \qquad (9.51b)$$

As an illustration, let us consider the case in which $EI = \text{const}$, $m = \text{const}$. In this case, the axial force can be obtained from Eq. (7.21) in the form

$$P(x) = \int_x^L m \Omega^2 \zeta \, d\zeta = \tfrac{1}{2} m \Omega^2 L^2 \left(1 - \frac{x^2}{L^2} \right) \qquad (9.52)$$

or

$$P_j(\xi) = \tfrac{1}{2} m \Omega^2 L^2 \left[1 - \left(\frac{h}{L} \right)^2 (j - \xi)^2 \right], \qquad 0 \le \xi \le 1 \qquad (9.53)$$

Introducing Eqs. (9.46) and (9.53) into Eqs. (9.51), we obtain the element stiffness matrices

$$K_j = \frac{EI}{h^3} \begin{bmatrix} 12 & 6 & -12 & 6 \\ & 4 & -6 & 2 \\ \text{Symm} & & 12 & -6 \\ & & & 4 \end{bmatrix}$$

$$+ \frac{1}{2} m \Omega^2 L n \left(\frac{1}{30} \left[1 - \left(\frac{j}{n} \right)^2 \right] \begin{bmatrix} 36 & -3 & -36 & 3 \\ & 4 & -3 & 1 \\ \text{Symm} & & 36 & -3 \\ & & & 4 \end{bmatrix} \right.$$

$$\left. - \frac{j}{30n} \begin{bmatrix} 36 & 0 & -36 & 6 \\ & 6 & 0 & -1 \\ \text{Symm} & & 36 & -6 \\ & & & 2 \end{bmatrix} + \frac{1}{210n^2} \begin{bmatrix} 72 & -6 & -72 & -15 \\ & 18 & 6 & -3 \\ \text{Symm} & & 72 & 15 \\ & & & 4 \end{bmatrix} \right)$$

$$j = 1, 2, \ldots, n \qquad (9.54a)$$

and mass matrices

$$M_j = \frac{mL}{420n} \begin{bmatrix} 156 & 22 & 54 & -13 \\ & 4 & 13 & -3 \\ & \text{Symm} & 156 & -22 \\ & & & 4 \end{bmatrix}, \quad j = 1, 2, \ldots, n \qquad (9.54b)$$

Because $w(0) = w'(0) = 0$, we must strike out the first two rows and columns from the matrices K_1 and M_1. The assembling process follows the pattern shown in (9.20), except that the order of the matrices K_j and M_j $(j = 1, 2, \ldots, n)$ is twice as large.

9.5   Two-dimensional domains

As indicated in Sec. 7.8, in two-dimensional boundary-value problems there is an additional factor to consider, namely, the shape of the boundary, a factor not present in one-dimensional boundary-value problems. Indeed, the shape of the boundary alone is often responsible for putting closed-form solutions out of reach. A smooth boundary can be taken as an indication that the solution itself may be smooth. But, if no closed-form solution is possible and an approximate solution by the finite element method is being sought, then a smooth boundary is likely to create difficulties in choosing elements to match the boundary. In fact, in many cases it is not feasible to divide the domain D exactly into n elements, and small slivers along the boundaries may not belong to any element, so that the finite element domain D^n may differ somewhat from the actual domain D. If we consider triangular elements, then the difference $D - D^n$ between a domain bounded by a smooth curve and one bounded by a polygon may be as indicated by the shaded areas in Fig. 9.8.

Two questions arise in two-dimensional boundary-value problems that did not arise in one-dimensional ones. They concern the choice of a numbering pattern for the nodal points and the choice of the shape of the elements. The first is relatively minor and is caused by the fact that the

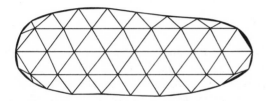

Figure 9.8

numbering of the nodal points cannot follow a straight uninterrupted line as in the one-dimensional case. One should be aware of the fact that the numbering controls the bandwidth of the stiffness and mass matrices, so that the numbering pattern should be such that the bandwidth is the smallest possible. On the other hand, the choice of the elements shape is a much more critical question and it requires an examination in depth.

In a classic paper, Courant (C3) suggested a numerical solution of the plane torsion problem for multi-connected domains by dividing the domains into small triangles over which the solution was assumed to be linear, thus anticipating the finite element method. Indeed, triangular elements have proved very versatile, owing to their ability to fill tightly a domain with smooth boundaries. Of course, this may imply the use of increasingly smaller elements as one approaches the boundary. Triangular elements are equally useful for cases in which the boundaries possess corners. However, in some cases quadrilateral elements hold the advantage, as they can reduce the number of elements to be used for a given domain.

The treatment of problems involving two-dimensional domains follows the same pattern as for one-dimensional domains, although some of the details may be different. In the case of a second-order problem, such as the membrane vibration of Sec. 7.8, the elements must be from the space κ_G^1, so that linear elements are admissible. Linear triangular elements have some nice features. One of them is that the plane $w(x, y) = a_1 + a_2 x + a_3 y$, defining the displacement in a given element, is determined uniquely by the values of the nodal displacements, namely, the values of w at the three vertices of the triangle. Moreover, the value of w along an element edge reduces to a linear function of a single variable, where the latter function is determined uniquely by the nodal displacements at the two end points of the edge, which implies that the nodal displacement at the third point has no effect on the value of w along the edge in question. As a result, *the continuity of w across the edge is guaranteed by continuity at the nodal points.*

The finite element solution can be written in the general form

$$w^n(x, y) = \sum_{j=1}^{n} a_j \phi_j(x, y) \qquad (9.55)$$

where $\phi_j(x, y)$ are trial functions. For linear elements, the trial functions have the form of *pyramid functions* (Fig. 9.9) and they represent the two-dimensional counterpart of the roof functions. Taking the height of the pyramid equal to unity, the coefficients a_j can once again be identified as nodal displacements.

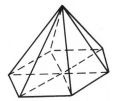

Figure 9.9

The calculation of the global stiffness and mass matrices is carried out by calculating first the element matrices and then assembling the results. In two-dimensional problems, we have the additional task of choosing a set of coordinates. First, we shall formulate the problem in terms of cartesian coordinates and then introduce the natural coordinates.

Let us consider the triangular element shown in Fig. 9.10 and denote the vectors from the origin 0 to the vertices 1, 2 and 3 by $\mathbf{r}_1$, $\mathbf{r}_2$ and $\mathbf{r}_3$, respectively. Moreover, let $P(x, y)$ be an arbitrary point inside the triangle and denote by $\mathbf{r}$ the radius vector from 0 to P. Then, the vectors from 1, 2 and 3 to P are simply $\mathbf{r}-\mathbf{r}_1$, $\mathbf{r}-\mathbf{r}_2$ and $\mathbf{r}-\mathbf{r}_3$, respectively. Because these three vectors lie in the same plane, we must have

$$c_1(\mathbf{r}-\mathbf{r}_1)+c_2(\mathbf{r}-\mathbf{r}_2)+c_3(\mathbf{r}-\mathbf{r}_3)=\mathbf{0} \tag{9.56}$$

so that, solving for $\mathbf{r}$, we obtain

$$\mathbf{r}=\frac{c_1\mathbf{r}_1+c_2\mathbf{r}_2+c_3\mathbf{r}_3}{c_1+c_2+c_3}=\sum_{i=1}^{3}\xi_i\mathbf{r}_i \tag{9.57}$$

where

$$\xi_i=\frac{c_i}{\sum_{j=1}^{3}c_j}, \qquad i=1,2,3 \tag{9.58}$$

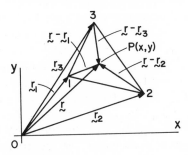

Figure 9.10

9 The finite element method

Equations (9.57–58) yield

$$x_1\xi_1 + x_2\xi_2 + x_3\xi_3 = x \qquad (9.59a)$$

$$y_1\xi_1 + y_2\xi_2 + y_3\xi_3 = y \qquad (9.59b)$$

$$\xi_1 + \xi_2 + \xi_3 = 1 \qquad (9.59c)$$

which can be written in the matrix form

$$\begin{bmatrix} 1 & 1 & 1 \\ x_1 & x_2 & x_3 \\ y_1 & y_2 & y_3 \end{bmatrix} \begin{bmatrix} \xi_1 \\ \xi_2 \\ \xi_3 \end{bmatrix} = \begin{bmatrix} 1 \\ x \\ y \end{bmatrix} \qquad (9.60)$$

The solution of Eq. (9.60) is simply

$$\begin{bmatrix} \xi_1 \\ \xi_2 \\ \xi_3 \end{bmatrix} = \begin{bmatrix} 1 & 1 & 1 \\ x_1 & x_2 & x_3 \\ y_1 & y_2 & y_3 \end{bmatrix}^{-1} \begin{bmatrix} 1 \\ x \\ y \end{bmatrix} \qquad (9.61)$$

or more explicitly

$$\xi_1(x, y) = \frac{1}{\Delta} [x_2 y_3 - x_3 y_2 + (y_2 - y_3)x + (x_3 - x_2)y]$$

$$\xi_2(x, y) = \frac{1}{\Delta} [x_3 y_1 - x_1 y_3 + (y_3 - y_1)x + (x_1 - x_3)y] \qquad (9.62)$$

$$\xi_3(x, y) = \frac{1}{\Delta} [x_1 y_2 - x_2 y_1 + (y_1 - y_2)x + (x_2 - x_1)y]$$

where

$$\Delta = (x_2 y_3 - x_3 y_2) + (x_3 y_1 - x_1 y_3) + (x_1 y_2 - x_2 y_1) \qquad (9.63)$$

The functions ξ_1, ξ_2 and ξ_3 have a very interesting geometric interpretation. We observe that for $x = x_1$, $y = y_1$, Eqs. (9.62) yield $\xi_1 = 1$, $\xi_2 = \xi_3 = 0$. Similarly, for $x = x_2$, $y = y_2$ and $x = x_3$, $y = y_3$, Eqs. (9.62) yield $\xi_1 = 0$, $\xi_2 = 1$, $\xi_3 = 0$ and $\xi_1 = \xi_2 = 0$, $\xi_3 = 1$, respectively. Hence, the vertices 1, 2 and 3 of the triangle are defined by the triplets $(1, 0, 0)$, $(0, 1, 0)$ and $(0, 0, 1)$, respectively. This suggests that the quantities ξ_i $(i = 1, 2, 3)$ can be used as coordinates, as shown in Fig. 9.11. For example, the edge 2, 3 is defined as $\xi_1 = 0$ and the vertex 1 as $\xi_1 = 0$. A line parallel to the edge 2, 3 is described by $\xi_1 = c = \text{const}$, where the constant c is proportional to the distance from 2, 3. The coordinates ξ_1, ξ_2, ξ_3 can be identified as the *natural coordinates* for the triangular element.

352

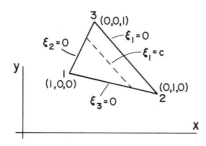

Figure 9.11

The quantities ξ_1, ξ_2, ξ_3 can be given another interpretation by noticing that $\Delta = 2A$, where A is the area of the triangular element. Letting $x_i = x$, $y_i = y$ ($i = 1, 2, 3$) in Eq. (9.63), in sequence, and comparing with Eqs. (9.62), we conclude that

$$\xi_i = A_i/A, \qquad i = 1, 2, 3 \tag{9.64}$$

where A_i is the area of the triangle formed by the point P with the edge opposite to the vertex i ($i = 1, 2, 3$) (Fig. 9.12). For this reason, ξ_1, ξ_2 and ξ_3 are also known as *area coordinates*.

The natural coordinates ξ_1, ξ_2, ξ_3 can be used to describe the element admissible functions. Indeed, letting

$$L_i = \xi_i, \qquad i = 1, 2, 3 \tag{9.65}$$

we observe that L_1, L_2 and L_3 simply define sections of pyramids, as shown in Figs. 9.13a, b and c respectively, so that they are really *linear interpolation* functions for the triangular element. Hence, denoting the nodal displacements by w_i ($i = 1, 2, 3$), the displacement at any point of the element e can be written in the form

$$w(\xi_1, \xi_2, \xi_3) = \sum_{i=1}^{3} L_i w_i = \mathbf{L}^T \mathbf{w}_e \tag{9.66}$$

The function $w(\xi_1, \xi_2, \xi_3)$ is shown in Fig. 9.13d.

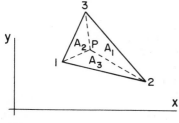

Figure 9.12

9   *The finite element method*

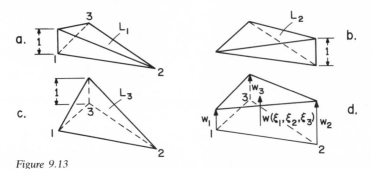

Figure 9.13

Next, we propose to calculate the stiffness and mass matrices for the membrane of Sec. 7.8. As suggested earlier, we shall calculate the element stiffness and mass matrices first and then assemble them. The energy inner product can be written as

$$[w, w] = \sum_{e=1}^{n} [w, w]_e \qquad (9.67)$$

where, for constant tension T,

$$[w, w]_e = T \int_{A_e} \left[\left(\frac{\partial w}{\partial x} \right)^2 + \left(\frac{\partial w}{\partial y} \right)^2 \right] dA_e \qquad (9.68)$$

is the contribution of the element e to the energy inner product, in which A_e is the element area. Moreover, we have

$$(\sqrt{m}w, \sqrt{m}w) = \sum_{e=1}^{n} (\sqrt{m}w, \sqrt{m}w)_e \qquad (9.69)$$

where, for constant mass density m,

$$(\sqrt{m}w, \sqrt{m}w)_e = m \int_{A_e} w^2 \, dA_e \qquad (9.70)$$

To evaluate the integrals (9.68) and (9.70), we shall express the integrands in terms of the natural coordinates. Using Eqs. (9.62), (9.65) and (9.66), we can write

$$\frac{\partial w}{\partial x} = \frac{\partial w}{\partial \xi_1} \frac{\partial \xi_1}{\partial x} + \frac{\partial w}{\partial \xi_2} \frac{\partial \xi_2}{\partial x} + \frac{\partial w}{\partial \xi_3} \frac{\partial \xi_3}{\partial x} = \frac{1}{2A_e} \left[\frac{\partial \mathbf{L}^T}{\partial \xi_1} (y_2 - y_3) \right.$$

$$\left. + \frac{\partial \mathbf{L}^T}{\partial \xi_2} (y_3 - y_1) + \frac{\partial \mathbf{L}^T}{\partial \xi_3} (y_1 - y_2) \right] \mathbf{w}_e$$

$$= \frac{1}{2A_e} [y_2 - y_3 \quad y_3 - y_1 \quad y_1 - y_2] \mathbf{w}_e \qquad (9.71a)$$

$$\frac{\partial w}{\partial y} = \frac{\partial w}{\partial \xi_1}\frac{\partial \xi_1}{\partial y} + \frac{\partial w}{\partial \xi_2}\frac{\partial \xi_2}{\partial y} + \frac{\partial w}{\partial \xi_3}\frac{\partial \xi_3}{\partial y} = \frac{1}{2A_e}\left[\frac{\partial \mathbf{L}^T}{\partial \xi_1}(x_3 - x_2)\right.$$

$$\left. + \frac{\partial \mathbf{L}^T}{\partial \xi_2}(x_1 - x_3) + \frac{\partial \mathbf{L}^T}{\partial \xi_3}(x_2 - x_1)\right]\mathbf{w}_e$$

$$= \frac{1}{2A_e}[x_3 - x_2 \quad x_1 - x_3 \quad x_2 - x_1]\mathbf{w}_e \tag{9.71b}$$

so that, introducing Eqs. (9.71) into Eq. (9.68), we obtain

$$[w, w]_e = \mathbf{w}_e^T K_e \mathbf{w}_e \tag{9.72}$$

where K_e is the element stiffness matrix

$$K_e = \frac{T}{4A_e^2}\int_{A_e}\left\{\begin{bmatrix}y_2 - y_3\\y_3 - y_1\\y_1 - y_2\end{bmatrix}\begin{bmatrix}y_2 - y_3\\y_3 - y_1\\y_1 - y_2\end{bmatrix}^T + \begin{bmatrix}x_3 - x_2\\x_1 - x_3\\x_2 - x_1\end{bmatrix}\begin{bmatrix}x_3 - x_2\\x_1 - x_3\\x_2 - x_1\end{bmatrix}^T\right\}dA_e \tag{9.73}$$

so that the integrand in (9.73) does not contain natural coordinates. On the other hand, introducing Eq. (9.66) into Eq. (9.70), we obtain

$$(\sqrt{m}w, \sqrt{m}w)_e = \mathbf{w}_e^T M_e \mathbf{w}_e \tag{9.74}$$

where M_e is the element mass matrix

$$M_e = m\int_{A_e}\begin{bmatrix}\xi_1\\\xi_2\\\xi_3\end{bmatrix}\begin{bmatrix}\xi_1\\\xi_2\\\xi_3\end{bmatrix}^T dA_e \tag{9.75}$$

The integrals (9.73) and (9.75) can be evaluated explicitly by means of the formula (H3, p. 84)

$$\int_{A_e}\xi_1^m\xi_2^n\xi_3^p\,dA_e = \frac{m!\,n!\,p!}{(m+n+p+2)!}\,2A_e \tag{9.76}$$

In the case of Eq. (9.73), $m = n = p = 0$, so that use of Eq. (9.76) yields

$$K_e = \frac{T}{4A_e}\left[\begin{array}{ccc}(y_2 - y_3)^2 + (x_3 - x_2)^2 & (y_2 - y_3)(y_3 - y_1) + (x_3 - x_2)(x_1 - x_3) \\ & (y_3 - y_1)^2 + (x_1 - x_3)^2 \\ \text{Symm} \end{array}\right.$$

$$\left.\begin{array}{c}(y_2 - y_3)(y_1 - y_2) + (x_3 - x_2)(x_2 - x_1)\\(y_3 - y_1)(y_1 - y_2) + (x_1 - x_3)(x_2 - x_1)\\(y_1 - y_2)^2 + (x_2 - x_1)^2\end{array}\right] \tag{9.77}$$

355

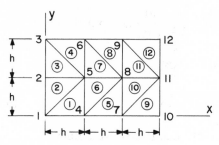

Figure 9.14

Moreover, using Eq. (9.76), Eq. (9.75) yields

$$M_e = \frac{mA_e}{12}\begin{bmatrix} 2 & 1 & 1 \\ 1 & 2 & 1 \\ 1 & 1 & 2 \end{bmatrix} \tag{9.78}$$

The next task is the assembly process. Before we can proceed with this task, however, we must specify the membrane configuration and boundary conditions, as well as the finite elements geometry. Assuming that the membrane is rectangular with all its sides free, we shall choose triangular elements and adopt the mesh and numbering scheme shown in Fig. 9.14. For convenience, we tabulate the nodal coordinates and element data, as shown in Tables I and II. Using the data from Tables I and II, we obtain the following element stiffness matrices

$$K_e = \frac{T}{2}\begin{bmatrix} 1 & -1 & 0 \\ -1 & 2 & -1 \\ 0 & -1 & 1 \end{bmatrix}, \qquad e = 1, 5, 9 \tag{9.79a}$$

$$K_e = \frac{T}{2}\begin{bmatrix} 1 & 0 & -1 \\ 0 & 1 & -1 \\ -1 & -1 & 2 \end{bmatrix}, \qquad e = 2, 4, 6, 8, 10, 12 \tag{9.79b}$$

$$K_e = \frac{T}{2}\begin{bmatrix} 2 & -1 & -1 \\ -1 & 1 & 0 \\ -1 & 0 & 1 \end{bmatrix}, \qquad e = 3, 7, 11 \tag{9.79c}$$

and the following element mass matrices

$$M_e = \frac{mh^2}{24}\begin{bmatrix} 2 & 1 & 1 \\ 1 & 2 & 1 \\ 1 & 1 & 2 \end{bmatrix}, \qquad e = 1, 2, \ldots, 12 \tag{9.80}$$

Table I			Table II			
Nodal Coordinates			Element Data			
Node Numbers	x	y	Element Number	Node Numbers		
1	0	0	1	1	4	5
2	0	h	2	1	5	2
3	0	$2h$	3	2	5	3
4	h	0	4	3	5	6
5	h	h	5	4	7	8
6	h	$2h$	6	4	8	5
7	$2h$	0	7	5	8	6
8	$2h$	h	8	6	8	9
9	$2h$	$2h$	9	7	10	11
10	$3h$	0	10	7	11	8
11	$3h$	h	11	8	11	9
12	$3h$	$2h$	12	9	11	12

The assembly process for two-dimensional domains is more involved than that for one-dimensional ones, because the nodal vectors do not have the same orderly structure. Nevertheless, this is only a bookkeeping problem, so that the global stiffness matrix can be shown to be

$$
K = \frac{T}{2}
\begin{bmatrix}
2 & -1 & 0 & -1 & & & & & & & & \\
-1 & 4 & -1 & 0 & -2 & & & & & & & \\
0 & -1 & 2 & 0 & 0 & -1 & & & & & & \\
-1 & 0 & 0 & 4 & -2 & 0 & -1 & & & & & \\
 & -2 & 0 & -2 & 8 & -2 & 0 & -2 & & & & \\
 & & -1 & 0 & -2 & 4 & 0 & 0 & -1 & & & \\
 & & & -1 & 0 & 0 & 4 & -2 & 0 & -1 & & \\
 & & & & -2 & 0 & -2 & 8 & -2 & 0 & -2 & \\
 & & & & & -1 & 0 & -2 & 4 & 0 & 0 & -1 \\
 & & & & & & -1 & 0 & 0 & 2 & -1 & 0 \\
 & & & & & & & -2 & 0 & -1 & 4 & -1 \\
 & & & & & & & & -1 & 0 & -1 & 2
\end{bmatrix}
$$

(9.81)

357

Similarly, the global mass matrix is

$$
M = \frac{mh^2}{24}
\begin{bmatrix}
4 & 1 & 0 & 1 & 2 & & & & & & & \\
1 & 4 & 1 & 0 & 2 & 0 & & & & & & \\
0 & 1 & 4 & 0 & 2 & 1 & 0 & & & & & \\
1 & 0 & 0 & 6 & 2 & 0 & 1 & 2 & & & & \\
2 & 2 & 2 & 2 & 12 & 2 & 0 & 2 & 0 & & & \\
0 & 1 & 0 & 2 & 6 & 0 & 2 & 1 & 0 & & & \\
& 0 & 1 & 0 & 0 & 6 & 2 & 0 & 1 & 2 & & \\
& & 2 & 2 & 2 & 2 & 12 & 2 & 0 & 2 & 0 & \\
& & & 0 & 1 & 0 & 2 & 6 & 0 & 2 & 1 & \\
& & & & 0 & 1 & 0 & 0 & 2 & 1 & 0 & \\
& & & & 2 & 2 & 2 & 1 & 8 & 1 & \\
& & & & & 0 & 1 & 0 & 1 & 2 &
\end{bmatrix}
\qquad (9.82)
$$

The missing entries in K and M are all zero.

A matrix is said to have *half-bandwidth* h if the entries (i, j) of the matrix are zero for all i and j satisfying the inequality $j > i + h$. Hence, the matrices K and M have half-bandwidth equal to three and four, respectively. Note that it would have been more expedient to list the matrices by their non-zero diagonals, but we chose to display the matrices in full, as this enables us to ascertain the singularity of K by simply adding up all its columns. The fact that K is singular should come as no surprise, because the membrane is free, so that a rigid-body translation is admissible. In fact, this provides one check for the correctness of K.

Next, let us consider quadratic interpolation functions for triangular elements. Because the natural coordinates ξ_1, ξ_2, ξ_3 must satisfy Eq. (9.59c), only two of the coordinates should be regarded as independent. Hence, let us express the interpolation functions in the form

$$
L_i(\xi_1, \xi_2, \xi_3) = c_{i1} + c_{i2}\xi_1 + c_{i3}\xi_2 + c_{i4}\xi_1^2 + c_{i5}\xi_2^2 + c_{i6}\xi_1\xi_2,
$$
$$
i = 1, 2, \ldots, 6 \qquad (9.83)
$$

where the dependence on ξ_3 is only implicit. Equations (9.83) indicate

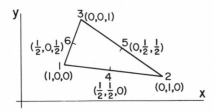

Figure 9.15

that we must have six nodes. We shall choose the nodes as shown in Fig. 9.15. The interpolation functions must satisfy

$$L_1(1, 0, 0) = 1, \qquad L_1(\tfrac{1}{2}, \tfrac{1}{2}, 0) = L_1(0, 1, 0) = L_1(0, \tfrac{1}{2}, \tfrac{1}{2})$$
$$= L_1(0, 0, 1) = L_1(\tfrac{1}{2}, 0, \tfrac{1}{2}) = 0$$

$$L_2(0, 1, 0) = 1, \qquad L_2(1, 0, 0) = L_2(\tfrac{1}{2}, \tfrac{1}{2}, 0) = L_2(0, \tfrac{1}{2}, \tfrac{1}{2})$$
$$= L_2(0, 0, 1) = L_2(\tfrac{1}{2}, 0, \tfrac{1}{2}) = 0$$

$$L_3(0, 0, 1) = 1, \qquad L_3(1, 0, 0) = L_3(\tfrac{1}{2}, \tfrac{1}{2}, 0) = L_3(0, 1, 0)$$
$$= L_3(0, \tfrac{1}{2}, \tfrac{1}{2}) = L_3(\tfrac{1}{2}, 0, \tfrac{1}{2}) = 0$$

$$L_4(\tfrac{1}{2}, \tfrac{1}{2}, 0) = 1, \qquad L_4(1, 0, 0) = L_4(0, 1, 0) = L_4(0, 0, 1)$$
$$= L_4(0, \tfrac{1}{2}, \tfrac{1}{2}) = L_4(\tfrac{1}{2}, 0, \tfrac{1}{2}) = 0$$

$$L_5(0, \tfrac{1}{2}, \tfrac{1}{2}) = 1, \qquad L_5(1, 0, 0) = L_5(\tfrac{1}{2}, \tfrac{1}{2}, 0) = L_5(0, 1, 0)$$
$$= L_5(0, 0, 1) = L_5(\tfrac{1}{2}, 0, \tfrac{1}{2}) = 0$$

$$L_6(\tfrac{1}{2}, 0, \tfrac{1}{2}) = 1, \qquad L_6(1, 0, 0) = L_6(\tfrac{1}{2}, \tfrac{1}{2}, 0) = L_6(0, 1, 0)$$
$$= L_6(0, \tfrac{1}{2}, \tfrac{1}{2}) = L_6(0, 0, 1) = 0$$

$$(9.84)$$

Introducing conditions (9.84) into Eqs. (9.83), in sequence, we obtain the interpolation functions

$$L_1 = -\xi_1 + 2\xi_1^2, \qquad L_2 = -\xi_2 + 2\xi_2^2, \qquad L_3 = -\xi_3 + 2\xi_3^2$$
$$L_4 = 4\xi_1\xi_2, \qquad L_5 = 4\xi_2\xi_3, \qquad L_6 = 4\xi_1\xi_3$$

$$(9.85)$$

The functions L_1 and L_4 are shown in Figs. 9.16a and b, respectively.

We shall not pursue higher-degree polynomials for triangular elements here. The interested reader is referred to the text by Huebner (H5, Sec. 5.8.1). Instead, we shall turn our attention to *quadrilateral elements*, and in particular to *rectangular elements*. The natural coordinates for rectangular elements are shown in Fig. 9.17. Because there are four nodes involved, the lowest degree polynomial possible is the *bilinear*, given by

$$L_i(\xi_1, \xi_2) = c_{i1} + c_{i2}\xi_1 + c_{i3}\xi_2 + c_{i4}\xi_1\xi_2 \qquad (9.86)$$

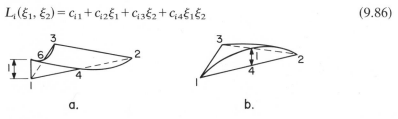

a.   b.

Figure 9.16

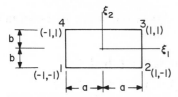

Figure 9.17

where the coefficients c_{ij} ($j = 1, 2, 3, 4$) are obtained for every interpolation function L_i ($i = 1, 2, 3, 4$) by imposing the conditions

$$
\begin{aligned}
L_1(-1, -1) &= 1, & L_1(1, -1) &= L_1(1, 1) = L_1(-1, 1) = 0 \\
L_2(1, -1) &= 1, & L_2(-1, -1) &= L_2(1, 1) = L_2(-1, 1) = 0 \\
L_3(1, 1) &= 1, & L_3(-1, -1) &= L_3(1, -1) = L_3(-1, 1) = 0 \\
L_4(-1, 1) &= 1, & L_4(-1, -1) &= L_4(1, -1) = L_4(1, 1) = 0
\end{aligned}
\tag{9.87}
$$

Introducing conditions (9.87) into Eq. (9.86), it is not difficult to show that the bilinear interpolation functions have the expressions

$$
\begin{aligned}
L_1 &= \tfrac{1}{4}(1 - \xi_1)(1 - \xi_2), & L_2 &= \tfrac{1}{4}(1 + \xi_1)(1 - \xi_2) \\
L_3 &= \tfrac{1}{4}(1 + \xi_1)(1 + \xi_2), & L_4 &= \tfrac{1}{4}(1 - \xi_1)(1 + \xi_2)
\end{aligned}
\tag{9.88}
$$

The function L_1 is shown in Fig. 9.18.

As an illustration, let us formulate the membrane problem in terms of bilinear rectangular elements. First, we wish to derive the element stiffness and mass matrices. To this end, we write the displacement w in the form

$$
w(\xi_1, \xi_2) = \mathbf{L}^T \mathbf{w}_e
\tag{9.89}
$$

where $\mathbf{L} = [L_1 \quad L_2 \quad L_3 \quad L_4]^T$ is the vector of interpolation functions with components given by Eqs. (9.88) and $\mathbf{w}_e = [w_1 \quad w_2 \quad w_3 \quad w_4]^T$ is the vector of nodal displacements. Considering the rectangular elements shown in Fig. 9.19, the relation between the cartesian coordinates x, y and local natural coordinates ξ_1, ξ_2 can be written as follows:

$$
\xi_1 = 1 - 2j + 2x/h, \quad (j-1)h \le x \le jh, \quad j = 1, 2, 3
\tag{9.90a}
$$

$$
\xi_2 = 1 - 2j + 2y/h, \quad (j-1)h \le y \le jh, \quad j = 1, 2
\tag{9.90b}
$$

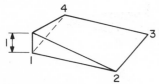

Figure 9.18

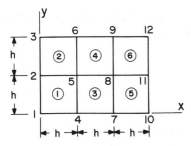

Figure 9.19

Using Eqs. (9.88–90), we obtain

$$\frac{\partial w}{\partial x} = \frac{2}{h}\frac{\partial w}{\partial \xi_1} = \frac{2}{h}\frac{\partial \mathbf{L}^T}{\partial \xi_1}\mathbf{w}_e$$

$$= \frac{1}{2h}\left[-(1-\xi_2) \quad 1-\xi_2 \quad 1+\xi_2 \quad -(1+\xi_2)\right]\mathbf{w}_e \qquad (9.91a)$$

$$\frac{\partial w}{\partial y} = \frac{2}{h}\frac{\partial w}{\partial \xi_2} = \frac{2}{h}\frac{\partial \mathbf{L}^T}{\partial \xi_2}\mathbf{w}_e$$

$$= \frac{1}{2h}\left[-(1-\xi_1) \quad -(1+\xi_1) \quad 1+\xi_1 \quad 1-\xi_1\right]\mathbf{w}_e \qquad (9.91b)$$

so that, introducing Eqs. (9.91) into Eq. (9.68), we obtain Eq. (9.72) in which

$$K_e = \frac{4T}{h^2}\int_{A_e}\left(\frac{\partial \mathbf{L}}{\partial \xi_1}\frac{\partial \mathbf{L}^T}{\partial \xi_1} + \frac{\partial \mathbf{L}}{\partial \xi_2}\frac{\partial \mathbf{L}^T}{\partial \xi_2}\right)dA_e$$

$$= \frac{T}{16}\int_{-1}^{1}\int_{-1}^{1}\left\{\begin{bmatrix} -(1-\xi_2) \\ 1-\xi_2 \\ 1+\xi_2 \\ -(1+\xi_2) \end{bmatrix}\begin{bmatrix} -(1-\xi_2) \\ 1-\xi_2 \\ 1+\xi_2 \\ -(1+\xi_2) \end{bmatrix}^T\right.$$

$$+ \begin{bmatrix} -(1-\xi_1) \\ -(1+\xi_1) \\ 1+\xi_1 \\ 1-\xi_1 \end{bmatrix}\begin{bmatrix} -(1-\xi_1) \\ -(1+\xi_1) \\ 1+\xi_1 \\ 1-\xi_1 \end{bmatrix}^T\left.\right\}d\xi_1\,d\xi_2$$

$$= \frac{T}{6}\begin{bmatrix} 4 & -1 & -2 & -1 \\ & 4 & -1 & -2 \\ \text{Symm} & & 4 & -1 \\ & & & 4 \end{bmatrix} \qquad (9.92)$$

is the element stiffness matrix. Moreover, introducing Eqs. (9.88–89) into Eq. (9.70), we obtain Eq. (9.74) in which

$$M_e = m \int_{A_e} \mathbf{L}\mathbf{L}^T \, dA_e$$

$$= \frac{mh^2}{64} \int_{-1}^{1} \int_{-1}^{1} \begin{bmatrix} (1-\xi_1)(1-\xi_2) \\ (1+\xi_1)(1-\xi_2) \\ (1+\xi_1)(1+\xi_2) \\ (1-\xi_1)(1+\xi_2) \end{bmatrix} \begin{bmatrix} (1-\xi_1)(1-\xi_2) \\ (1+\xi_1)(1-\xi_2) \\ (1+\xi_1)(1+\xi_2) \\ (1-\xi_1)(1+\xi_2) \end{bmatrix}^T d\xi_1 \, d\xi_2$$

$$= \frac{mh^2}{36} \begin{bmatrix} 4 & 2 & 1 & 2 \\ & 4 & 2 & 1 \\ & & 4 & 2 \\ \text{Symm} & & & 4 \end{bmatrix} \tag{9.93}$$

is the element mass matrix.

For the assembling process, we shall refer to the numbering scheme of Fig. 9.19 and list the element data, as shown in Table III. This permits us to derive the global stiffness matrix

$$K = \frac{T}{6} \begin{bmatrix} 4 & -1 & 0 & -1 & -2 \\ & 8 & -1 & -2 & -2 & -2 \\ & & 4 & 0 & -2 & -1 & 0 \\ & & & 8 & -2 & 0 & -1 & -2 \\ & & & & 16 & -2 & -2 & -2 & -2 \\ & & & & & 8 & 0 & -2 & -1 & 0 \\ & & \text{Symm} & & & & 8 & -2 & 0 & -1 & -2 \\ & & & & & & & 16 & -2 & -2 & -2 & -2 \\ & & & & & & & & 8 & 0 & -2 & -1 \\ & & & & & & & & & 4 & -1 & 0 \\ & & & & & & & & & & 8 & -1 \\ & & & & & & & & & & & 4 \end{bmatrix} \tag{9.94}$$

and we observe once again that the stiffness matrix is singular. Moreover,

Table III			
Element Data			
Element Number	Node Numbers		
1	1	4	5	2
2	2	5	6	3
3	4	7	8	5
4	5	8	9	6
5	7	10	11	8
6	8	11	12	9

the global mass matrix can be shown to be

$$M = \frac{mh^2}{36} \begin{bmatrix} 4 & 2 & 0 & 2 & 1 & & & & & & & \\ & 8 & 2 & 1 & 4 & 1 & & & & & & \\ & & 4 & 0 & 1 & 2 & 0 & & & & & \\ & & & 8 & 4 & 0 & 2 & 1 & & & & \\ & & & & 16 & 4 & 1 & 4 & 1 & & & \\ & & & & & 8 & 0 & 1 & 2 & 0 & & \\ & & \text{Symm} & & & & 8 & 4 & 0 & 2 & 1 & \\ & & & & & & & 16 & 4 & 1 & 4 & 1 \\ & & & & & & & & 8 & 0 & 1 & 2 \\ & & & & & & & & & 4 & 2 & 0 \\ & & & & & & & & & & 8 & 2 \\ & & & & & & & & & & & 4 \end{bmatrix} \tag{9.95}$$

Comparing the stiffness matrices obtained by using triangular and rectangular elements, we observe that the use of triangular elements resulted in a stiffness matrix with half-bandwidth equal to three, as opposed to the stiffness matrix obtained by using rectangular elements which was half-bandwidth equal to four. On the other hand, the latter matrix was easier to obtain. The mass matrices in both cases have half-bandwidth equal to four, and the work in deriving the matrices is about the same.

Higher-degree rectangular elements can be found in the text by Huebner (H5, p. 171).

Finally, one must mention elements with curved sides. Mathematically, one can use a coordinate transformation to map a curved boundary into a straight one. This can be done by means of the parametrical transformation $x = x(\xi_1, \xi_2)$, $y = y(\xi_1, \xi_2)$, where ξ_1 and ξ_2 represent curvilinear coordinates. One may describe the coordinate transformation by the same class of polynomials that are used as interpolation functions. Elements thus described are known as *isoparametric elements*. A discussion of isoparametric elements can be found in the text by Gallagher (Gl, Sec. 8.8).

9.6 Errors in the eigenvalues and eigenfunctions

We have expressed frequently special interest in the solution of the eigenvalue problem

$$Lw = \lambda mw, \qquad 0 < x < L \tag{9.96a}$$

$$B_i w = 0, \qquad i = 1, 2, \ldots, p, \qquad x = 0, L \tag{9.96b}$$

where L is a linear homogeneous self-adjoint differential operator of order $2p$. Through certain integrations by parts, the solution to the differential problem (9.96) has been reduced to the problem of rendering stationary the Rayleigh quotient

$$R(w) = \frac{[w, w]}{(\sqrt{m}w, \sqrt{m}w)} \tag{9.97}$$

Rayleigh's quotient has stationary values at the eigenfunctions w_i ($i = 1, 2, \ldots$), at which points Rayleigh's quotient is equal to the eigenvalues λ_i ($i = 1, 2, \ldots, n$), or

$$\lambda_i = R(w_i) = \frac{[w_i, w_i]}{(\sqrt{m}w_i, \sqrt{m}w_i)}, \qquad i = 1, 2, \ldots \tag{9.98}$$

Throughout this chapter, we have been concerned with finite element solutions of the above eigenvalue problem. Such solutions have been denoted by w^n, where w^n is a linear combination of admissible functions ϕ_i ($i = 1, 2, \ldots, n$) from the subspace S^n of the energy space κ_G^p. The approximate eigenvalues and eigenfunctions have been denoted by Λ_i^n and w_i^n, respectively.

Using the maximum minimum principle, it is possible to ascertain in a qualitative way how the approximate eigenvalues Λ_i^n relate to the actual eigenvalues λ_i. It turns out that more quantitative statements are also possible, and they concern not only eigenvalues but also eigenfunctions. Indeed, it can be shown that, for large n, the errors in the eigenvalues are given by

$$\Lambda_i^n - \lambda_i \simeq cn^{-2(k-p)}\lambda_i^{k/p}, \qquad i = 1, 2, \ldots, n \tag{9.99}$$

where c is a constant and $k - 1$ is the degree of the elements. Moreover, for large n, the errors in the eigenfunctions satisfy the inequalities

$$\|\sqrt{m}w_i^n - \sqrt{m}w_i\| \le c[n^{-k} + n^{-2(k-p)}]\lambda_i^{k/2p}, \qquad i = 1, 2, \ldots, n \qquad (9.100)$$

Proofs of (9.99–100) lie beyond the scope of this text. They can be found in the text by Strang and Fix (S2, Sec. 6.3).

There are several observations that can be made in connection with (9.99–100). The most obvious one is that errors can be reduced by increasing n. In fact, the errors approach zero as $n \to \infty$. Another observation is that errors increase with the mode number, so that at least one half of the approximate eigenvalues are unreliable (S2, p. 227). This provides a strong clue concerning truncation decisions. Indeed, the dimension of the space S^n should be at least twice as large as the number of accurate eigenvalues desired. Finally, using higher degree elements does not always help. It helps only for sufficiently large n.

9.7   Inconsistent mass matrices

Using the notation of Sec. 9.5, we can write the displacement of any point in a given element e in the form

$$w = \mathbf{L}^T\mathbf{w}_e \qquad (9.101)$$

where $\mathbf{L}$ is a vector of admissible functions and $\mathbf{w}_e$ is a vector of nodal displacements. In the finite element method, the admissible functions have the form of polynomials, known as interpolation functions. The mass weighted inner product has the expression

$$(\sqrt{m}w, \sqrt{m}w) = \sum_{e=1}^{n} \int_{A_e} mw^2 \, dA_e \qquad (9.102)$$

and can be written as

$$(\sqrt{m}w, \sqrt{m}w) = \sum_{e=1}^{n} \mathbf{w}_e^T M_e \mathbf{w}_e \qquad (9.103)$$

where

$$M_e = \int_{A_e} m\mathbf{L}\mathbf{L}^T \, dA_e \qquad (9.104)$$

is the element mass matrix. Similarly, one can write an expression for the element stiffness matrix K_e, based on the energy inner produce $[w, w]$. Then, the global mass and stiffness matrices M and K are obtained by assembling the element mass and stiffness matrices M_e and K_e, respectively. The discrete system eigenvalue problem has the usual form

$$Kw = \lambda^n Mw \qquad (9.105)$$

where w is the n-dimensional vector of the system nodal displacements.

It will prove of interest to examine the eigenvalue problem (9.105), and in particular the mass matrix, a little closer. Generally, the global mass matrix M is banded but not diagonal. As a result, the eigenvalue problem (9.105) is more difficult to solve than the one in which M is diagonal. For a positive definite matrix M, we would ordinarily perform a Cholesky decomposition (see Sec. 5.8) by writing

$$M = LL^T \qquad (9.106)$$

where L is a lower triangular matrix, and rewrite the eigenvalue problem (9.105) in the form

$$Ax = \lambda^n x \qquad (9.107)$$

where

$$A = L^{-1} K L^{-T} \qquad (9.108)$$

is a real symmetric matrix, in which $L^{-T} = (L^{-1})^T$, and

$$x = L^T w \qquad (9.109)$$

One disadvantage of the above procedure is that the matrix A is unlikely to be as sparse as the matrices K and M, so that the computational efficiency will suffer. For this and other reasons, it is not uncommon in the finite element practice to derive the global mass matrix M not by using Eq. (9.104) but by simply lumping the mass at the nodes, in a procedure similar to that described in Sec. 8.7. This procedure, of course, has the advantage that it yields a diagonal mass matrix, so that the matrix L for the Cholesky decomposition reduces to the simple form

$$L = M^{1/2} = \text{diag}\,[m_{ii}^{1/2}] \qquad (9.110)$$

Moreover, the bandedness of K is not affected. On the other hand, the practice of using a lumped mass matrix instead of a mass matrix derived by means of Eq. (9.104) represents a violation of the rules of the Rayleigh–Ritz method, resulting in an inconsistent formulation. For this reason, in the finite element terminology the lumped matrix is commonly referred to as an *inconsistent mass matrix*. Extending the terminology, one can refer to the mass matrix based on Eq. (9.104) as a *consistent mass matrix*, a term somewhat out of place because it would have no meaning if the Rayleigh–Ritz procedure were not violated by the introduction of the lumped mass matrix.

Notwithstanding the above argument, the lumped mass matrix is a fact of life, and it might be interesting to examine the effect on the system eigenvalues of the use of such an inconsistent mass matrix. The use of a lumped matrix tends to increase the value of the denominator of Rayleigh's quotient when compared with the one using a consistent mass matrix. As a result, the eigenvalues thus obtained tend to be lower than those obtained by a consistent formulation of the problem. Because the general idea of the Rayleigh–Ritz method is to lower the estimates of the eigenvalues, at first sight lumping appears attractive. This turns out not to be the case, however, because in an inconsistent formulation one can lose control of the computational accuracy, and in fact one can increase the error significantly (S2, p. 228). Moreover, with the current state of the art, the problem involved in working with a banded matrix M instead of a diagonal one is not very serious, provided the eigenvalue problem can be handled within the core of the computer. Problems requiring out-of-core solutions will be discussed in Ch. 10.

Lumping can create a problem of a different kind if some of the nodal coordinates are rotational coordinates, such as in bending vibration. Then, unless the lumped masses are assumed to possess rotary inertia, the mass matrix is likely to have zero diagonal entries corresponding to the rotational nodal displacements. In this case, it is advised to eliminate the rotational coordinates from the problem formulation in a process known as *static condensation*. This subject is also treated in Ch. 10.

10

Systems with a large number of degrees of freedom

10.1 Introduction

The advent of the digital computer has stimulated the development of increasingly powerful methods of analysis, such as the finite element method. But, application of the finite element method generally results in high-order stiffness and mass matrices, which in turn can place severe strains on various computational algorithms. In fact, the order may be so high (sometimes reaching into the tens of thousands) that rapid loss of accuracy and excessively large computer time render many computational algorithms unsatisfactory. Even ordinary operations such as matrix inversion or the evaluation of the determinant of a matrix can become a problem for high-order matrices. In fact, quite often in engineering practice matrix inversion is replaced by the solution of simultaneous algebraic equations (Sec. 5.2). In view of this, it is perhaps appropriate to distinguish between applied mathematics and 'applicable' mathematics, where the latter may be narrowly defined as the subject of applied mathematics that has survived the curse of high dimensionality. In applicable mathematics, the number of elementary operations must be kept to a minimum, which saves computer time and tends to reduce round-off errors.

In this chapter, we discuss two distinct approaches to the problem of high dimensionality. The first consists of reducing the order of the system by eliminating unnecessary or unwanted variables, a process known as *condensation*. The second approach consists of seeking a partial solution to the problem, such as a limited number of lower eigenvalues and eigenvectors, instead of the complete solution. In an ironic twist, the matrix iteration based on the power method, which has the drawback of losing accuracy with each additional mode computed, becomes once again respectable just because of its ability to produce a partial solution. In its modern version, however, one iterates to several modes simultaneously, in a process known as *subspace iteration*. Another method discussed in this chapter combines the ideas of subspace iteration with those of matrix iteration with shifts. The method is referred to here as the *method of sectioning.*

10.2   Static condensation

In Sec. 9.7, we have pointed out that it is common practice in the finite element method to use an inconsistent mass matrix in the form of a diagonal matrix of lumped masses. In problems in which the nodal displacements include rotations, such as in bending vibration, the associated entries in the mass matrix would have to be moments of inertia. But lumped mass matrices generally imply point masses and seldom include moments of inertia, unless there is a good reason to believe that rotary inertia effects are important. Hence, the diagonal entries associated with rotational displacements are generally zero. The fact that certain diagonal entries in the lumped mass matrix are zero is an indication that the corresponding displacements are not vital to the solution and can be eliminated from the problem formulation. The elimination process is known as *static condensation* and its net result is to reduce the order of the eigenvalue problem. It is very important in high-order problems, where it may take the difference between an in-core and an out-of-core solution. This difference is not guaranteed, however, because, although static condensation does reduce the order of the eigenvalue problem, it generally increases the bandwidth of the stiffness matrix and it can force an out-of-core solution once again. It is worth pointing out that static condensation involves no approximations and the condensation calculations are well amortized when several eigenvalues are being sought.

Let us consider the eigenvalue problem

$$K\mathbf{u} = \lambda M\mathbf{u} \tag{10.1}$$

where K is the symmetric stiffness matrix, M is the diagonal lumped mass matrix and $\mathbf{u}$ is the nodal displacement vector. Then, assuming that there are certain zero diagonal entries, we can rearrange M so that all the zero diagonal entries form a null submatrix. This enables us to partition the eigenvalue problem (10.1) as follows:

$$\begin{bmatrix} K_{11} & K_{12} \\ \hline K_{21} & K_{22} \end{bmatrix} \begin{bmatrix} \mathbf{u}_1 \\ \hline \mathbf{u}_2 \end{bmatrix} = \lambda \begin{bmatrix} M_{11} & 0 \\ \hline 0 & 0 \end{bmatrix} \begin{bmatrix} \mathbf{u}_1 \\ \hline \mathbf{u}_2 \end{bmatrix} \tag{10.2}$$

where the notation is obvious. Equation (10.2) can be separated into the two equations

$$K_{11}\mathbf{u}_1 + K_{12}\mathbf{u}_2 = \lambda M_{11}\mathbf{u}_1 \tag{10.3a}$$

$$K_{21}\mathbf{u}_1 + K_{22}\mathbf{u}_2 = \mathbf{0} \tag{10.3b}$$

369

Solving Eq. (10.3b) for $\mathbf{u}_2$, we have

$$\mathbf{u}_2 = -K_{22}^{-1}K_{21}\mathbf{u}_1 \tag{10.4}$$

so that, introducing Eq. (10.4) into Eq. (10.3a), we obtain the *condensed eigenvalue problem*

$$K_1\mathbf{u}_1 = \lambda M_1\mathbf{u}_1 \tag{10.5}$$

where

$$K_1 = K_{11} - K_{12}K_{22}^{-1}K_{21}, \qquad M_1 = M_{11} \tag{10.6}$$

are symmetric matrices of corresponding lower order.

The solution of Eq. (10.5) will yield a set of eigenvalues λ_i and associated eigenvectors $\mathbf{u}_{1i}$. Then, the remaining components of the nodal displacement vector $\mathbf{u}$, namely the components of $\mathbf{u}_2$, can be recovered by using Eq. (10.4).

10.3   Mass condensation

Quite often, the inertia forces associated with some of the displacements are known to be much smaller than those associated with others, so that the importance of the first class of displacements in an overall solution can be regarded as being relatively slight. Yet, in a computer solution of the eigenvalue problem, all displacements play equal part. Hence, the question arises as to whether it is not possible to eliminate from the problem formulation displacements of lesser importance, without affecting the results too greatly. In effect, this amounts to sacrificing the solution accuracy for the sake of a reduction in the computation effort. This is the essence of the *mass condensation* technique, sometimes referred to as an *eigenvalue economizer*. In the case of a very high-order system, if sufficient degrees of freedom are eliminated, then it may be possible to solve the problem in the computer core instead of out of core.

An early proponent of the technique, Irons (I1, I2) refers to the significant displacements as 'master' displacements and to the insignificant ones as 'slave' displacements. As displacements that tend to be slaves, he mentions the lengthwise displacements in a cantilever-like structures, displacements near a fixed point and rotational displacements, the latter on the basis that they are seldom used in plotting modal vectors. Note that rotational displacements are the ones eliminated in the static condensation of Sec. 10.2. Irons assumes that the potential energy has a minimum with respect to slave displacements, which is the same as saying that there are no applied forces associated with the slave displacements. By implication, inertia forces are included in the applied forces. A similar argument is advanced also by Przemiéniecki (P2), Guyan (G5) and Ramsden and Stoker (R2). Of course, the argument is only approximately valid.

Let us write the potential and kinetic energy for the system in the matrix form

$$V = \tfrac{1}{2}\mathbf{q}^T K \mathbf{q} \tag{10.7a}$$

$$T = \tfrac{1}{2}\dot{\mathbf{q}}^T M \dot{\mathbf{q}} \tag{10.7b}$$

and divide the displacement vectors $\mathbf{q}$ into the master displacement vector $\mathbf{q}_1$ and slave displacement vector $\mathbf{q}_2$, or

$$\mathbf{q} = [\mathbf{q}_1^T \,\vdots\, \mathbf{q}_2^T]^T \tag{10.8}$$

Then, the stiffness matrix K and mass matrix M can be partitioned accordingly, with the result

$$V = \frac{1}{2}\begin{bmatrix}\mathbf{q}_1 \\ \hline \mathbf{q}_2\end{bmatrix}^T \begin{bmatrix}K_{11} & \vdots & K_{12} \\ \hline K_{21} & \vdots & K_{22}\end{bmatrix}\begin{bmatrix}\mathbf{q}_1 \\ \hline \mathbf{q}_2\end{bmatrix} \tag{10.9a}$$

$$T = \frac{1}{2}\begin{bmatrix}\dot{\mathbf{q}}_1 \\ \hline \dot{\mathbf{q}}_2\end{bmatrix}^T \begin{bmatrix}M_{11} & \vdots & M_{12} \\ \hline M_{21} & \vdots & M_{22}\end{bmatrix}\begin{bmatrix}\dot{\mathbf{q}}_1 \\ \hline \dot{\mathbf{q}}_2\end{bmatrix} \tag{10.9b}$$

where $K_{21} = K_{12}^T$ and $M_{21} = M_{12}^T$. The condition that there be no applied forces in the direction of the slave displacements can be written symbolically in the form

$$\frac{\partial V}{\partial \mathbf{q}_2} = K_{21}\mathbf{q}_1 + K_{22}\mathbf{q}_2 = \mathbf{0} \tag{10.10}$$

where the equation implies static equilibrium in the direction of the slave displacements. Solving Eq. (10.10) for $\mathbf{q}_2$, we obtain

$$\mathbf{q}_2 = -K_{22}^{-1}K_{21}\mathbf{q}_1 \tag{10.11}$$

which can be used to eliminate $\mathbf{q}_2$ from the problem formulation.

Equation (10.11) can be regarded as a constraint equation, so that the complete displacement vector $\mathbf{q}$ can be expressed in terms of the master displacement vector $\mathbf{q}_1$ in the form

$$\mathbf{q} = C\mathbf{q}_1 \tag{10.12}$$

where C is a rectangular constraint matrix having the form

$$C = \begin{bmatrix}I \\ \hline -K_{22}^{-1}K_{21}\end{bmatrix} \tag{10.13}$$

10 Systems with a large number of degrees of freedom

in which I is a unit matrix of the same order as the dimension of $\mathbf{q}_1$. Introducing Eq. (10.12) into Eq. (10.7), we obtain

$$V = \tfrac{1}{2}\mathbf{q}_1^T K_1 \mathbf{q}_1 \tag{10.14a}$$

$$T = \tfrac{1}{2}\dot{\mathbf{q}}_1^T M_1 \dot{\mathbf{q}}_1 \tag{10.14b}$$

where the reduced stiffness and mass matrices are simply

$$K_1 = C^T K C = K_{11} - K_{12} K_{22}^{-1} K_{21} \tag{10.15a}$$

$$M_1 = C^T M C = M_{11} - K_{21}^T K_{22}^{-1} M_{21} - M_{12} K_{22}^{-1} K_{21} + K_{21}^T K_{22}^{-1} M_{22} K_{22}^{-1} K_{21} \tag{10.15b}$$

The matrix M_1 is generally known as the *condensed mass matrix*.

Comparing Eq. (10.11) with Eq. (10.4), we conclude that the mass condensation technique is equivalent to a static condensation technique. But, whereas static condensation involves no approximation, mass condensation does involve some approximation. The question arises as to what is being sacrificed as a result of the condensation process. To answer this question, let us consider the complete eigenvalue problem, which can be separated into

$$K_{11}\mathbf{q}_1 + K_{12}\mathbf{q}_2 = \lambda(M_{11}\mathbf{q}_1 + M_{12}\mathbf{q}_2) \tag{10.16a}$$

$$K_{21}\mathbf{q}_1 + K_{22}\mathbf{q}_2 = \lambda(M_{21}\mathbf{q}_1 + M_{22}\mathbf{q}_2) \tag{10.16b}$$

Solving Eq. (10.16b) for $\mathbf{q}_1$, we have

$$\mathbf{q}_2 = (K_{22} - \lambda M_{22})^{-1}(\lambda M_{21} - K_{21})\mathbf{q}_1 \tag{10.17}$$

so that, introducing Eq. (10.17) into Eq. (10.16a), we obtain

$$(K_{11} - K_{12}K_{22}^{-1}K_{21})\mathbf{q}_1 = \lambda(M_{11} - K_{12}K_{22}^{-1}M_{21} - M_{12}K_{22}^{-1}K_{21}$$
$$+ K_{12}K_{22}^{-1}M_{22}K_{22}^{-1}K_{21})\mathbf{q}_1 + \lambda^2(M_{12}K_{22}^{-1}M_{21} - M_{12}K_{22}^{-1}M_{22}K_{22}^{-1}K_{21}$$
$$- K_{12}K_{22}^{-1}M_{22}K_{22}^{-1}M_{21} + K_{12}K_{22}^{-1}M_{22}K_{22}^{-1}M_{22}K_{22}^{-1}K_{21})\mathbf{q}_1 + 0(\lambda^3) \tag{10.18}$$

Examining Eqs. (10.15) and (10.18), we conclude that the condensation used earlier in this section is tantamount to ignoring second- and higher-order terms in λ in Eq. (10.18), which can be justified if the coefficients of $\lambda^2, \lambda^3, \ldots$ are significantly smaller than the coefficient of λ. For this to be true, we must have the entries of M_{12} and M_{22} much smaller than the entries of K_{12} and K_{22}. Physically, this implies that *the slave displacements should be chosen from areas of high stiffness and low mass*. The effect on the system eigenvalues of retaining the term in λ^2 in the eigenvalue problem has been studied by Wright and Miles (W3). Of course, retention of this term is not advocated, as calculation with the term in λ^2 included is only an approximation to the original problem and yet it doubles the order of the condensed eigenvalue problem. Indeed, if $\mathbf{q}_1$ and $\mathbf{q}_2$ have the same dimension, the order of the condensed eigenvalue problem is the same as that of the original problem.

10.4   Simultaneous iteration

In Ch. 5, we have presented a large variety of computational algorithms for the solution of the eigenvalue problem for real symmetric matrices. The emphasis throughout Ch. 5 has been on a complete solution of the problem. However, in eigenvalue problems of high order, such as those resulting from the use of the finite element method, a complete solution of the problem may not be feasible. Moreover, because higher eigen-values and eigenvectors tend to be unreliable, one is well advised to seek only a limited number of lower eigenvalues and eigenvectors. In this regard, the matrix iteration by the power method comes immediately to mind, as the method permits iteration to one eigenvalue and eigenvector at a time, beginning with the lowest one and progressing to increasingly higher ones. The *simultaneous iteration method* is essentially a power method, but, as the name indicates, iteration is carried out to a given number of eigenvalues and eigenvectors simultaneously. The method was developed by Jennings (J1) and placed on a sounder mathematical foundation by Clint and Jennings (C2).

Let us consider the eigenvalue problem

$$A\mathbf{x}_i = \lambda_i\mathbf{x}_i, \qquad i = 1, 2, \ldots, n \tag{10.19}$$

where A is an $n \times n$ real symmetric matrix and x_i $(i = 1, 2, \ldots, n)$ is a real n-vector. The eigenvectors $\mathbf{x}_i$ are assumed to be normalized, so that they satisfy the orthonormality relation

$$\mathbf{x}_j^T\mathbf{x}_i = \delta_{ij}, \qquad i, j = 1, 2, \ldots, n \tag{10.20}$$

Next, let us consider m n-dimensional trial vectors $\mathbf{u}_1, \mathbf{u}_2, \ldots, \mathbf{u}_m$, $m < n$, and arrange them in the $n \times m$ matrix

$$U = [\mathbf{u}_1 \quad \mathbf{u}_2 \cdots \mathbf{u}_m] \tag{10.21}$$

Then, premultiplying U by A, we obtain

$$V = AU \tag{10.22}$$

where

$$V = [\mathbf{v}_1 \quad \mathbf{v}_2 \cdots \mathbf{v}_m] \tag{10.23}$$

373

is another $n \times m$ matrix. Clearly, the vectors $\mathbf{v}_1, \mathbf{v}_2, \ldots, \mathbf{v}_m$ are also n-dimensional.

The question arises as to the relation between the vectors $\mathbf{u}_1, \mathbf{u}_2, \ldots, \mathbf{u}_m$ and $\mathbf{v}_1, \mathbf{v}_2, \ldots, \mathbf{v}_m$. To answer this question, let us consider the $m \times m$ matrix

$$B = U^T V = U^T A U \qquad (10.24)$$

Although the matrix B is real and symmetric, it is generally not a diagonal matrix. A notable exception is the case in which the vectors $\mathbf{u}_1, \mathbf{u}_2, \ldots, \mathbf{u}_m$ are eigenvectors of A, in which case the vectors $\mathbf{v}_1, \mathbf{v}_2, \ldots, \mathbf{v}_m$ are proportional to the vectors $\mathbf{u}_1, \mathbf{u}_2, \ldots, \mathbf{u}_m$. Then, if the vectors $\mathbf{u}_1, \mathbf{u}_2, \ldots, \mathbf{u}_m$ are normalized so as to satisfy equations of the type (10.20), the matrix B is diagonal with its nonzero elements equal to the eigenvalues $\lambda_1, \lambda_2, \ldots, \lambda_m$, respectively. The matrix B can be diagonal also when the vectors $\mathbf{u}_1, \mathbf{u}_2, \ldots, \mathbf{u}_m$ are linear combinations of all the eigenvectors $\mathbf{x}_i$ ($i = 1, 2, \ldots, n$) so that the fact that B is diagonal is only a necessary but not a sufficient condition for the trial vectors $\mathbf{u}_1, \mathbf{u}_2, \ldots, \mathbf{u}_m$ to be eigenvectors.

For any arbitrary trial vectors $\mathbf{u}_1, \mathbf{u}_2, \ldots, \mathbf{u}_m$, the diagonal elements of B provide a measure of how far the trial vectors are from being orthogonal. For this reason, Clint and Jennings refer to the matrix B as an 'interaction matrix'.

The simultaneous iteration method can be described by the steps

$$U_{k+1} = A U_k \qquad k = 1, 2, \ldots \qquad (10.25)$$

However, such an iteration will cause the columns of the matrix U_{k+1} to tend to resemble closer and closer the first eigenvector, so that convergence to the higher eigenvectors is not possible. The problem can be circumvented by forcing the trial vectors to be orthogonal. Hence, orthogonalization will be performed in every iteration step. To this end, we can use the Gram-Schmidt orthogonalization process (see Sec. 1.6). Actually, we shall render the trial vectors orthonormal, a process that can be written in the symbolic form

$$\hat{U}_k = U_k L_k^T \qquad (10.26)$$

where $\hat{U}_k$ is a matrix of orthonormal trial vectors and L_k is a lower triangular matrix. Hence, the iteration process should really be expressed in the form

$$U_{k+1} = A \hat{U}_k, \qquad k = 1, 2, \ldots \qquad (10.27)$$

The iteration converges with the result

$$\lim_{k\to\infty} \hat{U}_{k+1} = \hat{X}^m, \qquad \lim_{k\to\infty} L_k^T = \Lambda^m \tag{10.28}$$

where $\hat{X}^m$ is an $n \times m$ matrix of the first m eigenvectors $\mathbf{x}_i$ and Λ^m is a diagonal matrix of the first m eigenvalues λ_i $(i = 1, 2, \ldots, m)$. Convergence characteristics are similar to those of the power method.

In the case in which the eigenvalue problem is in terms of two real symmetric matrices, or

$$K\mathbf{x}_i = \lambda_i M\mathbf{x}_i, \qquad i = 1, 2, \ldots, n \tag{10.29}$$

The iteration process has the form

$$KU_{k+1} = M\hat{U}_k, \qquad k = 1, 2, \ldots \tag{10.30}$$

Equation (10.30) represents m sets of n simultaneous algebraic equations, which can be solved by Gaussian elimination (see Sec. 5.2). Note that in this case $\hat{U}_{k+1}$ is obtained by orthonormalizing U_{k+1} with respect to M.

Equation (10.30) can be derived directly from the integral formulation of Sec. 7.10. Indeed, discretization of Eq. (7.215) yields

$$\mathbf{w}_{k+1} = \lambda AM\mathbf{w}_k, \qquad k = 1, 2, \ldots \tag{10.31}$$

where A is the flexibility matrix, M is the mass matrix and $\mathbf{w}$ is the displacement vector. Note that λ is a scalar resulting from the normalization of the vector $\mathbf{w}_{k+1}$. Simultaneous iteration simply implies the replacement of the vector $\mathbf{w}_k$ by a rectangular matrix. Now, if we recall that the flexibility matrix A is the inverse of the stiffness matrix K, then Eq. (10.31) leads directly to Eq. (10.30).

Example 10.1

Consider the eigenvalue problem of Example 5.7 and obtain the two lowest modes by simultaneous iteration.

The eigenvalue problem can be written in the form

$$KX = MX\Lambda \tag{a}$$

where

$$K = \begin{bmatrix} 2 & -1 & 0 \\ -1 & 3 & -2 \\ 0 & -2 & 2 \end{bmatrix}, \qquad M = \begin{bmatrix} 1 & 0 & 0 \\ 0 & 1 & 0 \\ 0 & 0 & 2 \end{bmatrix} \tag{b}$$

Let us use the initial trial matrix

$$U_1 = \begin{bmatrix} 1/\sqrt{23} & 1/\sqrt{7} \\ 2/\sqrt{23} & 2/\sqrt{7} \\ 3/\sqrt{23} & -1/\sqrt{7} \end{bmatrix} \tag{c}$$

Note that, whereas the columns of U_1 have been normalized so as to satisfy $\mathbf{u}_{1i}^T M \mathbf{u}_{1i} = 1$ ($i = 1, 2$), the vectors themselves are not orthogonal with respect to M. Introducing Eq. (c) into Eq. (10.30), we obtain

$$U_2 = \begin{bmatrix} 1.876626 & 0.377965 \\ 3.544738 & 0.377965 \\ 4.170280 & 0 \end{bmatrix} \tag{d}$$

Upon using the Gram-Schmidt orthonormalization process described in Sec. 1.6, we have

$$\hat{U}_2 = \begin{bmatrix} 0.263117 & 0.670821 \\ 0.497000 & 0.521749 \\ 0.584705 & -0.372677 \end{bmatrix}, \quad L_2 = \begin{bmatrix} 0.140207 & 0 \\ -0.089366 & 2.218531 \end{bmatrix} \tag{e}$$

Note that the orthonormalization process had to be modified to account for the fact that the vectors $\mathbf{u}_{21}$ and $\mathbf{u}_{22}$ must be orthonormal with respect to the matrix M and not in an ordinary sense.

The second and third iteration cycles yield

$$\hat{U}_3 = \begin{bmatrix} 0.268588 & 0.799851 \\ 0.500550 & 0.360189 \\ 0.581941 & -0.339487 \end{bmatrix}, \quad L_3 = \begin{bmatrix} 0.139139 & 0 \\ -0.015103 & 1.853671 \end{bmatrix} \tag{f}$$

and

$$\hat{U}_4 = \begin{bmatrix} 0.269065 & 0.847544 \\ 0.500744 & 0.282633 \\ 0.581733 & -0.317639 \end{bmatrix}, \quad L_4 = \begin{bmatrix} 0.139194 & 0 \\ -0.001111 & 1.766267 \end{bmatrix} \tag{g}$$

at which point convergence to the first eigenvalue has been achieved, as can be concluded from Example 5.7. The first column of $\hat{U}_4$ is also very close to the first eigenvector.

The eighth iteration cycle results in

$$\hat{U}_9 = \begin{bmatrix} 0.269108 & 0.877790 \\ 0.500758 & 0.223975 \\ 0.581731 & -0.299433 \end{bmatrix}, \qquad L_9 = \begin{bmatrix} 0.139194 & 0 \\ 0 & 1.745900 \end{bmatrix}$$

(h)

at which point we conclude that $\hat{U}_9$ and L_9 are reasonably close to $\hat{X}^2$ and Λ^2, respectively.

10.5   Subspace iteration

The simultaneous iteration described in Sec. 10.4 requires the orthogonalization of the trial vectors after every iteration step. To this end, we have used the Gram-Schmidt process, a process which often gives very inaccurate results in the sense that the computed vectors may be far from being orthogonal. Another orthogonalization process, known as the modified Gram-Schmidt, is mathematically equivalent but computationally superior (Sec. 1.6). A method proposed by Clint and Jennings (C2) involves a different orthogonalization of the iterated vector, rather than the Gram-Schmidt process. Although Clint and Jennings refer to their method as simultaneous iteration, the method has come to be known as *subspace iteration*. We shall discuss the method in the context of the more general symmetric eigenvalue problem, Eq. (10.29), instead of the standard one, Eq. (10.19).

Let us consider the simultaneous iteration described by

$$KU_{k+1} = M\hat{U}_k, \qquad k = 1, 2, \ldots \tag{10.32}$$

where $\hat{U}_k$ is a given $n \times m$ matrix whose columns are orthonormal with respect to M. Using Gaussian elimination, Eq. (10.32) yields the $n \times m$ matrix U_{k+1}. The next step consists of orthonormalizing U_{k+1}. To this end, let us consider the reduced eigenvalue problem

$$K_{k+1}P_{k+1} = M_{k+1}P_{k+1}\Lambda_{k+1}^m \tag{10.33}$$

where

$$K_{k+1} = U_{k+1}^T K U_{k+1} \tag{10.34a}$$

$$M_{k+1} = U_{k+1}^T M U_{k+1} \tag{10.34b}$$

are $m \times m$ real symmetric matrices, where $m < n$. Moreover, P_{k+1} is the matrix of eigenvectors and Λ_{k+1}^m the diagonal matrix of eigenvalues. We shall assume that P_{k+1} is orthonormal with respect to M_{k+1}. Then, for the next iteration step, we shall use the matrix

$$\hat{U}_{k+1} = U_{k+1} P_{k+1} \tag{10.35}$$

where U_{k+1} is orthonormal with respect to M. Indeed, by virtue of the fact that P_{k+1} is orthonormal with respect to M_{k+1}, we can use Eqs. (10.34b) and (10.35) and write

$$P_{k+1}^T M_{k+1} P_{k+1} = P_{k+1}^T U_{k+1}^T M U_{k+1} P_{k+1} = \hat{U}_{k+1}^T M \hat{U}_{k+1} = I \tag{10.36}$$

so that, if P_{k+1} is orthonormal with respect to M_{k+1}, then $\hat{U}_{k+1}$ is orthonormal with respect to M.

The iteration process converges, so that

$$\lim_{k \to \infty} \hat{U}_{k+1} = \hat{X}^m, \qquad \lim_{k \to \infty} \Lambda_{k+1}^m = \Lambda^m \tag{10.37}$$

The fact that each iteration cycle requires the solution of an eigenvalue problem may create some degree of skepticism concerning subspace iteration, as we replace one eigenvalue problem by another. It should be recalled, however, that the reduced eigenvalue problem, Eq. (10.33), is generally of considerably smaller order than the original eigenvalue problem, Eq. (10.29). Moreover, as k increases, the matrices K_{k+1} and M_{k+1} tend to become diagonal, thus making the task significantly easier. In addition, the solution of the reduced eigenvalue problem need not be highly accurate, as the results are fed back into an iterative process.

Practical aspects of the subspace iteration method have been discussed by Bathe and Wilson (B3).

Example 10.2

Solve the eigenvalue problem of Example 10.1 by subspace iteration.
The first step is the same as in Example 10.1, so that

$$U_2 = \begin{bmatrix} 1.876626 & 0.377965 \\ 3.544738 & 0.377965 \\ 4.170280 & 0 \end{bmatrix} \tag{a}$$

Now, instead of using the Gram-Schmidt process to orthonormalize U_2, we use Eqs. (10.34) and form the matrices

$$K_2 = U_2^T K U_2 = \begin{bmatrix} 7.086929 & 0.236433 \\ 0.236433 & 0.428573 \end{bmatrix}$$

$$M_2 = U_2^T M U_2 = \begin{bmatrix} 50.869364 & 2.049086 \\ 2.049086 & 0.285715 \end{bmatrix} \tag{b}$$

378

and solve the eigenvalue problem

$$K_2 P_2 = M_2 P_2 \Lambda_2^2 \tag{c}$$

The solution of Eq. (c) is

$$P_2 = \begin{bmatrix} 0.139498 & -0.090469 \\ 0.017506 & 2.218458 \end{bmatrix}, \quad \Lambda_2^2 = \begin{bmatrix} 0.139196 & 0 \\ 0 & 2.072345 \end{bmatrix}, \tag{d}$$

so that, using Eq. (10.35), we have

$$\hat{U}_2 = U_2 P_2 = \begin{bmatrix} 0.268402 & 0.668723 \\ 0.591101 & 0.517813 \\ 0.581746 & -0.377280 \end{bmatrix} \tag{e}$$

Introducing Eq. (e) into Eq. (10.32), we obtain the improved matrix

$$U_3 = \begin{bmatrix} 1.932996 & 0.431976 \\ 3.597590 & 0.195229 \\ 4.179337 & -0.182051 \end{bmatrix} \tag{f}$$

so that, using Eqs. (10.34), we can write

$$K_3 = U_3^T K U_3 = \begin{bmatrix} 7.184208 & 0.000196 \\ 0.000196 & 0.525907 \end{bmatrix}$$

$$M_3 = U_3^T M U_3 = \begin{bmatrix} 51.612844 & 0.015656 \\ 0.015656 & 0.290216 \end{bmatrix} \tag{g}$$

The solution of the eigenvalue problem

$$K_3 P_3 = M_3 P_3 \Lambda_3^2 \tag{h}$$

is

$$P_3 = \begin{bmatrix} 0.139194 & -0.000035 \\ 0.000989 & 1.856265 \end{bmatrix}, \quad \Lambda_3^2 = \begin{bmatrix} 0.139194 & 0 \\ 0 & 1.812155 \end{bmatrix} \tag{i}$$

so that, using Eq. (10.35), we obtain

$$\hat{U}_3 = U_3 P_3 = \begin{bmatrix} 0.269061 & 0.801862 \\ 0.500762 & 0.362397 \\ 0.581737 & -0.337935 \end{bmatrix} \tag{j}$$

and we note that the first mode has almost converged.

The completion of the computation is left as an exercise to the reader.

As a matter of interest, let us compare the results obtained here with

those obtained in Example 10.1. A comparison of the results corresponding to the first two iteration cycles indicates faster convergence for the
subspace iteration.

10.6   The method of sectioning

One criticism of the subspace iteration method is that its reliability
depends on the initially chosen matrix U_1. If, for example, one of the
desired eigenvectors, say x_i, is orthogonal to the space spanned by the
initial trial vectors $u_{11}, u_{12}, \ldots, u_{1m}$, then its existence could never be
known. Whereas this is not very likely and, moreover, in a computer
solution x_i will always creep into the space of the iterated vectors,
convergence can be very slow. Hence, it appears desirable to examine
some methods that are not sensititive to the choice of initial trial vectors.
One such method is the method based on Sturm's theorem in conjunction
with bisection presented in Sec. 5.12. The method is most efficient as an
in-core solution of small-banded systems of relatively low order. Essentially, the method seeks eigenvalues in a given interval. Then, the
eigenvectors are obtained by some other means, such as inverse iteration.
    Another method that looks for eigenvalues in a given interval is the
method of sectioning (J2). The method represents a combination of
several ideas, including iteration with shifts (see Sec. 3.2), eigenvalue
search in small subintervals and solution of a reduced eigenvalue problem. The method is particularly suitable for cases in which eigenvalues are
clustered.
    Let us consider the symmetric eigenvalue problem

$$K x_i = \lambda_i M x_i, \qquad i = 1, 2, \ldots, n \tag{10.38}$$

where M is positive definite. The object is to compute all the eigenvalues
λ_k and eigenvectors x_k such that $\alpha < \lambda_k < \beta$. We shall assume that there
are m such eigenvalues in the interval $\alpha < \lambda_k < \beta$ and that they are
ordered so that $\lambda_1 \leq \lambda_2 \leq \cdots \leq \lambda_m$. Then, we seek a set of m vectors
$y_1, y_2, \ldots, y_m$ which span the subspace E_n^m defined by $x_1, x_2, \ldots, x_m$.
Letting Y and X be the $n \times m$ matrices

$$Y = [y_1 \quad y_2 \cdots y_m], \qquad X = [x_1 \quad x_2 \cdots x_m] \tag{10.39}$$

then, for some $m \times m$ coefficient matrix Z, we have

$$Y = XZ \tag{10.40}$$

Assuming that Y has been found and that the matrices Y and X are orthonormal, in the sense that they satisfy

$$Y^TMY = X^TMX = I \tag{10.41}$$

it can be shown that the desired eigenvalues λ_k and the columns $\mathbf{z}_k$ of Z satisfy the reduced eigenvalue problem

$$\bar{K}\mathbf{z}_k = \lambda_k \mathbf{z}_k, \qquad k = 1, 2, \ldots, m \tag{10.42}$$

where

$$\bar{K} = Y^TKY \tag{10.43}$$

is an $m \times m$ matrix. In some respects, the procedure reminds one of the subspace iteration, but the resemblance is misleading. Indeed, the solution of the reduced eigenvalue problem (10.42) is the final step in the computation of the eigenvalues and not just some intermediate step in the iteration process. For this reason, the solution of Eq. (10.42) must be much more accurate than the solution of Eq. (10.33) in subspace iteration. The solution of Eq. (10.42) can be obtained by one of the methods discussed in Ch. 5. Hence, the main problem remaining is how to produce the vectors $\mathbf{y}_1, \mathbf{y}_2, \ldots, \mathbf{y}_m$, a problem discussed in the following.

Let us consider the matrix

$$H = K - \mu M, \qquad \alpha < \mu < \beta \tag{10.44}$$

where μ is some scalar chosen so that H is nonsingular. Then, beginning with some initial vector

$$\mathbf{y}^{(1)} = \sum_{i=1}^{n} a_i \mathbf{x}_i \tag{10.45}$$

the iteration process has the form

$$H\mathbf{y}^{(j+1)} = \mathbf{w}^{(j)}, \qquad j = 1, 2, \ldots \tag{10.46}$$

where

$$\mathbf{w}^{(j)} = M\mathbf{y}^{(j)}/\|\mathbf{y}^{(j)}\|, \qquad j = 2, 3, \ldots \tag{10.47}$$

in which $\|\mathbf{y}^{(j)}\|$ is the Euclidean norm of $\mathbf{y}^{(j)}$. Note that the solution of Eq. (10.46) can be obtained by Gaussian elimination (see Sec. 5.2).

Next, let us introduce the notation

$$\pi_j = \|\mathbf{y}^{(j)}\| \cdot \|\mathbf{y}^{(j-1)}\| \cdot \cdots \cdot \|\mathbf{y}^{(2)}\| \cdot \|\mathbf{y}^{(1)}\| \tag{10.48}$$

and

$$h = \lambda_k - \mu \tag{10.49}$$

381

where

$$|\lambda_k - \mu| = \min |\lambda_i - \mu| \neq 0, \qquad i = 1, 2, \ldots, n \tag{10.50}$$

Note that the magnitude of h is used because h can be negative. Then, it can be shown that

$$\pi_j \mathbf{w}^{(j)} = \frac{1}{h^{j-1}} \sum_{i=1}^{n} a_i \rho_i^{j-1} \mathbf{M} \mathbf{x}_i, \qquad j = 2, 3, \ldots \tag{10.51}$$

and

$$\pi_j \mathbf{y}^{(j+1)} = \frac{1}{h^j} \sum_{i=1}^{n} a_i \rho_i^j \mathbf{x}_i, \qquad j = 1, 2, \ldots \tag{10.52}$$

where

$$\rho_i = \frac{h}{\lambda_i - \mu}, \qquad i = 1, 2, \ldots, n \tag{10.53}$$

We observe that for

$$\sigma = \min (\beta - \mu, \mu - \alpha) \tag{10.54}$$

we have

$$|\rho_i| \leq \begin{cases} 1, & 1 \leq i \leq m \\ |h/\sigma| < 1, & m < i \leq n \end{cases} \tag{10.55}$$

This, together with estimates of h, permits us to find a set of vectors which span the subspace E_n^m. First, we choose μ at the center of the interval $\alpha \leq \mu \leq \beta$, or $\mu = (\alpha + \beta)/2$. Then, for a preselected number γ, $0 < \gamma < 1$, there are three possibilities:

i. The estimates of $|h|$ converge to a value greater than σ, in which case $m = 0$, so that there are no eigenvalues in the interval $\alpha \leq \mu \leq \beta$.
ii. The estimates of $|h|$ converge to a value less than $\gamma\sigma$, in which case we iterate until $\mathbf{y}^{(j)}$ is in E_n^m. Note that the iteration converges at the rate

$$|h/\sigma|^p < \gamma^p \tag{10.56}$$

iii. The estimates of $|h|$ converge to a value $\gamma\sigma < |h| < \sigma$, in which case there are no eigenvalues in the range $\mu - \gamma\sigma < \lambda < \mu + \gamma\sigma$. In this case, we investigate the two intervals $(\alpha, \mu - \gamma\sigma)$ and $(\mu + \gamma\sigma, \beta)$ separately. To this end, we use the same procedure but with different shifts μ.

It is shown by Jensen (J2) that the sequence of numbers $\|\mathbf{y}^{(j)}\|^{-1}$ converges to $|h|$.

For a given μ, the parameter γ controls the efficiency of the algorithm. It should be pointed out that convergence does not depend on γ, and the value of γ merely controls the length of the iteration cycle. Indeed, if γ is chosen as a small number and there is no eigenvalue in the neighborhood of μ, then this fact reveals itself immediately, thus avoiding slow convergence by simply stopping the iteration process and beginning a new one with a different μ. If γ is chosen as a large number, then shifting is minimized but convergence is likely to be slower, as in this case μ need not be so close to an eigenvalue.

The paper by Jensen (J2) contains a numerical example. According to Jensen, the method has been used to solve eigenvalue problems of order up to 3335, a very respectable number.

11

Substructure synthesis

11.1 General discussion

The search for new approaches to the analysis of complex structures has
led to the development of the finite element method. At about the same
time, however, other methods of analysis were being developed. One
such method, developed by Hurty (H6, H7) has come to be known as
component-mode synthesis. A similar method, developed by Gladwell
(G2), is generally referred to as *branch-mode analysis*, and in an extended form, due to Benfield and Hruda (B4), it is known as *component-
mode substitution*. In all three approaches the structure is regarded as a
collection of substructures and, in the spirit of Rayleigh–Ritz, the motion
of each substructure is represented by a linear combination of 'component modes'. The methods differ to some extent in the definition of the
modes, but in general one assumes that they are generated by solving
some form of substructure eigenvalue problem. But, as pointed out by
Meirovitch (M7) and Meirovitch and Hale (M8, H1, M10, M11), one
need use, in the spirit of Rayleigh–Ritz, only admissible functions. Although substructure modes are certainly suitable admissible functions,
they are only a relatively small subset of the much broader set of
admissible functions. To emphasize the mathematical requirement, and
play down the physical implication of the term component modes, we
shall refer to the method of regarding the structure as an assemblage of
substructures each represented by a finite set of suitable admissible
functions as *substructure synthesis*.

11.2 Component-mode synthesis

The interest lies in a dynamical formulation for a structure consisting of
an assemblage of substructures. Component-mode synthesis (H6, H7)
is a method specifically designed for this task. It derives the system
equations of motion by first deriving the equations of motion for each
individual substructure separately and then coupling the equations by

subjecting them to given constraints, where the latter reflect the interacting forces between the various substructures. Note that the term component refers to a given substructure or structural member. The motion of each substructure is described in terms of 'component modes', which are divided into three classes, namely, rigid-body modes, constraint modes and normal modes. The motion is represented by a linear combination of the component modes multiplied by time-dependent generalized coordinates. Integration over each substructure permits the derivation of the substructure Lagrange's equations. Then, the equations of motion for the entire structure are obtained in terms of the system generalized coordinates by assembling. The method is essentially a Rayleigh–Ritz type discretization procedure, combined with truncation at the substructure level. It is a way of simulating complex structures by only a limited number of degrees of freedom.

Let us consider a given substructure s and write the total displacement vector $\mathbf{u}_s(x, y, z, t)$ of an arbitrary point $P(x, y, z)$ on the substructure in the form

$$\mathbf{u}_s(x, y, z, t) = \mathbf{u}_s^R(x, y, z, t) + \mathbf{u}_s^C(x, y, z, t) + \mathbf{u}_s^N(x, y, z, t) \tag{11.1}$$

where $\mathbf{u}_s^R$ is a rigid-body displacement, $\mathbf{u}_s^C$ is a 'constraint displacement' and $\mathbf{u}_s^N$ is a displacement relative to the fixed constraints. The various displacements can be explained by referring to Fig. 11.1. Figure 11.1a shows the undisplaced, undeformed structural member, with a set of constraints indicated by arrows. The six numbered constraints 1–6 are selected as a statically determinate set, with the remaining constraints, labeled as i, j and k, considered as redundant. All the constraints are movable and they are the result of substructure s being attached to adjacent substructures that are themselves in motion. The solid line in Fig. 11.1b shows the member after it has undergone a rigid-body displacement, defined uniquely by the arbitrary displacements of the six statically determined constraints. In this displacement, point P undergoes the rigid-body displacement $\mathbf{u}_s^R$. The constraint displacement $\mathbf{u}_s^C$ represents a displacement resulting from motion of the redundant constraints relative to the rigid-body motion. This motion represents a linear combination of displacements obtained by letting each of the redundant constraints undergo an arbitrary displacement, as shown in Fig. 11.1b. The motion of the statically determinate and redundant constraints provide for the arbitrary displacement of all the movable constraints. In addition to these displacements, one must provide for the displacement of any point P on the substructure relative to the constraints. This motion is represented by $\mathbf{u}_s^N$ and is shown in Fig. 11.1c.

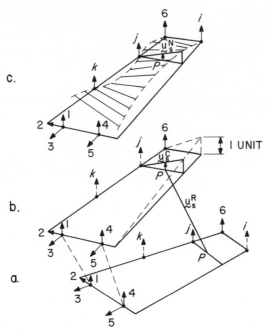

Figure 11.1

It is convenient to represent the three classes of displacements as series of space-dependent functions multiplied by time-dependent generalized coordinates, as follows:

$$\mathbf{u}_s^R = \Phi_s^R(x, y, z)\boldsymbol{\zeta}_s^R(t) \tag{11.2a}$$

$$\mathbf{u}_s^C = \Phi_s^C(x, y, z)\boldsymbol{\zeta}_s^C(t) \tag{11.2b}$$

$$\mathbf{u}_s^N = \Phi_s^N(x, y, z)\boldsymbol{\zeta}_s^N(t) \tag{11.2c}$$

where Φ_s^R, Φ_s^C and Φ_s^N are in general rectangular matrices and $\boldsymbol{\zeta}_s^R$, $\boldsymbol{\zeta}_s^C$ and $\boldsymbol{\zeta}_s^N$ are vectors. Ordinarily, Φ_s^R is a 3×6 matrix of rigid-body modes and $\boldsymbol{\zeta}_s^R$ is a six-dimensional vector of generalized coordinates associated with the rigid-body modes, unless the member s is constrained externally, in which case there may be less than six rigid-body modes. For no external constraints, the rigid-body modes can be conveniently taken as three translations and three rotations, which are kinematically equivalent to the six numbered displacements in Fig. 11.1. The matrix Φ_s^C of the so-called 'constraint modes' has three rows and as many columns as there are redundant constraints. Clearly, the generalized redundant constraint vector $\boldsymbol{\zeta}_s^C$ must have the same dimension as the number of columns of Φ_s^C.

Finally, Φ_s^N is a matrix of 'natural modes of vibration' and has three rows. In theory, there is an infinity of natural modes for a distributed elastic member, so that Φ_s^N has an infinite number of columns, from which it follows that the dimension of the natural mode vector ζ_s^N must also be infinite. In practice, only a finite number of natural modes is considered, so that ζ_s^N is finite-dimensional. Because the displacement $\mathbf{u}_s^N$ is measured relative to the constraints, the natural modes must be regarded as 'fixed-constraint' modes. It should be observed that the combined dimension of ζ_s^R and ζ_s^C represents the number of external (to the member) degrees of freedom, whereas the dimension of ζ_s^N represents the number of internal degrees of freedom. Introducing the matrix

$$\Phi_s = [\Phi_s^R \mathrel{\vdots} \Phi_s^C \mathrel{\vdots} \Phi_s^N] \tag{11.3}$$

and the vector

$$\zeta_s = [(\zeta_s^R)^T \mathrel{\vdots} (\zeta_s^C)^T \mathrel{\vdots} (\zeta_s^N)^T]^T \tag{11.4}$$

Eqs. (11.1) and (11.2) permit us to write

$$\mathbf{u}_s = \Phi_s \zeta_s \tag{11.5}$$

The kinetic energy associated with the member s has the general expression

$$T_s = \tfrac{1}{2} \int_{D_s} m_s \dot{\mathbf{u}}_s^T \dot{\mathbf{u}}_s \, dD_s \tag{11.6}$$

where m_s is the mass density and D_s is the domain of extension of the member. Introducing Eq. (11.5) into Eq. (11.6), we obtain the discretized form

$$T_s = \tfrac{1}{2} \dot{\zeta}_s^T M_s \dot{\zeta}_s \tag{11.7}$$

where

$$M_s = \int_{D_s} m_s \Phi_s^T \Phi_s \, dD_s \tag{11.8}$$

is the mass matrix for the substructure. Similarly, letting

$$\mathscr{F}_s = \tfrac{1}{2} \int_{D_s} c_s \dot{\mathbf{u}}_s^T \dot{\mathbf{u}}_s \, dD_s \tag{11.9}$$

be the Rayleigh dissipation function for the member s, in which c_s is a distributed damping coefficient, and using Eq. (11.5), we obtain

$$\mathscr{F}_s = \tfrac{1}{2} \dot{\zeta}_s^T C_s \dot{\zeta}_s \tag{11.10}$$

387

11   Substructure synthesis

where

$$C_s = \int_{D_s} c_s \Phi_s^T \Phi_s \, dD_s \tag{11.11}$$

is the member damping matrix. Moreover, using the notation of Sec. 7.4, we shall write the potential energy in the general form

$$V_s = \tfrac{1}{2}[\mathbf{u}_s, \mathbf{u}_s] \tag{11.12}$$

where $[\mathbf{u}_s, \mathbf{u}_s]$ is a three-dimensional energy integral of the type given by Eq. (7.73). Introducing Eq. (11.5) into Eq. (11.12), we can write

$$V_s = \tfrac{1}{2}\zeta_s^T K_s \zeta_s \tag{11.13}$$

where

$$K_s = [\Phi_s, \Phi_s] \tag{11.14}$$

is the substructure stiffness matrix. Finally, the virtual work for the member can be written as

$$\delta W_s = \int_{D_s} \mathbf{f}_s^T \, \delta \mathbf{u}_s \, dD_s \tag{11.15}$$

in which $\mathbf{f}_s$ is a distributed force vector, including external and constraint forces. Using Eq. (11.5), we obtain

$$\delta W_s = \mathbf{P}_s^T \, \delta \zeta_s \tag{11.16}$$

where

$$\mathbf{P}_s = \int_{D_s} \Phi_s^T \mathbf{f}_s \, dD_s \tag{11.17}$$

is a generalized force vector for the member, excluding damping forces.

The equations of motion for the substructure can be obtained by means of Lagrange's equations, which can be written in the symbolic form

$$\frac{d}{dt}\left(\frac{\partial T_s}{\partial \dot{\zeta}_s}\right) - \frac{\partial T_s}{\partial \zeta_s} + \frac{\partial \mathcal{F}_s}{\partial \dot{\zeta}_s} + \frac{\partial V}{\partial \zeta_s} = \mathbf{P}_s \tag{11.18}$$

Introducing Eqs. (11.7), (11.10) and (11.13) into Eq. (11.18), we obtain the substructure equations of motion

$$M_s \ddot{\zeta}_s(t) + C_s \dot{\zeta}_s(t) + K_s \zeta_s(t) = \mathbf{P}_s(t) \tag{11.19}$$

At this point, we begin the assembling process. Introducing the matrices

$$M^d = \text{block-diag}\,[M_s], \qquad C^d = \text{block-diag}\,[C_s],$$
$$K^d = \text{block-diag}\,[K_s], \qquad s = 1, 2, \ldots \tag{11.20}$$

388

as well as the vectors

$$\boldsymbol{\zeta}(t) = [\boldsymbol{\zeta}_1^T(t) \quad \boldsymbol{\zeta}_2^T(t) \quad \cdots \quad \boldsymbol{\zeta}_s^T(t) \quad \cdots]^T$$
$$\mathbf{P}(t) = [\mathbf{P}_1^T(t) \quad \mathbf{P}_2^T(t) \quad \cdots \quad \mathbf{P}_s^T(t) \quad \cdots]^T$$

(11.21)

we can write the equations of motion for the collection of substructures in the form

$$M^d\ddot{\boldsymbol{\zeta}}(t) + C^d\dot{\boldsymbol{\zeta}}(t) + K^d\boldsymbol{\zeta}(t) = \mathbf{P}(t)$$

(11.22)

Equation (11.22) represents a set of *disconnected* substructure equations, which explains the superscript *d*. The implication is that the vector $\boldsymbol{\zeta}(t)$ must contain a certain number of redundant components. These redundant components correspond to various constraints. The process of eliminating redundant coordinates is the same process that ties together the disconnected substructure equations. Assuming that the vector $\boldsymbol{\zeta}(t)$ has dimension m and that there are c constraints, then the number of independent generalized coordinates is $n = m - c$, where n is the number of degrees of freedom of the system. Denoting by $\mathbf{q}(t)$ the n-dimensional independent generalized coordinate vector, we can write the relation between $\boldsymbol{\zeta}(t)$ and $\mathbf{q}(t)$ in the matrix form

$$\boldsymbol{\zeta}(t) = \beta \mathbf{q}(t)$$

(11.23)

where β is an $m \times n$ transformation matrix, depending on the nature of the constraints. Introducing Eq. (11.23) into Eq. (11.22) and premultiplying by β^T, we obtain the *coupled* system equations of motion

$$M\ddot{\mathbf{q}}(t) + C\dot{\mathbf{q}}(t) + K\mathbf{q}(t) = \mathbf{Q}(t)$$

(11.24)

where

$$M = \beta^T M^d \beta, \qquad C = \beta^T C^d \beta, \qquad K = \beta^T K^d \beta$$

(11.25)

are real symmetric $n \times n$ matrices with obvious meaning and

$$\mathbf{Q}(t) = \beta^T \mathbf{P}(t)$$

(11.26)

is the n-dimensional generalized force vector for the system. Note that in the process of coupling the substructures, the constraint forces disappear, so that the vector $\mathbf{Q}(t)$ contains only the effect of external forces.

The matrix β reflects certain compatibility conditions at boundary points. For example, if a point is shared by substructures r and s, because the point has the same displacement whether it lies on substructure r or s, we have

$$\boldsymbol{u}_r = \boldsymbol{u}_s$$

(11.27)

where **u** represents the displacement vector. Moreover, if at the same point no angle change is possible, then

$$\boldsymbol{\theta}_r = \boldsymbol{\theta}_s \qquad (11.28)$$

where $\boldsymbol{\theta}$ represents a rotation vector, which carries the assumption of small rotations. The elimination of redundancies due to constraints is a process akin to static condensation (see Sec. 10.2). Indeed, if we write the constraint equations (11.27–28) in the general form

$$A\boldsymbol{\zeta} = \mathbf{0} \qquad (11.29)$$

where A is a $c \times m$ matrix, divide the vector $\boldsymbol{\zeta}$ into the n-dimensional vector **q** of independent variables and the c-dimensional vector $\boldsymbol{\zeta}_d$ of dependent variables, partition the matrix A as follows:

$$A = [A_1 \;\vdots\; A_2] \qquad (11.30)$$

where A_2 is a nonsingular $c \times c$ matrix and A_1 is an $n \times c$ matrix, then Eq. (11.29) can be rewritten as

$$A_1 \mathbf{q} + A_2 \boldsymbol{\zeta}_d = \mathbf{0} \qquad (11.31)$$

Equation (11.31) yields

$$\boldsymbol{\zeta}_d = -A_2^{-1} A_1 \mathbf{q} \qquad (11.32)$$

so that

$$\boldsymbol{\zeta} = \begin{bmatrix} \mathbf{q} \\ \overline{\boldsymbol{\zeta}_d} \end{bmatrix} = \begin{bmatrix} I \\ \overline{-A_2^{-1} A_1} \end{bmatrix} \mathbf{q} \qquad (11.33)$$

where I is a unit matrix of order n. Comparing Eqs. (11.23) and (11.33), we conclude that

$$\beta = \begin{bmatrix} I \\ \overline{-A_2^{-1} A_1} \end{bmatrix} \qquad (11.34)$$

The above procedure makes a sharp distinction between determinate and indeterminate constraints. In reality no such distinction exists and all boundary constraints should really receive equal treatment. This is the essence of an idea advanced by Craig and Bampton (C5), who suggest that all constraints can be regarded as boundary constraints and there is no need to identify rigid-body modes specifically. This is a particularly attractive idea, if we observe that both the rigid-body modes and the constraint modes were identified in Fig. 11.1b in the same manner, i.e., by boundary displacements. In view of this, it is only necessary to distinguish between external and internal degrees of freedom. The constraint modes are defined as the mode shapes due to successive unit displacements at the boundary points, all other boundary points being totally constrained. Craig and Bampton discuss the substructure in terms of a finite element model, instead of a distributed one, which permits a ready identification of the constraint displacements as nodal displacements at the boundaries. On the other hand, the interior modes are simply the normal modes of the substructure with totally constrained boundary nodal points. For details of the approach, the reader is advised to consult the paper by Craig and Bampton cited above.

The method described above is generally referred to as a 'fixed constraint mode' method, because the modes used to describe the motion corresponds to fixed constraints. To account for motion caused by concentrated loads at unconstrained points, in developing a computer program for the method, Bamford (B2) introduced another class of displacement modes, referred to as 'attachment modes'. The possibility of using unconstrained modes has been suggested by Goldman (G3) and by Hou (H4). Some ill-conditioning problems have been experienced in using unconstrained modes. The use of unconstrained modes has also been proposed by Dowell (D1), who, in addition, used Lagrange multipliers to enforce continuity. No convergence problems have been observed here.

An important aspect of the component synthesis is the selection of modes and its effect on the eigenvalue error. This problem has been addressed by Hurty (H8). In an attempt to optimize the eigenvalue error, Bajan et al (B1) have introduced the subsystem modes into the analysis in sequential groups, rather than simultaneously.

Truncation problems, and in particular the effect of modes not retained explicitly, have been discussed by Kuhar and Stahle (K2) and by Rubin (R4). The first paper presents a condensation scheme similar to the dynamic condensation of Sec. 10.3. Truncation problems have also been discussed by Hintz (H2), who identified Hurty's component mode synthesis as a static condensation (see Sec. 10.2).

11.3 Branch-mode analysis

Another method based on the substructure synthesis idea is known as *branch-mode analysis* (G2). The method was developed about the same time as the component-mode synthesis and it exhibits some of the same features, although differences exist. The systems considered have a chain-like configuration, in which every component is connected to two other components, one on each side. The method can be extended to cases in

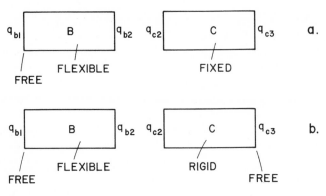

Figure 11.2

which several components are connected at one point. The components are grouped into subsets called *branches*. The motion of each branch is represented by *branch modes*, which are generally modes of vibration of the separate branches. The number of modes taken to represent the motion of a branch is left to the discretion of the analyst. The branch modes may be defined by letting only one of the components in the branch be flexible while the other are either fixed (Fig. 11.2a) or free (Fig. 11.2b), where in the latter case the component can vibrate as a rigid body. Note that compatibility requires that $q_{b2} = q_{c2}$, where in the first case both q_{b2} and q_{c2} are zero. Other types of components are possible. For example, the left end in Fig. 11.2b can be taken as fixed instead of free. Gladwell illustrates the method by means of a lumped system, although the method is not so restricted. He does not discuss redundant constraints.

Let us consider an n-degree-of-freedom system and write the kinetic and potential energy

$$T = \tfrac{1}{2}\dot{\mathbf{q}}^T M \dot{\mathbf{q}} \tag{11.35a}$$

$$V = \tfrac{1}{2}\mathbf{q}^T K \mathbf{q} \tag{11.35b}$$

Using Lagrange's equations in conjunction with Eqs. (11.35), one can derive n simultaneous ordinary differential equations, leading to an eigenvalue problem of order n. When n is very large, it is not feasible to retain all the degrees of freedom, so that the interest lies in a scheme for the reduction of the order of the eigenvalue problem from n to m, $m \ll n$, in a way that the m-degree-of-freedom system is capable of yielding good approximations to the modes of vibration of the original large-order system. To this end, the motion of every branch is represented by a reduced number of 'modes'. This is, of course, the same approach as the component-mode synthesis, and the main difference lies in the definition of the modes.

Assuming that the system consists of the branches $B, C, \ldots, S$ and that the motion of the branches is represented by $m_b, m_c, \ldots, m_s$ modes, then the number of degrees of freedom of the reduced system is $m = m_0 + m_b + m_c + \cdots + m_s$, where m_0 is the number of rigid-body modes of the entire system. Letting $X_0, X_b, X_c, \ldots, X_s$ be $n \times m_0, n \times m_b, n \times m_c, \ldots, n \times m_s$ matrices of rigid-body modes and branch modes and introducing the $n \times m$ matrix

$$X = [X_0 \mid X_b \mid X_c \mid \cdots \mid X_s] \tag{11.36}$$

The n-dimensional configuration vector $\mathbf{q}$ can be written in the form

$$\mathbf{q} = X\mathbf{u} \tag{11.37}$$

where $\mathbf{u}$ represents the m-dimensional reduced configuration vector. Introducing Eq. (11.37) into Eqs. (11.35), we obtain

$$T = \tfrac{1}{2}\dot{\mathbf{u}}^T M^* \dot{\mathbf{u}} \tag{11.38a}$$

$$V = \tfrac{1}{2}\mathbf{u}^T K^* \mathbf{u} \tag{11.38b}$$

in which

$$M^* = X^T M X \tag{11.39a}$$

$$K^* = X^T K X \tag{11.39b}$$

are reduced mass and stiffness matrices. The corresponding eigenvalue problem is

$$K^* U = M^* U \Lambda^m \tag{11.40}$$

where Λ^m is a diagonal matrix of estimated eigenvalues and U is the matrix of the eigenvectors of the reduced problem. Denoting by $\mathbf{u}_i$ $(i = 1, 2, \ldots, m)$ the eigenvectors of the reduced problem, the estimated eigenvectors of the original problem are obtained by means of Eq. (11.37) in the form

$$\mathbf{q}_i = X\mathbf{u}_i, \qquad i = 1, 2, \ldots, m \tag{11.41}$$

Of course, estimates of only the m lowest system modes are obtained.

Because there are m_0 rigid-body modes, m_0 of the eigenvalues are zero. Hence, the problem can be further reduced by eliminating the rigid-body modes, in a process similar to the static condensation. To this end, let us partition the matrix X as follows:

$$X = [X_0 \mid X_a] \tag{11.42}$$

where

$$X_a = [X_b \mid X_c \mid \cdots \mid X_s] \tag{11.43}$$

is an $n \times (m - m_0)$ matrix of branch modes. Then, the matrices M^* and K^* can be partitioned accordingly, with the results,

$$M^* = \begin{bmatrix} M_{00} & M_{0a} \\ M_{a0} & M_{aa} \end{bmatrix} \tag{11.44a}$$

$$K^* = \begin{bmatrix} K_{00} & K_{0a} \\ K_{a0} & K_{aa} \end{bmatrix} \tag{11.44b}$$

where

$$M_{00} = X_0^T M X_0, \quad M_{0a} = X_0^T M X_a = M_{a0}^T, \quad M_{aa} = X_a^T M X_a \tag{11.45a}$$

$$K_{00} = X_0^T K X_0, \quad K_{0a} = X_0^T K X_a = K_{a0}^T, \quad K_{aa} = X_a^T K X_a \tag{11.45b}$$

Similarly, let us partition the matrices U and Λ^m as follows:

$$U = \begin{bmatrix} U_{00} & U_{0a} \\ U_{a0} & U_{aa} \end{bmatrix} \tag{11.46a}$$

$$\Lambda^m = \begin{bmatrix} 0 & 0 \\ 0 & \Lambda^{m-m_0} \end{bmatrix} \tag{11.46b}$$

where Λ^{m-m_0} is the diagonal matrix of nonzero eigenvalues. Introducing Eqs. (11.44–46) into Eq. (11.41), we obtain

$$0 = (M_{00} U_{0a} + M_{0a} U_{aa}) \Lambda^{m-m_0} \tag{11.47a}$$

$$K_{aa} U_{aa} = (M_{a0} U_{0a} + M_{aa} U_{aa}) \Lambda^{m-m_0} \tag{11.47b}$$

In addition, we have $K_{aa} U_{a0} = 0$, which is satisfied identically by virtue of the fact that U_{a0} is part of the matrix representing the rigid-body modes. Equation (11.47a) can be solved for U_{0a}, with the result

$$U_{0a} = -M_{00}^{-1} M_{0a} U_{aa} \tag{11.48}$$

which, when introduced into Eq. (11.47b), yields the further reduced eigenvalue problem

$$K_{aa} U_{aa} = (M_{aa} - M_{a0} M_{00}^{-1} M_{0a}) U_{aa} \Lambda^{m-m_0} \tag{11.49}$$

The question remains as to how to produce the transformation matrix X to be used in Eq. (11.37). This matrix is obtained by solving branch eigenvalue problems. For example, in the case of Fig. 11.2a, the matrix X_b is obtained by solving the eigenvalue problem associated with the branch B with its left end free and its right end fixed and taking a limited number of these modes. The paper by Gladwell cited above presents a numerical example.

A problem can arise in the treatment of fixed-fixed systems. In this case, it is shown by Gladwell that the system must be divided into a minimum of three branches.

11.4 Component-mode substitution

The component-mode substitution (B4) resembles both Hurty's component-mode synthesis and Gladwell's branch-mode analysis. Whereas the general ideas are the same, substantive differences exist. The method differs from the component-mode synthesis in that the component modes need not be constrained and can be free-free. More importantly, however, the component-mode substitution does not require that the generalized coordinates of the static constraint modes appear in the final formulation, thus reducing the order of the simulation. Connections are handled by using component stiffness matrices. Static constraint modes are used when component modes contain a fixed interface. Such constraint modes are eliminated by using 'branch components', where a branch component is defined as a component whose motion is defined relative to interfaces. A component containing a fixed interface is called a constrained-branch component. This permits the use of constraint modes for the connection, as well as the elimination of the generalized coordinates associated with the constraint modes. In many respects, component-mode substitution is closer to the branch-mode analysis. The branches are defined in a similar way and it requires the solution of eigenvalue problems for branch components. For example, if a chain-like branch has n components, then it is necessary to solve $2n - 1$ eigenvalue problems. However, these eigenvalue problems are of much lower order than that of the overall problem.

The efficiency of the method can be improved by applying stiffness and inertial loadings to the free interface coordinates of the component under consideration. These loadings represent reduced stiffness and mass properties of the system. Both the reduced-stiffness and reduced-mass properties are formed by a reduction process akin to static condensation (see Sec. 10.2).

Let us consider a system consisting of two components, a and b, and denote by $\bar{\mathbf{q}}_a$ and $\bar{\mathbf{q}}_b$ the interface displacement vectors and by $\hat{\mathbf{q}}_a$ and $\hat{\mathbf{q}}_b$

11 Substructure synthesis

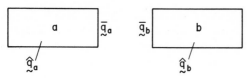

Figure 11.3

the interior displacement vectors (Fig. 11.3), so that the component displacement vectors are

$$\mathbf{q}_a = [\bar{\mathbf{q}}_a^T \vdots \hat{\mathbf{q}}_a^T], \qquad \mathbf{q}_b = [\bar{\mathbf{q}}_b^T \vdots \hat{\mathbf{q}}_b^T]^T \tag{11.50}$$

Moreover, we shall denote by $\bar{\mathbf{p}}_a$, $\hat{\mathbf{p}}_a$, $\bar{\mathbf{p}}_b$ and $\hat{\mathbf{p}}_b$ the associated force vectors, so that

$$\mathbf{p}_a = [\bar{\mathbf{p}}_a^T \vdots \hat{\mathbf{p}}_a^T]^T, \qquad \mathbf{p}_b = [\bar{\mathbf{p}}_b^T \vdots \hat{\mathbf{p}}_b^T]^T \tag{11.51}$$

are the component force vectors. The relationship between the forces and displacements for component b can be written in the form

$$\mathbf{p}_b = k_b \mathbf{q}_b \tag{11.52}$$

where k_b is the component stiffness matrix. Equation (11.52) can be partitioned as follows

$$\begin{bmatrix} \bar{\mathbf{p}}_b \\ \hline \hat{\mathbf{p}}_b \end{bmatrix} = \begin{bmatrix} k_{b11} & \vdots & k_{b12} \\ \hline k_{b21} & \vdots & k_{b22} \end{bmatrix} \begin{bmatrix} \bar{\mathbf{q}}_b \\ \hline \hat{\mathbf{q}}_b \end{bmatrix} \tag{11.53}$$

For no interior forces, $\hat{\mathbf{p}}_b = \mathbf{0}$, we obtain the equation

$$k_{b21}\bar{\mathbf{q}}_b + k_{b22}\hat{\mathbf{q}}_b = \mathbf{0} \tag{11.54}$$

which permits us to write

$$\hat{\mathbf{q}}_b \big|_{\hat{\mathbf{p}}_b = \mathbf{0}} = \hat{\mathbf{q}}_{cb} = T_{cb}\bar{\mathbf{q}}_b \tag{11.55}$$

where

$$T_{cb} = -k_{b22}^{-1}k_{b21} \tag{11.56}$$

so that

$$\mathbf{q}_b \big|_{\hat{\mathbf{p}}_b = \mathbf{0}} = \begin{bmatrix} \bar{\mathbf{q}}_b \\ \hline \hat{\mathbf{q}}_{cb} \end{bmatrix} = T_{rb}\bar{\mathbf{q}}_b \tag{11.57}$$

where T_{rb} is the transformation matrix

$$T_{rb} = \begin{bmatrix} I \\ \hline T_{cb} \end{bmatrix} \tag{11.58}$$

Equations (11.54–58) are recognized as describing the static condensation procedure of Sec. 10.2.

The potential energy of component b is

$$V_b = \tfrac{1}{2}\mathbf{q}_b^T k_b \mathbf{q}_b \qquad (11.59)$$

Using Eq. (11.57), the potential energy can be written in the constrained form

$$V_{cb} = \tfrac{1}{2}\bar{\mathbf{q}}_b^T \bar{k}_b \bar{\mathbf{q}}_b \qquad (11.60)$$

where

$$\bar{k}_b = T_{rb}^T k_b T_{rb} \qquad (11.61)$$

is referred to as the reduced stiffness matrix. Similarly, if the kinetic energy of component b is

$$T_b = \tfrac{1}{2}\dot{\mathbf{q}}_b^T m_b \dot{\mathbf{q}}_b \qquad (11.62)$$

then the constrained kinetic energy is

$$T_{cb} = \tfrac{1}{2}\bar{\mathbf{q}}_b^T \bar{m}_b \bar{\mathbf{q}}_b \qquad (11.63)$$

where

$$\bar{m}_b = T_{rb}^T m_b T_{rb} \qquad (11.64)$$

The total potential energy for uncoupled components a and b is

$$V_T = \tfrac{1}{2}\mathbf{q}_a^T k_a \mathbf{q}_a + \tfrac{1}{2}\mathbf{q}_b^T k_b \mathbf{q}_b \qquad (11.65)$$

Because the two components are coupled together by the interface constraints, we have

$$\bar{\mathbf{q}}_a = \bar{\mathbf{q}}_b \qquad (11.66)$$

and note that $\bar{\mathbf{q}}_a$ and $\bar{\mathbf{q}}_b$ are assumed to be expressed in the same system of coordinates. This generally involves a transformation from one local system of coordinates to another, where the transformation matrix is the matrix of direction cosines. Using Eq. (11.66), the coordinates transformation coupling components a and b is

$$\bar{\mathbf{q}}_b = T_L \mathbf{q}_a \qquad (11.67)$$

where

$$T_L = [I \mid 0] \qquad (11.68)$$

Hence, inserting Eqs. (11.60) and (11.67) into Eq. (11.65), we obtain the total potential energy for the constrained system

$$V^a = \tfrac{1}{2}\mathbf{q}_a^T k^a \mathbf{q}_a \tag{11.69}$$

where

$$k^a = k_a + T_L^T \bar{k}_b T_L \tag{11.70}$$

is the constrained stiffness matrix. In a similar fashion, the total kinetic energy for the constrained system is

$$T^a = \tfrac{1}{2}\dot{\mathbf{q}}_a^T m^a \dot{\mathbf{q}}_a \tag{11.71}$$

where

$$m^a = m_a + T_L^T \bar{m}_b T_L \tag{11.72}$$

is the mass matrix for the constrained system.

The equations of motion for the constrained system have the compact matrix form

$$m^a \ddot{\mathbf{q}}_a + k^a \mathbf{q}_a = \mathbf{0} \tag{11.73}$$

Letting Φ_a be the modal matrix of the system, obtained by solving the eigenvalue problem associated with Eq. (11.73), the displacement vector $\mathbf{q}_a$ can be written as

$$\mathbf{q}_a(t) = \begin{bmatrix} \bar{\mathbf{q}}_a \\ \hline \hat{\mathbf{q}}_a \end{bmatrix} = \Phi_a \boldsymbol{\xi}_a(t) = \begin{bmatrix} \bar{\Phi}_a \\ \hline \hat{\Phi}_a \end{bmatrix} \boldsymbol{\xi}_a(t) \tag{11.74}$$

where $\boldsymbol{\xi}_a(t)$ is the generalized displacement vector.

In the general case, in which there are more than two components involved, component b can be regarded as representing all other components of the system in the branch coupled together.

Next, let us assume that the noninterface displacement of component b can be expressed as a linear combination of constraint modes and fixed-constraint modes, in the same way as they were defined in Sec. 11.2. The constraint displacement $\hat{\mathbf{q}}_{cb}$ is essentially a linear combination of the constraint modes, so that the columns of the matrix T_{cb}, Eq. (11.56), represent the constraint modes. We shall denote the noninterface displacement vector relative to the constraint displacement by $\hat{\mathbf{q}}_{nb}$, so that

$$\hat{\mathbf{q}}_b = \hat{\mathbf{q}}_{cb} + \hat{\mathbf{q}}_{nb} \tag{11.75}$$

Using Eq. (11.55), we can write

$$\begin{bmatrix} \mathbf{q}_a \\ \hline \mathbf{q}_b \end{bmatrix} = \begin{bmatrix} \bar{\mathbf{q}}_a \\ \hat{\mathbf{q}}_a \\ \hline \bar{\mathbf{q}}_b \\ \hat{\mathbf{q}}_b \end{bmatrix} = T_1^c \begin{bmatrix} \bar{\mathbf{q}}_a \\ \hat{\mathbf{q}}_a \\ \hline \bar{\mathbf{q}}_b \\ \hat{\mathbf{q}}_{nb} \end{bmatrix} \tag{11.76}$$

where

$$
T_1^c = \begin{bmatrix} I & 0 & 0 & 0 \\ 0 & I & 0 & 0 \\ 0 & 0 & I & 0 \\ 0 & 0 & T_{cb} & I \end{bmatrix} \tag{11.77}
$$

Recalling that $\bar{\mathbf{q}}_a = \bar{\mathbf{q}}_b$, we can eliminate $\bar{\mathbf{q}}_b$ from the vector on the right side of Eq. (11.76), with the result

$$
\begin{bmatrix} \bar{\mathbf{q}}_a \\ \hat{\mathbf{q}}_a \\ \bar{\mathbf{q}}_b \\ \hat{\mathbf{q}}_{nb} \end{bmatrix} = T_2^c \begin{bmatrix} \bar{\mathbf{q}}_a \\ \hat{\mathbf{q}}_a \\ \hat{\mathbf{q}}_{nb} \end{bmatrix} \tag{11.78}
$$

where

$$
T_2^c = \begin{bmatrix} I & 0 & 0 \\ 0 & I & 0 \\ I & 0 & 0 \\ 0 & 0 & I \end{bmatrix} \tag{11.79}
$$

Hence, Eq. (11.76) can be rewritten as

$$
\begin{bmatrix} \bar{\mathbf{q}}_a \\ \hat{\mathbf{q}}_a \\ \bar{\mathbf{q}}_b \\ \hat{\mathbf{q}}_b \end{bmatrix} T_3^c \begin{bmatrix} \bar{\mathbf{q}}_a \\ \hat{\mathbf{q}}_a \\ \hat{\mathbf{q}}_{nb} \end{bmatrix} \tag{11.80}
$$

where

$$
T_3^c = T_1^c T_2^c \tag{11.81}
$$

The displacement $\hat{\mathbf{q}}_{nb}$ is a superposition of fixed-constraint modes for component b. The fixed-interface equations of motion have the matrix form

$$
m_{b22}\hat{\mathbf{q}}_b + k_{b22}\hat{\mathbf{q}}_b = \mathbf{0} \tag{11.82}
$$

having the solution

$$
\hat{\mathbf{q}}_b(t)\,\big|_{\bar{\mathbf{q}}_b = \mathbf{0}} = \hat{\mathbf{q}}_{nb}(t) = \hat{\Phi}_b^c \boldsymbol{\xi}_b^c(t) \tag{11.83}
$$

where $\hat{\Phi}_b^c$ is the matrix of the fixed-constraint modes, obtained by solving the eigenvalue problem associated with Eq. (11.82). Using Eqs. (11.74) and (11.83), we can write

$$\begin{bmatrix} \mathbf{q}_a \\ \hline \hat{\mathbf{q}}_a \\ \hline \hat{\mathbf{q}}_{nb} \end{bmatrix} = T_4^c \begin{bmatrix} \boldsymbol{\xi}_a \\ \hline \boldsymbol{\xi}_b^c \end{bmatrix} \tag{11.84}$$

where

$$T_4^c = \begin{bmatrix} \bar{\Phi}_a & \vdots & 0 \\ \hline \hat{\Phi}_a & \vdots & 0 \\ \hline 0 & \vdots & \hat{\Phi}_b^c \end{bmatrix} \tag{11.85}$$

Finally, we wish to relate the uncoupled displacements of components a and b to the generalized coordinate vectors $\boldsymbol{\xi}_a$ and $\boldsymbol{\xi}_b^c$. Combining Eqs. (11.80) and (11.84), we have simply

$$\begin{bmatrix} \mathbf{q}_a \\ \hline \mathbf{q}_b \end{bmatrix} = \begin{bmatrix} \bar{\mathbf{q}}_a \\ \hline \hat{\mathbf{q}}_a \\ \hline \bar{\mathbf{q}}_b \\ \hline \hat{\mathbf{q}}_b \end{bmatrix} = T_5^c \begin{bmatrix} \boldsymbol{\xi}_a \\ \hline \boldsymbol{\xi}_b^c \end{bmatrix} \tag{11.86}$$

where

$$T_5^c = T_3^c T_4^c = \begin{bmatrix} \bar{\Phi}_a & \vdots & 0 \\ \hline \hat{\Phi}_a & \vdots & 0 \\ \hline \bar{\Phi}_a & \vdots & 0 \\ \hline T_{cb}\Phi_a & \vdots & \hat{\Phi}_b^c \end{bmatrix} \tag{11.87}$$

We are now in the position of deriving the system modes. Indeed, the uncoupled system potential energy is

$$V_T = \frac{1}{2} \begin{bmatrix} \mathbf{q}_a \\ \hline \mathbf{q}_b \end{bmatrix}^T \begin{bmatrix} k_a & \vdots & 0 \\ \hline 0 & \vdots & k_b \end{bmatrix} \begin{bmatrix} \mathbf{q}_a \\ \hline \mathbf{q}_b \end{bmatrix} \tag{11.88}$$

so that, using Eq. (11.86), we obtain the potential energy for the coupled system

$$V_{ab} = \frac{1}{2} \begin{bmatrix} \boldsymbol{\xi}_a \\ \hline \boldsymbol{\xi}_b^c \end{bmatrix}^T K^c \begin{bmatrix} \boldsymbol{\xi}_a \\ \hline \boldsymbol{\xi}_b^c \end{bmatrix} \tag{11.89}$$

where

$$K^c = (T_5^c)^T \begin{bmatrix} k_a & \vdots & 0 \\ \hline 0 & \vdots & k_b \end{bmatrix} T_5^c \tag{11.90}$$

Similarly, the kinetic energy for the coupled system is

$$T_{ab} = \frac{1}{2}\begin{bmatrix} \dot{\xi}_a \\ \hline \dot{\xi}_b^c \end{bmatrix}^T M^c \begin{bmatrix} \dot{\xi}_a \\ \hline \dot{\xi}_b^c \end{bmatrix} \tag{11.91}$$

where

$$M^c = (T_5^c)^T \begin{bmatrix} m_a & \vdots & 0 \\ \hline 0 & \vdots & m_b \end{bmatrix} T_5^c \tag{11.92}$$

The system modes are obtained by solving the eigenvalue problem

$$K^c \Phi_m^c = M^c \Phi_m^c \Lambda_m^c \tag{11.93}$$

where Φ_m^c is the system modal matrix and Λ_m^c is the diagonal matrix of natural frequencies squared. Hence, to obtain the system modes, it is necessary to solve three eigenvalue problems, namely, those associated with Eqs. (11.73) and (11.82) and that described by Eq. (11.93).

The above procedure shows how to compute the system modes in terms of constrained component-branch modes. A similar procedure can be derived to compute the system modes in terms of free-free component modes. For details, the reader is referred to the paper by Benfield and Hruda (B4).

11.5 Substructure synthesis

The methods discussed in Secs. 11.2–11.4 have two things in common: They all regard a structure as an assemblage of substructures and the motion of every substructure is represented as a superposition of component modes. The main difference lies in the kind of modes to be used. In fact, the merits of using free-free modes, as opposed to constrained modes, have been examined in detail. Nevertheless, there seems to be a consensus that the modes should satisfy a certain substructure eigenvalue problem.

The implied requirement that the component modes be functions or vectors satisfying some substructure eigenvalue problem, however, can place undue hardship on the analyst. Indeed, producing eigenfunctions for a substructure can be quite a difficult task in itself. Moreover, even if substructure eigenfunctions can be produced, they are likely to be difficult to work with. However, this hardship is not really necessary from a mathematical point of view and can be avoided. It has been generally assumed that any substructure synthesis approach represents some form of a Rayleigh–Ritz method, which implies that some stationarity principle must exist for the structure. In view of this, *the only requirement to be imposed on the functions used to represent the substructure motion is that they belong to a complete set of admissible functions.* Quite often, admissible functions can be chosen in the form of low-degree polynomials, which are very attractive computationally. A substructure synthesis method based on admissible functions has been developed by Meirovitch (M7) and Meirovitch and Hale (M8, H1, M10, M11) and is the subject of this section. The various substructure syntheses based on component modes discussed in Secs. 11.2–4 can be regarded as special cases of the one discussed in this section.

One must distinguish between structures in which adjacent substructures are connected at a single point and those in which they are connected at an infinite number of points, such as in the case of internal boundaries consisting of lines and surfaces. Cases in which internal boundaries actually consist of a finite number of points are very rare, although cases in which actual boundaries consisting of lines and surfaces are represented by a finite number of points are quite frequent. The general idea of representing substructure motion by admissible functions is not affected by the type of internal boundaries, so that the main difference between the cases of single-point and infinitely-many-points boundaries lies in the satisfaction of the geometric compatibility conditions. In the following, we shall discuss both cases.

i. *Single-point internal boundaries* (M7, M8)

In the case in which adjacent substructures are connected at a single point, the geometric compatibility can be satisfied automatically by a suitable kinematical procedure. It is convenient to describe the motion in terms of rigid-body motions of local reference frames for the substructures. Then, deformations are defined simply as motions relative to the moving reference frames. This approach is capable of accommodating rotating substructures, which makes it ideally suited for the treatment of gyroscopic systems, such as helicopters and spacecraft with spinning parts.

Let us consider the system shown in Fig. 11.4 and concentrate our attention on the description of the motion of a central substructure C and of a typical appended substructure A. It will prove convenient to introduce an inertial system XYZ with the origin at 0, as well as to identify a reference frame $x_C y_C z_C$ with the substructure C. The origin of $x_C y_C z_C$ can be taken at some convenient point C on the substructure, such as the mass center of the substructure when in undeformed state. This choice is not mandatory, however, and any other convenient point can be used. To describe the motion of substructure C, let us denote the radius vector

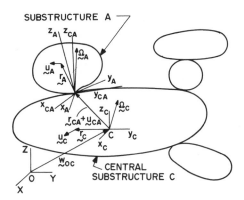

Figure 11.4

from 0 to the origin C by $\mathbf{w}_{0C}$, the nominal position of any material point on the substructure relative to $x_C y_C z_C$ by $\mathbf{r}_C$, the displacement of the point relative to $x_C y_C z_C$ by $\mathbf{u}_C$ and the angular velocity of the frame $x_C y_C z_C$ by $\mathbf{\Omega}_C$. Then, if $\mathbf{w}_C$ denotes the position of the point in question relative to the inertial space, the absolute velocity vector of that point is

$$\dot{\mathbf{w}}_C = T_{C0}\dot{\mathbf{w}}_{0C} - (\tilde{r}_C + \tilde{u}_C)\mathbf{\Omega}_C + \dot{\mathbf{u}}_C \tag{11.94}$$

where T_{C0} is the matrix of direction cosines between the inertial axes XYZ and the local axes $x_C y_C z_C$ and $\tilde{r}_C + \tilde{u}_C$ is a skew symmetric matrix with entries equal to the components of $\mathbf{r}_C + \mathbf{u}_C$. Note that (M3, Sec. 3.2) $(\tilde{r}_C + \tilde{u}_C)\mathbf{\Omega}_C$ is simply the matrix expression of the vector cross product $(\mathbf{r}_C + \mathbf{u}_C) \times \mathbf{\Omega}_C$. As a matter of interest, the velocity vectors $\dot{\mathbf{w}}_{0C}$ and $\mathbf{\Omega}_C$ can be recognized as representing the rigid-body motions of substructure C and the vector $\mathbf{u}_C$ can be interpreted as representing the deformation at any point on the substructure. Similarly, introducing a reference frame $x_A y_A z_A$ for the substructure A with the origin at point A connecting the two substructures, the absolute velocity of any point on the substructure A is

$$\dot{\mathbf{w}}_A = T_{AC}\dot{\mathbf{w}}_{CA} - (\tilde{r}_A + \tilde{u}_A)\mathbf{\Omega}_A + \dot{\mathbf{u}}_A \tag{11.95}$$

where the notation is obvious. We note that $\dot{\mathbf{w}}_{CA}$ is simply $\dot{\mathbf{w}}_C$ evaluated at the connecting point A. The angular velocity $\mathbf{\Omega}_A$ of the reference frame $x_A y_A z_A$ has the expression

$$\mathbf{\Omega}_A = T_{AC}\mathbf{\Omega}_C + L_A\dot{\mathbf{\beta}}_{CA} + \mathbf{\omega}_A \tag{11.96}$$

in which $\dot{\mathbf{\beta}}_{CA}$ is an elastic angular velocity due to the deformation of the substructure C at the connecting point A and $\mathbf{\omega}_A$ is the angular velocity of the reference $x_A y_A z_A$ relative to a reference frame $x_{CA} y_{CA} z_{CA}$ attached to C at point A, where we note that the angular velocity of the frame $x_{CA} y_{CA} z_{CA}$ includes both Ω_C and $\dot{\beta}_{CA}$. The matrix L_A is simply the matrix of direction cosines between axes $x_A y_A z_A$ and $x_{CA} y_{CA} z_{CA}$, and it can involve explicit time dependence. The coordinate transformations defined by the matrices T_{AC} and L_A guarantee that both $\dot{\mathbf{w}}_A$ and Ω_A are in terms of components along the local axes $x_A y_A z_A$.

Equation (11.96) for the angular velocity Ω_A of the frame $x_A y_A z_A$ is sufficiently general to accommodate a number of cases of interest. If substructure A is attached rigidly to the central substructure C, then we simply let $\omega_A = \mathbf{0}$. If substructure A is hinged, then ω_A represents the rotation of the substructure A relative to substructure C. This case arises generally when substructure A possesses uniform angular velocity relative to substructure C, such as in the case of a helicopter rotor blade. In this case, it is common to restrict the secular rotation of substructure A so as to take place about the spin axis alone. Angular motions about axes normal to the spin axis may also be permitted, but they are generally nonsecular in character.

The above procedure is ideally suited to chain-like structures. Indeed, if several substructures form a chain, then a substructure B attached to A bears the same relation to A as A bears to C. In this case, $\dot{\mathbf{w}}_B$ and Ω_B can be obtained by simply replacing A by B and C by A in Eqs. (11.95) and (11.96). The procedure is equally applicable to the case in which several members are connected at the same point. The helicopter rotor, where several blades can be regarded as being attached to the drive shaft at the same point, represents such an example.

The next step is to derive the system Lagrange's equations of motion. To this end, it is necessary to determine the functional dependence of the Lagrangian. If we regard A as a typical peripheral substructure, and if summation over all substructures of the same type is implied, then the system kinetic energy can be written in the general form

$$T = \frac{1}{2} \int_{m_C} \dot{\mathbf{w}}_C^T \dot{\mathbf{w}}_C \, dm_C + \frac{1}{2} \int_{m_A} \dot{\mathbf{w}}_A^T \dot{\mathbf{w}}_A \, dm_A \qquad (11.97)$$

where m_C and m_A are associated masses and $\dot{\mathbf{w}}_C$ and $\dot{\mathbf{w}}_A$ are given by Eqs. (11.95) and (11.96), respectively. Assuming that in equilibrium substructure A rotated about axis z_A with the uniform angular velocity ω_A relative to substructure C, while all other motions are zero, we can write

$$\Omega_C = \Theta_C \dot{\theta}_C \qquad (11.98a)$$

$$\omega_A = \omega_A \mathbf{1}_A + \Theta_A \dot{\theta}_A \qquad (11.98b)$$

where $\mathbf{l}_A$ is the vector of direction cosines between z_A when substructure A is in undeformed state and $x_{CA}y_{CA}z_{CA}$ and Θ_C and Θ_A are 3×3 matrices depending on the rotations θ_{Ci} and θ_{Ai} ($i = 1, 2, \ldots, n$), respectively. Note that, whereas in general θ_{Ci} and θ_{Ai} can be arbitrarily large and cannot be represented by vectors, the angular rates can and are indeed represented by the vectors $\dot{\boldsymbol{\theta}}_C$ and $\dot{\boldsymbol{\theta}}_A$, respectively. It follows from the above that the functional dependence of the kinetic energy is

$$T = T(\dot{\mathbf{w}}_{0C}, \theta_{Ci}, \dot{\boldsymbol{\theta}}_C, \mathbf{u}_C, \dot{\mathbf{u}}_C, \mathbf{u}_{CA}, \dot{\mathbf{u}}_{CA}, \theta_{Ai}, \dot{\boldsymbol{\theta}}_A, \mathbf{u}_A, \dot{\mathbf{u}}_A, t) \tag{11.99}$$

Assuming that the potential energy is due entirely to elastic effects, including restraints between substructures, we shall write it in the general form

$$V = U(\theta_{A1}, \theta_{A2}) + \tfrac{1}{2}[\mathbf{u}_C, \mathbf{u}_C] + \tfrac{1}{2}[\mathbf{u}_A, \mathbf{u}_A] \tag{11.100}$$

where U represents the potential energy due to torsional springs at point A restraining the angular motion about axes normal to z_A. The square brackets in (11.100) represent the usual energy inner products.

Equations (11.99–100) can be used to derive general Lagrange's equations of motion. The equations will have a hybrid form, in the sense that some of the differential equations will be ordinary and some will be partial. Moreover, the equations will be nonlinear and possibly possess time-dependent coefficients. Assuming small motions in the neighborhood of a given equilibrium state and considering only rotating members with inertial symmetry, the equations of motion can be reduced to linear equations with time-independent coefficients. In addition, it must be recognized that some of the coordinates are ignorable (M3, Sec. 2.10) and can be eliminated. The resulting system of equations are still hybrid.

The system can be discretized by a Rayleigh–Ritz approach. To this end, we represent the elastic motions as follows:

$$\mathbf{u}_C = \Phi_C \boldsymbol{\eta}_C \tag{11.101a}$$

$$\mathbf{u}_A = \Phi_A \boldsymbol{\eta}_A \tag{11.101b}$$

where Φ_C and Φ_A are $3\times N_C$ and $3\times N_A$ matrices of space-dependent admissible functions for substructures C and A, respectively, and $\boldsymbol{\eta}_C$ and $\boldsymbol{\eta}_A$ are associated N_C-dimensional and N_A-dimensional vectors of time-dependent generalized coordinates. The term admissible is to be interpreted here in the sense that the functions must satisfy the geometric boundary conditions imposed by the kinematical procedure. Then, introducing the generalized configuration vector

$$\mathbf{q} = [\theta_{C1} \quad \theta_{C2} \quad \theta_{A1} \quad \theta_{A2} \quad \boldsymbol{\eta}_C^T \quad \boldsymbol{\eta}_A^T]^T \tag{11.102}$$

405

the kinetic energy can be written as

$$T = \tfrac{1}{2}\dot{\mathbf{q}}^T M \dot{\mathbf{q}} + \mathbf{q}^T F \dot{\mathbf{q}} + \tfrac{1}{2}\mathbf{q}^T K_T \mathbf{q} \tag{11.103}$$

where, in symbolic form,

$$M = \frac{\partial^2 T}{\partial \dot{\mathbf{q}}^2}, \qquad F = \frac{\partial^2 T}{\partial \dot{\mathbf{q}} \, \partial \mathbf{q}}, \qquad K_T = \frac{\partial^2 T}{\partial \mathbf{q}^2} \tag{11.104}$$

are square matrices of constant coefficients. Similarly, the potential energy reduces to

$$V = \tfrac{1}{2}\mathbf{q}^T K_V \mathbf{q} \tag{11.105}$$

where K_V is a block-diagonal matrix of the form

$$K_V = \text{block-diag}\,[0 \mathbin{\vert} \partial^2 U/\partial \boldsymbol{\theta}_A^2 \mathbin{\vert} [\Phi_C, \Phi_C] \mathbin{\vert} [\Phi_A, \Phi_A]] \tag{11.106}$$

The system Lagrange's equations can be written in the symbolic form

$$\frac{\mathrm{d}}{\mathrm{d}t}\left(\frac{\partial L}{\partial \dot{\mathbf{q}}}\right) - \frac{\partial L}{\partial \mathbf{q}} = \mathbf{0} \tag{11.107}$$

where $L = T - V$ is the system Lagrangian. Using Eqs. (11.103) and (11.105), the equations of motion can be shown to be

$$M\ddot{\mathbf{q}} + G\dot{\mathbf{q}} + K\mathbf{q} = \mathbf{0} \tag{11.108}$$

where M is a symmetric mass matrix,

$$G = F^T - F \tag{11.109}$$

is a skew symmetric gyroscopic matrix and

$$K = K_V - K_T \tag{11.110}$$

is a symmetric stiffness matrix, including elastic and centrifugal effects.

Equation (11.108) describes a typical linear gyroscopic system, discussed on various occasions throughout this book. It will be recalled that a stationarity principle and an inclusion principle exist for the system. Of course, the above formulation remains valid also when there are no rotating substructures. In this case, the system simply ceases to be gyroscopic.

The question of how to select admissible functions has been addressed in Ch. 8. It should be clear that *the finite element method is entirely consistent with this approach*, as the interpolation functions of the finite element method can be regarded as special classes of admissible functions to be used to represent the motion of substructures.

ii. *Infinitely-many-points internal boundaries* (H1, M10, M11)

The substructure synthesis presented earlier in this section was based on a kinematical procedure guaranteeing automatically the satisfaction of the geometric compatibility conditions at internal boundaries. The procedure is suitable for single-point connected substructures, such as in chain-like structures. In the case of continuous two- and three-dimensional substructures, however, internal boundaries consist of lines and surfaces, and hence of an infinity of points. In such cases, the kinematical procedure described earlier is no longer possible and the geometric compatibility must be ensured by other means. The method to be described here consists of regarding the structure as a disconnected assemblage of substructures, with the motion of each substructure being represented by a finite series of admissible functions. Then the structure is forced to act as a whole by invoking the geometric compatibility conditions, which reduces to the imposition of constraints on the disconnected displacement vector. The method contains the various component-mode syntheses presented so far as special cases.

It should be clear from the onset that it is impossible to satisfy the geometric compatibility conditions at the entire infinity of points on internal boundaries. This would require the satisfaction of an infinite set of constraint equations, which is inconsistent with a model reduction procedure based on a finite set of admissible functions. Hence, the internal geometric compatibility must be violated to a certain extent. This raises the question of whether a substructure synthesis that does not satisfy the internal compatibility conditions exactly is a Rayleigh–Ritz method at all. The answer is a qualified yes, as shown in the following.

Let us consider the case in which the actual geometric compatibility conditions at internal boundaries cannot be satisfied exactly, so that the disconnected structure is joined together by imposing geometric compatibility conditions that are only approximations of the actual ones. It will prove convenient to regard the structure that is no longer disconnected, but whose internal boundary conditions are only approximations of the actual ones, as a fictitious structure lying between the totally disconnected structure and the actual structure and to refer to it as an 'intermediate structure'. Then, we can state that *the Rayleigh–Ritz method is valid for the intermediate structure.*

The imposition of the approximate geometric compatibility conditions can be done in various ways. We shall do it here by the weighted residual method. In this case, the intermediate structure represents a mathematical contrivance defined by the type of weighting functions used and their number. In the limit, as the number of weighting functions increases, the intermediate structure must approach the actual structure.

Let us consider two typical adjacent substructures r and s and assume that the energy inner product for each of the substructures contain derivatives of the displacements through order p. This implies that the geometric compatibility conditions contain derivatives ranging from zero to $p-1$. For example, for $p = 2$, geometric compatibility is guaranteed by the satisfaction of

$$\mathbf{u}_r = \mathbf{u}_s \qquad\qquad\qquad (11.111a)$$

$$\partial\mathbf{u}_r/\partial\mathbf{n}_r = \partial\mathbf{u}_s/\partial\mathbf{n}_s \qquad\qquad\qquad (11.111b)$$

at every point of the internal boundary S_{rs} shared by substructures r and s, where $\partial\mathbf{u}_r/\partial\mathbf{n}_r$ denotes the partial derivative of the displacement vector $\mathbf{u}_r$ with respect to the outward normal $\mathbf{n}_r$ of the internal boundary on substructure r. Clearly, for a general boundary and for general sets of substructure admissible functions, Eqs. (11.111) can be satisfied only approximately. To this end, we consider the weighted residual method in conjunction with the finite number M_{rs} of weighting functions g_{rsi} ($i = 1, 2, \ldots, M_{rs}$). Hence, we set the weighted average error equal to zero on S_{rs}, or

$$\int_{S_{rs}} g_{rsi}(\mathbf{u}_r - \mathbf{u}_s)\, dS_{rs} = \mathbf{0}, \qquad i = 1, 2, \ldots, M_{rs} \qquad (11.112a)$$

$$\int_{S_{rs}} g_{rsi}\left(\frac{\partial\mathbf{u}_r}{\partial\mathbf{n}_r} - \frac{\partial\mathbf{u}_s}{\partial\mathbf{n}_s}\right) dS_{rs} = \mathbf{0}, \qquad i = 1, 2, \ldots, M_{rs} \qquad (11.112b)$$

Equations (11.112) represent the constraint equations that join the disconnected substructures together to form the intermediate structure. As the number M_{rs} of weighting functions tends to infinity, the geometric compatibility conditions (11.111) will be satisfied at every point of S_{rs}, provided the weighting functions are from a complete set and they are such that the integrals in Eqs. (11.112) are defined and that Eqs. (11.112) themselves are linearly independent. Hence, as $M_{rs} \to \infty$ the intermediate structure approaches the actual structure. For practical reasons, however, M_{rs} must be taken as finite. It is interesting to note that in the special case in which g_{rsi} are spatial Dirac delta functions, Eqs. (11.112) guarantee the satisfaction of Eqs. (11.111) at a finite number M_{rs} of points on S_{rs}, so that in this case the intermediate structure is obtained from the actual structure by replacing the actual boundaries by a finite set of discrete points.

The question remains as to the nature of the eigensolution of the intermediate structure. To this end, we recall that Eqs. (11.112) play the role of constraint equations imposed on the disconnected structure. As in any Rayleigh–Ritz method, constraints tend to raise the computed eigenvalues. In this case, however, *the computed eigenvalues of the intermediate structure provide upper bounds for the eigenvalues of the disconnected structure* and not for the eigenvalues of the actual structure. Hence, the possibility exists that the computed eigenvalues of the intermediate structure are lower than the corresponding ones of the actual structure. The computed eigenvalues can still be made to converge to those of the actual structure, but it is necessary to consider two limiting processes, one in which the number of substructure admissible functions is increased and the other in which the number of internal boundary weighting functions is increased. Mathematical justification for above statements, as well as a numerical example, can be found in H1. The method has been extended to structures consisting of discrete substructures in M10 and generalized in M11.

Bibliography

B1. Bajan, R. L., Feng, C. C. and Jaszlics, I. J., "Vibration Analysis of Complex Structural Systems by Modal Substitution," *Shock and Vibration Bulletin*, No. 39, Part 3, 1969, pp. 99–105.

B2. Bamford, R. M., "A Modal Combination Program for Dynamic Analysis of Structures," TM33-290, Jet Propulsion Laboratory, Pasadena, Calif., July 1967.

B3. Bathe, K. J. and Wilson, W. L., "Large Eigenvalue Problems in Dynamic Analysis," *Journal of the Engineering Mechanics Division*, ASCE, Vol. 98, 1972, pp. 1471–1485.

B4. Benfield, W. A. and Hruda, R. F., "Vibration Analysis of Structures by Component Mode Substitution," *AIAA Journal*, Vol. 9, No. 7, 1971, pp. 1255–1261.

B5. Bolotin, V. V., *The Dynamic Stability of Elastic Systems*, Holden-Day, Inc., San Francisco, Calif., 1964.

B6. Bowman, F., *Introduction to Bessel Functions*, Dover Publications, Inc., New York, 1958.

C1. Caughey, T. L., "Classical Normal Modes in Damped Linear Dynamic Systems," *Journal of Applied Mechanics*, Vol. 27, 1960, pp. 269–271.

C2. Clint, M. and Jennings, A., "The Evaluation of Eigenvalue and Eigenvectors of Real Symmetric Matrices by Simultaneous Iteration," *The Computer Journal*, Vol. 13, No. 1, 1970, pp. 76–80.

C3. Courant, R., "Variational Methods for the Solution of Problems of Equilibrium and Vibrations," *Bulletin of the American Mathematical Society*, Vol. 49, Jan. 1943, pp. 1–23.

C4. Courant, R. and Hilbert, D., *Methods of Mathematical Physics*, Vol. 1, Interscience Publishers, Inc., New York 1953.

C5. Craig, R. R. Jr. and Bampton, M. C. C., "Coupling of Substructures for Dynamic Analysis," *AIAA Journal*, Vol. 6, No. 7, 1968, pp. 1313–1319.

D1. Dowell, E. H., "Free Vibration of an Arbitrary Structure in Terms of Component Modes," *Journal of Applied Mechanics*, Vol. 39, No. 3, 1972, pp. 727–732.

D2. Dowell, E. H. et al, *A Modern Course in Aeroelasticity*, Sijthoff & Noordhoff International Publishers, Alphen aan den Rijn, The Netherlands, 1978.

F1. Finlayson, B. A., *The Method of Weighted Residuals and Variational Principles*, Academic Press, New York, 1972.

F2. Francis, J. G. F., "The QR Transformation, Parts I and II," *The Computer Journal*, Vol. 4, 1961, pp. 265–271, 332–345.

F3. Franklin, J. N., *Matrix Theory*, Prentice-Hall, Inc., Englewood Cliffs, N.J., 1968.

F4. Friedman, B., *Principles and Techniques of Applied Mathematics*, John Wiley & Sons, Inc., New York, 1956.

F5. Fung, Y. C., *Theory of Aeroelasticity*, Dover Publications, Inc., New York, 1969.

G1. Gallagher, R. H., *Finite Element Analysis*, Prentice-Hall, Inc., Englewood Cliffs, N.J., 1975.

G2. Gladwell, G. M. L., "Branch Mode Analysis of Vibrating Systems," *Journal of Sound and Vibration*, Vol. 1, 1964, pp. 41–59.

G3. Goldman, R. L., "Vibration Analysis by Dynamic Partitioning," *AIAA Journal*, Vol. 7, No. 6, pp. 1152–1154.

G4. Gould, S. H., *Variational Methods for Eigenvalue Problems* (Second Edition), University of Toronto Press, Toronto, Canada, 1966.

G5. Guyan, R. J., "Reduction of Stiffness and Mass Matrices," *AIAA Journal*, Vol. 3, No. 2, 1965, p. 380.

H1. Hale, A. L. and Meirovitch, L., "A General Substructure Synthesis Method for the Dynamic Simulation of Complex Structures," *Journal of Sound and Vibration*, Vol. 69, No. 2, 1980, pp. 309–326.

H2. Hintz, R. M., "Analytical Methods in Component Modal Synthesis," *AIAA Journal*, Vol. 13, No. 8, 1975, pp. 1007–1016.

H3. Holland, I. and Bell, K. (Editors), *Finite Element Methods in Stress Analysis*, Tapir Press, Trondheim, Norway, 1969.

H4. Hou, S. N., "Review of a Modal Synthesis Technique and a New Approach," *Shock and Vibration Bulletin*, No. 40, Part, 4, 1969, pp. 25–30.

H5. Huebner, K. H., *The Finite Element Method for Engineers*, John Wiley & Sons, Inc., New York, 1975.

H6. Hurty, W. C., "Vibrations of Structural Systems by Component-Mode Synthesis," *Journal of the Engineering Mechanics Division*, ASCE, Vol. 86, Aug. 1960, pp. 51–69.

H7. Hurty, W. C., "Dynamic Analysis of Structural Systems Using Component Modes," *AIAA Journal*, Vol. 3, No. 4, 1965, pp. 678–685.

H8. Hurty, W. C., "A Criterion for Selecting Realistic Natural Modes of a Structure," TM33–364, Jet Propulsion Laboratory, Pasadena, Calif., Jan. 1967.

H9. Huseyin, K., *Vibrations and Stability of Multiple Parameter Systems*, Sijthoff & Noordhoff International Publishers, Alphen aan den Rijn, The Netherlands, 1978.

I1. Irons, B., "Structural Eigenvalue Problems: Elimination of Unwanted Variables," *AIAA Journal*, Vol. 3, No. 5, 1965, pp. 962–963.

I2. Irons, B., "Eigenvalue Economisers in Vibration Problems," *Journal of the Royal Aeronautical Society*, Vol. 67, Aug. 1963, pp. 526–528.

J1. Jennings, A., "A Direct Iteration Method of Obtaining Latent Roots and Vectors of a Symmetric Matrix," *Proceedings of the Cambridge Philosophical Society*, Vol. 63, 1967, pp. 755–765.

J2. Jensen, P. S., "The Solution of Large Symmetric Eigenproblems by Sectioning," *SIAM Journal of Numerical Analysis*, Vol. 9, No. 4, 1972, pp. 534–545.

K1. Kublanovskaya, V. N., "On Some Algorithm for the Solution of the Complete Eigenvalue Problem," *Z. Vycisl. Mat. i Mat. Fiz.*, Vol. 1, 1961, pp. 555–570; *USSR Comput. Math. Math. Phys.*, Vol. 3, pp. 637–657.

K2. Kuhar, E. J. and Stahle, C. V., "Dynamic Transformation Method for Modal Synthesis," *AIAA Journal*, Vol. 12, No. 5, 1974, pp. 672–678.

L1. Lancaster, P., *Lambda-matrices and Vibrating Systems*, Pergamon Press, New York, 1966.

L2. Likins, P. W., "Stability Analysis of Mechanical Systems with Constraint Damping," *AIAA Journal*, Vol. 5, No. 11, 1967, pp. 2091–2094.

M1. Martin, H. C. and Carey, G. F., *Introduction to Finite Element Analysis*, McGraw-Hill Book Co., New York, 1973.

M2. Meirovitch, L., *Analytical Methods in Vibrations*, The Macmillan Co., New York, 1967.

M3. Meirovitch, L., *Methods of Analytical Dynamics*, McGraw-Hill Book Co., New York, 1970.

M4. Meirovitch, L., "A New Method of Solution of the Eigenvalue

Problem for Gyroscopic Systems," *AIAA Journal*, Vol. 12, No. 10, 1974, pp. 1337–1342.

M5. Meirovitch, L., "A Modal Analysis for the Response of Linear Gyroscopic Systems," *Journal of Applied Mechanics*, Vol. 42, No. 2, 1975, pp. 446–450.

M6. Meirovitch, L., *Elements of Vibration Analysis*, McGraw-Hill Book Co., New York, 1975.

M7. Meirovitch, L., "A Stationarity Principle for the Eigenvalue Problem for Rotating Structures," *AIAA Journal*, Vol. 14, No. 10, 1976, pp. 1387–1394.

M8. Meirovitch, L., and Hale, A. L., "Synthesis and Dynamic Characteristics of Large Structures with Rotating Substructures," *Proceedings of the IUTAM Symposium on the Dynamics of Multibody Systems*. (Editor: K. Magnus), Springer-Verlag, Berlin, 1978, pp. 231–244.

M9. Meirovitch, L. and Ryland, G., "Response of Slightly Damped Gyroscopic Systems," *Journal of Sound and Vibration*, Vol. 67, No. 1, 1979, pp. 1–19.

M10. Meirovitch, L. and Hale, A. L., "A General Dynamic Synthesis for Structures with Discrete Substructures," Paper No. 80-0798, AIAA/ASME/ASCE/AHS 21st Structures, Structural Dynamics & Materials Conference, Seattle, Washington, May 12–14, 1980.

M11. Meirovitch, L. and Hale, A. L., "On the Substructure Synthesis Method," An International Conference on Recent Advances in Structural Dynamics, Southampton, England, 7–11 July, 1980.

M12. Melosh, R. J. "Basis for Derivation of Matrices for the Direct Stiffness Method," *AIAA Journal*, Vol. 1, No. 7, 1963, pp. 1631–1636.

M13. Mikhlin, S. G., *Variational Methods in Mathematical Physics*, Pergamon Press, New York, 1964.

M14. Mikhlin, S. G. and Smolitskiy, K. L., *Approximate Methods for Solution of Differential and Integral Equations*, American Elsevier Publishing Company, Inc., New York, 1967.

M15. Mingori, D. L., "A Stability Theorem for Mechanical Systems with Constraint Damping," *Journal of Applied Mechanics*, Vol. 37, 1970, pp. 253–258.

M16. Murdoch, D. C., *Linear Algebra*, John Wiley & Sons, Inc., New York, 1970.

N1. Noble, B. and Daniel, J. W., *Applied Linear Algebra* (Second Edition), Prentice-Hall, Inc., Englewood Cliffs, N.J., 1977.

P1. Paidoussis, M. P. and Issid, N. T., "Dynamics Stability of Pipes

Conveying Fluid," *Journal of Sound and Vibration*, Vol. 33, No. 3, 1974, pp. 267–294.

P2. Przemieniecki, J. S., *Theory of Matrix Structural Analysis*, McGraw-Hill Book Co., New York, 1968.

R1. Ralston, A., *A First Course in Numerical Analysis*, McGraw-Hill Book Co., New York, 1965.

R2. Ramsden, J. N. and Stoker, J. R., "Mass Condensation: A Semi-Automated Method for Reducing the Size of Vibration Problems," *International Journal for Numerical Methods in Engineering*, Vol. 1, 1969, pp. 333–349.

R3. Rayleigh (Lord), *Theory of Sound*, Vol. 1, Dover Publications, Inc., New York, 1945 (First American Edition of the 1894 Edition).

R4. Rubin, S., "Improved Component-Mode Representation for Structural Dynamic Analysis." *AIAA Journal*, Vol. 13, No. 8, 1975, pp. 995–1006.

R5. Rutishauser, H., "Solution of Eigenvalue Problems with the LR-Transformation," *Nat. Bur. Standards Appl. Math. Ser.*, Vol. 49, 1958, pp. 47–81.

R6. Ryland, G. and Meirovitch, L., "Response of Vibrating Systems with Perturbed Parameters," *Journal of Guidance and Control*, Vol. 3, No. 4, 1980, pp. 298-303

S1. Stewart, G. W., *Introduction to Matrix Computations*, Academic Press, New York, 1973.

S2. Strang, G. and Fix, G. J., *An Analysis of the Finite Element Method*, Prentice-Hall, Inc., Englewood Cliffs, N.J., 1973.

T1. Turner, M. J., Clough, R. W., Martin, H. C. and Topp, L. J., "Stiffness and Deflection Analysis of Complex Structures," *Journal of Aeronautical Sciences*, Vol. 23, 1956, pp. 805–823.

W1. Wilkinson, J. H., *The Algebraic Eigenvalue Problem*, Oxford University Press, London, 1965.

W2. Wilkinson, J. H. and Reinsch, G., *Linear Algebra*, Springer-Verlag, New York, 1971.

W3. Wright, G. C. and Miles, G. A., "An Economical Method for Determining the Smallest Eigenvalues of Large Linear Systems," *International Journal for Numerical Methods in Engineering*, Vol. 3, 1971, pp. 25–33.

Z1. Ziegler, H., *Principles of Structural Stability*, Blaisdell Publishing Company, Waltham, Massachusetts, 1968.

Suggested problems

Chapter 1

1.1 Check whether the vectors

$$\mathbf{x}_1 = \begin{bmatrix} 1 \\ 1 \\ 1 \\ 1 \end{bmatrix}, \quad \mathbf{x}_2 = \begin{bmatrix} 1 \\ -1 \\ 1 \\ 1 \end{bmatrix}, \quad \mathbf{x}_3 = \begin{bmatrix} 1 \\ 0 \\ 2 \\ 1 \end{bmatrix}, \quad \mathbf{x}_4 = \begin{bmatrix} 1 \\ 1 \\ -1 \\ 2 \end{bmatrix}$$

are independent.

1.2 Select a basis for L^3 from the vectors

$$\mathbf{x}_1 = \begin{bmatrix} 1 \\ 1 \\ 1 \end{bmatrix}, \quad \mathbf{x}_2 = \begin{bmatrix} 2 \\ -1 \\ 1 \end{bmatrix}, \quad \mathbf{x}_3 = \begin{bmatrix} 0 \\ 3 \\ 1 \end{bmatrix}, \quad \mathbf{x}_4 = \begin{bmatrix} 1 \\ 1 \\ -1 \end{bmatrix}$$

1.3 Solve the problem of Example 1.3 by the ordinary Gram-Schmidt process.

1.4 Determine whether the matrix

$$A = \begin{bmatrix} 1 & 1 & 1 & 1 \\ 1 & -1 & 0 & 1 \\ 1 & 1 & 2 & -1 \\ 1 & 1 & 1 & 2 \end{bmatrix}$$

is singular or not by calculating the value of its determinant.

1.5 Calculate the inverse of the matrix A of Problem 1.4.

1.6 Determine the rank of the matrix

$$A = \begin{bmatrix} 1 & 2 & 0 & 1 \\ 1 & -1 & 3 & 1 \\ 1 & 1 & 1 & -1 \end{bmatrix}$$

1.7 Use the approach of Example 1.8 to determine a set of vectors spanning the null space of all solutions of the equation

$$A\mathbf{x} = \mathbf{0}$$

where A is the matrix of Problem 1.6.

Chapter 2

2.1 The system shown in Fig. P2.1 consists of two rigid uniform links connected to one another and to a rotating drum by means of torsional springs. The angular velocity of the drum is $\Omega = $ const. Denote by θ_1 and θ_2 the angular displacements of the links relative to axes x, y, where the axes are embedded in the drum, and derive the equations of motion for small motions about equilibrium.

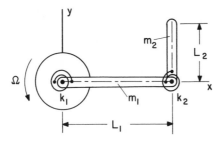

Figure P 2.1

2.2 Derive the equations of motion for the system shown in Fig. P2.2 first by using x_i $(i = 1, 2, 3, 4)$ as generalized coordinates and then by using

$$z_1 = x_1, \quad z_2 = x_2 - x_1, \quad z_3 = x_3 - x_2, \quad z_4 = x_4 - x_3$$

Assume that the springs exhibit linear behavior and that they are unstretched in equilibrium.

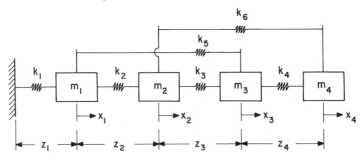

Figure P 2.2

416

2.3 An *n*-story building is assumed to consist of rigid concrete beams supported by massless elastic columns fixed at both ends, as shown in Fig. P2.3. Derive the equations describing the horizontal translation of the beams. The masses of the beams are denoted by M_i and the bending stiffness of the columns by $\frac{1}{2}EI_i$ $(i = 1, 2, \ldots, n)$. The columns are of height Hl.

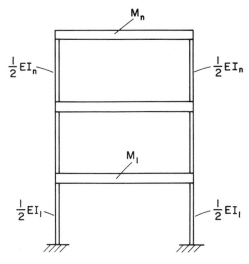

Figure P2.3

2.4 A massless elastic circular bar with a mass M attached at midspan is spinning with the constant angular velocity Ω, as shown in Fig. P2.4. Let X, Y be a set of inertial axes and x, y a set of axes embedded in the bar and derive the equations of motion of M in terms of the displacement components x and y along the shaft

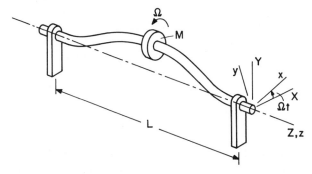

Figure P2.4

417

axes. Assume that M is acted upon by elastic restoring forces proportional to the displacements x and y, as well as by internal and external damping forces opposing the motion, where the internal damping forces are proportional to $\dot{x}$ and $\dot{y}$ and the external damping forces are proportional to $\dot{X}$ and $\dot{Y}$. The constants of proportionality are k, c and h, respectively.

2.5 Consider the system of Problem 2.1, let

$$L_1 = 2L, L_2 = L$$

$$M_1 = M_2 = M$$

$$k_1 = 2k, k_2 = k$$

and derive the eigenvalue problem.

2.6 Consider the system of Problem 2.2, let

$$k_1 = k_2 = 2k, k_3 = k_4 = k_5 = k_6 = k$$

$$m_1 = 2m, m_2 = m_3 = m_4 = m$$

and derive the two eigenvalue problems, the first corresponding to the coordinates x_i $(i = 1, 2, 3, 4)$ and the second corresponding to the coordinates z_i $(i = 1, 2, 3, 4)$. Expand the characteristic equation for both cases and draw conclusions.

2.7 Consider the system of Problem 2.3, let $M_i = M$ and $EI_i = EI$ $(i = 1, 2, \ldots, n)$ and derive the eigenvalue problem.

2.8 Derive the eigenvalue problem for the system of Problem 2.4.

Chapter 3

3.1 Consider the eigenvalue problem described by Eq. (3.1), where

$$A = \begin{bmatrix} 2 & 3 \\ 1 & 2 \end{bmatrix}$$

derive the characteristic polynomial and obtain the eigenvalues λ_1 and λ_2. Then compute the eigenvectors $\mathbf{x}_1$ and $\mathbf{x}_2$ by solving the algebraic equations $A\mathbf{x}_i = \lambda_i \mathbf{x}_i$ $(i = 1, 2)$.

3.2 Consider the matrix A of Problem 3.1 and the linear transformation (3.8), where

$$P = \begin{bmatrix} 1 & 1 \\ 0 & 1 \end{bmatrix}$$

and compute the matrix B according to Eq. (3.12). Verify that the characteristic polynomial associated with the matrix B is the same as that associated with A, compute the eigenvectors $\mathbf{y}_1$ and $\mathbf{y}_2$ of the matrix B and show the relation of $\mathbf{y}_1$ and $\mathbf{y}_2$ to the eigenvectors $\mathbf{x}_1$ and $\mathbf{x}_2$ of the matrix A.

3.3 Verify relations (3.23) and (3.24) for the matrix A of Problem 3.1.

3.4 Let $n = 2$ in Problem 2.7, solve the eigenvalue problem and verify the orthogonality of the eigenvectors. Plot the eigenvectors.

3.5 Derive the equation of the ellipse corresponding to the system of Problem 3.4. Solve the eigenvalue problem by finding the principal axes of the ellipse.

3.6 Obtain the left eigenvectors of the matrix A of Problem 3.1 and show that they satisfy the biorthogonality relations (3.107).

Chapter 4

4.1 Consider the system of Problem 2.5, let $k = 10mL^2\Omega^2$ and use Rayleigh's quotient to estimate the lowest eigenvalue. Show that there are two distinct trial vectors that can lead to the same approximation.

4.2 Consider the system of Problem 2.6 and use Rayleigh's quotient to estimate the lowest eigenvalue. Show that the static displacement pattern, obtained by applying on each mass a force proportional to the value of the mass, provides a satisfactory trial vector.

4.3 Consider the matrix

$$A = \begin{bmatrix} 2 & -1 & 0 \\ -1 & 3 & -\sqrt{2} \\ 0 & -\sqrt{2} & 1 \end{bmatrix}$$

and solve the eigenvalue problem. Then, solve the eigenvalue problem for the 2×2 matrix obtained by removing the third row and column and verify the inclusion principle.

4.4 Obtain the eigenvalues of the matrix

$$A = \begin{bmatrix} 5 & -1 & 0 \\ -1 & 2 & -1 \\ 0 & -1 & 5 \end{bmatrix}$$

and show that they are consistent with Gerschgorin's theorems.

4.5 Use the approach of Sec. 4.9 and produce a first-order perturbation theory for the eigensolution of

$$K\mathbf{u} = \lambda M\mathbf{u}$$

where

$$K = K_0 + K_1, \qquad M = M_0 + M_1$$

in which K, M, K_0 and M_0 are all positive definite real symmetric matrices and K_1 and M_1 are 'small' relative to K_0 and M_0.

4.6 Use the approach of Sec. 4.9 and produce a second-order perturbation theory for the eigensolution of

$$A\mathbf{x} = \lambda\mathbf{x}$$

where

$$A = A_0 + A_1$$

in which A and A_0 are real symmetric matrices and A_1 is 'small' relative to A_0.

4.7 Compute a second-order perturbation solution for the eigenvalue problem of Example 4.6.

Chapter 5

5.1 Solve the set of simultaneous algebraic equations

$$5x_1 - 4x_2 - x_3 + 2x_4 = 17$$

$$-x_1 + 2x_2 - 3x_3 - x_4 = 5$$

$$2x_1 - 3x_2 + x_3 + 4x_4 = 14$$

$$3x_1 + 4x_2 - 2x_3 + 5x_4 = 36$$

by the Gaussian elimination method.

5.2 The eigenvalues of the matrix

$$A = \begin{bmatrix} 2.5 & -1 & 0 \\ -1 & 5 & -\sqrt{2} \\ 0 & -\sqrt{2} & 10 \end{bmatrix}$$

are known to be

$$\lambda_1 = 2.119322, \qquad \lambda_2 = 5, \qquad \lambda_3 = 10.380678$$

Find the associated eigenvectors.

5.3 Solve the eigenvalue problem

$$\omega^2 M\mathbf{u} = K\mathbf{u}$$

where

$$K = k\begin{bmatrix} 4 & -2 & 0 & 0 \\ -2 & 3 & -1 & 0 \\ 0 & -1 & 2 & -1 \\ 0 & 0 & -1 & 1 \end{bmatrix}, \qquad M = m\begin{bmatrix} 3 & 0 & 0 & 0 \\ 0 & 2 & 0 & 0 \\ 0 & 0 & 1 & 0 \\ 0 & 0 & 0 & 1 \end{bmatrix}$$

Use the matrix iteration by the power method in conjunction with Hotelling's deflation.

5.4 Consider the matrix

$$A = \begin{bmatrix} 4 & -2 & 6 & 4 \\ -2 & 2 & -1 & 3 \\ 6 & -1 & 22 & 13 \\ 4 & 3 & 13 & 46 \end{bmatrix}$$

Use Sylvester's criterion to establish that the matrix is positive definite and then decompose the matrix into

$$A = LL^T$$

where L is a lower triangular matrix, by Cholesky's method.

5.5 Solve the eigenvalue problem of Problem 5.3 by the Jacobi method.

5.6 The eigenvalue problem of Problem 5.3 can be written in the form

$$Av = \lambda v$$

where

$$A = M^{1/2}K^{-1}M^{1/2}, \qquad v = M^{1/2}u, \qquad \lambda = 1/\omega^2$$

Use Givens' method to tridiagonalize the matrix A.

5.7 Tridiagonalize the matrix A of Problem 5.6 by Householder's method.

5.8 Determine the eigenvalues of the matrix A of Problem 5.6 by the method based on Sturm's theorem.

5.9 Determine the eigenvalues of the matrix A of Problem 5.6 by the QR method.

5.10 Determine the eigenvectors of the tridiagonal matrix associated with the matrix A of Problem 5.6 by inverse iteration. Then, make the proper transformations to obtain the actual eigenvectors u_i $(i = 1, 2, 3, 4)$.

Chapter 6

6.1 Start with Eq. (6.12), use a transformation of the variable of integration and prove Eq. (6.13).

6.2 Consider the system shown in Fig. 6.2a and let $c = 0$, $m = k = 1$. Then, obtain the response to the force $F(t)$ shown in Fig. P6.1 and determine F_0 and T so that $x(t) \equiv 0$ for $t > T$. The initial conditions are zero.

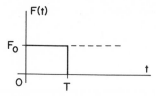

Figure P6.1

6.3 Determine the response of the system shown in Fig. 6.2a to the force $F(t)$ depicted in Fig. P6.2.

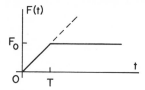

Figure P6.2

6.4 Solve Problem 6.3 by regarding the system as a discrete-time system.

6.5 Obtain the response of the system shown in Fig. 5.1 to the excitation

$$F_1(t) = F_3(t) = 0, \qquad F_2(t) = -F_4(t) = F_0 u(t)$$

where $u(t)$ is the unit step function. The initial conditions are zero. Interpret the result. (Note that the system eigenvalues were computed in Example 5.11 and that the first eigenvector was computed in Example 5.13.)

6.6 Determine the response of the system of Problem 2.7 to a horizontal displacement of the foundation having the form of the triangular pulse shown in Fig. P6.3. Let $n = 3$.

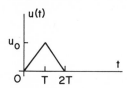

Figure P6.3

6.7 Show that the complex eigenvalue problem (6.44) can be reduced to the real eigenvalue problem (6.45–46).

6.8 Use modal analysis and derive the response of the system of

422

Problem 4.1 to the torque

$$T_1(t) = \hat{T}\,\delta(t)$$

applied to link 1.

6.9 Solve Problem 6.5 by regarding the system as a discrete-time system.

6.10 A bead of mass m is free to slide along a circular hoop of radius R, while the hoop is rotating about a vertical axis with the constant angular velocity Ω (see Fig. P6.4). Derive the differential equation of motion, identify the equilibrium points and test the stability of motion in the neighbourhood of the equilibrium points.

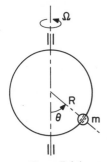

Figure P6.4

Chapter 7

7.1 A bar of length L is hinged at point 0 to a shaft rotating with the constant angular velocity Ω (Fig. P7.1). The bar has mass per unit length $m(x)$ and bending stiffness $EI(x)$. Denote the angle between axis x and the shaft axis by $\theta(t)$ and the transverse bending displacement by $w(x, t)$ and derive expressions for the kinetic and

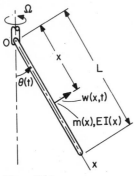

Figure P7.1

potential energy. The angle θ can be arbitrarily large but the elastic displacement can be assumed to be small.

7.2 A uniform cable of mass per unit length ρ hangs freely from a ceiling, as shown in Fig. P7.2. Derive the boundary value problem for the lateral vibration of the cable.

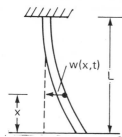

Figure P7.2

7.3 A bar of mass per unit length $m(x)$ and bending stiffness $EI(x)$ lies on an elastic foundation of stiffness per unit length $k(x)$. The bar is free at both ends (Fig. P7.3). Derive the boundary value problem for the bending vibration of the bar.

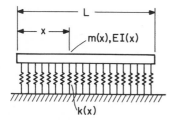

Figure P7.3

7.4 A bar of mass per unit length $m(x)$ and bending stiffness $EI(x)$ is fixed at the end $x = 0$ and hinged at the end $x = L$ in a way that the rotational motion is restrained through a torsional spring of stiffness k_T. Derive the boundary value problem for the bending vibration of the bar for the case in which a distributed force $f(x, t)$ is acting upon the system (Fig. P7.4).

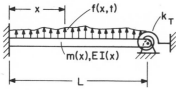

Figure P7.4

424

7.5 A circular shaft of polar mass moment of inertia per unit length $I(x)$ and torsional stiffness $GJ(x)$ is fixed at $x = 0$ and has a rigid disc of mass moment of inertia I_D attached at $x = L$ (Fig. P7.5). Derive the boundary value problem for torsional vibration.

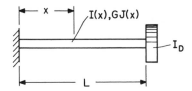

Figure P7.5

7.6 Let the whirling elastic bar of Problem 2.4 have the mass per unit length $m = $ const and bending stiffness $EI = $ const and derive the boundary value problem. Ignore the mass moments of inertia of the bar and of the rigid mass about their own axis.

7.7 A bar of mass per unit length $m(x)$ and bending stiffness $EI(x)$ is free at both ends. At the end $x = 0$, however, there is a constant axial force P, as shown in Fig. P7.6. Derive the boundary value problem for the bending vibration of the system. (Hint: Assume that the axial force is given by $P(x) = m(x)a$, where a is the constant acceleration of the bar in the axial direction.)

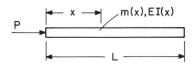

Figure P7.6

7.8 Use Hamilton's principle and derive the equations of motion for the system of Problem 7.1. The system of equations is hybrid as it consists of one ordinary differential equation and one partial differential equation, where the latter is supplemented by certain boundary conditions. (As a check, note that by letting $\theta(t) = \pi/2 = $ const the boundary value problem for $w(x, t)$ must reduce to that of Sec. 7.2, where in the latter the rotary inertia is ignored.)

7.9 Derive the eigenvalue problem for the system of Problem 7.3. Identify the differential operator L and check whether the system is self-adjoint and positive definite.

7.10 Repeat Problem 7.9 for the system of Problem 7.4.

7.11 Repeat Problem 7.9 for the system of Problem 7.5.

7.12 Let $m(x) = m = $ const, $EI(x) = EI = $ const in the eigenvalue problem of Problem 7.10 and derive the characteristic equation.

7.13 Derive the eigenvalue problem for the system of Problem 7.2 and prove that the eigenfunctions are orthogonal without solving the eigenvalue problem. Then, solve the eigenvalue problem and plot the first three modes. (Hint: the differential equation of motion can be reduced to a Bessel equation by a suitable transformation.)

7.14 Consider the system of Problem 7.8, let $m(x) = m = $ const, $EI(x) = EI = $ const and (1) determine the quilibrium position $\theta = \theta_0 = $ const, $w = w_0(x)$, (2) let $\theta(t) = \theta_0 + \theta_1(t)$, $w(x, t) = w_0(x) + w_1(x, t)$ where $\theta_1(t)$ and $w_1(x, t)$ are small displacements from the equilibrium position and derive the linearized equations of motion and (3) derive the eigenvalue problem.

7.15 Consider the eigenvalue problem of Problem 7.9, let $m(x) = m = $ const, $EI(x) = EI = $ const, $k(x) = k = $ const and solve the eigenvalue problem. Compare the eigensolution with the eigensolution for the unrestrained free–free bar $(k = 0)$ and draw conclusions.

7.16 Consider the eigenvalue problem of Problem 7.10, let $m(x) = m = $ const, $EI(x) = EI = $ const, $k_T = 0.2 EI/L$ and obtain the first four natural frequencies and natural modes. Plot the modes.

7.17 A uniform bar in bending is hinged at $x = 0$ and free at $x = L$. Solve the eigenvalue problem and plot the first four modes. Ignore the effect of gravity.

7.18. Let $I(x) = I = $ const, $GJ(x) = GJ = $ const in Problem 7.11 and derive the characteristic equation. Then, let $I_D = 0.5 IL$, obtain the first four roots of the characteristic equations and plot the first four modes.

7.19 Derive the characteristic equation for a uniform circular membrane supported at the boundary $r = a$ by a uniformly distributed spring of stiffness k.

7.20 Use the approach of Sec. 4.2 to prove inequalities (7.197) and (7.198).

7.21 Consider the eigenvalue problem of Problem 7.13, use the trial function

$$u(x) = \cos \frac{\pi x}{2L}$$

and verify inequality (7.197). Compare $R(u = \cos \pi x/2L)$ with the first eigenvalue obtained in Problem 7.13 and draw conclusions.

7.22 Write down the influence function for a uniform simply supported bar in bending. Write the eigenvalue problem in integral form and use as a trial function the admissible function

$$w_0(x) = x(L - x)$$

to generate an improved trial function. Draw conclusions as to the nature of the improved trial function.

7.23 A uniform string clamped at both ends is plucked initially according to

$$w(x, 0) = w_0(x) = \begin{cases} 3cx/2L, & 0 < x < 2L/3 \\ 3c(1 - x/L), & 2L/3 < x < L \end{cases}$$

Derive an expression for the subsequent motion.

7.24 The system of Problem 7.15 is subjected to the force

$$p(x, t) = \hat{P}\delta(x - L/3)\delta(t)$$

where $\delta(t)$ is the unit impulse. Obtain the system response. The initial conditions are zero. Draw conclusions as to the modes participation in the response.

7.25 The cable of Problem 7.13 is acted upon by the lateral force

$$p(x, t) = p_0\delta(x - L/2)\mathscr{u}(t)$$

at $x = L/2$, where $\mathscr{u}(t)$ is the unit step function. Derive the system response. The initial conditions are zero.

7.26 The circular membrane of Problem 7.19 is acted upon by the force

$$p(r, \theta, t) = \hat{p}_0\delta(r - a/2) \sin \theta \delta(t)$$

Derive the system response. The initial conditions are zero.

Chapter 8

8.1 Consider the cable of Problem 7.2 and obtain an approximate eigensolution by the Rayleigh-Ritz method. Use as comparison functions the eigenfunctions of a cable with the same end conditions but with the tension T constant throughout the cable. Use $n = 3, 4$ and compare the computed eigensolution with the exact one obtained in Problem 7.13. Also verify the inclusion principle.

8.2 Obtain an approximate solution of the eigenvalue problem for the system of Problem 7.18 by the assumed-modes method in conjunction with the admissible functions $\phi_i = \sin (2i - 1)\pi x/2L$ $(i = 1, 2, \ldots, n)$. Compare the solution for $n = 4$ to the 'exact' solution obtained in Problem 7.18 and draw conclusions.

8.3 Obtain approximate solutions to the eigenvalue problem for the system of Problem 7.16 by the Rayleigh-Ritz method in two ways: first by using the admissible functions $\phi_i = x^2 \sin i\pi x/L$ $(i = 1, 2, \ldots, n)$ and then by using the comparison functions $\phi_i = x^2 \sin \beta_i x$ $(i = 1, 2, \ldots, n)$, where β_i are such that ϕ_i satisfy the natural boundary condition at $x = L$. Let $n = 4$ and compare the approximate solutions thus obtained with the 'exact' solution obtained in Problem 7.16. Draw conclusions.

8.4 The cantilever bar shown in Fig. P8.1 has rectangular cross section of unit width and height varying according to

$$h(x) = \frac{3h}{2}\left(1 - \frac{2x}{3L}\right)$$

Let ρ be the mass per unit volume and E the modulus of elasticity and obtain an approximate eigensolution for the bending vibration of the bar by the assumed modes method. Solve the problem in two different ways: first by using as admissible functions the eigenfunctions of the uniform cantilever bar and then by using the admissible functions $\phi_i = x^{i+1}$ $(i = 1, 2, \ldots, n)$. Let $n = 4, 5, \ldots, 10$, tabulate and compare the computed natural frequencies and draw conclusions.

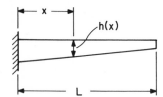

Figure P 8.1

8.5 Consider the eigenvalue problem defined by Eqs. (7.132–133), let $m(x) = m = $ const, $EI(x) = EI = $ const and use Rayleigh's quotient to prove that the lowest natural frequency is always higher than Ω.

8.6 Consider the system of Problem 8.5 and use the assumed modes method in conjunction with a set of admissible functions consisting of the eigenfunctions of a uniform nonrotating cantilever bar to formulate the eigenvalue problem. Solve the eigenvalue problem for $n = 3, 4, \ldots, 10$ using the values $\Omega^2 = EI/mL^4$, $9EI/mL^4$ and $25EI/mL^4$. Then, plot the computed natural frequencies vs n for every value of Ω and verify the conclusion of Problem 8.5. Also verify the inclusion principle.

8.7 Obtain an approximate solution of the eigenvalue problem of Problem 7.14 by the Rayleigh-Ritz method for the case in which $\Omega^2 = EI/mL^4$. Use as comparison functions the first four eigenfunctions obtained in Problem 7.17.

8.8 Formulate the eigenvalues problem for the system of Problem 7.6. Then, let $\Omega^2 = 0.5EI/mL^4$, $M = 0.25\ mL$, ignore both internal and external damping and solve the eigenvalue problem by the Rayleigh-Ritz method. Use three admissible functions in the form of the first three eigenfunctions of a uniform bar in bending fixed at both ends.

8.9 Formulate the eigenvalue problem for the system of Problem 7.16 by the Galerkin method. Use the same comparison functions as those used in Problem 8.3 and draw conclusions.

8.10 Formulate the eigenvalue problem for the system of Problem 7.16 by the collocation method. Use the same comparison functions as those used in Problem 8.3, solve the eigenvalue problem, compare the results with those obtained in Problem 8.3 and draw conclusions.

8.11 Formulate the eigenvalue problem for the system of Problem 7.16 by the least squares method. Use the same comparison functions as those used in Problem 8.3, solve the eigenvalue problem, compare results with those obtained in Problems 8.3 and 8.10 and draw conclusions.

8.12 A rectangular membrane clamped at the boundaries $x = \pm a$, $y = \pm b$ is subjected to uniform tension and has the mass per unit area

$$\rho(x, y) = \rho_0(1 + x^2/a^2)(1 + y^2/b^2)$$

Select a set of admissible functions and formulate the eigenvalue problem by the Rayleigh-Ritz method. Solve the eigenvalue problem by using sixteen admissible functions.

8.13 Solve the eigenvalue problem for the system of Problem 8.4 by the lumped-parameter method (Sec. 8.7) for $n = 10$ and compare the computed natural frequencies with the corresponding ones obtained in Problem 8.4.

8.14 Formulate the response problem for a non-self-adjoint system by the Galerkin method.

Chapter 9

9.1 Solve the eigenvalue problem of Problem 8.1 by the finite element method using (i) linear elements, (ii) quadratic elements and (iii) cubic elements. Determine the number of elements necessary in each of the three cases to produce a computed lowest frequency agreeing within five significant figures with that obtained in Problem 8.1 with $n = 4$.

9.2 Formulate the eigenvalue problem for the system of Problem 7.17 by the finite element method using Hermite cubics.

9.3 Formulate the eigenvalue problem for the system of Problem 8.4 by the finite element method using Hermite cubics. Solve the eigenvalue problem corresponding to ten elements.

9.4 Formulate the eigenvalue problem for the system of Problem 8.7 by the finite element method using Hermite cubics. Solve the eigenvalue problem corresponding to ten elements.

9.5 Formulate the eigenvalue problem for the system of Problem 8.8 by the finite element method using Hermite cubics for the case in which $\Omega^2 = 25EI/mL^4$. Solve the eigenvalue problem corresponding to ten elements.

9.6 Formulate the eigenvalue problem for the truss shown in Fig. P9.1 by the finite element method. The truss is hinged at all connecting points and its members can be regarded as uniform rods in axial vibration, all having the same cross-sectional area and mass density.

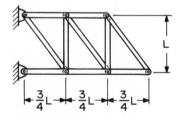

Figure P9.1

9.7 Let the columns in the system of Problem 2.7 have mass per unit length $m = $ const. Formulate the eigenvalue problem by the finite element method. Divide the columns into the same number of elements. The axial forces in the columns can be ignored and the left columns can be assumed to undergo the same motion as the right columns.

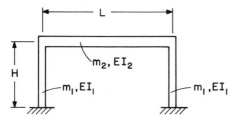

Figure P9.2

9.8 Formulate the eigenvalue problem for the system of Fig. P9.2 by the finite element method.

9.9 Formulate the eigenvalue problem for the system of Problem 8.12 by the finite element method.

9.10 Formulate the membrane eigenvalue problem of Sec. 9.5 by using the interpolation functions given by Eqs. (9.85).

9.11 Derive the equations of motion for the system of Problem 7.4 by the finite element method. Assume that the bar is uniform, $m(x) = m = $ const, $EI(x) = EI = $ const.

Chapter 11

11.1 Formulate the eigenvalue problem for the system of Problem 9.7 by the substructure synthesis method.

11.2 Formulate the eigenvalue problem for the system of Problem 9.8 by the substructure synthesis method.

Author index

Subject index

Subject index

Positive, semidefinite system: (*Cont.*)
   discrete, 44
   distributed, see continuous
Potential energy:
   discrete systems, 30
   distributed systems, 235
Potential energy density, 231
Principal coordinates, 201
Proportional damping, 211, 327
Pyramid functions, 350

QL method, 167
QR method, 162–172
Quadrilateral elements, 359

Rayleigh dissipation function, 30
Rayleigh energy method, 292
Rayleigh principle, 74–84
Rayleigh quotient:
   continuous systems, 274
   discrete systems, 74
      conservative gyroscopic, 81
      conservative nongyroscopic, 80
   distributed systems, see continuous
Rayleigh-Ritz method, 286–298
Rayleigh theorem, 84–88
Rectangular elements, 359
Response:
   continuous systems, 281–284
   discrete systems,
      conservative gyroscopic, 204–210
      conservative nongyroscopic, 199–204
      damped nongyroscopic, 210–217
      general dynamical, 217–221
   discrete-time systems, 192–199, 221–223
   discretized systems,
      damped nongyroscopic, 326–327
      undamped gyroscopic, 324–326
      undamped nongyroscopic, 322–324
   distributed systems, see continuous systems
   single-degree-of-freedom systems, 186–192
Ritz system, 289
   eigenfunctions, 290
   eigenvalues, 290
Rigid-body modes, 44
Roof functions, 332
Rotary inertia, 248
Rotation matrix, 139

Sampler, 194
Sampling period, 194
Sectioning, method of, 380
Self-adjoint eigenvalue problem:
   algebraic, 72
   differential, 248
Self-adjoint systems:
   continuous, 248
   discrete, 72
   distributed, see continuous
Shape functions, 335
Shift in eigenvalues, 52, 166
Significant behavior, 225
Similarity transformation, 53
Simultaneous iteration, 373
Space:
   configuration, 31
   phase, 32
   state, 32, 45
   vector, see vector space
Spatial Dirac delta function, 282
Square summable functions, 245
Stability of motion, 223–228
Stable equilibrium, 225, 226
State vector, 32, 45
Static condensation, 369
Stiffness coefficients:
   elastic, 34, 288
   geometric, 34
Stiffness matrix, 35, 289
   complex, 212
Structural damping, 212
Structural damping factor, 212
Sturm sequence, 158
Sturm theorem, 159
Subdomains, method of, 308
Subspace iteration, 377
Superposition principle, 185
Sylvester's criterion, 95
Synchronous motion:
   continuous systems, 239
   discrete systems, 41
System:
   conservative, 41
   damped, 47
   gyroscopic, 40
   linear, 35
   linearized, 33
   natural, 40

438